Ford Explorer
Mazda Navajo &
Mercury Mountaineer
Automotive
Repair
Manual

**by Jay Storer
and John H Haynes**
Member of the Guild of Motoring Writers

Models covered:
All Ford Explorer, Mazda Navajo and
Mercury Mountaineer models
1991 through 2001

(5E7 - 36024)
(2021)

ABCDE
FGHIJ

AUTOMOTIVE
PARTS &
ACCESSORIES
ASSOCIATION MEMBER

Haynes Publishing Group
Sparkford Nr Yeovil
Somerset BA22 7JJ England

Haynes North America, Inc
861 Lawrence Drive
Newbury Park
California 91320 USA

About this manual

Its purpose

The purpose of this manual is to help you get the best value from your vehicle. It can do so in several ways. It can help you decide what work must be done, even if you choose to have it done by a dealer service department or a repair shop; it provides information and procedures for routine maintenance and servicing; and it offers diagnostic and repair procedures to follow when trouble occurs.

We hope you use the manual to tackle the work yourself. For many simpler jobs, doing it yourself may be quicker than arranging an appointment to get the vehicle into a shop and making the trips to leave it and pick it up. More importantly, a lot of money can be saved by avoiding the expense the shop must pass on to you to cover its labor and overhead costs. An added benefit is the sense of satisfaction and accomplishment that you feel after doing the job yourself.

Using the manual

The manual is divided into Chapters. Each Chapter is divided into numbered Sections, which are headed in bold type between horizontal lines. Each Section consists of consecutively numbered paragraphs.

At the beginning of each numbered Section you will be referred to any illustrations which apply to the procedures in that Section. The reference numbers used in illustration captions pinpoint the pertinent Section and the Step within that Section. That is, illustration 3.2 means the illustration refers to Section 3 and Step (or paragraph) 2 within that Section.

Procedures, once described in the text, are not normally repeated. When it's necessary to refer to another Chapter, the reference will be given as Chapter and Section number. Cross references given without use of the word "Chapter" apply to Sections and/or paragraphs in the same Chapter. For example, "see Section 8" means in the same Chapter.

References to the left or right side of the vehicle assume you are sitting in the driver's seat, facing forward.

Even though we have prepared this manual with extreme care, neither the publisher nor the author can accept responsibility for any errors in, or omissions from, the information given.

NOTE

A **Note** provides information necessary to properly complete a procedure or information which will make the procedure easier to understand.

CAUTION

A **Caution** provides a special procedure or special steps which must be taken while completing the procedure where the Caution is found. Not heeding a Caution can result in damage to the assembly being worked on.

WARNING

A **Warning** provides a special procedure or special steps which must be taken while completing the procedure where the Warning is found. Not heeding a Warning can result in personal injury.

Acknowledgements

We are grateful to the Ford Motor Company for assistance with technical information and certain illustrations. Technical writers who contributed to this project include Alan Ahlstrand and Jeff Kibler.

© Haynes North America, Inc. 1991, 1992, 1993, 1998, 1999, 2000, 2001

With permission from J.H. Haynes & Co. Ltd.

A book in the Haynes Automotive Repair Manual Series

Printed in the U.S.A.

ISBN 1 56392 442 0

Library of Congress Control Number: 2001095061

Contents

Haynes photographer, mechanic and author with 1991 Explorer

Introduction to the Ford Explorer, Mazda Navajo, and Mercury Mountaineer

The models covered by this manual are available in a variety of trim options. In the 1991 through 1994 model years, a Mazda Navajo model was available, which was mechanically identical to the Ford Explorer. Beginning with 1997 models, the Lincoln-Mercury Division of Ford Motor Company offered the Mercury Mountaineer model, which was mechanically identical to the Ford Explorer models for 1997 and 1998.

The 4.0L pushrod V6 engine has been available for all the years covered by this manual, with the optional SOHC V6 becoming available in 1997, and the 5.0L V8 engine becoming an option beginning with 1996 models.

Chassis layout is conventional, with the engine mounted at the front and the power being transmitted through either a manual or automatic transmission to a driveshaft and solid rear axle. On 4WD models a transfer case also transmits power to the front axle by way of a driveshaft. Transmissions used are a five-speed overdrive manual transmission and both four-speed and five-speed overdrive automatic transmissions. The V8 engine is available only with an automatic transmission.

1994 and earlier vehicles use twin I-beam front suspension with coil springs and radius arms. 1995 and later models use upper and lower control arms and torsion

bars. All models use semi-elliptical leaf springs at the rear. Some models have an optional Automatic Ride Control system that controls ride height and firmness with a computer, sensors and air-powered shock absorbers.

All models are equipped with power assisted front disc brakes, and most models have drum rear brakes. Some models have rear disc brakes. 1994 and earlier models are equipped with a rear-wheel anti-lock brake system, while 1995 and later models are equipped with a four-wheel anti-lock brake system.

Vehicle identification numbers

Vehicle identification numbers

Modifications are a continuing and unpublicized process in vehicle manufacturing. Since spare parts manuals and lists are compiled on a numerical basis, the individual vehicle numbers are necessary to correctly identify the component required.

Vehicle Identification Number (VIN)

This very important identification number is located on a plate attached to the left side dashboard just inside the windshield **(see illustration)**. The VIN also appears on the Vehicle Certificate of Title and Registration. It contains information such as where and when the vehicle was manufactured, engine type, the model year and the body style.

Vehicle Certification Label

The Vehicle Certification label (VC label, also referred to as the Truck Safety Compliance Certification label) is attached to the front of the left (driver's side) door pillar **(see illustration)**. The upper half of the label contains the name of the manufacturer, the month and year of production, the Gross Vehicle Weight Rating (GVWR), the Gross Axle Weight Rating (GAWR) and the certification statement.

The VC label also contains the VIN number, which is used for warranty identification of the vehicle, and provides such information as manufacturer, type of restraint system, body type, engine, transmission, model year and vehicle serial number.

Engine numbers

Labels containing the engine code, calibration and serial numbers, as well as plant name, can be found on the valve cover, as well as stamped on the engine itself **(see illustration)**. The engine code is also found as the eighth digit of the VIN code. The letter X indicates the 4.0L pushrod V6, the letter E indicates the 4.0L SOHC V6, and the letter P indicates the 5.0L V8.

Manual transmission numbers

The manual transmission identification number and serial numbers can be found on a tag on the left or right side of the transmission main case **(see illustration)**.

Automatic transmission numbers

Tags with the automatic transmission serial number, build date and other information are attached with a bolt, usually at the extension housing **(see illustration)**.

Vehicle Emissions Control Information (VECI) label

This label is found under the hood (see Chapter 6).

The VIN number is visible through the windshield on the driver's side

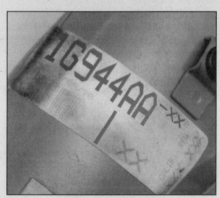

The engine identification label is usually affixed to the valve cover

A – FORD MOTOR COMPANY PART NUMBER
B – SERIAL NUMBER
C – BAR CODE (INVENTORY PURPOSES)

The manual transmission identification label is affixed to the side of the transmission case

The vehicle certification label, also known as the Truck Safety Compliance Certification label, is located on the left front door pillar

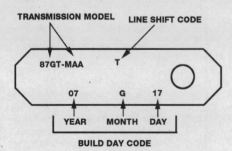

TRANSMISSION MODEL   LINE SHIFT CODE

87GT-MAA   T

07   G   17

YEAR   MONTH   DAY

BUILD DAY CODE

The automatic transmission identification label is attached with a bolt, usually to the extension housing

The transfer case identification tag is retained by a bolt at the rear of the transfer case

The differential identification tag is bolted to the differential cover

Buying parts

Replacement parts are available from many sources, which generally fall into one of two categories - authorized dealer parts departments and independent retail auto parts stores. Our advice concerning these parts is as follows:

Retail auto parts stores: Good auto parts stores will stock frequently needed components which wear out relatively fast, such as clutch components, exhaust systems, brake parts, tune-up parts, etc. These stores often supply new or reconditioned parts on an exchange basis, which can save a considerable amount of money. Discount auto parts stores are often very good places to buy materials and parts needed for general vehicle maintenance such as oil, grease, filters, spark plugs, belts, touch-up paint, bulbs, etc. They also usually sell tools and general accessories, have convenient hours, charge lower prices and can often be found not far from home.

Authorized dealer parts department: This is the best source for parts which are unique to the vehicle and not generally available elsewhere (such as major engine parts, transmission parts, trim pieces, etc.).

Warranty information: If the vehicle is still covered under warranty, be sure that any replacement parts purchased - regardless of the source - do not invalidate the warranty!

To be sure of obtaining the correct parts, have engine and chassis numbers available and, if possible, take the old parts along for positive identification.

Maintenance techniques, tools and working facilities

Maintenance techniques

There are a number of techniques involved in maintenance and repair that will be referred to throughout this manual. Application of these techniques will enable the home mechanic to be more efficient, better organized and capable of performing the various tasks properly, which will ensure that the repair job is thorough and complete.

Fasteners

Fasteners are nuts, bolts, studs and screws used to hold two or more parts together. There are a few things to keep in mind when working with fasteners. Almost all of them use a locking device of some type, either a lockwasher, locknut, locking tab or thread adhesive. All threaded fasteners should be clean and straight, with undamaged threads and undamaged corners on the hex head where the wrench fits. Develop the habit of replacing all damaged nuts and bolts with new ones. Special locknuts with nylon or fiber inserts can only be used once. If they are removed, they lose their locking ability and must be replaced with new ones.

Rusted nuts and bolts should be treated with a penetrating fluid to ease removal and prevent breakage. Some mechanics use turpentine in a spout-type oil can, which works quite well. After applying the rust penetrant, let it work for a few minutes before trying to loosen the nut or bolt. Badly rusted fasteners may have to be chiseled or sawed off or removed with a special nut breaker, available at tool stores.

If a bolt or stud breaks off in an assembly, it can be drilled and removed with a special tool commonly available for this purpose. Most automotive machine shops can perform this task, as well as other repair procedures, such as the repair of threaded holes that have been stripped out.

Flat washers and lockwashers, when removed from an assembly, should always be replaced exactly as removed. Replace any damaged washers with new ones. Never use a lockwasher on any soft metal surface (such as aluminum), thin sheet metal or plastic.

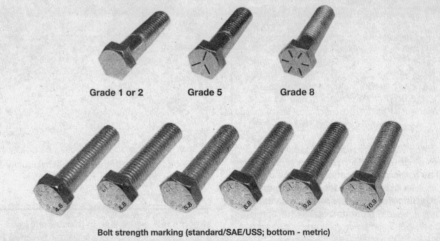

Grade 1 or 2   Grade 5   Grade 8

Bolt strength marking (standard/SAE/USS; bottom - metric)

Grade	Identification	Grade	Identification
Hex Nut Grade 5	3 Dots	Hex Nut Property Class 9	Arabic 9
Hex Nut Grade 8	6 Dots	Hex Nut Property Class 10	Arabic 10

Standard hex nut strength markings

Metric hex nut strength markings

Class 10.9   Class 9.8   Class 8.8

Metric stud strength markings

00-1 HAYNES

Fastener sizes

For a number of reasons, automobile manufacturers are making wider and wider use of metric fasteners. Therefore, it is important to be able to tell the difference between standard (sometimes called U.S. or SAE) and metric hardware, since they cannot be interchanged.

All bolts, whether standard or metric, are sized according to diameter, thread pitch and length. For example, a standard 1/2 - 13 x 1 bolt is 1/2 inch in diameter, has 13 threads per inch and is 1 inch long. An M12 - 1.75 x 25 metric bolt is 12 mm in diameter, has a thread pitch of 1.75 mm (the distance between threads) and is 25 mm long. The two bolts are nearly identical, and easily confused, but they are not interchangeable.

In addition to the differences in diameter, thread pitch and length, metric and standard bolts can also be distinguished by examining the bolt heads. To begin with, the distance across the flats on a standard bolt head is measured in inches, while the same dimension on a metric bolt is sized in millimeters (the same is true for nuts). As a result, a standard wrench should not be used on a metric bolt and a metric wrench should not be used on a standard bolt. Also, most standard bolts have slashes radiating out from the center of the head to denote the grade or strength of the bolt, which is an indication of the amount of torque that can be applied to it. The greater the number of slashes, the greater the strength of the bolt. Grades 0 through 5 are commonly used on automobiles. Metric bolts have a property class (grade) number, rather than a slash, molded into their heads to indicate bolt strength. In this case, the higher the number, the stronger the bolt. Property class numbers 8.8, 9.8 and 10.9 are commonly used on automobiles.

Strength markings can also be used to distinguish standard hex nuts from metric hex nuts. Many standard nuts have dots stamped into one side, while metric nuts are marked with a number. The greater the number of dots, or the higher the number, the greater the strength of the nut.

Metric studs are also marked on their ends according to property class (grade). Larger studs are numbered (the same as metric bolts), while smaller studs carry a geometric code to denote grade.

It should be noted that many fasteners, especially Grades 0 through 2, have no distinguishing marks on them. When such is the case, the only way to determine whether it is standard or metric is to measure the thread pitch or compare it to a known fastener of the same size.

Standard fasteners are often referred to as SAE, as opposed to metric. However, it should be noted that SAE technically refers to a non-metric fine thread fastener only. Coarse thread non-metric fasteners are referred to as USS sizes.

Since fasteners of the same size (both standard and metric) may have different strength ratings, be sure to reinstall any bolts, studs or nuts removed from your vehicle in their original locations. Also, when replacing a fastener with a new one, make sure that the new one has a strength rating equal to or greater than the original.

Metric thread sizes

	Ft-lbs	Nm
M-6	6 to 9	9 to 12
M-8	14 to 21	19 to 28
M-10	28 to 40	38 to 54
M-12	50 to 71	68 to 96
M-14	80 to 140	109 to 154

Pipe thread sizes

1/8	5 to 8	7 to 10
1/4	12 to 18	17 to 24
3/8	22 to 33	30 to 44
1/2	25 to 35	34 to 47

U.S. thread sizes

1/4 - 20	6 to 9	9 to 12
5/16 - 18	12 to 18	17 to 24
5/16 - 24	14 to 20	19 to 27
3/8 - 16	22 to 32	30 to 43
3/8 - 24	27 to 38	37 to 51
7/16 - 14	40 to 55	55 to 74
7/16 - 20	40 to 60	55 to 81
1/2 - 13	55 to 80	75 to 108

Standard (SAE and USS) bolt dimensions/grade marks

G   Grade marks (bolt strength)
L   Length (in inches)
T   Thread pitch (number of threads per inch)
D   Nominal diameter (in inches)

Metric bolt dimensions/grade marks

P   Property class (bolt strength)
L   Length (in millimeters)
T   Thread pitch (distance between threads in millimeters)
D   Diameter

Tightening sequences and procedures

Most threaded fasteners should be tightened to a specific torque value (torque is the twisting force applied to a threaded component such as a nut or bolt). Overtightening the fastener can weaken it and cause it to break, while undertightening can cause it to eventually come loose. Bolts, screws and studs, depending on the material they are made of and their thread diameters, have specific torque values, many of which are noted in the Specifications at the beginning of each Chapter. Be sure to follow the torque recommendations closely. For fasteners not assigned a specific torque, a general torque value chart is presented here as a guide. These torque values are for dry (unlubricated) fasteners threaded into steel or cast iron (not aluminum). As was previously mentioned, the size and grade of a fastener determine the amount of torque that can safely be applied to it. The figures listed here are approximate for Grade 2 and Grade 3 fasteners. Higher grades can tolerate higher torque values.

Fasteners laid out in a pattern, such as cylinder head bolts, oil pan bolts, differential cover bolts, etc., must be loosened or tightened in sequence to avoid warping the com-

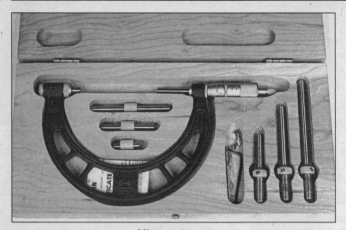

Micrometer set

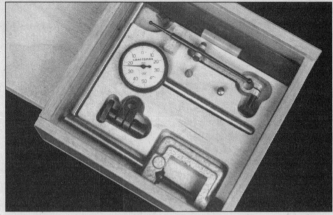

Dial indicator set

ponent. This sequence will normally be shown in the appropriate Chapter. If a specific pattern is not given, the following procedures can be used to prevent warping.

Initially, the bolts or nuts should be assembled finger-tight only. Next, they should be tightened one full turn each, in a criss-cross or diagonal pattern. After each one has been tightened one full turn, return to the first one and tighten them all one-half turn, following the same pattern. Finally, tighten each of them one-quarter turn at a time until each fastener has been tightened to the proper torque. To loosen and remove the fasteners, the procedure would be reversed.

Component disassembly

Component disassembly should be done with care and purpose to help ensure that the parts go back together properly. Always keep track of the sequence in which parts are removed. Make note of special characteristics or marks on parts that can be installed more than one way, such as a grooved thrust washer on a shaft. It is a good idea to lay the disassembled parts out on a clean surface in the order that they were removed. It may also be helpful to make sketches or take instant photos of components before removal.

When removing fasteners from a component, keep track of their locations. Sometimes threading a bolt back in a part, or putting the washers and nut back on a stud, can prevent mix-ups later. If nuts and bolts cannot be returned to their original locations, they should be kept in a compartmented box or a series of small boxes. A cupcake or muffin tin is ideal for this purpose, since each cavity can hold the bolts and nuts from a particular area (i.e. oil pan bolts, valve cover bolts, engine mount bolts, etc.). A pan of this type is especially helpful when working on assemblies with very small parts, such as the carburetor, alternator, valve train or interior dash and trim pieces. The cavities can be marked with paint or tape to identify the contents.

Whenever wiring looms, harnesses or connectors are separated, it is a good idea to identify the two halves with numbered pieces of masking tape so they can be easily reconnected.

Gasket sealing surfaces

Throughout any vehicle, gaskets are used to seal the mating surfaces between two parts and keep lubricants, fluids, vacuum or pressure contained in an assembly.

Many times these gaskets are coated with a liquid or paste-type gasket sealing compound before assembly. Age, heat and pressure can sometimes cause the two parts to stick together so tightly that they are very difficult to separate. Often, the assembly can be loosened by striking it with a soft-face hammer near the mating surfaces. A regular hammer can be used if a block of wood is placed between the hammer and the part. Do not hammer on cast parts or parts that could be easily damaged. With any particularly stubborn part, always recheck to make sure that every fastener has been removed.

Avoid using a screwdriver or bar to pry apart an assembly, as they can easily mar the gasket sealing surfaces of the parts, which must remain smooth. If prying is absolutely necessary, use an old broom handle, but keep in mind that extra clean up will be necessary if the wood splinters.

After the parts are separated, the old gasket must be carefully scraped off and the gasket surfaces cleaned. Stubborn gasket material can be soaked with rust penetrant or treated with a special chemical to soften it so it can be easily scraped off. A scraper can be fashioned from a piece of copper tubing by flattening and sharpening one end. Copper is recommended because it is usually softer than the surfaces to be scraped, which reduces the chance of gouging the part. Some gaskets can be removed with a wire brush, but regardless of the method used, the mating surfaces must be left clean and smooth. If for some reason the gasket surface is gouged, then a gasket sealer thick enough to fill scratches will have to be used during reassembly of the components. For most applications, a non-drying (or semi-drying) gasket sealer should be used.

Hose removal tips

Warning: *If the vehicle is equipped with air conditioning, do not disconnect any of the A/C hoses without first having the system depressurized by a dealer service department or a service station.*

Hose removal precautions closely parallel gasket removal precautions. Avoid scratching or gouging the surface that the hose mates against or the connection may leak. This is especially true for radiator hoses. Because of various chemical reactions, the rubber in hoses can bond itself to the metal spigot that the hose fits over. To remove a hose, first loosen the hose clamps that secure it to the spigot. Then, with slip-joint pliers, grab the hose at the clamp and rotate it around the spigot. Work it back and forth until it is completely free, then pull it off. Silicone or other lubricants will ease removal if they can be applied between the hose and the outside of the spigot. Apply the same lubricant to the inside of the hose and the outside of the spigot to simplify installation.

As a last resort (and if the hose is to be replaced with a new one anyway), the rubber can be slit with a knife and the hose peeled from the spigot. If this must be done, be careful that the metal connection is not damaged.

If a hose clamp is broken or damaged, do not reuse it. Wire-type clamps usually weaken with age, so it is a good idea to replace them with screw-type clamps whenever a hose is removed.

Tools

A selection of good tools is a basic requirement for anyone who plans to maintain and repair his or her own vehicle. For the owner who has few tools, the initial investment might seem high, but when compared to the spiraling costs of professional auto maintenance and repair, it is a wise one.

To help the owner decide which tools are needed to perform the tasks detailed in this manual, the following tool lists are offered: *Maintenance and minor repair, Repair/overhaul* and *Special.*

The newcomer to practical mechanics

Dial caliper

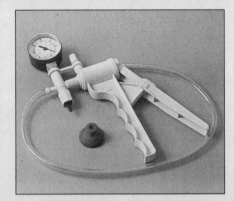

Hand-operated vacuum pump

Timing light

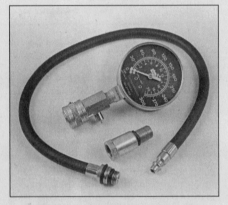

Compression gauge with spark plug hole adapter

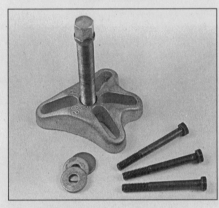

Damper/steering wheel puller

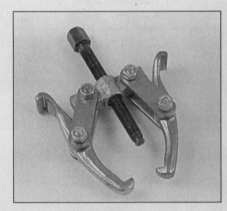

General purpose puller

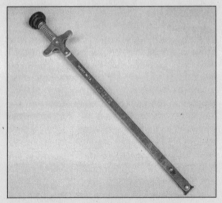

Hydraulic lifter removal tool

Valve spring compressor

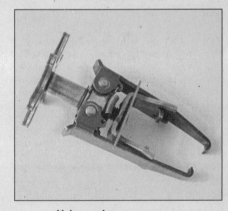

Valve spring compressor

Ridge reamer

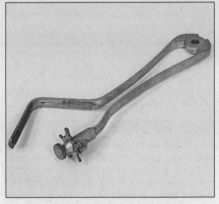

Piston ring groove cleaning tool

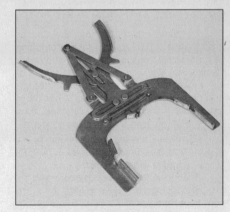

Ring removal/installation tool

Ring compressor

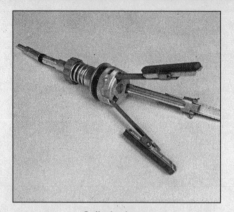

Cylinder hone

Brake hold-down spring tool

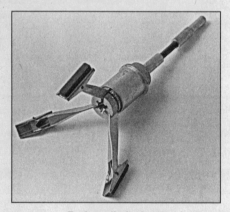

Brake cylinder hone

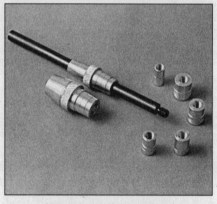

Clutch plate alignment tool

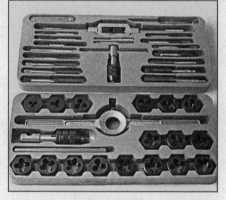

Tap and die set

should start off with the *maintenance and minor repair* tool kit, which is adequate for the simpler jobs performed on a vehicle. Then, as confidence and experience grow, the owner can tackle more difficult tasks, buying additional tools as they are needed. Eventually the basic kit will be expanded into the *repair and overhaul* tool set. Over a period of time, the experienced do-it-yourselfer will assemble a tool set complete enough for most repair and overhaul procedures and will add tools from the special category when it is felt that the expense is justified by the frequency of use.

Maintenance and minor repair tool kit

The tools in this list should be considered the minimum required for performance of routine maintenance, servicing and minor repair work. We recommend the purchase of combination wrenches (box-end and open-end combined in one wrench). While more expensive than open end wrenches, they offer the advantages of both types of wrench.

> *Combination wrench set (1/4-inch to 1 inch or 6 mm to 19 mm)*
> *Adjustable wrench, 8 inch*
> *Spark plug wrench with rubber insert*
> *Spark plug gap adjusting tool*
> *Feeler gauge set*
> *Brake bleeder wrench*

> *Standard screwdriver (5/16-inch x 6 inch)*
> *Phillips screwdriver (No. 2 x 6 inch)*
> *Combination pliers - 6 inch*
> *Hacksaw and assortment of blades*
> *Tire pressure gauge*
> *Grease gun*
> *Oil can*
> *Fine emery cloth*
> *Wire brush*
> *Battery post and cable cleaning tool*
> *Oil filter wrench*
> *Funnel (medium size)*
> *Safety goggles*
> *Jackstands (2)*
> *Drain pan*

Note: *If basic tune-ups are going to be part of routine maintenance, it will be necessary to purchase a good quality stroboscopic timing light and combination tachometer/dwell meter. Although they are included in the list of special tools, it is mentioned here because they are absolutely necessary for tuning most vehicles properly.*

Repair and overhaul tool set

These tools are essential for anyone who plans to perform major repairs and are in addition to those in the maintenance and minor repair tool kit. Included is a comprehensive set of sockets which, though expensive, are invaluable because of their versatil-

ity, especially when various extensions and drives are available. We recommend the 1/2-inch drive over the 3/8-inch drive. Although the larger drive is bulky and more expensive, it has the capacity of accepting a very wide range of large sockets. Ideally, however, the mechanic should have a 3/8-inch drive set and a 1/2-inch drive set.

> *Socket set(s)*
> *Reversible ratchet*
> *Extension - 10 inch*
> *Universal joint*
> *Torque wrench (same size drive as sockets)*
> *Ball peen hammer - 8 ounce*
> *Soft-face hammer (plastic/rubber)*
> *Standard screwdriver (1/4-inch x 6 inch)*
> *Standard screwdriver (stubby - 5/16-inch)*
> *Phillips screwdriver (No. 3 x 8 inch)*
> *Phillips screwdriver (stubby - No. 2)*
> *Pliers - vise grip*
> *Pliers - lineman's*
> *Pliers - needle nose*
> *Pliers - snap-ring (internal and external)*
> *Cold chisel - 1/2-inch*
> *Scribe*
> *Scraper (made from flattened copper tubing)*
> *Centerpunch*
> *Pin punches (1/16, 1/8, 3/16-inch)*
> *Steel rule/straightedge - 12 inch*

Allen wrench set (1/8 to 3/8-inch or
* 4 mm to 10 mm)*
A selection of files
Wire brush (large)
Jackstands (second set)
Jack (scissor or hydraulic type)

Note: *Another tool which is often useful is an electric drill with a chuck capacity of 3/8-inch and a set of good quality drill bits.*

Special tools

The tools in this list include those which are not used regularly, are expensive to buy, or which need to be used in accordance with their manufacturer's instructions. Unless these tools will be used frequently, it is not very economical to purchase many of them. A consideration would be to split the cost and use between yourself and a friend or friends. In addition, most of these tools can be obtained from a tool rental shop on a temporary basis.

This list primarily contains only those tools and instruments widely available to the public, and not those special tools produced by the vehicle manufacturer for distribution to dealer service departments. Occasionally, references to the manufacturer's special tools are included in the text of this manual. Generally, an alternative method of doing the job without the special tool is offered. However, sometimes there is no alternative to their use. Where this is the case, and the tool cannot be purchased or borrowed, the work should be turned over to the dealer service department or an automotive repair shop.

Valve spring compressor
Piston ring groove cleaning tool
Piston ring compressor
Piston ring installation tool
Cylinder compression gauge
Cylinder ridge reamer
Cylinder surfacing hone
Cylinder bore gauge
Micrometers and/or dial calipers
Hydraulic lifter removal tool
Balljoint separator
Universal-type puller
Impact screwdriver
Dial indicator set
Stroboscopic timing light (inductive
* pick-up)*
Hand operated vacuum/pressure pump
Tachometer/dwell meter
Universal electrical multimeter
Cable hoist
Brake spring removal and installation
* tools*
Floor jack

Buying tools

For the do-it-yourselfer who is just starting to get involved in vehicle maintenance and repair, there are a number of options available when purchasing tools. If maintenance and minor repair is the extent of the work to be done, the purchase of individual tools is satisfactory. If, on the other hand, extensive work is planned, it would be a good idea to purchase a modest tool set from one of the large retail chain stores. A set can usually be bought at a substantial savings over the individual tool prices, and they often come with a tool box. As additional tools are needed, add-on sets, individual tools and a larger tool box can be purchased to expand the tool selection. Building a tool set gradually allows the cost of the tools to be spread over a longer period of time and gives the mechanic the freedom to choose only those tools that will actually be used.

Tool stores will often be the only source of some of the special tools that are needed, but regardless of where tools are bought, try to avoid cheap ones, especially when buying screwdrivers and sockets, because they won't last very long. The expense involved in replacing cheap tools will eventually be greater than the initial cost of quality tools.

Care and maintenance of tools

Good tools are expensive, so it makes sense to treat them with respect. Keep them clean and in usable condition and store them properly when not in use. Always wipe off any dirt, grease or metal chips before putting them away. Never leave tools lying around in the work area. Upon completion of a job, always check closely under the hood for tools that may have been left there so they won't get lost during a test drive.

Some tools, such as screwdrivers, pliers, wrenches and sockets, can be hung on a panel mounted on the garage or workshop wall, while others should be kept in a tool box or tray. Measuring instruments, gauges, meters, etc. must be carefully stored where they cannot be damaged by weather or impact from other tools.

When tools are used with care and stored properly, they will last a very long time. Even with the best of care, though, tools will wear out if used frequently. When a tool is damaged or worn out, replace it. Subsequent jobs will be safer and more enjoyable if you do.

How to repair damaged threads

Sometimes, the internal threads of a nut or bolt hole can become stripped, usually from overtightening. Stripping threads is an all-too-common occurrence, especially when working with aluminum parts, because aluminum is so soft that it easily strips out.

Usually, external or internal threads are only partially stripped. After they've been cleaned up with a tap or die, they'll still work. Sometimes, however, threads are badly damaged. When this happens, you've got three choices:

1) *Drill and tap the hole to the next suitable oversize and install a larger diameter bolt, screw or stud.*
2) *Drill and tap the hole to accept a threaded plug, then drill and tap the plug to the original screw size. You can also buy a plug already threaded to the original size. Then you simply drill a hole to*
the specified size, then run the threaded plug into the hole with a bolt and jam nut. Once the plug is fully seated, remove the jam nut and bolt.
3) *The third method uses a patented thread repair kit like Heli-Coil or Slimsert. These easy-to-use kits are designed to repair damaged threads in straight-through holes and blind holes. Both are available as kits which can handle a variety of sizes and thread patterns. Drill the hole, then tap it with the special included tap. Install the Heli-Coil and the hole is back to its original diameter and thread pitch.*

Regardless of which method you use, be sure to proceed calmly and carefully. A little impatience or carelessness during one of these relatively simple procedures can ruin your whole day's work and cost you a bundle if you wreck an expensive part.

Working facilities

Not to be overlooked when discussing tools is the workshop. If anything more than routine maintenance is to be carried out, some sort of suitable work area is essential.

It is understood, and appreciated, that many home mechanics do not have a good workshop or garage available, and end up removing an engine or doing major repairs outside. It is recommended, however, that the overhaul or repair be completed under the cover of a roof.

A clean, flat workbench or table of comfortable working height is an absolute necessity. The workbench should be equipped with a vise that has a jaw opening of at least four inches.

As mentioned previously, some clean, dry storage space is also required for tools, as well as the lubricants, fluids, cleaning solvents, etc. which soon become necessary.

Sometimes waste oil and fluids, drained from the engine or cooling system during normal maintenance or repairs, present a disposal problem. To avoid pouring them on the ground or into a sewage system, pour the used fluids into large containers, seal them with caps and take them to an authorized disposal site or recycling center. Plastic jugs, such as old antifreeze containers, are ideal for this purpose.

Always keep a supply of old newspapers and clean rags available. Old towels are excellent for mopping up spills. Many mechanics use rolls of paper towels for most work because they are readily available and disposable. To help keep the area under the vehicle clean, a large cardboard box can be cut open and flattened to protect the garage or shop floor.

Whenever working over a painted surface, such as when leaning over a fender to service something under the hood, always cover it with an old blanket or bedspread to protect the finish. Vinyl covered pads, made especially for this purpose, are available at auto parts stores.

Jacking and towing

Warning: *On models equipped with Automatic Ride Control, the system service switch must be turned OFF (see Chapter 10) before jacking or towing.*

Jacking

When lifting the vehicle with either the supplied jack or a floor jack, block the wheel that is diagonally opposite the one being lifted to prevent the vehicle from moving. **Warning:** *If equipped with an under-chassis mounted spare tire, remove the tire or tire carrier from the rack before the vehicle is raised in order to a void sudden weight release from the chassis.*

When using the vehicle jack, the manufacturer recommends positioning the jack under the axles, shock absorber or jacking bracket, as close to the wheel to be removed as possible. **Caution:** *When raising the rear axle, don't place the jack under the differential, this could cause the rear cover to leak. Place the jack under the axle tubes.*

When raising the vehicle for service, always place jackstands under the frame rails. NEVER work under the vehicle when it is supported only by a jack!

Towing

Equipment specifically designed for towing should be used. It should be attached to the frame members of the vehicle, not the bumpers or brackets. Safety is a major consideration when towing and all applicable state and local laws must be obeyed. A safety chain system must be used at all times. Remember that power steering and power brakes will not work with the engine off.

It is recommended 2WD vehicles be towed from the rear, with the rear wheels off the ground. If it's absolutely necessary, 2WD vehicles can be towed from the front with the front wheels off the ground, provided that speeds do not exceed 35 mph and the distance is less than 50 miles; the transmission may be damaged if the speed/mileage limitations are exceeded. Four-wheel drive vehicles must <u>not</u> be towed with all four wheels on the ground. They must be towed with the front wheels raised and the rear wheels on a towing dolly or placed on a flat bed trailer or truck.

When the vehicle is towed with the rear wheels raised, the steering wheel must be

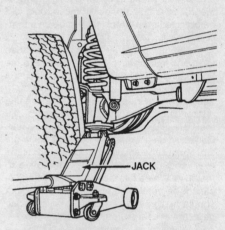

Front axle jacking location under the axle arm (1991 to 1994 models)

clamped in the straight ahead position with a special device designed for use during towing. **Warning:** *Don't use the steering lock to keep the front wheels pointed straight ahead. It's not strong enough for this purpose.*

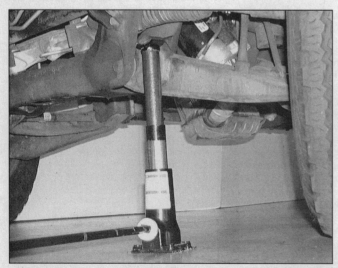

Front suspension jacking point, under the lower control arm tab (1995 and later models)

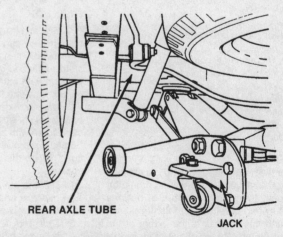

Rear jacking locations - Position the jack under the rear axle tube, not under the differential

Booster battery (jump) starting

Observe these precautions when using a booster battery to start a vehicle:

a) *Before connecting the booster battery, make sure the ignition switch is in the Off position.*
b) *Turn off the lights, heater and other electrical loads.*
c) *Your eyes should be shielded. Safety goggles are a good idea.*
d) *Make sure the booster battery is the same voltage as the dead one in the vehicle.*
e) *The two vehicles MUST NOT TOUCH each other!*
f) *Make sure the transaxle is in Neutral (manual) or Park (automatic).*
g) *If the booster battery is not a maintenance-free type, remove the vent caps and lay a cloth over the vent holes.*

Connect the red jumper cable to the positive (+) terminals of each battery **(see illustration)**.

Connect one end of the black jumper cable to the negative (-) terminal of the booster battery. The other end of this cable should be connected to a good ground on the vehicle to be started, such as a bolt or bracket on the body.

Start the engine using the booster battery, then, with the engine running at idle speed, disconnect the jumper cables in the reverse order of connection.

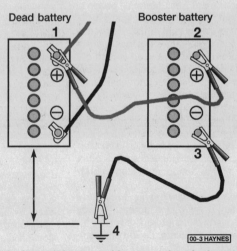

Make the booster battery cable connections in the numerical order shown (note that the negative cable of the booster battery is NOT attached to the negative terminal of the dead battery)

Automotive chemicals and lubricants

A number of automotive chemicals and lubricants are available for use during vehicle maintenance and repair. They include a wide variety of products ranging from cleaning solvents and degreasers to lubricants and protective sprays for rubber, plastic and vinyl.

Cleaners

Carburetor cleaner and choke cleaner is a strong solvent for gum, varnish and carbon. Most carburetor cleaners leave a dry-type lubricant film which will not harden or gum up. Because of this film it is not recommended for use on electrical components.

Brake system cleaner is used to remove grease and brake fluid from the brake system, where clean surfaces are absolutely necessary. It leaves no residue and often eliminates brake squeal caused by contaminants.

Electrical cleaner removes oxidation, corrosion and carbon deposits from electrical contacts, restoring full current flow. It can also be used to clean spark plugs, carburetor jets, voltage regulators and other parts where an oil-free surface is desired.

Demoisturants remove water and moisture from electrical components such as alternators, voltage regulators, electrical connectors and fuse blocks. They are non-conductive, non-corrosive and non-flammable.

Degreasers are heavy-duty solvents used to remove grease from the outside of the engine and from chassis components. They can be sprayed or brushed on and, depending on the type, are rinsed off either with water or solvent.

Lubricants

Motor oil is the lubricant formulated for use in engines. It normally contains a wide variety of additives to prevent corrosion and reduce foaming and wear. Motor oil comes in various weights (viscosity ratings) from 5 to 80. The recommended weight of the oil depends on the season, temperature and the demands on the engine. Light oil is used in cold climates and under light load conditions. Heavy oil is used in hot climates and where high loads are encountered. Multi-viscosity oils are designed to have characteristics of both light and heavy oils and are available in a number of weights from 5W-20 to 20W-50.

Gear oil is designed to be used in differentials, manual transmissions and other areas where high-temperature lubrication is required.

Chassis and wheel bearing grease is a heavy grease used where increased loads and friction are encountered, such as for wheel bearings, balljoints, tie-rod ends and universal joints.

High-temperature wheel bearing grease is designed to withstand the extreme temperatures encountered by wheel bearings in disc brake equipped vehicles. It usually contains molybdenum disulfide (moly), which is a dry-type lubricant.

White grease is a heavy grease for metal-to-metal applications where water is a problem. White grease stays soft under both low and high temperatures (usually from -100 to +190-degrees F), and will not wash off or dilute in the presence of water.

Assembly lube is a special extreme pressure lubricant, usually containing moly, used to lubricate high-load parts (such as main and rod bearings and cam lobes) for initial start-up of a new engine. The assembly lube lubricates the parts without being squeezed out or washed away until the engine oiling system begins to function.

Silicone lubricants are used to protect rubber, plastic, vinyl and nylon parts.

Graphite lubricants are used where oils cannot be used due to contamination problems, such as in locks. The dry graphite will lubricate metal parts while remaining uncontaminated by dirt, water, oil or acids. It is electrically conductive and will not foul electrical contacts in locks such as the ignition switch.

Moly penetrants loosen and lubricate frozen, rusted and corroded fasteners and prevent future rusting or freezing.

Heat-sink grease is a special electrically non-conductive grease that is used for mounting electronic ignition modules where it is essential that heat is transferred away from the module.

Sealants

RTV sealant is one of the most widely used gasket compounds. Made from silicone, RTV is air curing, it seals, bonds, waterproofs, fills surface irregularities, remains flexible, doesn't shrink, is relatively easy to remove, and is used as a supplementary sealer with almost all low and medium temperature gaskets.

Anaerobic sealant is much like RTV in that it can be used either to seal gaskets or to form gaskets by itself. It remains flexible, is solvent resistant and fills surface imperfections. The difference between an anaerobic sealant and an RTV-type sealant is in the curing. RTV cures when exposed to air, while an anaerobic sealant cures only in the absence of air. This means that an anaerobic sealant cures only after the assembly of parts, sealing them together.

Thread and pipe sealant is used for sealing hydraulic and pneumatic fittings and vacuum lines. It is usually made from a Teflon compound, and comes in a spray, a paint-on liquid and as a wrap-around tape.

Chemicals

Anti-seize compound prevents seizing, galling, cold welding, rust and corrosion in fasteners. High-temperature ant-seize, usually made with copper and graphite lubricants, is used for exhaust system and exhaust manifold bolts.

Anaerobic locking compounds are used to keep fasteners from vibrating or working loose and cure only after installation, in the absence of air. Medium strength locking compound is used for small nuts, bolts and screws that may be removed later. High-strength locking compound is for large nuts, bolts and studs which aren't removed on a regular basis.

Oil additives range from viscosity index improvers to chemical treatments that claim to reduce internal engine friction. It should be noted that most oil manufacturers caution against using additives with their oils.

Gas additives perform several functions, depending on their chemical makeup. They usually contain solvents that help dissolve gum and varnish that build up on carburetor, fuel injection and intake parts. They also serve to break down carbon deposits that form on the inside surfaces of the combustion chambers. Some additives contain upper cylinder lubricants for valves and piston rings, and others contain chemicals to remove condensation from the gas tank.

Miscellaneous

Brake fluid is specially formulated hydraulic fluid that can withstand the heat and pressure encountered in brake systems. Care must be taken so this fluid does not come in contact with painted surfaces or plastics. An opened container should always be resealed to prevent contamination by water or dirt.

Weatherstrip adhesive is used to bond weatherstripping around doors, windows and trunk lids. It is sometimes used to attach trim pieces.

Undercoating is a petroleum-based, tar-like substance that is designed to protect metal surfaces on the underside of the vehicle from corrosion. It also acts as a sound-deadening agent by insulating the bottom of the vehicle.

Waxes and polishes are used to help protect painted and plated surfaces from the weather. Different types of paint may require the use of different types of wax and polish. Some polishes utilize a chemical or abrasive cleaner to help remove the top layer of oxidized (dull) paint on older vehicles. In recent years many non-wax polishes that contain a wide variety of chemicals such as polymers and silicones have been introduced. These non-wax polishes are usually easier to apply and last longer than conventional waxes and polishes.

Conversion factors

Length (distance)

Inches (in)	X	25.4	= Millimetres (mm)	X	0.0394 = Inches (in)
Feet (ft)	X	0.305	= Metres (m)	X	3.281 = Feet (ft)
Miles	X	1.609	= Kilometres (km)	X	0.621 = Miles

Volume (capacity)

Cubic inches (cu in; in³)	X	16.387	= Cubic centimetres (cc; cm³)	X	0.061 = Cubic inches (cu in; in³)
Imperial pints (Imp pt)	X	0.568	= Litres (l)	X	1.76 = Imperial pints (Imp pt)
Imperial quarts (Imp qt)	X	1.137	= Litres (l)	X	0.88 = Imperial quarts (Imp qt)
Imperial quarts (Imp qt)	X	1.201	= US quarts (US qt)	X	0.833 = Imperial quarts (Imp qt)
US quarts (US qt)	X	0.946	= Litres (l)	X	1.057 = US quarts (US qt)
Imperial gallons (Imp gal)	X	4.546	= Litres (l)	X	0.22 = Imperial gallons (Imp gal)
Imperial gallons (Imp gal)	X	1.201	= US gallons (US gal)	X	0.833 = Imperial gallons (Imp gal)
US gallons (US gal)	X	3.785	= Litres (l)	X	0.264 = US gallons (US gal)

Mass (weight)

Ounces (oz)	X	28.35	= Grams (g)	X	0.035 = Ounces (oz)
Pounds (lb)	X	0.454	= Kilograms (kg)	X	2.205 = Pounds (lb)

Force

Ounces-force (ozf; oz)	X	0.278	= Newtons (N)	X	3.6 = Ounces-force (ozf; oz)
Pounds-force (lbf; lb)	X	4.448	= Newtons (N)	X	0.225 = Pounds-force (lbf; lb)
Newtons (N)	X	0.1	= Kilograms-force (kgf; kg)	X	9.81 = Newtons (N)

Pressure

Pounds-force per square inch (psi; lbf/in²; lb/in²)	X	0.070	= Kilograms-force per square centimetre (kgf/cm²; kg/cm²)	X	14.223 = Pounds-force per square inch (psi; lbf/in²; lb/in²)
Pounds-force per square inch (psi; lbf/in²; lb/in²)	X	0.068	= Atmospheres (atm)	X	14.696 = Pounds-force per square inch (psi; lbf/in²; lb/in²)
Pounds-force per square inch (psi; lbf/in²; lb/in²)	X	0.069	= Bars	X	14.5 = Pounds-force per square inch (psi; lbf/in²; lb/in²)
Pounds-force per square inch (psi; lbf/in²; lb/in²)	X	6.895	= Kilopascals (kPa)	X	0.145 = Pounds-force per square inch (psi; lbf/in²; lb/in²)
Kilopascals (kPa)	X	0.01	= Kilograms-force per square centimetre (kgf/cm²; kg/cm²)	X	98.1 = Kilopascals (kPa)

Torque (moment of force)

Pounds-force inches (lbf in; lb in)	X	1.152	= Kilograms-force centimetre (kgf cm; kg cm)	X	0.868 = Pounds-force inches (lbf in; lb in)
Pounds-force inches (lbf in; lb in)	X	0.113	= Newton metres (Nm)	X	8.85 = Pounds-force inches (lbf in; lb in)
Pounds-force inches (lbf in; lb in)	X	0.083	= Pounds-force feet (lbf ft; lb ft)	X	12 = Pounds-force inches (lbf in; lb in)
Pounds-force feet (lbf ft; lb ft)	X	0.138	= Kilograms-force metres (kgf m; kg m)	X	7.233 = Pounds-force feet (lbf ft; lb ft)
Pounds-force feet (lbf ft; lb ft)	X	1.356	= Newton metres (Nm)	X	0.738 = Pounds-force feet (lbf ft; lb ft)
Newton metres (Nm)	X	0.102	= Kilograms-force metres (kgf m; kg m)	X	9.804 = Newton metres (Nm)

Vacuum

Inches mercury (in. Hg)	X	3.377	= Kilopascals (kPa)	X	0.2961 = Inches mercury
Inches mercury (in. Hg)	X	25.4	= Millimeters mercury (mm Hg)	X	0.0394 = Inches mercury

Power

Horsepower (hp)	X	745.7	= Watts (W)	X	0.0013 = Horsepower (hp)

Velocity (speed)

Miles per hour (miles/hr; mph)	X	1.609	= Kilometres per hour (km/hr; kph)	X	0.621 = Miles per hour (miles/hr; mph)

Fuel consumption*

Miles per gallon, Imperial (mpg)	X	0.354	= Kilometres per litre (km/l)	X	2.825 = Miles per gallon, Imperial (mpg)
Miles per gallon, US (mpg)	X	0.425	= Kilometres per litre (km/l)	X	2.352 = Miles per gallon, US (mpg)

Temperature

Degrees Fahrenheit = (°C x 1.8) + 32

Degrees Celsius (Degrees Centigrade; °C) = (°F - 32) x 0.56

*It is common practice to convert from miles per gallon (mpg) to litres/100 kilometres (l/100km), where mpg (Imperial) x l/100 km = 282 and mpg (US) x l/100 km = 235

Safety first!

Regardless of how enthusiastic you may be about getting on with the job at hand, take the time to ensure that your safety is not jeopardized. A moment's lack of attention can result in an accident, as can failure to observe certain simple safety precautions. The possibility of an accident will always exist, and the following points should not be considered a comprehensive list of all dangers. Rather, they are intended to make you aware of the risks and to encourage a safety conscious approach to all work you carry out on your vehicle.

Essential DOs and DON'Ts

DON'T rely on a jack when working under the vehicle. Always use approved jackstands to support the weight of the vehicle and place them under the recommended lift or support points.

DON'T attempt to loosen extremely tight fasteners (i.e. wheel lug nuts) while the vehicle is on a jack - it may fall.

DON'T start the engine without first making sure that the transmission is in Neutral (or Park where applicable) and the parking brake is set.

DON'T remove the radiator cap from a hot cooling system - let it cool or cover it with a cloth and release the pressure gradually.

DON'T attempt to drain the engine oil until you are sure it has cooled to the point that it will not burn you.

DON'T touch any part of the engine or exhaust system until it has cooled sufficiently to avoid burns.

DON'T siphon toxic liquids such as gasoline, antifreeze and brake fluid by mouth, or allow them to remain on your skin.

DON'T inhale brake lining dust - it is potentially hazardous (see *Asbestos* below).

DON'T allow spilled oil or grease to remain on the floor - wipe it up before someone slips on it.

DON'T use loose fitting wrenches or other tools which may slip and cause injury.

DON'T push on wrenches when loosening or tightening nuts or bolts. Always try to pull the wrench toward you. If the situation calls for pushing the wrench away, push with an open hand to avoid scraped knuckles if the wrench should slip.

DON'T attempt to lift a heavy component alone - get someone to help you.

DON'T rush or take unsafe shortcuts to finish a job.

DON'T allow children or animals in or around the vehicle while you are working on it.

DO wear eye protection when using power tools such as a drill, sander, bench grinder, etc. and when working under a vehicle.

DO keep loose clothing and long hair well out of the way of moving parts.

DO make sure that any hoist used has a safe working load rating adequate for the job.

DO get someone to check on you periodically when working alone on a vehicle.

DO carry out work in a logical sequence and make sure that everything is correctly assembled and tightened.

DO keep chemicals and fluids tightly capped and out of the reach of children and pets.

DO remember that your vehicle's safety affects that of yourself and others. If in doubt on any point, get professional advice.

Asbestos

Certain friction, insulating, sealing, and other products - such as brake linings, brake bands, clutch linings, torque converters, gaskets, etc. - may contain asbestos. Extreme care must be taken to avoid inhalation of dust from such products, since it is hazardous to health. If in doubt, assume that they do contain asbestos.

Fire

Remember at all times that gasoline is highly flammable. Never smoke or have any kind of open flame around when working on a vehicle. But the risk does not end there. A spark caused by an electrical short circuit, by two metal surfaces contacting each other, or even by static electricity built up in your body under certain conditions, can ignite gasoline vapors, which in a confined space are highly explosive. Do not, under any circumstances, use gasoline for cleaning parts. Use an approved safety solvent.

Always disconnect the battery ground (-) cable at the battery before working on any part of the fuel system or electrical system. Never risk spilling fuel on a hot engine or exhaust component. It is strongly recommended that a fire extinguisher suitable for use on fuel and electrical fires be kept handy in the garage or workshop at all times. Never try to extinguish a fuel or electrical fire with water.

Fumes

Certain fumes are highly toxic and can quickly cause unconsciousness and even death if inhaled to any extent. Gasoline vapor falls into this category, as do the vapors from some cleaning solvents. Any draining or pouring of such volatile fluids should be done in a well ventilated area.

When using cleaning fluids and solvents, read the instructions on the container carefully. Never use materials from unmarked containers.

Never run the engine in an enclosed space, such as a garage. Exhaust fumes contain carbon monoxide, which is extremely poisonous. If you need to run the engine, always do so in the open air, or at least have the rear of the vehicle outside the work area.

If you are fortunate enough to have the use of an inspection pit, never drain or pour gasoline and never run the engine while the vehicle is over the pit. The fumes, being heavier than air, will concentrate in the pit with possibly lethal results.

The battery

Never create a spark or allow a bare light bulb near a battery. They normally give off a certain amount of hydrogen gas, which is highly explosive.

Always disconnect the battery ground (-) cable at the battery before working on the fuel or electrical systems.

If possible, loosen the filler caps or cover when charging the battery from an external source (this does not apply to sealed or maintenance-free batteries). Do not charge at an excessive rate or the battery may burst.

Take care when adding water to a non maintenance-free battery and when carrying a battery. The electrolyte, even when diluted, is very corrosive and should not be allowed to contact clothing or skin.

Always wear eye protection when cleaning the battery to prevent the caustic deposits from entering your eyes.

Household current

When using an electric power tool, inspection light, etc., which operates on household current, always make sure that the tool is correctly connected to its plug and that, where necessary, it is properly grounded. Do not use such items in damp conditions and, again, do not create a spark or apply excessive heat in the vicinity of fuel or fuel vapor.

Secondary ignition system voltage

A severe electric shock can result from touching certain parts of the ignition system (such as the spark plug wires) when the engine is running or being cranked, particularly if components are damp or the insulation is defective. In the case of an electronic ignition system, the secondary system voltage is much higher and could prove fatal.

Troubleshooting

Contents

This section provides an easy reference guide to the more common problems which may occur during the operation of your vehicle. These problems and possible causes are grouped under various components or systems; i.e. Engine, Cooling System, etc., and also refer to the Chapter and/or Section which deals with the problem.

Remember that successful troubleshooting is not a mysterious black art practiced only by professional mechanics. It's simply the result of a bit of knowledge combined with an intelligent, systematic approach to the problem. Always work by a process of elimination, starting with the simplest solution and working through to the most complex - and never overlook the obvious. Anyone can forget to fill the gas tank or leave the lights on overnight, so don't assume that you are above such oversights.

Finally, always get clear in your mind why a problem has occurred and take steps to ensure that it doesn't happen again. If the electrical system fails because of a poor connection, check all other connections in the system to make sure that they don't fail as well. If a particular fuse continues to blow, find out why - don't just go on replacing fuses. Remember, failure of a small component can often be indicative of potential failure or incorrect functioning of a more important component or system.

Many systems on modern vehicles are controlled by the vehicle computer, and some models have extra computers for special features, such as the airbag system, Automatic Ride Control, or automatic climate control. Much of today's diagnostic work is done with expensive scan tools not available to the home mechanic. When common-sense troubleshooting doesn't turn up an obvious problem, the vehicle should be taken to a dealer service department for electronic scan-tool troubleshooting.

Engine

1 Engine will not rotate when attempting to start

1 Battery terminal connections loose or corroded. Check the cable terminals at the battery; tighten cable clamp and/or clean off corrosion as necessary (Chapter 1).
2 Battery discharged or faulty. If the cable ends are clean and tight on the battery posts, turn the key to the On position and switch on the headlights or windshield wipers. If they won't run, the battery is discharged.
3 Automatic transmission not engaged in park (P) or Neutral (N).
4 Broken, loose or disconnected wires in the starting circuit. Inspect all wires and connectors at the battery, starter solenoid and

ignition switch (on steering column).
5 Starter motor pinion jammed in driveplate ring gear. Remove starter (Chapter 5) and inspect pinion and driveplate (Chapter 2).
6 Starter solenoid faulty (Chapter 5).
7 Starter motor faulty (Chapter 5).
8 Ignition switch faulty (Chapter 12).
9 Engine seized. Try to turn the crankshaft with a large socket and breaker bar on the pulley bolt.
10 Starter relay faulty (Chapter 5)
11 Transmission Range (TR) sensor out of adjustment or defective (Chapter 6)

2 Engine rotates but will not start

1 Fuel tank empty.
2 Battery discharged (engine rotates slowly).
3 Battery terminal connections loose or corroded.
4 Fuel not reaching fuel injectors. Check for clogged fuel filter or lines and defective fuel pump. Also make sure the tank vent lines aren't clogged (Chapter 4). If you have had an impact, check the fuel pump inertia switch (Chapter 4).
5 Low cylinder compression. Check as described in Chapter 2.
6 Water in fuel. Drain tank and fill with new fuel.
7 Wet or damaged ignition components (Chapters 1 and 5).
8 Worn, faulty or incorrectly gapped spark plugs (Chapter 1).
9 Broken, loose or disconnected wires at the ignition coil or faulty coil (Chapter 5).
10 Timing chain failure or wear affecting valve timing (Chapter 2).
11 Fuel injection or engine control systems failure (Chapters 4 and 6).

3 Starter motor operates without turning engine

1 Starter pinion sticking. Remove the starter (Chapter 5) and inspect.
2 Starter pinion or driveplate teeth worn or broken. Remove the inspection cover and inspect.

4 Engine hard to start when cold

1 Battery discharged or low. Check as described in Chapter 1.
2 Fuel not reaching the fuel injectors. Check the fuel filter, lines and fuel pump (Chapters 1 and 4).
3 Defective spark plugs (Chapter 1).
4 Defective engine coolant temperature

sensor (Chapter 6).
5 Fuel injection or engine control systems malfunction (Chapters 4 and 6).

5 Engine hard to start when hot

1 Air filter dirty (Chapter 1).
2 Fuel not reaching the fuel injection (see Section 4). Check for a vapor lock situation, brought about by clogged fuel tank vent lines.
3 Bad engine ground connection.
4 Fuel injection or engine control systems malfunction (Chapters 4 and 6).

6 Starter motor noisy or engages roughly

1 Pinion or driveplate teeth worn or broken. Remove the inspection cover on the left side of the engine and inspect.
2 Starter motor mounting bolts loose or missing.

7 Engine starts but stops immediately

1 Loose or damaged wire harness connections at distributor, coil or alternator.
2 Intake manifold vacuum leaks. Make sure all mounting bolts/nuts are tight and all vacuum hoses connected to the manifold are attached properly and in good condition.
3 Insufficient fuel pressure (Chapter 4).
4 Fuel injection or engine control systems malfunction (Chapters 4 and 6).

8 Engine 'lopes' while idling or idles erratically

1 Vacuum leaks. Check mounting bolts at the intake manifold for tightness. Make sure that all vacuum hoses are connected and in good condition. Use a stethoscope or a length of fuel hose held against your ear to listen for vacuum leaks while the engine is running. A hissing sound will be heard. A soapy water solution will also detect leaks. Check the intake manifold gasket surfaces.
2 Leaking EGR valve or plugged PCV valve (Chapters 1 and 6).
3 Air filter clogged (Chapter 1).
4 Fuel pump not delivering sufficient fuel (Chapter 4).
5 Leaking head gasket. Perform a cylinder compression check (Chapter 2).
6 Timing chain(s) worn (Chapter 2).
7 Camshaft lobes worn (Chapter 2).
8 Valves burned or otherwise leaking (Chapter 2).
9 Ignition timing out of adjustment (Chapter 5).
10 Ignition system not operating properly

(Chapters 1 and 5).
11 Fuel injection or engine control systems malfunction (Chapters 4 and 6).

9 Engine misses at idle speed

1 Spark plugs faulty or not gapped properly (Chapter 1).
2 Faulty spark plug wires (Chapter 1).
3 Wet or damaged ignition components (Chapter 5).
4 Short circuits in ignition, coil or spark plug wires.
5 Sticking or faulty emissions systems (Chapter 6).
6 Clogged fuel filter and/or foreign matter in fuel. Remove the fuel filter (Chapter 1) and inspect.
7 Vacuum leaks at intake manifold or hose connections. Check as described in Section.
8 Low or uneven cylinder compression. Check as described in Chapter 2.
9 Fuel injection or engine control systems malfunction (Chapters 4 and 6).

10 Excessively high idle speed

1 Sticking throttle linkage (Chapter 4).
2 Vacuum leaks at intake manifold or hose connections. Check as described in Section 8.
3 Fuel injection or engine control systems malfunction (Chapters 4 and 6).

11 Battery will not hold a charge

1 Alternator drivebelt defective or not adjusted properly (Chapter 1).
2 Battery cables loose or corroded (Chapter 1).
3 Alternator not charging properly (Chapter 5).
4 Loose, broken or faulty wires in the charging circuit (Chapter 5).
5 Short circuit causing a continuous drain on the battery.
6 Battery defective internally.

12 Alternator light stays on

1 Fault in alternator or charging circuit (Chapter 5).
2 Alternator drivebelt defective or not properly adjusted (Chapter 1).

13 Alternator light fails to come on when key is turned on

1 Faulty bulb (Chapter 12).
2 Defective alternator (Chapter 5).
3 Fault in the printed circuit, dash wiring or bulb holder (Chapter 12).

14 Engine misses throughout driving speed range

1 Fuel filter clogged and/or impurities in the fuel system. Check fuel filter (Chapter 1) or clean system (Chapter 4).
2 Faulty or incorrectly gapped spark plugs (Chapter 1).
3 Incorrect ignition timing (Chapter 5).
4 Defective spark plug wires (Chapter 1).
5 Emissions system components faulty (Chapter 6).
6 Low or uneven cylinder compression pressures. Check as described in Chapter 2.
7 Weak or faulty ignition coil(s) (Chapter 5).
8 Weak or faulty ignition system (Chapter 5).
9 Vacuum leaks at intake manifold or vacuum hoses. (see Section 8).
10 Dirty or clogged fuel injector(s) (Chapter 4).
11 Leaky EGR valve (Chapter 6).
12 Fuel injection or engine control systems malfunction (Chapters 4 and 6).

15 Hesitation or stumble during acceleration

1 Ignition system not operating properly (Chapter 5).
2 Dirty or clogged fuel injector(s) (Chapter 4).
3 Low fuel pressure. Check for proper operation of the fuel pump and for restrictions in the fuel filter and lines (Chapter 4).
4 Fuel injection or engine control systems malfunction (Chapters 4 and 6).

16 Engine stalls

1 Idle speed incorrect (Chapter 4).
2 Fuel filter clogged and/or water and impurities in the fuel system (Chapter 1).
3 Damaged or wet distributor cap and wires.
4 Emissions system components faulty (Chapter 6).
5 Faulty or incorrectly gapped spark plugs (Chapter 1). Also check the spark plug wires (Chapter 1).
6 Vacuum leak at the intake manifold or vacuum hoses. Check as described in Section 8.
7 Fuel injection or engine control systems malfunction (Chapters 4 and 6).

17 Engine lacks power

1 Incorrect ignition timing (Chapter 5).
2 Faulty or incorrectly gapped spark plugs (Chapter 1).
3 Air filter dirty (Chapter 1).

4 Faulty ignition coil(s) (Chapter 5).
5 Brakes binding (Chapters 1 and 10).
6 Automatic transmission fluid level incorrect, causing slippage (Chapter 1).
7 Fuel filter clogged and/or impurities in the fuel system (Chapters 1 and 4).
8 EGR system not functioning properly (Chapter 6).
9 Use of sub-standard fuel. Fill tank with proper octane fuel.
10 Low or uneven cylinder compression pressures. Check as described in Chapter 2.
11 Vacuum leak at intake manifold or vacuum hoses (check as described in Section 8).
12 Dirty or clogged fuel injector(s) (Chapters 1 and 4).
13 Fuel injection or engine control systems malfunction (Chapters 4 and 6).
14 Restricted exhaust system (Chapter 4).

18 Engine backfires

1 EGR system not functioning properly (Chapter 6).
2 Ignition timing incorrect (Chapter 5).
3 Vacuum leak (refer to Section 8).
4 Damaged valve springs or sticking valves (Chapter 2).
5 Vacuum leak at the intake manifold or vacuum hoses (see Section 8).

19 Engine surges while holding accelerator steady

1 Vacuum leak at the intake manifold or vacuum hoses (see Section 8).
2 Restricted air filter (Chapter 1).
3 Fuel pump or pressure regulator defective (Chapter 4).
4 Fuel injection or engine control systems malfunction (Chapters 4 and 6).

20 Pinging or knocking engine sounds when engine is under load

1 Incorrect grade of fuel. Fill tank with fuel of the proper octane rating.
2 Ignition timing incorrect (Chapter 5).
3 Carbon build-up in combustion chambers. Remove cylinder head(s) and clean combustion chambers (Chapter 2).
4 Incorrect spark plugs (Chapter 1).
5 Fuel injection or engine control systems malfunction (Chapters 4 and 6).
6 Restricted exhaust system (Chapter 4).

21 Engine diesels (continues to run) after being turned off

1 Idle speed too high (Chapter 4).

2   Ignition timing incorrect (Chapter 5).
3   Incorrect spark plug heat range (Chapter 1).
4   Vacuum leak at the intake manifold or vacuum hoses (see Section 8).
5   Carbon build-up in combustion chambers. Remove the cylinder head(s) and clean the combustion chambers (Chapter 2).
6   Valves sticking (Chapter 2).
7   EGR system not operating properly (Chapter 6).
8   Fuel injection or engine control systems malfunction (Chapters 4 and 6).
9   Check for causes of overheating (Section 27).

22   Low oil pressure

1   Improper grade of oil.
2   Oil pump worn or damaged (Chapter 2).
3   Engine overheating (refer to Section 27).
4   Clogged oil filter (Chapter 1).
5   Clogged oil strainer (Chapter 2).
6   Oil pressure gauge not working properly (Chapter 2).

23   Excessive oil consumption

1   Loose oil drain plug.
2   Loose bolts or damaged oil pan gasket (Chapter 2).
3   Loose bolts or damaged front cover gasket (Chapter 2).
4   Front or rear crankshaft oil seal leaking (Chapter 2).
5   Loose bolts or damaged valve cover gasket (Chapter 2).
6   Loose oil filter (Chapter 1).
7   Loose or damaged oil pressure switch (Chapter 2).
8   Pistons and cylinders excessively worn (Chapter 2).
9   Piston rings not installed correctly on pistons (Chapter 2).
10   Worn or damaged piston rings (Chapter 2).
11   Intake and/or exhaust valve oil seals worn or damaged (Chapter 2).
12   Worn valve stems or guides.
13   Worn or damaged valves/guides (Chapter 2).
14   Faulty or incorrect PCV valve allowing too much crankcase airflow.

24   Excessive fuel consumption

1   Dirty or clogged air filter element (Chapter 1).
2   Incorrect ignition timing (Chapter 5).
3   Incorrect idle speed (Chapter 4).
4   Low tire pressure or incorrect tire size (Chapter 10).
5   Inspect for binding brakes.
6   Fuel leakage. Check all connections,

lines and components in the fuel system (Chapter 4).
7   Dirty or clogged fuel injectors (Chapter 4).
8   Fuel injection or engine control systems malfunction (Chapters 4 and 6).
9   Thermostat stuck open or not installed.
10   Improperly operating transmission.

25   Fuel odor

1   Fuel leakage. Check all connections, lines and components in the fuel system (Chapter 4).
2   Fuel tank overfilled. Fill only to automatic shut-off.
3   Charcoal canister filter in Evaporative Emissions Control system clogged (Chapter 1).
4   Vapor leaks from Evaporative Emissions Control system lines (Chapter 6).

26   Miscellaneous engine noises

1   A strong dull noise that becomes more rapid as the engine accelerates indicates worn or damaged crankshaft bearings or an unevenly worn crankshaft. To pinpoint the trouble spot, remove the spark plug wire from one plug at a time and crank the engine over. If the noise stops, the cylinder with the removed plug wire indicates the problem area. Replace the bearing and/or service or replace the crankshaft (Chapter 2).
2   A similar (yet slightly higher pitched) noise to the crankshaft knocking described in the previous paragraph, that becomes more rapid as the engine accelerates, indicates worn or damaged connecting rod bearings (Chapter 2). The procedure for locating the problem cylinder is the same as described in Paragraph 1.
3   An overlapping metallic noise that increases in intensity as the engine speed increases, yet diminishes as the engine warms up indicates abnormal piston and cylinder wear (Chapter 2). To locate the problem cylinder, use the procedure described in Paragraph 1.
4   A rapid clicking noise that becomes faster as the engine accelerates indicates a worn piston pin or piston pin hole. This sound will happen each time the piston hits the highest and lowest points in the stroke (Chapter 2). The procedure for locating the problem piston is described in Paragraph 1.
5   A metallic clicking noise coming from the water pump indicates worn or damaged water pump bearings or pump. Replace the water pump with a new one (Chapter 3).
6   A rapid tapping sound or clicking sound that becomes faster as the engine speed increases indicates "valve tapping." This can be identified by holding one end of a section of hose to your ear and placing the other end at different spots along the valve cover. The

point where the sound is loudest indicates the problem valve. If the pushrod and rocker arm components are in good shape, you likely have a collapsed valve lifter. Changing the engine oil and adding a high viscosity oil treatment will sometimes cure a stuck lifter problem. If the problem persists, the lifters, pushrods and rocker arms must be removed for inspection (Chapter 2).
7   A steady metallic rattling or rapping sound coming from the area of the timing chain cover indicates a worn, damaged or out-of-adjustment timing chain. Service or replace the chain and related components (Chapter 2).

Cooling system

27   Overheating

1   Insufficient coolant in system (Chapter 1).
2   Drivebelt defective or not adjusted properly (Chapter 1).
3   Radiator core blocked or radiator grille dirty and restricted (Chapter 3).
4   Thermostat faulty (Chapter 3).
5   Cooling fan not functioning properly (Chapter 3).
6   Radiator cap not maintaining proper pressure.
7   Defective water pump (Chapter 3).
8   Improper grade of engine oil.
9   Inaccurate temperature gauge (Chapter 12).

28   Overcooling

1   Thermostat faulty (Chapter 3).
2   Inaccurate temperature gauge (Chapter 12).

29   External coolant leakage

1   Deteriorated or damaged hoses. Loose clamps at hose connections (Chapter 1).
2   Water pump seals defective. If this is the case, water will drip from the weep hole in the water pump body (Chapter 3).
3   Leakage from radiator core or header tank. This will require the radiator to be professionally repaired (see Chapter 3 for removal procedures).
4   Leakage from the coolant reservoir.
5   Engine drain plugs or water jacket freeze plugs leaking (Chapters 1 and 2).
6   Leak from coolant temperature switch (Chapter 3).
7   Leak from damaged gaskets or small cracks (Chapter 2).

8   Leak from oil cooler or oil cooler adapter housing.

30   Internal coolant leakage

Note: *Internal coolant leaks can usually be detected by examining the oil. Check the dipstick and inside the rocker arm cover for water deposits and an oil consistency like that of a milkshake.*
1   Leaking cylinder head gasket. Have the system pressure tested or remove the cylinder head (Chapter 2) and inspect.
2   Cracked cylinder bore or cylinder head. Dismantle engine and inspect (Chapter 2).
3   Loose cylinder head bolts (tighten as described in Chapter 2).

31   Abnormal coolant loss

1   Overfilling system (Chapter 1).
2   Coolant boiling away due to overheating (see causes in Section 27).
3   Internal or external leakage (see Sections 29 and 30).
4   Faulty radiator cap. Have the cap pressure tested.
5   Cooling system being pressurized by engine compression. This could be due to a cracked head or block or leaking head gasket(s). Have the system tested for the presence of combustion gas in the coolant.

32   Poor coolant circulation

1   Inoperative water pump. A quick test is to pinch the top radiator hose closed with your hand while the engine is idling, then release it. You should feel a surge of coolant if the pump is working properly (Chapter 3).
2   Restriction in cooling system. Drain, flush and refill the system (Chapter 1). If necessary, remove the radiator (Chapter 3) and have it reverse flushed or professionally cleaned.
3   Loose water pump drivebelt (Chapter 1).
4   Thermostat sticking (Chapter 3).
5   Insufficient coolant (Chapter 1).

33   Corrosion

1   Excessive impurities in the water. Distilled water is recommended.
2   Insufficient antifreeze solution (refer to Chapter 1 for the proper ratio of water to antifreeze).
3   Infrequent flushing and draining of system. Regular flushing of the cooling system should be carried out at the specified intervals as described in (Chapter 1).

Clutch

34   Fails to release (pedal pressed to the floor - shift lever does not move freely in and out of Reverse)

1   Leak in the clutch hydraulic system. Check the master cylinder, slave cylinder and lines (Chapter).
2   Clutch plate warped or damaged (Chapter 8).

35   Clutch slips (engine speed increases with no increase in vehicle speed)

1   Clutch plate oil soaked or lining worn. Remove clutch (Chapter) and inspect.
2   Clutch plate not seated.
3   Pressure plate worn (Chapter 8).

36   Grabbing (chattering) as clutch is engaged

1   Oil on clutch plate lining. Remove (Chapter 8) and inspect. Correct any leakage source.
2   Worn or loose engine or transmission mounts. These units move slightly when the clutch is released. Inspect the mounts and bolts (Chapter 2).
3   Worn splines on clutch plate hub. Remove the clutch components (Chapter 8) and inspect.
4   Warped pressure plate or flywheel. Remove the clutch components and inspect.

37   Squeal or rumble with clutch fully engaged (pedal released)

1   Release bearing binding on transmission bearing retainer. Remove clutch components (Chapter 8) and check bearing. Remove any burrs or nicks; clean and relubricate bearing retainer before installing.

38   Squeal or rumble with clutch fully disengaged (pedal depressed)

1   Worn, defective or broken release bearing (Chapter 8).
2   Worn or broken pressure plate springs (or diaphragm fingers) (Chapter 8).

39   Clutch pedal stays on floor when disengaged

1   Linkage or release bearing binding.

Inspect the linkage or remove the clutch components as necessary.
2   Make sure proper pedal stop (bumper) is installed.

Manual transmission

Note: *All the following references are in Chapter 7A, unless noted.*

40   Noisy in Neutral with engine running

1   Input shaft bearing worn.
2   Damaged main drive gear bearing.
3   Worn countershaft bearings.
4   Worn or damaged countershaft endplay shims.

41   Noisy in all gears

1   Any of the above causes, and/or:
2   Insufficient lubricant (see the checking procedures in Chapter 1).

42   Noisy in one particular gear

1   Worn, damaged or chipped gear teeth for that particular gear.
2   Worn or damaged synchronizer for that particular gear.

43   Slips out of high gear

1   Transmission loose on clutch housing.
2   Dirt between the transmission case and engine or misalignment of the transmission.

44   Difficulty in engaging gears

1   Clutch not releasing completely (see clutch adjustment in Chapter 1).
2   Loose, damaged or out-of-adjustment shift linkage. Make a thorough inspection, replacing parts as necessary.

45   Oil leakage

1   Excessive amount of lubricant in the transmission (see Chapter 1 for correct checking procedures). Drain lubricant as required.
2   Transmission oil seal or vehicle speed sensor O-ring in need of replacement.

Automatic transmission

Note: *Due to the complexity of the automatic transmission, it's difficult for the home mechanic to properly diagnose and service this component without an expensive scan tool. For problems other than the following, the vehicle should be taken to a dealer service department or a transmission shop.*

46   General shift mechanism problems

1   Common problems which may be attributed to a misadjusted shift cable are:

a) *Engine starting in gears other than Park or Neutral.*
b) *Indicator on shifter pointing to a gear other than the one actually being selected.*
c) *Vehicle moves when in Park.*

2   Refer to Chapter 7B to check the shift cable and or the neutral start switch/Transmission Range sensor adjustment.

47   Transmission will not downshift with accelerator pedal pressed to the floor

Since these transmissions are electronically controlled, your dealer or a professional shop with the proper equipment will have to diagnose the probable cause.

48   Transmission slips, shifts rough, is noisy or has no drive in forward or reverse gears

There are many probable causes for the above problems, but before taking the vehicle to a repair shop, check the level and condition of the fluid as described in Chapter 1. Correct fluid level as necessary or change the fluid and filter if needed. If the problem persists, have a professional diagnose the problem.

49   Fluid leakage

1   Automatic transmission fluid is a deep red color. Fluid leaks should not be confused with engine oil, which can easily be blown by air flow to the transmission.

2   To pinpoint a leak, first remove all built-up dirt and grime from around the transmission. Degreasing agents and/or steam cleaning will achieve this. With the underside clean, drive the vehicle at low speeds so air flow will not blow the leak far from its source. Raise the vehicle and determine where the leak is coming from. Common areas of leakage are:

a) *Pan: Tighten the mounting bolts and/or replace the pan gasket as necessary (see Chapter 7).*
b) *Filler pipe: Replace the rubber seal where the pipe enters the transmission case.*
c) *Transmission oil lines: Tighten the connectors where the lines enter the transmission case and/or replace the lines.*
d) *Vent pipe: Transmission overfilled and/or water in fluid (see checking procedures, Chapter 1).*
e) *Speedometer connector: Replace the O-ring where the speedometer sensor enters the transmission case (Chapter 7).*

Transfer case

50   Transfer case is difficult to shift into the desired range

1   Speed may be too great to permit engagement. Stop the vehicle and shift into the desired range.
2   Shift linkage loose, bent or binding on a manual shift transfer case. Check the linkage for damage or wear and replace or lubricate as necessary (Chapter 7).
3   Defective circuit and or range switch on electric shift transfer case (Chapter 7).
4   If the vehicle has been driven on a paved surface for some time, the driveline torque can make shifting difficult. Stop and shift into two-wheel drive on paved or hard surfaces.
5   Insufficient or incorrect grade of lubricant. Drain and refill the transfer case with the specified lubricant. (Chapter 1).
6   Worn or damaged internal components. Disassembly and overhaul of the transfer case may be necessary (Chapter 7).

51   Transfer case noisy in all gears

Insufficient or incorrect grade of lubricant. Drain and refill (Chapter 1).

52   Noisy or jumps out of four-wheel drive Low range

1   Transfer case not fully engaged. Stop the vehicle, shift into Neutral and then engage 4L.
2   Shift linkage loose, worn or binding. Tighten, repair or lubricate linkage as necessary.
3   Shift fork cracked, inserts worn or fork binding on the rail. See your dealer for a new or rebuilt unit.

53   Lubricant leaks from the vent or output shaft seals

1   Transfer case is overfilled. Drain to the proper level (Chapter 1).
2   Vent is clogged or jammed closed. Clear or replace the vent.
3   Output shaft seal incorrectly installed or damaged. Replace the seal and check contact surfaces for nicks and scoring.

Driveshaft

54   Oil leak at seal end of driveshaft

Defective transmission or transfer case oil seal. See Chapter 7 for replacement procedures. While this is done, check the splined yoke for burrs or a rough condition which may be damaging the seal. Burrs can be removed with crocus cloth or a fine whetstone.

55   Knock or clunk when the transmission is under initial load (just after transmission is put into gear)

1   Loose or disconnected rear suspension components. Check all mounting bolts, nuts and bushings (Chapter 10).
2   Loose driveshaft bolts. Inspect all bolts and nuts and tighten them to the specified torque.
3   Worn or damaged universal joint bearings. Check for wear (Chapter 8).

56   Metallic grinding sound consistent with vehicle speed

Pronounced wear in the universal joint bearings. Check as described in Chapter 8.

57   Vibration

Note: *Before assuming that the driveshaft is at fault, make sure the tires are perfectly balanced and perform the following test.*

1   Install a tachometer inside the vehicle to monitor engine speed as the vehicle is driven. Drive the vehicle and note the engine speed at which the vibration (roughness) is most pronounced. Now shift the transmission to a different gear and bring the engine speed to the same point.
2   If the vibration occurs at the same engine speed (rpm) regardless of which gear the transmission is in, the driveshaft is NOT at fault since the driveshaft speed varies.

3   If the vibration decreases or is eliminated when the transmission is in a different gear at the same engine speed, refer to the following probable causes.
4   Bent or dented driveshaft. Inspect and replace as necessary (Chapter 8).
5   Undercoating or built-up dirt, etc. on the driveshaft. Clean the shaft thoroughly and recheck.
6   Worn universal joint bearings. Remove and inspect (Chapter 8).
7   Driveshaft and/or companion flange out of balance. Check for missing weights on the shaft. Remove the driveshaft (see Chapter 8) and reinstall 180-degrees from original position, then retest. Have the driveshaft professionally balanced if the problem persists.

Axles

58   Noise

1   Road noise. No corrective procedures available.
2   Tire noise. Inspect tires and check tire pressures (Chapter 1).
3   Rear wheel bearings loose, worn or damaged (Chapter 8).

59   Vibration

See probable causes under Driveshaft. Proceed under the guidelines listed for the driveshaft. If the problem persists, check the rear wheel bearings by raising the rear of the vehicle and spinning the rear wheels by hand. Listen for evidence of rough (noisy) bearings. Remove and inspect (Chapter 8).

60   Oil leakage

1   Pinion seal damaged (Chapter 8).
2   Axleshaft oil seals damaged (Chapter 8).
3   Differential inspection cover leaking. Tighten the bolts or replace the gasket as required (Chapters 1 and 8).

Brakes

Note: *Before assuming that a brake problem exists, make sure that the tires are in good condition and inflated properly (see Chapter 1), that the front end alignment is correct and that the vehicle is not loaded with weight in an unequal manner.*

61   Vehicle pulls to one side during braking

1   Defective, damaged or oil contaminated

disc brake pads or shoes on one side. Inspect as described in Chapter 9.
2   Excessive wear of brake shoe or pad material or drum/disc on one side. Inspect and correct as necessary.
3   Loose or disconnected front suspension components. Inspect and tighten all bolts to the specified torque (Chapter 10).
4   Defective drum brake or caliper assembly. Remove the drum or caliper and inspect for a stuck piston or other damage (Chapter 9).
5   Inadequate lubrication of front brake caliper slide rails. Remove caliper and lubricate slide rails (Chapter 9).

62   Noise (high-pitched squeal with the brakes applied)

1   Disc brake pads worn out. The noise comes from the wear sensor rubbing against the disc (does not apply to all vehicles) or the actual pad backing plate itself if the material is completely worn away. Replace the pads with new ones immediately (Chapter 9). If the pad material has worn completely away, the brake discs should be inspected for damage as described in Chapter 9.
2   Linings contaminated with dirt or grease. Replace pads or shoes.
3   Incorrect linings. Replace with correct linings.

63   Excessive brake pedal travel

1   Partial brake system failure. Inspect the entire system (Chapter 9) and correct as required.
2   Insufficient fluid in the master cylinder. Check (Chapter 1), add fluid and bleed the system if necessary (Chapter 9).
3   Rear brakes not adjusting properly. Make a series of starts and stops while the vehicle is in Reverse. If this does not correct the situation, remove the drums and inspect the self-adjusters (Chapter 9).
4   Problem with the anti-lock brake system (Chapter 9).

64   Brake pedal feels spongy when depressed

1   Air in the hydraulic lines. Bleed the brake system (Chapter 9).
2   Faulty flexible hoses. Inspect all system hoses and lines. Replace parts as necessary.
3   Master cylinder mounting bolts/nuts loose.
4   Master cylinder defective (Chapter 9).
5   Problem with the anti-lock brake system (Chapter 9).

65   Excessive effort required to stop vehicle

1   Power brake booster not operating properly (Chapter 9).
2   Excessively worn linings or pads. Inspect and replace if necessary (Chapter 9).
3   One or more caliper pistons or wheel cylinders seized or sticking. Inspect and rebuild as required (Chapter 9).
4   Brake linings or pads contaminated with oil or grease. Inspect and replace as required (Chapter 9).
5   Problem with the anti-lock brake system (Chapter 9).

66   Pedal travels to the floor with little resistance

1   Little or no fluid in the master cylinder reservoir caused by leaking wheel cylinder(s), leaking caliper piston(s), loose, damaged or disconnected brake lines. Inspect the entire system and correct as necessary.
2   Worn master cylinder seals (Chapter 9).
3   Problem with the anti-lock brake system (Chapter 9).

67   Brake pedal pulsates during brake application

1   Caliper improperly installed. Remove and inspect (Chapter 9).
2   Disc or drum defective. Remove (Chapter 9) and check for excessive lateral runout and parallelism. Have the disc or drum resurfaced or replace it with a new one.

Suspension and steering systems

68   Vehicle pulls to one side

1   Tire pressures uneven (Chapter 1).
2   Defective tire (Chapter 1).
3   Excessive wear in suspension or steering components (Chapter 10).
4   Front end in need of alignment.
5   Front brakes dragging. Inspect the brakes as described in Chapter 9.

69   Shimmy, shake or vibration

1   Tire or wheel out-of-balance or out-of-round. Have professionally balanced.
2   Loose, worn or out-of-adjustment front wheel bearings (Chapter 1).
3   Shock absorbers and/or suspension components worn or damaged (Chapter 10).

70 Excessive pitching and/or rolling around corners or during braking

1   Defective shock absorbers. Replace as a set (Chapter 10).
2   Broken or weak springs and/or suspension components. Inspect as described in Chapter 10.

71 Abnormal ride height/quality

If the vehicle rides too softly, too harshly, or abnormally high or low, at one or both ends, have the Automatic Ride Control system checked at a dealership.

72 Excessively stiff steering

1   Lack of fluid in power steering fluid reservoir (Chapter 1).
2   Incorrect tire pressures (Chapter 1).
3   Lack of lubrication at steering joints (Chapter 1).
4   Front end out of alignment.
5   Lack of power assistance.

73 Excessive play in steering

1   Loose front wheel bearings (Chapters 1 and 10).
2   Excessive wear in suspension or steering components (Chapter 10).
3   Steering gearbox damaged or out of adjustment (Chapter 10).

74 Lack of power assistance

1   Steering pump drivebelt faulty or not adjusted properly (Chapter 1).
2   Fluid level low (Chapter 1).
3   Hoses or lines restricted. Inspect and replace parts as necessary.
4   Air in power steering system. Bleed the system (Chapter 10).

75 Excessive tire wear (not specific to one area)

1   Incorrect tire pressures (Chapter 1).
2   Tires out-of-balance. Have professionally balanced.
3   Wheels damaged. Inspect and replace as necessary.

4   Suspension or steering components excessively worn (Chapter 10).

76 Excessive tire wear on outside edge

1   Inflation pressures incorrect (Chapter 1).
2   Excessive speed in turns.
3   Front end alignment incorrect. Have professionally aligned.
4   Suspension arm bent or twisted (Chapter 10).

77 Excessive tire wear on inside edge

1   Inflation pressures incorrect (Chapter 1).
2   Front end alignment incorrect. Have professionally aligned.
3   Loose or damaged steering components (Chapter 10).

78 Tire tread worn in one place

1   Tires out-of-balance.
2   Damaged or buckled wheel. Inspect and replace if necessary.
3   Defective tire (Chapter 1).

Chapter 1
Tune-up and routine maintenance

Contents

Specifications

Recommended lubricants and fluids

Note: *Listed here are the manufacturers recommendations at the time this manual was written. Manufacturers occasionally upgrade their fluid and lubricant specifications, so check with your auto parts store for current recommendations.*

Engine oil
    Type .... API grade SG
    Viscosity .... See accompanying chart
Power steering fluid type
    1991 to 1995 .... Type F automatic transmission fluid
    1996 and later .... Mercon type automatic transmission fluid
Brake fluid type .... DOT 3 heavy duty brake fluid
Automatic transmission fluid type .... Mercon V-type automatic transmission fluid

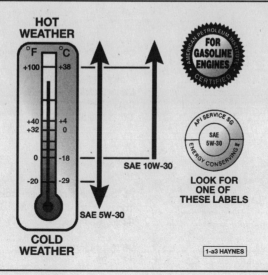

Recommended engine oil viscosity

1-a3 HAYNES

Recommended lubricants and fluids (continued)

Manual transmission lubricant type	
Mitsubishi transmission	SAE 80W EP gear lubricant
Mazda transmission	Mercon type automatic transmission fluid
Transfer case lubricant type	Mercon type automatic transmission fluid
Coolant type	50/50 mixture of ethylene glycol-based antifreeze and water
Front wheel bearing grease	NLGI No. 2 lithium base grease containing polyethylene and molybdenum disulfide
Automatic locking front hubs	NLGI No. 2 lithium base grease containing polyethylene and molybdenum disulfide
Front spindle and thrust bearings (4WD)	NLGI No. 2 lithium base grease containing polyethylene and molybdenum disulfide
Caliper slide rail grease	Disc brake caliper slide rail grease
Chassis grease	NLGI No. 2 lithium base grease containing polyethylene and molybdenum disulfide
Differential lubricant*	Synthetic SAE 75W-140 rear axle lubricant

Traction-Lok axles add 4 oz. of friction modifier when oil is changed.

Capacities (approximate)

Engine oil	
With filter change	5 quarts
Without filter change	4 quarts
Cooling system*	
V6 models	7.8 to 13.2 quarts
V8 models	12.8 to 15.7 quarts
Transfer case	
Warner 13-54	2.5 pints
1996 and later 4WD	1.5 quarts
1996 and later AWD	1.3 quarts
Front axle differential	3.5 pints
Rear axle differential	
1994 and earlier	5.3 pints
1995 and later	5.5 to 5.8 pints
Automatic transmission (drain and refill)	
5R55E	4.0 quarts
4R70W	5.0 quarts
Manual transmission	
Mazda	5.6 pints
Mitsubishi	4.8 pints

Capacity may vary +/- 15% due to equipment variations. Most service refills take only 80% of listed capacity because some coolant remains in the engine.

Brakes

Disc brake pad thickness (minimum)	1/8-inch
Drum brake shoe lining thickness (minimum)	1/16-inch above rivet heads

Ignition system

Spark plug	
Type	
V6 engines	
1991 through 1993	Motorcraft AWSF-42C or equivalent
1994 through 1995	Motorcraft AWSF-42PP
1996 and later	Motorcraft AGSF-22PP
V8 engine	Motorcraft AWSF-32EE
Gap	0.054-inch
Ignition timing	Not adjustable

Fuel system

Idle speed	Not adjustable

Torque specifications

	Ft-lbs (unless otherwise indicated)
Wheel lug nuts	100
Spark plugs	84 to 168 in-lbs
Oil pan drain plug	
4.0L pushrod V6	96 to 144 in-lbs
4.0L SOHC V6 and 5.0L V8	17 to 22

Manual transmission drain plug.. 30 to 43
Transfer case drain plug .... 84 to 204 in-lbs
Automatic transmission pan bolts .... 108 to 132 in-lbs
Differential cover bolts
   Front .... 20 to 25
   Rear .... 28 to 38
Front hub adjusting nut (2WD models)
   Step 1 .... 17 to 25
   Step 2 .... Back off 1/2-turn
   Step 3 .... 18 to 20 in-lbs
Front hub adjusting nut (1994 and earlier 4WD models)
   Manual locking hubs
      Step 1.. 35
      Step 2.. Back off 1/4-turn
      Step 3.. 16 in-lbs
      Step 4 (outer locknut).. 150
      Endplay .... 0 to 0.003 inch
      Hub turning torque (maximum) .... 25 in-lbs
   Automatic locking hubs
      Step 1.. 35
      Step 2.. Back-off 1/4-turn
      Step 3.. 16 in-lbs
      Endplay .... 0 to 0.003 inch
      Hub turning torque (maximum) .... 25 in-lbs

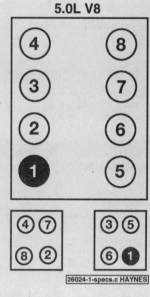

5.0L V8

4.0L PUSHROD V6

4.0L SOHC V6

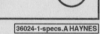

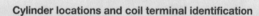

36024-1-specs.A HAYNES

36024-1-specs.b HAYNES

36024-1-specs.c HAYNES

Cylinder locations and coil terminal identification

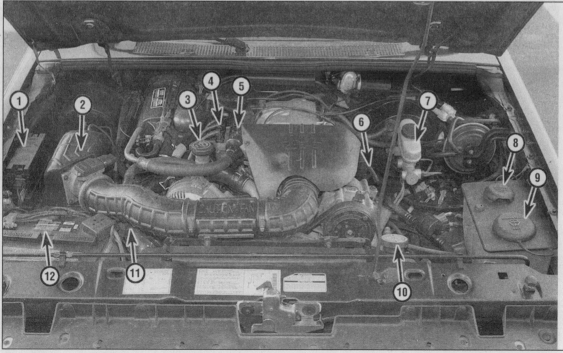

Typical engine compartment layout - 4.0L pushrod V6

1 Relay center
2 Air cleaner housing
3 Engine oil filler cap
4 Ignition coil pack
5 Spark plug wires
6 Engine oil dipstick
7 Brake master cylinder reservoir
8 Engine coolant reservoir
9 Windshield washer reservoir
10 Radiator cap
11 Upper radiator hose
12 Battery

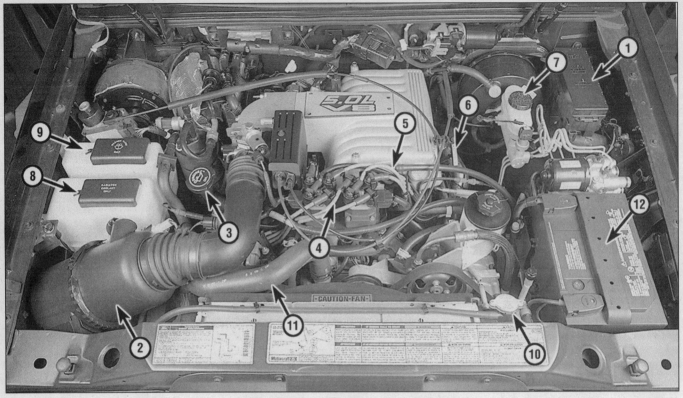

Typical engine compartment layout - 5.0L V8

1	Relay center	5	Spark plug wires	8	Engine coolant reservoir	110	Radiator cap
2	Air cleaner housing	6	Engine oil dipstick	9	Windshield washer	11	Upper radiator hose
3	Engine oil filler cap	7	Brake master cylinder		reservoir	12	Battery
4	Ignition coil pack		reservoir				

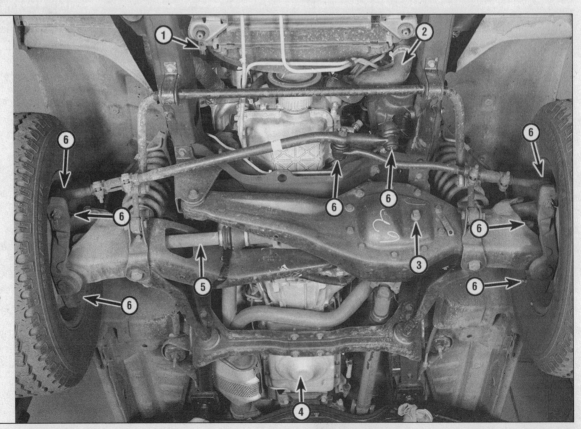

Typical engine compartment underside components (1994 and earlier 4WD models)

1 Radiator drain
2 Lower radiator hose
3 Front differential check/fill plug
4 Transmission pan
5 Front driveaxle
6 Grease fittings

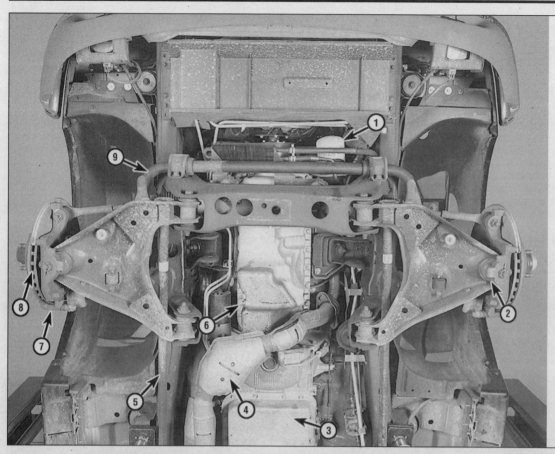

Typical engine compartment underside components (1996 and later 2WD models)

1 Oil filter
2 Lower ball joint
3 Automatic transmission fluid pan
4 Exhaust system
5 Torsion bar
6 Engine oil drain plug
7 Brake caliper
8 Brake disc
9 Stabilizer bar

1

Typical rear underside components

1 Rear shock absorber
2 Leaf spring assembly
3 Exhaust hanger
4 Muffler
5 Rear differential

Ford Explorer, Mazda Navajo, Mercury Mountaineer Maintenance schedule

The following maintenance intervals are based on the assumption that the vehicle owner will be doing the maintenance or service work, as opposed to having a dealer service department do the work. Although the time/mileage intervals are loosely based on factory recommendations, most have been shortened to ensure, for example, that such items as lubricants and fluids are checked/changed at intervals that promote maximum engine/driveline service life. Also, subject to the preference of the individual owner interested in keeping his or her vehicle in peak condition at all times, and with the vehicle's ultimate resale in mind, many of the maintenance procedures may be performed more often than recommended in the following schedule. We encourage such owner initiative.

Because off-road use necessitates more frequent maintenance, a separate schedule is included for vehicles used off road.

When the vehicle is new it should be serviced initially by a factory authorized dealer service department to protect the factory warranty. In many cases the initial maintenance check is done at no cost to the owner (check with your dealer service department for more information).

Every 250 miles or weekly, whichever comes first

Check the engine oil level (Section 3)
Check the engine coolant level (Section 3)
Check the brake fluid level (Section 3)
Check the clutch fluid level (Section 3)
Check the windshield washer fluid level (Section 3)
Check the tires and tire pressures (Section 4)
Perform owner safety checks (Section 5)

Every 3,000 miles or 3 months, whichever comes first

All items listed above, plus . . .
Change the engine oil and oil filter (Section 6)
Check the power steering fluid level (Section 7)
Check the automatic transmission fluid level (Section 8)
Rotate the tires (Section 9)

Every 6,000 miles or 6 months, whichever comes first

All the items listed above, plus . . .
Inspect/replace the underhood hoses (Section 10)
Check/adjust the drivebelt (Section 11)
Check/service the battery (Section 12)
Inspect/lubricate automatic transmission shift linkage (Section 14)
Lubricate the chassis (Section 15)
Rotate the tires (Section 9)

Every 12,000 miles or 12 months, whichever comes first

All items listed above, plus . . .
Check/replenish the manual transmission lubricant (Section 16)
Check the differential lubricant level (Section 17)
Check the transfer case lubricant level (Section 18)
Replace the air filter (Section 19)
Check/replace the PCV valve (Section 20)
Check the fuel system (Section 21)
Inspect the cooling system (Section 23)
Inspect the exhaust system (Section 24)

Inspect the brakes and lubricate caliper friction points (Section 26)
Check/lubricate the front wheel bearings (2WD models) (Section 27)
Check hub lock, spindle and front wheel bearing lubrication (1994 and earlier 4WD models) (Section 28)
Inspect/replace the windshield wiper blades (Section 29)

Every 24,000 miles or 24 months, whichever comes first

All items listed above plus . . .
Service the cooling system (drain, flush and refill) (Section 30)
Check/replace the spark plug wires (Section 13)
Lubricate automatic transmission linkage (Section 14)

Every 30,000 miles or 30 months, whichever comes first

Change the transfer case lubricant (Section 18)
Lubricate the driveshaft slip yokes (Section 31)
Lubricate right front driveaxle slip yoke (1994 and earlier 4WD models) (Section 31)
Check/replace the fuel filter (Section 22)
Change the automatic transmission fluid and filter (Section 8)

Every 60,000 miles or 60 months, whichever comes first

Replace platinum type spark plugs (Section 13)
Change differential lubricant (Section 17)

Off-road operation

If the vehicle is driven off road, perform the following maintenance items every 1,000 miles. If the vehicle is driven in mud or water, perform the items daily.
Inspect the brakes (Section 26)
Inspect the front wheel bearings (2WD models) (Section 27)
Check hub lock, spindle and front wheel bearing lubrication (1994 and earlier 4WD models) (Section 28)
Inspect exhaust system (Section 24)
Lubricate driveshaft slip yoke and U-joints (if equipped with grease fittings) (Sections 15 and 31)

1 Introduction

This Chapter is designed to help the home mechanic maintain his or her vehicle with the goals of maximum performance, economy, safety and reliability in mind.

Included is a master maintenance schedule (page 3), followed by procedures dealing specifically with each item on the schedule. Visual checks, adjustments, component replacement and other helpful items are included. Refer to the accompanying illustrations of the engine compartment and the underside of the vehicle for the locations of various components. Servicing the vehicle, in accordance with the mileage/time maintenance schedule and the step-by-step procedures will result in a planned maintenance program that should produce a long and reliable service life. Keep in mind that it is a comprehensive plan, so maintaining some items but not others at specified intervals will not produce the same results.

As you service the vehicle, you will discover that many of the procedures can - and should - be grouped together because of the nature of the particular procedure you're performing or because of the close proximity of two otherwise unrelated components to one another.

For example, if the vehicle is raised for chassis lubrication, you should inspect the exhaust, suspension, steering and fuel systems while you're under the vehicle. When you're rotating the tires, it makes good sense to check the brakes since the wheels are already removed. Finally, let's suppose you have to borrow or rent a torque wrench. Even if you only need it to tighten the spark plugs, you might as well check the torque of as many critical fasteners as time allows.

The first step in this maintenance program is to prepare yourself before the actual work begins. Read through all the procedures you're planning to do, then gather up all the parts and tools needed. If it looks like you might run into problems during a particular job, seek advice from a mechanic or an experienced do-it-yourselfer.

Caution: *If the vehicle is equipped with Automatic Ride Control (ARC), make sure the air suspension switch is turned to the OFF position before the vehicle is raised, towed or jump-started to prevent damage to the system components (see Chapter 10).*

2 Tune-up general information

The term tune-up is used in this manual to represent a combination of individual operations rather than one specific procedure.

If, from the time the vehicle is new, the routine maintenance schedule is followed closely and frequent checks are made of fluid levels and high wear items, as suggested throughout this manual, the engine will be kept in relatively good running condition and the need for additional work will be minimized.

More likely than not, however, there will be times when the engine is running poorly due to a lack of regular maintenance. This is even more likely if a used vehicle, which has not received regular and frequent maintenance checks, is purchased. In such cases, an engine tune-up will be needed outside of the regular maintenance intervals.

The first step in any tune-up or diagnostic procedure to help correct a poor running engine is a cylinder compression check. A compression check (see Chapter 2, Part D) will help determine the condition of internal engine components and should be used as a guide for tune-up and repair procedures. If, for instance, a compression check indicates serious internal engine wear, a conventional tune-up will not improve the performance of the engine and would be a waste of time and money. Because of its importance, the compression check should be done by someone with the right equipment and the knowledge to use it properly.

The following procedures are those most often needed to bring as generally poor running engine back into a proper state of tune.

Minor tune-up

Clean, inspect and test the battery (see Section 12)
Check all engine related fluids (see Section 3)
Check the drivebelts (see Section 11)
Replace the spark plugs (see Section 13)
Inspect the spark plug wires (see Section 13)
Check the PCV valve (see Section 20)
Check the air filter (see Section 19)
Check the cooling system (see Section 23)
Check all underhood hoses (see Section 10)

Major tune-up

All items listed under minor tune-up, plus . . .
Check the ignition system (see Chapter 5)
Check the charging system (see Chapter 5)
Check the fuel system (see Chapter 4)
Replace the spark plug wires (see Section 13)

3 Fluid level checks

Note: *The following are fluid level checks to be performed on a 250 mile or weekly basis. Additional fluid level checks can be found in specific maintenance procedures which follow. Regardless of intervals, be alert to fluid leaks under the vehicle which would indicate a fault to be corrected immediately.*

1   Fluids are an essential part of the lubrication, cooling, brake and windshield washer systems. Because the fluids gradually become depleted and/or contaminated dur-

3.4a Remove the dipstick, wipe it clean, then reinsert it all the way before withdrawing it for an accurate oil level check

ing normal operation of the vehicle, they must be periodically replenished. See Recommended lubricants and fluids at the beginning of this Chapter before adding fluid to any of the following components. **Note:** *The vehicle must be on level ground when fluid levels are checked.*

Engine oil

Refer to illustrations 3.4a, 3.4b and 3.6

2   Engine oil is checked with a dipstick, which is located on the side of the engine (refer to the underhood illustration at the front of this Chapter for dipstick location). The dipstick extends through a metal tube down into the oil pan.

3   The engine oil should be checked before the vehicle has been driven, or about 15 minutes after the engine has been shut off. If the oil is checked immediately after driving the vehicle, some of the oil will remain in the upper part of the engine, resulting in an inaccurate reading on the dipstick.

4   Pull the dipstick out of the tube **(see illustration)** and wipe all of the oil away from the end with a clean rag or paper towel. Insert the clean dipstick all the way back into the tube and pull it out again. Note the oil at the end of the dipstick. At its highest point, the oil should be above the ADD mark, in the SAFE range **(see illustration)**.

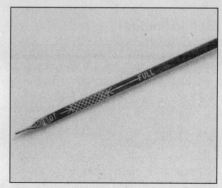

3.4b The oil level should appear between the ADD 1 QT and SAFE marks; don't overfill the crankcase

3.6 To add oil, remove the filler cap from the valve cover

3.9 The coolant recovery reservoir is combined with the windshield washer fluid reservoir (there are separate compartments for the two fluids)

5   It takes one quart of oil to raise the level from the ADD mark to the FULL or MAX mark on the dipstick. Do not allow the level to drop below the ADD mark or oil starvation may cause engine damage. Conversely, overfilling the engine (adding oil above the FULL or MAX mark) may cause oil fouled spark plugs, oil leaks or oil seal failures.

6   To add oil, remove the filler cap located on the valve cover **(see illustration)**. After adding oil, wait a few minutes to allow the level to stabilize, then pull the dipstick out and check the level again. Add more oil if required. Install the filler cap and tighten it by hand only.

7   Checking the oil level is an important preventive maintenance step. A consistently low oil level indicates oil leakage through damaged seals, defective gaskets or past worn rings or valve guides. The condition of the oil should also be noted. If the oil looks milky in color or has water droplets in it, the cylinder head gasket(s) may be blown or the head(s) or block may be cracked. The engine should be repaired immediately. Whenever you check the oil level, slide your thumb and index finger up the dipstick before wiping off the oil. If you see small dirt or metal particles clinging to the dipstick, the oil should be changed (see Section 6).

Engine coolant

Refer to illustration 3.9
Warning: *Do not allow antifreeze to come in contact with your skin or painted surfaces of the vehicle. Rinse off spills immediately with plenty of water. Antifreeze is highly toxic if ingested. Never leave antifreeze lying around in an open container or in puddles on the floor; children and pets are attracted by it's sweet smell and may drink it. Check with local authorities about disposing of used antifreeze. Many communities have collection centers which will see that antifreeze is disposed of safely. Never dump used antifreeze on the ground or pour it into drains.*

Note: *Non-toxic antifreeze is now manufactured and available at local auto parts stores, but even this type should be disposed of properly.*

8   All vehicles covered by this manual are equipped with a pressurized coolant recovery system. A plastic coolant reservoir located at the front of the engine compartment is connected by a hose to the radiator filler neck. If the engine overheats, coolant escapes through a valve in the radiator cap and travels through the hose into the reservoir. As the engine cools, the coolant is automatically drawn back into the cooling system to maintain the correct level.

9   The coolant level in the reservoir **(see illustration)** should be checked regularly. **Warning:** *Do not remove the radiator cap to check the coolant level when the engine is warm!* The level in the reservoir varies with the temperature of the engine. When the engine is cold, the coolant level should be at or slightly above the COLD FULL mark on the reservoir. Once the engine has warmed up, the level should be at or near the FULL HOT mark. If it isn't, allow the engine to cool, then remove the cap from the reservoir and add a 50/50 mixture of ethylene glycol based antifreeze and water. Don't use rust inhibitors or additives.

10   Drive the vehicle and recheck the coolant level. If only a small amount of coolant is required to bring the system up to the proper level, water can be used. However, repeated additions of water will dilute the antifreeze and water solution. In order to maintain the proper ratio of antifreeze and water, always top up the coolant level with the correct mixture. An empty plastic milk jug or bleach bottle makes an excellent container for mixing coolant.

11   If the coolant level drops consistently, there may be a leak in the system. Inspect the radiator, hoses, filler cap, drain plugs and water pump (see Section 23). If no leaks are noted, have the radiator cap pressure tested by a service station.

12   If you have to remove the radiator cap, wait until the engine has cooled completely, then wrap a thick cloth around the cap and turn it to the first stop. If coolant or steam escapes, let the engine cool down longer, then remove the cap.

13   Check the condition of the coolant as well. It should be relatively clear. If it's brown or rust colored, the system should be drained, flushed and refilled. Even if the coolant appears to be normal, the corrosion inhibitors wear out, so it must be replaced at the specified intervals.

Brake and clutch fluid

Refer to illustration 3.16
Warning: *Brake fluid can harm your eyes and damage painted surfaces, so use extreme caution when handling or pouring it. Do not use brake fluid that has been standing open or is more than one year old. Brake fluid absorbs moisture from the air, which can cause a dangerous loss of brake effectiveness. Use only the specified type of brake fluid. Mixing different types (such as DOT 3 or 4 and DOT 5) can cause brake failure.*

3.16 Check the brake fluid level by looking through the translucent plastic reservoir

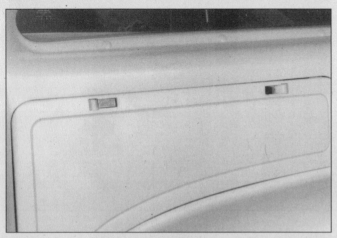

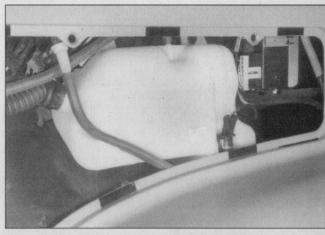

3.23a To check the rear windshield washer fluid level on early models, lift the tabs and remove the access panel in the luggage compartment . . .

3.23b . . . to expose the reservoir

14   The brake master cylinder is mounted at the left (driver's side) rear corner of the engine compartment. The clutch fluid reservoir (used on models with manual transmissions) is mounted adjacent to it.

15   To check the clutch fluid level, observe the level through the translucent reservoir. The level should be at or near the step molded into the reservoir. If the level is low, remove the reservoir cap to add the specified fluid. Clutch fluid is contained in a separate reservoir next to the brake master cylinder - clean the rubber cap before returning it to the reservoir.

16   The brake fluid level is checked by looking through the plastic reservoir mounted on the master cylinder. The fluid level should be between the MAX and MIN lines on the reservoir **(see illustration)**. If the fluid level is low, wipe the top of the reservoir and the cap with a clean rag to prevent contamination of the system as the cap is unscrewed. Top up with the recommended brake fluid, but do not overfill.

17   While the reservoir cap is off, check the master cylinder reservoir for contamination. If rust deposits, dirt particles or water droplets are present, the system should be drained and refilled by a dealer service department or repair shop.

18   After filling the reservoir to the proper level, make sure the cap is seated to prevent fluid leakage and/or contamination.

19   The fluid level in the master cylinder will drop slightly as the disc brake pads wear. A very low level may indicate worn brake pads. Check for wear (see Section 26).

20   If the brake fluid level drops consistently, check the entire system for leaks immediately. Examine all brake lines, hoses and connections, along with the calipers, wheel cylinders and master cylinder (see Section 26).

21   When checking the fluid level, if you discover one or both reservoirs empty or nearly empty, the brake or clutch hydraulic system should be checked for leaks and bled (see Chapters 8 and 9).

Windshield washer fluid

Refer to illustrations 3.23a, 3.23b and 3.23c

22   Fluid for the front windshield washer system is stored in a plastic reservoir in the engine compartment. The front windshield washer reservoir is combined with the coolant reservoir (there are separate compartments for the two different fluids) **(see illustration 3.9)**.

23   On early models, fluid for the rear windshield washer system is stored in a reservoir in the left quarter panel. To check the fluid level, remove the access cover from the left side of the luggage compartment **(see illustrations)**. To add fluid, locate the filler tube cap on the quarter panel above the tail light **(see illustration)**. Open the cap, add fluid and close the cap. On 1997 and later models, the fluid for the rear washer is stored in the main washer compartment in the engine compartment.

24   In milder climates, plain water can be used in the reservoir, but it should be kept no more than 2/3 full to allow for expansion if the water freezes. In colder climates, use windshield washer system antifreeze, available at any auto parts store, to lower the freezing

point of the fluid. This comes in concentrated or pre-mixed form. If you purchase concentrated antifreeze, mix the antifreeze with water in accordance with the manufacturer's directions on the container. **Caution:** *Do not use cooling system antifreeze - it will damage the vehicle's paint.*

4   Tire and tire pressure checks

Refer to illustrations 4.2, 4.3, 4.4a, 4.4b and 4.8

1   Periodic inspection of the tires may save you the inconvenience of being stranded with a flat tire. It can also provide you with vital information regarding possible problems in the steering and suspension systems before major damage occurs.

2   Tires are equipped with 1/2-inch wide bands that will appear when tread depth reaches 1/16-inch, but they don't appear until the tires are worn out. Tread wear can be monitored with a simple, inexpensive device known as a tread depth indicator **(see illustration)**.

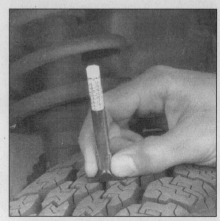

3.23c To fill the reservoir, lift the cap on the quarter panel above the tail light and pour washer fluid into the hole

4.2 Use a tire tread depth indicator to monitor tire wear - they are available at auto parts stores and service stations and cost very little

UNDERINFLATION

CUPPING

Cupping may be caused by:
- Underinflation and/or mechanical irregularities such as out-of-balance condition of wheel and/or tire, and bent or damaged wheel.
- Loose or worn steering tie-rod or steering idler arm.
- Loose, damaged or worn front suspension parts.

OVERINFLATION

INCORRECT TOE-IN OR EXTREME CAMBER

FEATHERING DUE TO MISALIGNMENT

4.3 This chart will help you determine the condition of the tires, the probable cause(s) of abnormal wear and the corrective action necessary

3   Note any abnormal tire wear (see illustration). Tread pattern irregularities such as cupping, flat spots and more wear on one side that the other are indications of front end alignment and/or balance problems. If any of these conditions are noted, take the vehicle to a tire shop or service station to correct the problem.

4   Look closely for cuts, punctures and embedded nails or tacks. Sometimes a tire will hold air pressure for a short time or leak down very slowly after a nail has embedded itself in the tread. If a slow leak persists, check the valve stem core to make sure it is tight (see illustration). Examine the tread for an object that may have embedded itself in the tire or for a "plug" that may have begun to leak (radial tire punctures are repaired with a plug that is installed in the puncture). If a puncture is suspected, it can be easily verified by spraying a solution of soapy water onto the puncture (see illustration). The soapy solution will bubble if there is a leak. Unless the puncture is unusually large, a tire shop or service station can usually repair the tire.

5   Carefully inspect the inner sidewall of each tire for evidence of brake fluid leakage. If you see any, inspect the brakes immediately.

6   Correct air pressure adds miles to the lifespan of the tires, improves mileage and enhances overall ride quality. Tire pressure cannot be accurately estimated by looking at

a tire, especially if it's a radial. A tire pressure gauge is essential. Keep an accurate gauge in the glove compartment. The pressure gauges attached to the nozzles of air hoses at gas stations are often inaccurate.

7   Always check tire pressure when the tires are cold. Cold, in this case, means the vehicle has not been driven over a mile in the three hours preceding a tire pressure check. A pressure rise of four to eight pounds is not uncommon once the tires are warm.

8   Unscrew the valve cap protruding from

4.4a If a tire loses air on a steady basis, check the valve core first to make sure it's snug (special inexpensive wrenches are commonly available at auto parts stores)

the wheel or hubcap and push the gauge firmly onto the valve stem (see illustration). Note the reading on the gauge and compare the figure to the recommended tire pressure shown in your owner's manual or on the tire placard on the passenger side door or door pillar. Be sure to reinstall the valve cap to keep dirt and moisture out of the valve stem mechanism. Check all four tires and, if necessary, add enough air to bring them to the recommended pressure.

9   Don't forget to keep the spare tire

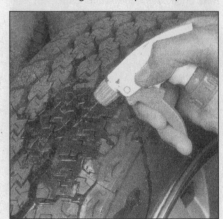

4.4b If the valve core is tight, raise the corner of the vehicle with the low tire and spray a soapy water solution onto the tread as the tire is turned slowly - leaks will cause small bubbles to appear

4.8 To extend the life of the tires, check the air pressure at least once a week with an accurate gauge (don't forget the spare!)

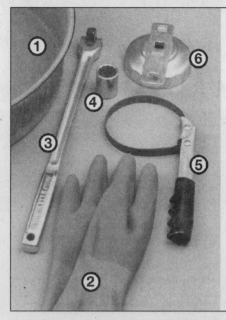

6.2 These tools are required when changing the engine oil and filter

1 *Drain pan* - It should be fairly shallow in depth, but wide to prevent spills
2 *Rubber gloves* - When removing the drain plug and filter, you will get oil on your hands (the gloves will prevent burns)
3 *Breaker bar* - Sometimes the oil drain plug is tight and a long breaker bar is needed to loosen it
4 *Socket* - To be used with the breaker bar or a ratchet (must be the correct size to fit the drain plug - six-point preferred)
5 *Filter wrench* - This is a metal band-type wrench, which requires clearance around the filter to be effective
6 *Filter wrench* - This type fits on the bottom of the filter and can be turned with a ratchet or breaker bar (different size wrenches are available for different types of filters)

inflated to the specified pressure (refer to your owner's manual or the placard attached to the door pillar). Note that the pressure recommended for temporary (mini) spare tires is higher than for the tires on the vehicle.

5 Owner safety checks

1 Most of these checks can be easily performed while the vehicle is being driven, simply by paying attention to the specified items. The checks are intended to make the vehicle owner aware of potential safety problems before they occur.
2 Check the seat belts for wear, fraying and cuts. Make sure the buckles latch securely and that the automatic retractors function correctly. Do not try to repair seat belts; always replace them if any problems are found.
3 Make sure the ignition key cannot be removed when the transmission is in any gear other than Park (automatic) or Reverse (manual). Make sure the steering column locks when the key is removed from the ignition. It may be necessary to rotate the steering wheel slightly to lock the steering column.
4 Check the parking brake. The easiest way to do this is to park on a steep hill, set the parking brake and note whether it keeps the vehicle from rolling.
5 If equipped with an automatic transmission, also check the Park mechanism. Place the transmission in Park, release the parking brake and note whether the transmission holds the vehicle from rolling. If the vehicle rolls while in Park, the transmission should be taken to a qualified shop for repairs.
6 If equipped with an automatic transmission, note whether the shift indicator shows the proper gear. If it doesn't, refer to Chapter 7 for linkage adjustment procedures.
7 If equipped with an automatic transmission, make sure the vehicle starts only in Park or Neutral. If it starts in any other gear, refer to Chapter 7 for switch adjustment procedures.
8 If equipped with a manual transmission,

the starter should operate only when the clutch pedal is pressed to the floor. If the starter operates when it shouldn't, refer to Chapter 8 for clutch/starter interlock switch service.
9 Make sure the brakes do not pull to one side while stopping. The brake pedal should feel firm, but excessive effort should not be required to stop the vehicle. If the pedal sinks too low, if you have to pump it more than once to get a firm pedal, or if pedal effort is too high, refer to Chapter 9 for repair procedures. A squealing sound from the front brakes may be caused by the pad wear indicators. Refer to Chapter 9 for pad replacement procedures.
10 Rearview mirrors should be clean and undamaged. They should hold their position when adjusted.
11 Sun visors should hold their position when adjusted. They should remain securely out of the way when lifted off the windshield.
12 Make sure the defroster blows heated air onto the windshield. If it doesn't, refer to Chapter 3 for heating system service.
13 The horn should sound with a clearly audible tone every time it is operated. If not, refer to Chapter 12.
14 Make sure the windows are clean and undamaged.
15 Turn on the lights, then walk around the car and make sure they work. Check headlights in both the high beam and low beam positions. Check the turn indicators for one side of the vehicle, then for the other side. If possible, have an assistant watch the brake lights while you push the pedal. If no assistant is available, the brake lights can be checked by backing up to a wall or garage door, then pressing the pedal. There should be three distinct patches of red light when the brake pedal is pressed.
16 Make sure locks operate smoothly when the key is turned. Lubricate locks if necessary with Lock Lubricant or an equivalent product.

Make sure all latches hold securely.
17 On models with airbags, each time the vehicle is started, the airbag warning light on the instrument panel should glow for a few seconds, then go out. During this period the airbag system is performing a self-check. If the light doesn't go out or flashes while driving, return the vehicle to your dealer for diagnosis. If the light blinks in a repeated pattern while driving, note the number of flashes. There will be a two-digit code, represented by several flashes (the first digit, i.e. if there are three flashes, the first digit is three), then a pause and another series to indicate the second digit. This code will be helpful to your dealer service department, especially for intermittent problems with the airbag system.

6 Engine oil and filter change

Refer to illustrations 6.2, 6.7 and 6.16
1 Frequent oil changes are the most important preventive maintenance procedures that can be done by the home mechanic. As engine oil ages, it becomes diluted and contaminated, which leads to premature engine wear.
2 Make sure that you have all the necessary tools before you begin this procedure (see illustration). You should also have plenty of rags or newspapers handy for mopping up oil spills.
3 Start the engine and allow it to reach normal operating temperature - oil and sludge will flow more easily when warm. If new oil, a filter or tools are needed, use the vehicle to go get them and warm up the engine oil at the same time. Park on a level surface and shut off the engine when it's warmed up. Remove the oil filler cap from the valve cover.
4 Access to the oil drain plug and filter will be improved if the vehicle can be lifted on a hoist, driven onto ramps or supported by jackstands. **Warning:** *DO NOT work under a*

6.7 Remove the oil pan drain plug with a box-end wrench or socket - an open-end or adjustable wrench shouldn't be used, since it may round off the corners of the drain plug

6.16 Lubricate the oil filter gasket with clean engine oil before installing the filter on the engine

7.2 Location of the power steering pump (arrow) - typical model shown

vehicle supported only by a bumper, hydraulic or scissors-type jack - always use jackstands!

5   Raise the vehicle and support it on jackstands. Make sure it is safely supported! **Caution:** *If the vehicle is equipped with Automatic Ride Control (ARC), make sure the air suspension switch is turned to the OFF position before the vehicle is raised to prevent damage to the system components (see Chapter 10).*

6   If you haven't changed the oil on this vehicle before, get under it and locate the drain plug and the oil filter. The exhaust components will be hot as you work, so note how they are routed to avoid touching them.

7   Being careful not to touch the hot exhaust components, position a drain pan under the plug in the bottom of the engine. Clean the area around the plug, then remove the plug **(see illustration)**. It's a good idea to wear an old glove while unscrewing the plug the final few turns to avoid being scalded by hot oil. It will also help to hold the drain plug against the threads as you unscrew it, then pull it away from the drain hole suddenly. This will place your arm out of the way of the hot oil, as well as reducing the chances of dropping the drain plug into the drain pan.

8   It may be necessary to move the drain pan slightly as oil flow slows to a trickle. Inspect the old oil for the presence of metal particles.

9   After all the oil has drained, wipe off the drain plug with a clean rag. Any small metal particles clinging to the plug would immediately contaminate the new oil.

10   Reinstall the plug and tighten it securely, but don't strip the threads.

11   Move the drain pan into position under the oil filter.

12   Loosen the oil filter by turning it counterclockwise with a filter wrench. Any standard filter wrench will work.

13   Sometimes the oil filter is screwed on so tightly that it can't be loosened. If it is, punch a metal bar or long screwdriver directly

through it, as close to the engine as possible, and use it as a T-bar to turn the filter. Be prepared for oil to spurt out of the canister as it's punctured.

14   Once the filter is loose, use your hands to unscrew it from the block. Just as the filter is detached from the block, immediately tilt the open end up to prevent oil inside the filter from spilling out.

15   Using a clean rag, wipe off the mounting surface on the block. Also, make sure that none of the old gasket remains stuck to the mounting surface. It can be removed with a scraper if necessary.

16   Compare the old filter with the new one to make sure they are the same type. Smear some engine oil on the rubber gasket of the new filter and screw it into place **(see illustration)**. Overtightening the filter will damage the gasket, so don't use a filter wrench. Most filter manufacturers recommend tightening the filter by hand only. Normally, they should be tightened 3/4-turn after the gasket contacts the block, but be sure to follow the directions on the filter or container.

17   Remove all tools and materials from under the vehicle, being careful not to spill the oil in the drain pan, then lower the vehicle.

18   Add new oil to the engine through the oil filler cap in the rocker arm cover. Use a funnel to prevent oil from spilling onto the top of the engine. Pour four quarts of fresh oil into the engine. Wait a few minutes to allow the oil to drain into the pan, then check the level on the dipstick (see Section 3 if necessary). If the oil level is in the SAFE range, install the filler cap.

19   Start the engine and run it for about a minute. While the engine is running, look under the vehicle and check for leaks at the oil pan drain plug and around the oil filter. If either one is leaking, stop the engine and tighten the plug or filter slightly.

20   Wait a few minutes, then recheck the level on the dipstick. Add oil as necessary to bring the level into the SAFE range.

21   During the first few trips after an oil change, make it a point to check frequently for leaks and proper oil level.

22   The old oil drained from the engine can-

not be reused in its present state and should be recycled. Oil reclamation centers, auto repair shops and gas stations will normally accept the oil for recycling. After the oil has cooled, it can be drained into a container (plastic jugs, bottles, milk cartons, etc.) for transport to a disposal site.

7   Power steering fluid level check

Refer to illustrations 7.2 and 7.5

1   Check the power steering fluid level periodically to avoid steering system problems, such as damage to the pump. **Caution:** *DO NOT hold the steering wheel against either stop (extreme left or right turn) for more than five seconds. If you do, the power steering pump could be damaged.*

2   The power steering pump, located at the left front corner of the engine on all models, is equipped with a twist-off cap with an integral fluid level dipstick **(see illustration)**.

3   Park the vehicle on level ground and apply the parking brake.

4   Run the engine until it has reached normal operating temperature. With the engine at idle, turn the steering wheel back-and-forth several times to get any air out of the steering system. Shut the engine off, remove the cap by turning it counterclockwise, wipe the dipstick clean and reinstall the cap (make sure it is seated).

5   Remove the cap again and note the fluid level. It must be between the two lines designating the FULL HOT range **(see illustration)** (be sure to use the proper temperature range on the dipstick when checking the fluid level - the FULL COLD lines on the reverse side of the dipstick are only usable when the engine is cold).

6   Add small amounts of fluid until the level is correct. **Caution:** *Do not overfill the pump. If too much fluid is added, remove the excess with a clean syringe or suction pump.*

7   Check the power steering hoses and connections for leaks and wear (see Section 10).

8   Check the condition and tension of the serpentine drivebelt (see Section 11).

7.5 Check the power steering fluid level with the engine at normal operating temperature - fluid should not go above the Full Hot mark

8.5 Make sure the area around the transmission dipstick is clean (arrow), then pull it out of the tube - fluid level is checked with the engine idling

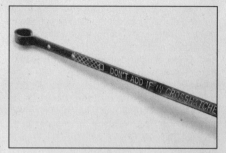

8.6 Follow the directions stamped on the transmission dipstick to get an accurate reading

8 Automatic transmission fluid check/change

Fluid level check

Refer to illustrations 8.5 and 8.6
Caution: *The use of transmission fluid other than the type listed in this Chapter's Specifications could result in transmission malfunctions or failure.*

1   The automatic transmission fluid should be carefully maintained. Low fluid level can lead to slipping or loss of drive, while overfilling can cause foaming and loss of fluid. Either condition can cause transmission damage.

2   Since transmission fluid expands as it heats up, the fluid level should only be checked when the transmission is warm (at normal operating temperature). If the vehicle has just been driven over 20 miles, the transmission can be considered warm. **Caution:** *If the vehicle has just been driven for a long time at high speed or in city traffic, in hot weather, or if it has been pulling a trailer, an accurate fluid level reading cannot be obtained. Allow the transmission to cool down for about 30 minutes. You can also check the transmission fluid level when the transmission is cold. If the vehicle has not been driven for over five hours and the fluid is*

about room temperature (70 to 95-degrees F), the transmission is cold. However, the fluid level is normally checked with the transmission warm to ensure accurate results.

3   Immediately after driving the vehicle, park it on a level surface, set the parking brake and start the engine. While the engine is idling, depress the brake pedal and move the selector lever through all the gear ranges, beginning and ending in Park.

4   Locate the automatic transmission dipstick tube in the engine compartment (see the illustration at the front of this chapter for dipstick location).

5   With the engine still idling, pull the dipstick away from the tube **(see illustration)**, wipe it off with a clean rag, push it all the way back into the tube and withdraw it again, then note the fluid level.

6   If the transmission is cold, the level should be in the room temperature range on the dipstick (between the two circles); if it's warm, the fluid level should be in the operating temperature range (between the two lines) **(see illustration)**. If the level is low, add the specified automatic transmission fluid through the dipstick tube - use a clean funnel to prevent spills.

7   Add just enough of the recommended fluid to fill the transmission to the proper level. It takes about one pint to raise the level

from the low mark to the high mark when the fluid is hot, so add the fluid a little at a time and keep checking the level until it's correct.

8   The condition of the fluid should also be checked along with the level. If the fluid is black or a dark reddish-brown color, or if it smells burned, it should be changed (see below). If you are in doubt about its condition, purchase some new fluid and compare the two for color and smell.

Fluid and filter change

Refer to illustrations 8.14, 8.16 and 8.19
9   At the specified intervals, the transmission fluid should be drained and replaced. Since the fluid will remain hot long after driving, perform this procedure only after the engine has cooled down completely.

10   Before beginning work, purchase the specified transmission fluid (see *Recommended lubricants and fluids* at the front of this Chapter), a new filter and gasket. Never reuse the old filter or gasket!

11   Other tools necessary for this job include jackstands to support the vehicle in a raised position, a drain pan capable of holding at least eight quarts, newspapers and clean rags.

12   Raise the vehicle and support it securely on jackstands. DO NOT crawl under the vehicle when it is supported only by a jack! Place the drain pan beneath the transmission. **Caution:** *If the vehicle is equipped with Automatic Ride Control (ARC), make sure the air suspension switch is turned to the OFF position before the vehicle is raised to prevent damage to the system components (see Chapter 10).*

13   Pry loose the two clips that secure the catalytic converter heat shield to the edge of the transmission fluid pan. **Note:** *It's impossible to remove the forward clip completely from the pan. For this reason, the pan must be removed toward the driver's side of the vehicle.*

14   With the drain pan in place, remove the mounting bolts from the front and sides of the transmission fluid pan **(see illustration)**.

15   Loosen the rear pan bolts approximately four turns.

16   Carefully pry the transmission pan loose with a screwdriver. Let the pan hang down so the fluid can drain **(see illustration)**. Don't damage the pan or transmission gasket surfaces or leaks could develop.

8.14 Remove the bolts around the front and sides of the transmission . . .

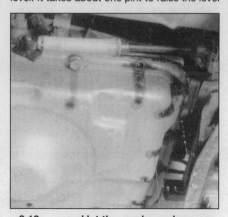

8.16 . . . and let the pan hang down so the fluid can drain

17   Remove the remaining bolts, pan and gasket. Carefully clean the gasket surface of the transmission to remove all traces of the old gasket and sealant.

18   Drain the fluid from the transmission pan, clean it with solvent and dry it with compressed air. On most models, there will be a magnet on the bottom of the pan. Remove and clean it and the pan with solvent or lacquer thinner, then replace the magnet in its original position. Do not be alarmed if there is some fine metal sludge around the magnet when you first inspect it. This is normal between fluid changes.

19   Remove the filter from the mount inside the transmission **(see illustration)**.

20   Install a new filter and gasket. Tighten the mounting bolt securely.

21   Make sure the gasket surface on the transmission pan is clean, then install a new gasket. Put the pan in place against the transmission and install the bolts. Working around the pan, tighten each bolt a little at a time until the torque listed in this Chapter's Specifications is reached. Don't overtighten the bolts!

22   Lower the vehicle and add automatic transmission fluid through the filler tube. **Caution:** *Refer to the Specifications at the front of this chapter for the correct amount and type of transmission fluid. Use of the wrong type or the wrong amount can cause transmission damage.*

23   With the transmission in Park and the parking brake set, run the engine at a fast idle, but don't race it.

24   Move the gear selector through each range and back to Park. Check the fluid level. Add fluid if needed to reach the correct level, and check again after 15 minutes of driving.

25   Check under the vehicle for leaks after the first few trips.

8.19 Once the pan is removed the filter is accessible

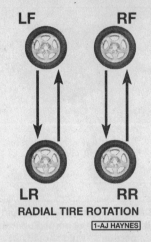

RADIAL TIRE ROTATION

1-AJ HAYNES

9.2 Tire rotation diagram

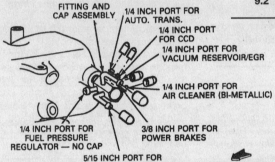

10.6 Several vacuum hoses are connected to one fitting on the intake manifold (typical)

9   Tire rotation

Refer to illustration 9.2

1   The tires should be rotated at the specified intervals and whenever uneven wear is noticed. Since the vehicle will be raised and the tires checked anyway, check the brakes also (see Section 26). **Note:** *Even if you don't rotate the tires, at least check the lug nut tightness.*

2   It is recommended that the tires be rotated in a specific pattern **(see illustration)**.

3   Refer to the information in Jacking and towing at the front of this manual for the proper procedure to follow when raising the vehicle and changing a tire. If the brakes must be checked, don't apply the parking brake as stated.

4   The vehicle must be raised on a hoist or supported on jackstands to get all four tires off the ground. Make sure the vehicle is safely supported! **Caution:** *If the vehicle is equipped with Automatic Ride Control (ARC), make sure the air suspension switch is turned to the OFF position before the vehicle is*

raised to prevent damage to the system components (see Chapter 10).

5   After the rotation procedure is finished, check and adjust the tire pressures as necessary and be sure to check the lug nut tightness.

10   Underhood hose check and replacement

Caution: *Replacement of air conditioning hoses must be left to a dealer service department or air conditioning shop that has the equipment to depressurize the system safely. Never disconnect air conditioning hoses or components until the system has been depressurized.*

General

1   High temperatures under the hood can cause deterioration of the rubber and plastic hoses used for engine, accessory and emission systems operation. Periodic inspection should be made for cracks, loose clamps, material hardening and leaks.

2   Information specific to the cooling system can be found in Section 23.

3   Most (but not all) hoses are secured to the fitting with clamps. Where clamps are used, check to be sure they haven't lost their tension, allowing the hose to leak. If clamps aren't used, make sure the hose has not expanded and/or hardened where it slips over the fitting, allowing it to leak.

PCV system hose

4   To reduce hydrocarbon emissions, crankcase blow-by gas is vented through the PCV valve to the intake manifold via a rubber hose on most models. The blow-by gases mix with incoming air in the intake manifold before being burned in the combustion chambers.

5   Check the PCV hose for cracks, leaks and other damage. Disconnect it from the valve cover and the intake manifold and check the inside for obstructions. If it's clogged, clean it out with solvent.

Vacuum hoses

Refer to illustration 10.6

6   It's quite common for vacuum hoses, especially those in the emissions system, to be color coded or identified by colored stripes molded into them. Various systems require hoses with different wall thickness, collapse resistance and temperature resistance. When replacing hoses, be sure the new ones are made of the same material. A number of hoses connect to a vacuum fitting on the intake manifold **(see illustration)**.

7   Often the only effective way to check a hose is to remove it completely from the vehicle. If more than one hose is removed, be sure to label the hoses and fittings to ensure correct installation.

8   When checking vacuum hoses, be sure to include any plastic T-fittings in the check. Inspect the fittings for cracks and the hose where it fits over each fitting for distortion, which could cause leakage.

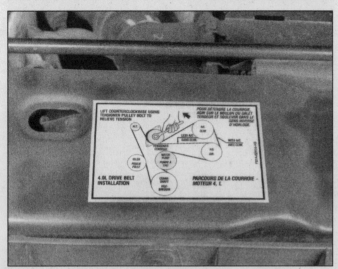

11.2 The belt's pathway is illustrated on a decal on the radiator support (typical)

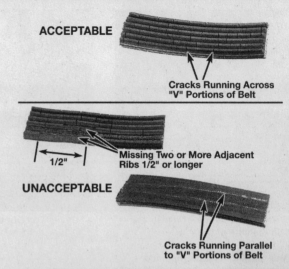

11.3 Small cracks in the underside of a V-ribbed belt are acceptable - lengthwise cracks, or missing pieces that cause the belt to make noise, are cause for replacement

9   A small piece of vacuum hose can be used as a stethoscope to detect vacuum leaks. Hold one end of the hose to your ear and probe around vacuum hoses and fittings, listening for the "hissing" sound characteristic of a vacuum leak. **Warning:** *When probing with the vacuum hose stethoscope, be careful not to come into contact with moving engine components such as the drivebelt, cooling fan, etc.*

Fuel hoses

Warning: *There are certain precautions which must be taken when servicing or inspecting fuel system components. Work in a well ventilated area and do not allow open flames (cigarettes, appliance pilot lights, etc.) or bare light bulbs near the work area. Mop up any spills immediately and do not store fuel-soaked rags where they could ignite.*

10   The fuel lines are usually under pressure, so if any fuel lines are to be disconnected be prepared to catch spilled fuel. **Warning:** *You must relieve the fuel system pressure before servicing the fuel lines. Refer to Chapter 4 for the fuel system pressure relief procedure.*

11   Check all rubber fuel lines for deterioration and chafing. Check especially for cracks in areas where the hose bends and just before fittings, such as where a hose attaches to the fuel pump, fuel filter or fuel injection system.

12   High quality fuel line should be used for fuel line replacement. Never, under any circumstances, use unreinforced vacuum line, clear plastic tubing or water hose for fuel lines.

13   Spring-type clamps are commonly used on fuel lines. These clamps often lose their tension over a period of time, and can be "sprung" during removal. Replace all spring-type clamps with screw clamps whenever a hose is replaced.

Metal lines

14   Sections of metal line are often used for fuel line between the fuel pump and fuel injection system. Check carefully to make sure the line isn't bent, crimped or cracked.

15   If a section of metal fuel line must be replaced, use seamless steel tubing only, since copper and aluminum tubing do not have the strength necessary to withstand the vibration caused by the engine.

16   Check the metal brake lines where they enter the master cylinder and brake proportioning unit (if used) for cracks in the lines and loose fittings. Any sign of brake fluid leakage calls for an immediate thorough inspection of the brake system.

Nylon fuel lines

17   Nylon fuel lines are used at several points in the fuel injection system. These lines require special materials and methods for repair. Refer to Chapter 4 for details.

Power steering hoses

18   Check the power steering hoses for leaks, loose connections and worn clamps. Tighten loose connections. Worn clamps or leaky hoses should be replaced.

11   Drivebelt check, adjustment and replacement

1   The accessory drivebelt is located at the front of the engine. The single serpentine belt drives the water pump, alternator, power steering pump and air conditioning compressor. The condition and tension of the drivebelt is critical to the operation of the engine and accessories. Excessive tension causes bearing wear, while insufficient tension produces slippage, noise, component vibration and belt failure. Because of their compo-

sition and the high stress to which they are subjected, drivebelts stretch and continue to deteriorate as they get older. As a result, they must be periodically checked. The serpentine belt has an automatic tensioner and requires no adjustment for the life of the belt.

Check

Refer to illustrations 11.2 and 11.3

2   These vehicles use a single V-ribbed belt to drive all of the accessories. This is known as a "serpentine" belt because of the winding path it follows between various drive, accessory and idler pulleys **(see illustration)**.

3   With the engine off, open the hood and locate the drivebelt at the front of the engine. With a flashlight, check the belt for separation of the rubber plies from each side of the core, a severed core, separation of the ribs from the rubber and torn or worn ribs. Also check for fraying and glazing, which gives the belt a shiny appearance. Cracks in the rib side of V-ribbed belts are acceptable, as are small chunks missing from the ribs. If a V-ribbed belt has lost chunks bigger than 1/2-inch from two adjacent ribs, or if the missing chunks cause belt noise, the belt should be replaced **(see illustration)**. Both sides of the belt should be inspected, which means you'll have to twist it to check the underside. Use your fingers to feel the belt where you can't see it. If any of the above conditions are evident, replace the belt as described below.

Adjustment

4   Tension is set by an automatic tensioner. Manual adjustment is not required.

Replacement

Refer to illustrations 11.5a and 11.5b

5   To replace a serpentine belt, rotate the tensioner counterclockwise to release tension off the belt **(see illustrations)**. Slip the belt off the pulleys.

1

11.5a Belt wear indicator marks are located on the side of the tensioner body - when the belt reaches the maximum wear mark it must be replaced

11.5b To remove the belt, rotate the tensioner bolt against the spring force to release tension off the belt

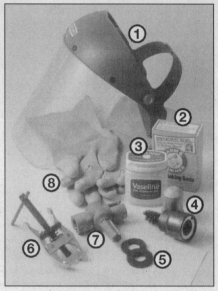

12.1 Tools and materials required for battery maintenance

1   *Face shield/safety goggles - When removing corrosion with a brush, the acidic particles can easily fly up into your eyes*
2   *Baking soda - A solution of baking soda and water can be used to neutralize corrosion*
3   *Petroleum jelly - A layer of this on the battery posts will help prevent corrosion*
4   *Battery post/cable cleaner - This wire brush cleaning tool will remove all traces of corrosion from the battery posts and cable clamps*
5   *Treated felt washers - Placing one of these on each post, directly under the cable clamps, will help prevent corrosion*
6   *Puller - Sometimes the cable clamps are very difficult to pull off the posts, even after the nut/bolt has been completely loosened. This tool pulls the clamp straight up and off the post without damage.*
7   *Battery post/cable cleaner - Here is another cleaning tool which is a slightly different version of number 4 above, but it does the same thing*
8   *Rubber gloves - Another safety item to consider when servicing the battery; remember that's acid inside the battery!*

12.4 On replacement batteries that are not sealed, remove the cell caps to check the water level in the battery - if the level is low, add distilled water only

12.8a Battery terminal corrosion usually appears as light, fluffy powder

6   Hold the tensioner in the released position. Install a new belt and make sure it is routed correctly. Be sure the ribs of the new belt engage the pulley ribs correctly. Release the tensioner.

12   Battery check, maintenance and charging

Check and maintenance

Refer to illustrations 12.1, 12.4, 12.8a, 12.8b, 12.8c and 12.8d
Warning: *Certain precautions must be followed when checking and servicing the battery. Hydrogen gas, which is highly flammable, is always present in the battery cells, so keep lighted tobacco and all other flames and sparks away from it. The electrolyte inside the battery is actually dilute sulfuric acid, which will cause injury if splashed on your skin or in your eyes. It will also ruin clothes and painted surfaces. When removing the battery cables, always detach the negative cable first and hook it up last!*
1   Battery maintenance is an important procedure which will help ensure that you are

not stranded because of a dead battery. Several tools are required for this procedure **(see illustration)**.
2   Before servicing the battery, always turn the engine and all accessories off and disconnect the cable from the negative terminal of the battery.
3   A sealed (sometimes called maintenance free) battery is standard equipment. The cell caps cannot be removed, no electrolyte checks are required and water cannot be added to the cells. However, if an aftermarket battery has been installed and it is a type that requires regular maintenance, the following procedures can be used.
4   Check the electrolyte level in each of the battery cells **(see illustration)**. It must be above the plates. There's usually a split-ring indicator in each cell to indicate the correct level. If the level is low, add distilled water only, then install the cell caps. **Caution:** *Overfilling the cells may cause electrolyte to spill over during periods of heavy charging, causing corrosion and damage to nearby components.*
5   If the positive terminal and cable clamp on your vehicle's battery is equipped with a rubber protector, make sure that it's not torn or damaged. It should completely cover the terminal.
6   The external condition of the battery

should be checked periodically. Look for damage such as a cracked case.
7   Check the tightness of the battery cable clamps to ensure good electrical connections and inspect the entire length of each cable, looking for cracked or abraded insulation and frayed conductors.
8   If corrosion (visible as white, fluffy deposits) is evident, remove the cables from the terminals, clean them with a battery brush and reinstall them **(see illustrations)**. Corrosion can be kept to a minimum by

installing specially treated washers available at auto parts stores or by applying a layer of petroleum jelly or grease to the terminals and cable clamps after they are assembled.

9   Make sure that the battery carrier is in good condition and that the hold-down clamp bolt is tight. If the battery is removed (see Chapter 5 for the removal and installation procedure), make sure that no parts remain in the bottom of the carrier when it's reinstalled. When reinstalling the hold-down clamp, don't overtighten the bolt.

10   Corrosion on the carrier, battery case and surrounding areas can be removed with a solution of water and baking soda. Apply the mixture with a small brush, let it work, then rinse it off with plenty of clean water.

11   Any metal parts of the vehicle damaged by corrosion should be coated with a zinc-based primer, then painted.

12   Additional information on the battery and jump-starting can be found in Chapter 5 and the front of this manual.

Charging

13   Remove all of the cell caps (if equipped) and cover the holes with a clean cloth to prevent spattering electrolyte. Disconnect the negative battery cable and hook the battery charger leads to the battery posts (positive to positive, negative to negative), then plug in the charger. Make sure it is set at 12-volts if it has a selector switch.

14   If you're using a charger with a rate higher than two amps, check the battery regularly during charging to make sure it doesn't overheat. If you're using a trickle charger, you can safely let the battery charge overnight after you've checked it regularly for the first couple of hours.

15   If the battery has removable cell caps, measure the specific gravity with a hydrometer every hour during the last few hours of the charging cycle. Hydrometers are available inexpensively from auto parts stores - follow the instructions that come with the hydrometer. Consider the battery charged when there's no change in the specific gravity reading for two hours and the electrolyte in the cells is gassing (bubbling) freely. The specific gravity reading from each cell should be very close to the others. If not, the battery probably has a bad cell(s).

16   Some batteries with sealed tops have built-in hydrometers on the top that indicate the state of charge by the color displayed in the hydrometer window. Normally, a bright-colored hydrometer indicates a full charge and a dark hydrometer indicates the battery still needs charging. Check the battery manufacturer's instructions to be sure you know what the colors mean.

17   If the battery has a sealed top and no built-in hydrometer, you can hook up a digital voltmeter across the battery terminals to check the charge. A fully charged battery should read 12.6-volts or higher.

18   Further information on the battery and jump starting can be found in Chapter 5 and at the front of this manual.

12.8b Removing the cable from a battery post with a wrench - sometimes a special battery pliers is required for this procedure if corrosion has caused deterioration of the nut hex (always remove the ground cable first and hook it up last!)

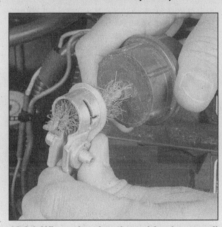

12.8d When cleaning the cable clamps, all corrosion must be removed (the inside of the clamp is tapered to match the taper on the post, so don't remove too much material)

13   Spark plug and wire check and replacement

Spark plugs

Refer to illustrations 13.2, 13.5a, 13.5b, 13.6, and 13.10

1   The spark plugs are located on the sides of the engine.

2   In most cases, the tools necessary for spark plug replacement include a spark plug socket which fits into a ratchet (spark plug sockets are padded inside to prevent damage to the porcelain insulators on the new plugs and to hold the plugs in the socket during removal and installation), various extensions and a gap gauge to check and adjust the gaps on the new plugs **(see illustration)**. A special plug wire removal tool is available for separating the wire boots from the spark plugs, but it isn't absolutely necessary. A torque wrench should be used to tighten the new plugs.

12.8c Regardless of the type of tool used on the battery posts, a clean, shiny surface should be the result

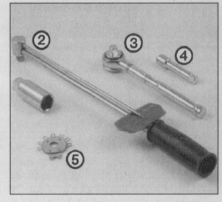

13.2 Tools required for changing spark plugs

1   *Spark plug socket - This will have special padding inside to protect the spark plug's porcelain insulator*

2   *Torque wrench - Although not mandatory, using this tool is the best way to ensure the plugs are tightened properly*

3   *Ratchet - Standard hand tool to fit the spark plug socket*

4   *Extension - Depending on model and accessories, you may need special extensions and universal joints to reach one or more of the plugs*

5   *Spark plug gap gauge - This gauge for checking the gap comes in a variety of styles. Make sure the gap for your engine is included.*

3   The best approach when replacing the spark plugs is to purchase the new ones in advance, adjust them to the proper gap and replace the plugs one at a time. When buying the new spark plugs, be sure to obtain the correct type for your particular engine. This information can be found on the *Vehicle Emission Control Information* label located under the hood, in the vehicle owner's manual and in this Chapter's Specifications. If differences exist between the plug specified on the emissions label and in the owner's manual, assume the emissions label is correct.

1

13.5a Spark plug manufacturers recommend using a wire type gauge when checking the gap - if the wire does not slide between the electrodes with a slight drag, adjustment is required

13.5b To change the gap, bend the *side* electrode only, as indicated by the arrows, and be very careful not to crack or chip the porcelain insulator surrounding the center electrode

Note: *These vehicles use platinum-tipped spark plugs. The factory-specified replacement interval for this type of plug is 60,000 miles.*

4   Allow the engine to cool completely before attempting to remove any of the plugs. While you are waiting for the engine to cool, check the new plugs for defects and adjust the gaps.

5   The gap is checked by inserting the proper thickness gauge between the electrodes at the tip of the plug **(see illustration)**. The gap between the plugs should be the same as the one specified on the *Vehicle Emissions Control Information* label or in this Chapter's Specifications. The gauge wire should just slide between the electrodes with a slight amount of drag. If the gap is incorrect, use the adjuster on the gauge body to bend the curved side electrode slightly until the specified gap is obtained **(see illustration)**. If the side electrode is not exactly over the center electrode, bend it with the adjuster until it is. Check for cracks in the porcelain insulator (if any are found, the plug should not be used).

6   With the engine cool, remove the spark plug wire from one spark plug. Pull only on the boot at the end of the wire - do not pull

on the wire. A plug wire removal tool should be used if available **(see illustration)**.

7   If compressed air is available, use it to blow any dirt or foreign material away from the spark plug hole. A common bicycle pump will also work. The idea here is to eliminate the possibility of debris falling into the cylinder as the spark plug is removed.

8   Place the spark plug socket over the plug and remove it from the engine by turning in a counterclockwise direction.

9   Compare the spark plug to those shown in the spark plug condition chart on the inside of the back cover to get an indication of the general running condition of the engine.

10   Thread one of the new plugs into the hole until you can no longer turn it with your fingers, then tighten it with a torque wrench (if available) or the ratchet. It's a good idea to slip a short length of rubber hose over the end of the plug to use as a tool to thread it into place, particularly if the cylinder head is made of aluminum **(see illustration)**. The hose will grip the plug well enough to turn it, but will start to slip if the plug begins to cross-thread in the hole - this will prevent damaged threads and the accompanying repair costs.

11   Before pushing the spark plug wire onto

the end of the plug, inspect it following the procedures outlined below.

12   Attach the plug wire to the new spark plug, again using a twisting motion on the boot until it is seated on the spark plug.

13   Repeat the procedure for the remaining spark plugs, replacing them one at a time to prevent mixing up the spark plug wires.

Spark plug wires

Refer to illustration 13.21

Note: *Every time a spark plug wire is detached from a spark plug or the coil, silicone dielectric compound (a white grease available at auto parts stores) should be applied to the inside of each boot before reconnection. Use a small standard screwdriver to coat the entire inside surface of each boot with a thin layer of the compound.*

14   The spark plug wires should be checked and, if necessary, replaced at the same time new spark plugs are installed.

15   The easiest way to identify bad wires is to make a visual check while the engine is running. In a dark, well-ventilated garage, start the engine and look at each plug wire. Be careful not to come into contact with any moving engine parts. If there is a break in the wire, you will see arcing or a small spark at the damaged area. If arcing is noticed, make a note to obtain new wires.

16   The spark plug wires should be inspected one at a time, beginning with the spark plug for the number one cylinder (the cylinder closest to the radiator on the right bank), to prevent confusion. Clearly label each original plug wire with a piece of tape marked with the correct number. The plug wires must be reinstalled in the correct order to ensure proper engine operation **(see the illustrations at the beginning of this Chapter)**.

17   Disconnect the spark plug wire from the first spark plug. A removal tool can be used **(see illustration 13.6)**, or you can grab the wire boot, twist it slightly and pull the wire free. Do not pull on the wire itself, only on the rubber boot.

18   Push the wire and boot back onto the end of the spark plug. It should fit snugly. If it doesn't, detach the wire and boot once more and use a pair of pliers to carefully crimp the metal connector inside the wire boot until it does.

19   Using a clean rag, wipe the entire length of the wire to remove built-up dirt and grease.

20   Once the wire is clean, check for burns, cracks and other damage. Do not bend the wire sharply or you might break the conductor.

21   Disconnect the wire from the ignition coil pack **(see illustration)**. Again, pull only on the rubber boot. Check for corrosion and a tight fit. Replace the wire in the coil pack.

22   Inspect each of the remaining spark plug wires, making sure that each one is securely fastened at the coil pack and spark plug when the check is complete.

23   If new spark plug wires are required,

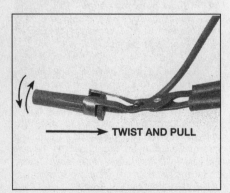

13.6 When removing the spark plug wires, pull only on the boot and twist it back-and-forth

TWIST AND PULL

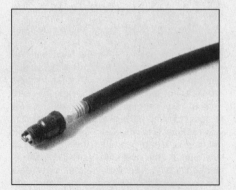

13.10 A length of 3/8-inch ID rubber hose will save time and prevent damaged threads when installing the spark plugs

13.21 Squeeze the two tabs together to release a spark plug wire from the coil-pack

14.3 Lubricate the automatic transmission shift cable at the points indicated

14.4 Carefully pry the kickdown cable end off the ballstud to lubricate the stud

purchase a set for your specific engine model. Pre-cut wire sets with the boots already installed are available. Remove and replace the wires one at a time to avoid mix-ups in the firing order.

14 Automatic transmission shift linkage lubrication

Refer to illustrations 14.3 and 14.4

1   Open the hood and locate the shift cable pivot point on the steering column. Also locate the cable end and pivot point on the transmission lever (raise the vehicle and support it securely on jackstands, if necessary).

2   Clean the cable ends and pivot points.

3   Carefully pry the cable ends off the pivot points and lubricate the pivot points with multi-purpose grease **(see illustration)**.

4   On earlier models, carefully pry the cable end off the kickdown lever ballstud, then lubricate the stud with multi-purpose grease **(see illustration)**. Reconnect the cable.

15 Chassis lubrication

Refer to illustrations 15.1, 15.6a, 15.6b, 15.6c, 15.13, 15.14a and 15.14b

1   Refer to *Recommended lubricants and fluids* at the front of this Chapter to obtain the necessary grease, etc. You'll also need a grease gun **(see illustration)**. Occasionally plugs will be installed rather than grease fittings. If so, grease fittings will have to be purchased and installed.

2   Look under the vehicle and locate the grease fittings or plugs in the tie-rod ends. If there are plugs, remove them and buy grease fittings, which will thread into the component. A dealer or auto parts store will be able to supply the correct fittings. Straight, as well as angled, fittings are available.

3   For easier access under the vehicle, raise it with a jack and place jackstands under the frame. Make sure the vehicle is safely supported - DO NOT crawl under the vehicle when it is supported only by the jack!

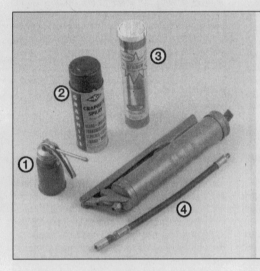

15.1 Materials required for chassis and body lubrication

1   Engine oil - Light engine oil in a can like this can be used for door and hood hinges

2   Graphite spray - Used to lubricate lock cylinders

3   Grease - Grease, in a variety of types and weights, is available for use in a grease gun. Check the Specifications for your requirements.

4   Grease gun - A common grease gun, shown here with a detachable hose and nozzle, is needed for chassis lubrication. After use, clean it thoroughly!

If the wheels are to be removed at this interval for tire rotation or brake inspection, loosen the lug nuts slightly while the vehicle is still on the ground.

4   Before beginning, force a little grease out of the nozzle to remove any dirt from the end of the gun. Wipe the nozzle clean with a rag.

5   With the grease gun and plenty of clean rags, crawl under the vehicle.

6   Wipe the tie-rod end grease fitting nipple clean and push the nozzle firmly over it.

Squeeze the trigger on the grease gun to force grease into the component **(see illustrations)**. They should be lubricated until the rubber seal is firm to the touch. Don't pump too much grease into the fitting as it could rupture the seal. If grease escapes around the grease gun nozzle, the nipple is clogged or the nozzle is not completely seated on the fitting. Resecure the gun nozzle to the fitting and try again. If necessary, replace the fitting with a new one.

15.6a Lubricate the tie-rod ends (one on each side of the vehicle) . . .

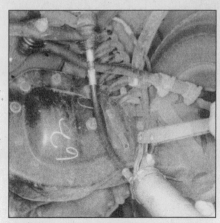

15.6b . . . the Pitman arm connection to the steering linkage . . .

15.6c . . . and the steering linkage cross rod

7   Wipe the excess grease from the components and the grease fitting. Repeat the procedure for the remaining fitting(s). **Note:** *Some models have grease fittings in the driveshaft U-joints.*

8   Open the hood and smear a little all-purpose grease on the hood latch mechanism. Have an assistant pull the hood release lever from inside the vehicle as you lubricate the cable at the latch.

9   Lubricate all the hinges (door, hood, etc.) with engine oil to keep them in proper working order.

10   The key lock cylinders can be lubricated with spray graphite or silicone lubricant, which is available at auto parts stores.

11   Lubricate the door weatherstripping with silicone spray. This will reduce chafing and retard wear.

12   Lubricate the parking brake linkage. Note that two different types of grease are required. Use multi-purpose grease on the linkage, adjuster assembly and connectors; use speedometer cable lubricant on parts of the cable that touch other parts of the vehi-

cle. Lubricate the cable twice, once with the parking brake set and once with it released.

13   On manual transmission equipped models, lubricate the clutch pedal/linkage pivot points **(see illustration)**.

14   Remove the protective cover from the throttle lever ballstud **(see illustration)**. Carefully unsnap the throttle linkage from the ballstud **(see illustration)**. Lubricate the ballstud with multi-purpose grease, then reconnect the linkage and install the protective cover.

---

16   Manual transmission lubricant level check and change

Note: *The transmission lubricant level and quality should not deteriorate under normal driving conditions. However, it's recommended that you check the level occasionally. The most convenient time would be when the vehicle is raised for another reason, such as an engine oil change.*

Level check

Refer to illustration 16.1

1   The transmission has a check/fill plug which must be removed to check the lubricant level **(see illustration)**. If the vehicle is raised to gain access to the plug, be sure to support it safely on jackstands - DO NOT crawl under a vehicle which is supported only by a jack! **Caution:** *If the vehicle is equipped with Automatic Ride Control (ARC), make sure the air suspension switch is turned to the OFF position before the vehicle is raised to prevent damage to the system components (see Chapter 10).*

2   Remove the plug from the transmission and use your little finger to reach inside the housing and feel the lubricant level. It should be at or very near the bottom of the plug hole.

3   If it isn't, add the recommended lubricant through the plug hole with a syringe or squeeze bottle.

4   Install and tighten the plug securely and check for leaks after the first few miles of driving.

Lubricant change

5   Manual transmission lubricant does not normally need changing during the life of the vehicle, but if you wish to do so, place a drain pan beneath the drain plug. Remove the filler plug, then remove the drain plug and let the lubricant drain into a pan. Let the lubricant drain for 10 minutes or more, then reinstall the filler plug and tighten securely.

6   Fill the transmission to the bottom of the filler plug hole with recommended lubricant.

7   Install and tighten the filler plug securely and check for leaks after the first few miles of driving.

---

17   Differential lubricant level check and change

Note: *This procedure applies to both the front and rear differential.*

Level check

Refer to illustrations 17.2a and 17.2b

1   The differential has a check/fill plug which must be removed to check the lubricant level. If the vehicle is raised to gain access to the plug, be sure to support it safely on jackstands - DO NOT crawl under the vehicle when it's supported only by the jack! **Caution:** *If the vehicle is equipped with Automatic Ride Control (ARC), make sure the air suspension switch is turned to the OFF position before the vehicle is raised to prevent damage to the system components (see Chapter 10).* **Note:** *On 1997 and later models, the differential is "lubricated for life", and does not need to be changed, except under severe usage or if the differential has been submerged in water (stream-crossing or boat-launching).*

2   Remove the lubricant check/fill plug from the differential **(see illustrations)**. Use a 3/8-inch drive ratchet and short extension without a socket to unscrew the plug from the rear differential; use an open end wrench to unscrew the plug from the front differential. On 1995 and later 4WD models, the front dif-

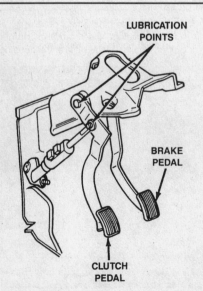

15.13 Clutch linkage lubrication points

15.14a The throttle linkage protective cover is secured by two screws (and a clip on the underside, which isn't visible in this photo)

15.14b After the cover is removed, carefully pry the linkage from the stud, grease the stud and reconnect the linkage

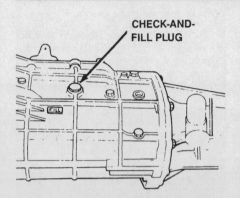

CHECK-AND-
FILL PLUG

16.1 Manual transmission filler plug

17.2a The check/fill plug for the rear differential is located in the differential housing, facing the front of the vehicle

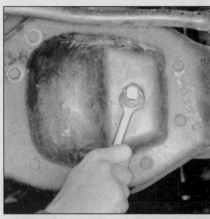

17.2b The check/fill plug for the front differential on 4WD models is located in the axle housing - use an open-end wrench to loosen it (1994 and earlier model shown)

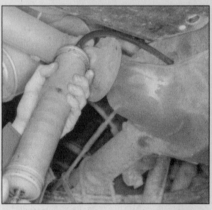

17.7 A suction pump can be used to remove the old lubricant from all differentials - it's the only method for differentials that don't have a drain plug

ferential fill plug is on the driver's side, just behind the left driveaxle.

3   Use your little finger as a dipstick to make sure the lubricant level is even with the bottom of the plug hole. If not, use a syringe to add the recommended lubricant until it just starts to run out of the opening. On some models a tag is located in the area of the plug which gives information regarding lubricant type, particularly on models equipped with a limited slip differential.

4   Install the plug and tighten it securely.

Lubricant change

Refer to illustrations 17.7, 17.8, 17.11 and 17.12

5   If it is necessary to change the differential lubricant, remove the check/fill plug **(see illustration 17.2a or 17.2b)**, then drain the differential. Some differentials can be drained by removing the drain plug, while on some rear differentials it's necessary to remove the cover plate on the differential housing. As an alternative, a hand suction pump can be used to remove the differential lubricant through the filler hole. This is the only way to drain front axles which don't have a drain plug. If you remove the cover plate, obtain a tube of silicone sealant to be used when reinstalling the differential cover.

6   If equipped with a drain plug, remove

the plug and allow the differential lubricant to drain completely. After the lubricant has drained, install the plug and tighten it securely.

7   If a suction pump is being used, insert the flexible hose **(see illustration)**. Work the hose down to the bottom of the differential housing and pump the lubricant out.

8   If the differential is being drained by removing the cover plate, remove all of the bolts except the two near the top. Loosen the remaining two bolts and use them to keep the cover loosely attached. Allow the lubricant to drain into the pan, then completely remove the cover **(see illustration)**.

9   Using a lint-free rag, clean the inside of the cover and the accessible areas of the differential housing. As this is done, check for chipped gears and metal particles in the lubricant, indicating that the differential should be more thoroughly inspected and/or repaired.

10   Clean all old gasket material from the cover and differential housing.

11   Apply a thin film of RTV sealant to the cover mating surface, then run a thick bead all the way around inside the cover bolt holes **(see illustration)**.

12   Place the cover on the differential hous-

ing and install the bolts. Tighten the bolts securely in a criss-cross pattern **(see illustration)**. Don't overtighten them or the cover may be distorted and leaks may develop.

13   On all models, use a hand pump, syringe or funnel to fill the differential housing

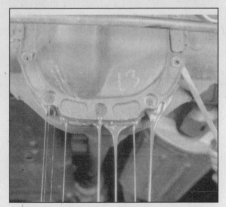

17.8 The rear differential can be drained by removing the cover if you don't have a suction pump

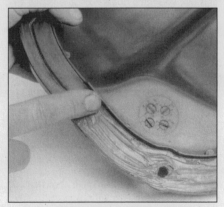

17.11 If you drain the differential by removing the cover, apply a thin film of RTV sealant to the differential cover just before installation

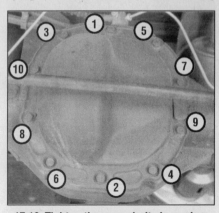

17.12 Tighten the cover bolts in a criss-cross pattern - don't overtighten the bolts, or the cover may be distorted, causing it to leak

1

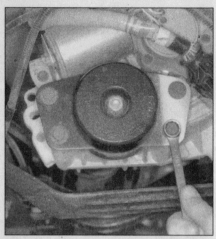

18.2 If the vehicle has a transfer case damper, unbolt it to gain access to the fill and drain plugs

18.6 If the vehicle's skid plate will obstruct removal of the drain plug, unbolt it from the frame rail to gain access to the plug

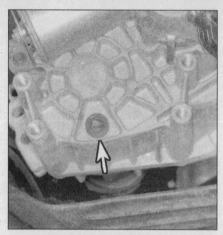

18.7 Location of the transfer case drain plug

19.2a Loosen the clamp (arrow) and disconnect the air outlet tube and hose

19.2b Unplug the electrical connector for the Mass Air Flow sensor

19.2c Label and disconnect the vacuum lines at the rear of the cover, remove the cover retaining screws and slide the cover sideways out of the retaining tab slots

with the specified lubricant until it's level with the bottom of the plug hole.

14   Install the check/fill plug and tighten it securely.

18   Transfer case lubricant level check and change

Level check

Refer to illustration 18.2

1   The transfer case has a check/fill plug which must be removed to check the lubricant level. If the vehicle is raised to gain access to the plug, be sure to support it safely on jackstands - DO NOT crawl under a vehicle which is supported only by a jack! **Caution:** *If the vehicle is equipped with Automatic Ride Control (ARC), make sure the air suspension switch is turned to the OFF position before the vehicle is raised to prevent damage to the system components (see Chapter 10).*

2   Remove the transfer case damper (if equipped) to gain access to the fill plug **(see illustration)**.

3   Remove the plug from the transfer case and use your little finger to reach inside the housing and feel the lubricant level. It should be at or very near the bottom of the plug hole.

4   If it isn't, add the recommended lubricant through the plug hole with a syringe or squeeze bottle.

5   Install and tighten the plug securely and check for leaks after the first few miles of driving.

Lubricant change

Refer to illustration 18.6 and 18.7

6   To change the lubricant, first note whether the transfer case skid plate will be in the way when the lubricant is drained. If it will be, remove it **(see illustration)**.

7   Remove the fill plug first. This will speed draining. Remove the drain plug **(see illustration)** and let the lubricant drain.

8   Clean the drain plug threads, then rein-

stall the plug after the lubricant has finished draining.

9   Fill the transaxle to the bottom of the filler plug threads with the recommended lubricant listed in this Chapter's Specifications.

10   Install the filler plug and tighten it securely.

19   Air filter replacement

1994 and earlier models

Refer to illustrations 19.2a, 19.2b, 19.2c, 19.3, 19.5a and 19.5b

1   Purchase a new filter element.

2   Disconnect the PCV hose from the air outlet tube, then disconnect the air outlet tube from the air cleaner cover. Disconnect the electrical connector from the Mass Air Flow (MAF) sensor. Label and disconnect the vacuum lines at the rear of the cover **(see illustrations)**.

19.3 Lift the cover off and take the element out of the housing

19.5a If necessary, squeeze the hot air hose retainers and push the hose out of the housing . . .

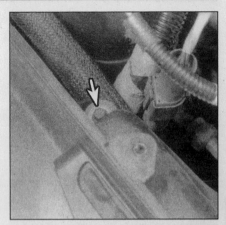

19.5b . . . then remove the mounting screw and lift the housing out

19.9a Release the clamps (arrows) on the air filter housing - 4.0L SOHC V6 shown

19.9b On the 5.0L V8, release the single round clamp (arrow) on the air cleaner housing

1995 and later models

Refer to illustrations 19.9a, 19.9b and 19.10

9   Loosen the screw-type clamp or locking clamp on the air filter housing **(see illustrations)**.

10   Pull the air filter out of the housing **(see illustration)**. Wipe the housing out with a damp rag and install the new filter element.

11   Assembly is the reverse of the removal procedure.

20   Positive Crankcase Ventilation (PCV) valve check and replacement

Refer to illustrations 20.1 and 20.2

Note: *To maintain the efficiency of the PCV system, clean the hoses and check the PCV valve at the intervals recommended in the maintenance schedule. For additional information on the PCV system, refer to Chapter 6.*

1   Locate the PCV valve **(see illustration)**. On 5.0L V8 engines, the PCV valve is located at the back of the intake manifold.

2   To check the valve, first pull it out of the grommet. Shake the valve **(see illustration)**. It should rattle, indicating that it is not clogged with deposits. If the valve does not

3   Remove the cover retaining screws. Disengage the cover tabs from the slots and lift the cover off **(see illustration)**.

4   Remove the filter element.

5   Wipe the inside of the air cleaner housing with a clean cloth. If it's necessary to remove the housing, squeeze the hot air hose and push it out of the housing. Remove the mounting screw and lift the housing out **(see**

illustrations).

6   Place the new air filter element in the housing. If the element is marked TOP be sure the marked side faces up.

7   Reinstall the cover and retaining screws. Don't overtighten the screws!

8   Reconnect the vacuum hoses, air outlet tube and electrical connector.

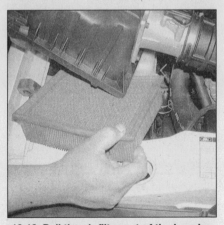

19.10 Pull the air filter out of the housing

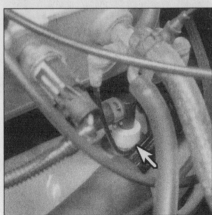

20.1 The PCV valve (arrow) is typically mounted in the valve cover near the brake booster - 4.0L pushrod V6

20.2 Shake the valve to test it - if it doesn't rattle, the valve is stuck and should be replaced

rattle, replace it with a new one. If it does rattle, reinstall it.

3   Start the engine and allow it to idle, then disconnect the PCV hose. If vacuum is felt, the PCV valve system is working properly (see Chapter 6 for additional PCV system information).

4   If no vacuum is felt, the oil filler cap, hoses or valve cover gasket may be leaking or the PCV valve may be bad. Check for vacuum leaks at the valve, filler cap and all hoses.

5   Pull straight up on the valve to remove it. Check the rubber grommet for cracks and distortion. If it's damaged, replace it.

6   If the valve is clogged, the hose is also probably plugged. Remove the hose and clean with solvent.

7   After cleaning the hose, inspect it for damage, wear and deterioration. Make sure it fits snugly on the fittings.

8   If necessary, install a new PCV valve. **Note:** *The elbow (if equipped) is not part of the PCV valve. A new valve will not include the elbow. The original must be transferred to the new valve. If a new elbow is purchased, it may be necessary to soak it in warm water for up to an hour to slip it onto the new valve. Do not attempt to force the elbow onto the valve or it will break.*

9   Install the clean PCV system hose. Make sure that the PCV valve and hose are secure.

21   Fuel system check

Warning: *Certain precautions should be observed when inspecting or servicing the fuel system components. Work in a well ventilated area and don't allow open flames (cigarettes, appliance pilot lights, etc.) near the work area. Mop up spills immediately. Do not store fuel soaked rags where they could ignite. It is a good idea to keep a dry chemical (Class B) fire extinguisher near the work area any time the fuel system is being serviced.*

1   If you smell gasoline while driving or after the vehicle has been sitting in the sun, inspect the fuel system immediately.

2   Remove the fuel filler cap and inspect it for damage and corrosion. The gasket should have an unbroken sealing imprint. If the gasket is damaged or corroded, install a new cap.

3   Inspect the fuel feed and return lines for cracks. Make sure that the connections between the fuel lines and the fuel injection system and between the fuel lines and the in-line fuel filter are tight. **Warning:** *The fuel system pressure must be relieved before servicing fuel system components. The fuel system pressure relief procedure is outlined in Chapter 4.*

4   Since some components of the fuel system - the fuel tank and some of the fuel feed and return lines, for example - are underneath the vehicle, they can be inspected more easily with the vehicle raised on a hoist. If that's not possible, raise the vehicle and support it

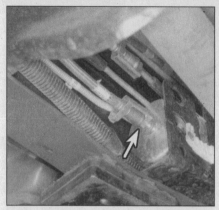

22.1 The inline fuel filter is mounted in the left frame rail (typical)

on jackstands. **Caution:** *If the vehicle is equipped with Automatic Ride Control (ARC), make sure the air suspension switch is turned to the OFF position before the vehicle is raised to prevent damage to the system components (see Chapter 10).*

5   With the vehicle raised and safely supported, inspect the gas tank and filler neck for punctures, cracks or other damage. The connection between the filler neck and the tank is particularly critical. Sometimes a rubber filler neck will leak because of loose clamps or deteriorated rubber. Inspect all fuel tank mounting brackets and straps to be sure the tank is securely attached to the vehicle. **Warning:** *Do not, under any circumstances, try to repair a fuel tank (except rubber components). A welding torch or any open flame can easily cause fuel vapors inside the tank to explode.*

6   Carefully check all rubber hoses and metal or nylon lines leading away from the fuel tank. Check for loose connections, deteriorated hoses, crimped lines and other damage. Repair or replace damaged sections as necessary (see Chapter 4).

22   Fuel filter replacement

Refer to illustration 22.1

Warning: *Gasoline is extremely flammable, so extra safety precautions must be observed when working on any part of the fuel system. Do not smoke and don't allow open flames or bare light bulbs near the vehicle. Also, don't perform fuel system maintenance procedures in a garage where a natural gas type appliance, such as a water heater or clothes dryer, with a pilot light is present!*

1   The fuel filter is mounted within the left frame rail **(see illustration)**. Periodic fuel filter replacement isn't required and the manufacturer states that the inline filter used on fuel injected models should last the life of the vehicle. Replace the filter only if it becomes clogged.

2   Obtain a new fuel filter before starting. **Warning:** *Be sure the new filter is specifically designed for your engine. Fuel injection sys-*

23.3 The radiator cap seals and the sealing surfaces in the radiator filler neck should be checked for built-up corrosion - the radiator cap should be replaced if the seals are brittle or deteriorated

tem filters are built to withstand high pressure, and as a result, often cost more than filters meant for use in carbureted systems. Filters meant for carbureted systems may burst due to the high pressure. Also, be sure the new filter includes replacement hairpin clips (if used). The manufacturer recommends against reusing the clips. **Warning:** *Before removing the fuel filter, the fuel system pressure must be relieved. See Chapter 4.*

3   Position the front end of the vehicle higher than the rear to prevent fuel siphoning. **Caution:** *If the vehicle is equipped with Automatic Ride Control (ARC), make sure the air suspension switch is turned to the OFF position before the vehicle is raised to prevent damage to the system components (see Chapter 10).* Remove the gas cap, then reinstall it after relieving the fuel system pressure.

4   Inspect the hose fittings at both ends of the filter to see if they're clean. If more than a light coating of dust is present, clean the fittings before proceeding.

5   Disconnect the push-connect fittings from the filter (see Chapter 4). **Note:** *On some models, you may first have to remove several bolts and a shield protecting the filter.*

6   Note which way the arrow on the filter is pointing - the new filter must be installed the same way. Loosen the clamp screw and detach the filter from the bracket.

7   Install the new filter in the bracket with the arrow pointing in the right direction and tighten the clamp screw securely. On some models, the filter simply snaps into a spring clip, or a pair of clips.

8   Carefully connect each hose to the filter (see Chapter 4).

9   Start the engine and check for fuel leaks.

23   Cooling system check

Refer to illustrations 23.3, 23.4a and 23.4b

1   Many major engine failures can be attributed to a faulty cooling system. If the vehicle is equipped with an automatic trans-

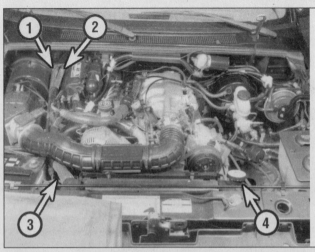

23.4a Radiator and heater hose locations (typical) - the lower radiator hose, not shown, is connected to the bottom of the radiator

1 Heater return
2 Heater supply
3 Upper radiator hose
4 Coolant recovery hose

Check for a chafed area that could fail prematurely.

Check for a soft area indicating the hose has deteriorated inside.

Overtightening the clamp on a hardened hose will damage the hose and cause a leak.

Check each hose for swelling and oil-soaked ends. Cracks and breaks can be located by squeezing the hose.

mission, the cooling system also plays an important role in prolonging transmission life because it cools the fluid.

2   The engine should be cold for the cooling system check, so perform the following procedure before the vehicle is driven for the day or after it has been shut off for at least three hours.

3   Remove the radiator cap **(see illustration)** and clean it thoroughly, inside and out, with clean water. Also clean the filler neck on the radiator. The presence of rust or corrosion in the filler neck means the coolant should be changed (see Section 30). The coolant inside the radiator should be relatively clean and transparent. If it's rust colored, drain the system and refill with new coolant.

4   Carefully check the radiator hoses and smaller diameter heater hoses **(see illustrations)**. Inspect each coolant hose along its entire length, replacing any hose which is cracked, swollen or deteriorated. Cracks will show up better if the hose is squeezed. Pay close attention to hose clamps that secure the hoses to cooling system components. Hose clamps can pinch and puncture hoses, resulting in coolant leaks.

5   Make sure all hose connections are tight. A leak in the cooling system will usually show up as white or rust colored deposits on the area adjoining the leak. If wire-type clamps are used on the hoses, it may be a good idea to replace them with screw-type clamps.

6   Clean the front of the radiator and air conditioning condenser with compressed air, if available, or a soft brush. Remove all bugs, leaves, etc. embedded in the condenser fins. Be extremely careful not to damage the cooling fins or cut your fingers on them.

7   If the coolant level has been dropping consistently and no leaks are detectable, have the radiator cap and cooling system pressure checked at a service station.

24 Exhaust system check

Refer to illustrations 24.4a, 24.4b and 24.4c

1   With the engine cold (at least three

hours after the vehicle has been driven), check the complete exhaust system from the engine to end of the tailpipe. Ideally, the inspection should be done with the vehicle on a hoist to permit unrestricted access. If a hoist isn't available, raise the vehicle and support it securely on jackstands. **Caution:** *If the vehicle is equipped with Automatic Ride Control (ARC), make sure the air suspension switch is turned to the OFF position before the vehicle is raised to prevent damage to the system components (see Chapter 10).*

2   Check the exhaust pipes and connections for evidence of leaks, severe corrosion and damage. Make sure that all brackets and hangers are in good condition and are tight.

3   At the same time, inspect the underside of the body for holes, corrosion, open seams, etc. which may allow exhaust gases to enter the passenger compartment. Seal all body openings with silicone or body putty.

4   Rattles and other noises can often be traced to the exhaust system, especially the mounts, hangers and heat shields. Try to move the pipes, muffler and catalytic converter **(see illustrations)**. If the components can come in contact with the body or suspension parts, secure the exhaust system with new mounts.

23.4b Hoses, like drivebelts, have a habit of failing at the worst possible time - to prevent the inconvenience of a blown radiator or heater hose, inspect them carefully as shown here

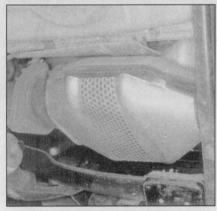

24.4a A catalytic converter is mounted beside the transmission - on automatic transmission models, the heat shield should be securely mounted in its clips

24.4b Also check the heat shields around the rear portion of the catalytic converter assembly

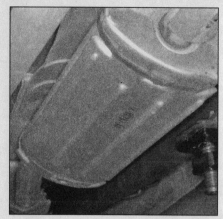

24.4c The main muffler (shown) should be securely attached to its brackets, as should the rear muffler

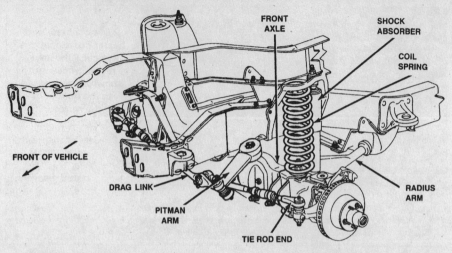

FRONT AXLE

SHOCK ABSORBER

COIL SPRING

FRONT OF VEHICLE

DRAG LINK

PITMAN ARM

TIE ROD END

RADIUS ARM

25.9a Steering and front suspension components - 1994 and earlier 4WD models

DRAG LINK

I-BEAM AXLE

SHOCK ABSORBER

COIL SPRING

RADIUS ARM

SPINDLE

FRONT OF VEHICLE

STABILIZER BAR

PITMAN ARM

TIE ROD END

BALL JOINTS

25.9b Steering and front suspension components - 1994 and earlier 2WD models

5   Check the running condition of the engine by inspecting inside the end of the tailpipe. The exhaust deposits here are an indication of engine state-of-tune. If the pipe is black and sooty or coated with white deposits, the engine may need a tune-up, including a thorough fuel system inspection.

25   Steering and suspension check

Note: *The steering linkage and suspension components should be checked periodically. Worn or damaged suspension and steering linkage components can result in excessive and abnormal tire wear, poor ride quality and vehicle handling, and reduced fuel economy. For detailed illustrations of the steering and suspension components, refer to Chapter 10.*

25.9c Steering and front suspension components - 1995 and later 2WD models

1   Frame
2   Upper control arm
3   Not used
4   Torsion bar
5   Nut
6   Bolt
7   Front wheel spindle
8   Not used
9   Nut
10 Cotter pin
11 Nut
12 Front shock absorber
13 Tie-rod end
14 Front stabilizer bar link
15 Lower control arm
16 Not used
17 Front stabilizer bar

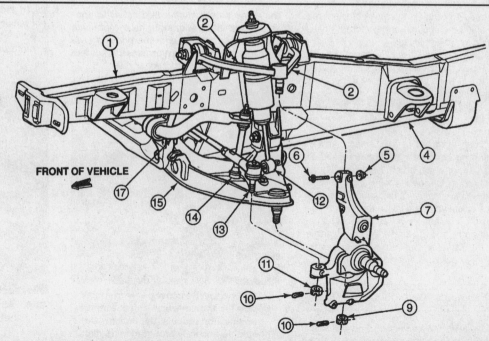

FRONT OF VEHICLE

25.9d Steering and front suspension components - 1995 and later 4WD models

1   Upper control arm
2   Frame
3   Nut
4   Front wheel hub and spindle
5   Not used
6   Hub washer and nut assembly
7   Nut
8   Cotter pin
9   Cotter pin
10 Nut
11 Front wheel driveshaft
12 Not used
13 Lower control arm

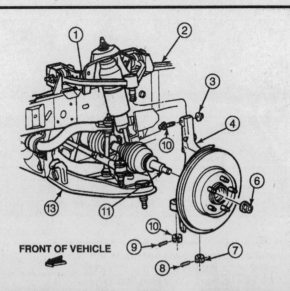

FRONT OF VEHICLE

25.10a To check the suspension balljoints, try to move the lower edge of each front tire in-and-out while watching/feeling for movement at the top of the tire . . .

1

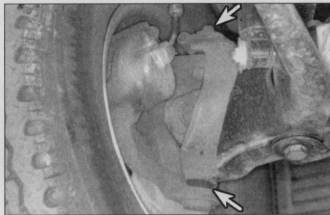

25.10b . . . if there's movement, repeat the test and look for looseness at the balljoints

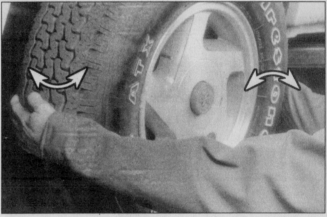

25.11a To check the steering gear and idler arm mounts and tie-rod connections for play, grasp each front tire like this and try to move it back-and-forth . . .

Shock absorber check

1   Park the vehicle on level ground, turn the engine off and set the parking brake. Check the tire pressures.

2   Push down at one corner of the vehicle, then release it while noting the movement of the body. It should stop moving and come to rest in a level position with one or two bounces.

3   If the vehicle continues to move up-and-down or if it fails to return to its original position, a worn or weak shock absorber is probably the reason.

4   Repeat the above check at each of the three remaining corners of the vehicle.

5   Raise the vehicle and support it on jack-stands. **Caution:** *If the vehicle is equipped with Automatic Ride Control (ARC), make sure the air suspension switch is turned to the OFF position before the vehicle is raised to prevent damage to the system components (see Chapter 10).*

6   Check the shock absorbers for evidence of fluid leakage. A light film of fluid is no cause for concern. Make sure that any fluid noted is from the shocks and not from any other source. If leakage is noted, replace the

shocks as a set.

7   Check the shock absorbers to be sure that they are securely mounted and undamaged. Check the upper mounts for damage and wear. If damage or wear is noted, replace the shock absorbers as a set.

8   If the shock absorbers must be replaced, refer to Chapter 10 for the procedure.

Steering and suspension check

Refer to illustrations 25.9a, 25.9b, 25.9c, 25.9d, 25.10a, 25.10b, 25.11a and 25.11b

9   Visually inspect the steering system components for damage and distortion **(see illustrations)**. Look for leaks and damaged seals, boots and fittings.

10 Clean the lower end of the steering knuckle. Have an assistant grasp the lower edge of the tire and move the wheel in-and-out **(see illustration)** while you look for movement at the steering knuckle-to-axle arm balljoints **(see illustration)**. If there is any movement, the balljoint(s) must be replaced.

11 Grasp each front tire at the front and rear edges, push in at the front, pull out at the rear and feel for play in the steering linkage

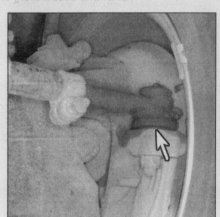

25.11b . . . if play is noted, check the steering gear mounts and make sure that they're tight; look for looseness at the tie-rod ends (shown) and the steering linkage connections

(see illustrations). If any freeplay is noted, check the steering gear mounts and the tie-rod balljoints for looseness. If the steering gear mounts are loose, tighten them. If the tie-rods are loose, the balljoints may be worn

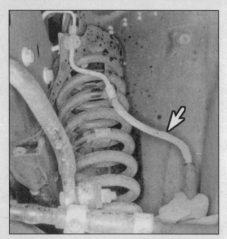

26.7a The brake hoses at the front of the vehicle (arrow) should be inspected and replaced if they show any defects

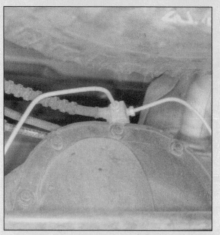

26.7b The rear brake hose meets the metal brake lines at a junction block on the axle housing - they should also be inspected and replaced if they show any defects

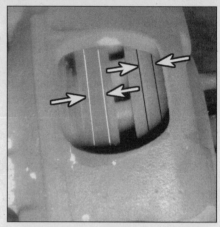

26.11 The lining thickness of the front disc brake pads (arrows) can be checked through the caliper inspection hole

(check to make sure the nuts are tight). Additional steering and suspension system illustrations can be found in Chapter 10.

Front wheel bearing check

12   Refer to Section 27 (2WD) or Section 28 (1994 and earlier 4WD models) for the wheel bearing check, repack and adjustment procedure. 1995 and later 4WD models use a sealed hub and bearing assembly, if play is noted in the hub/bearing assembly or if the bearing is noisy, refer to Chapter 10 for the replacement procedure.

26   Brake system check

Warning: *Dust produced by lining wear and deposited on brake components may contain asbestos, which is hazardous to your health. DO NOT blow it out with compressed air and DO NOT inhale it! DO NOT use gasoline or solvents to remove the dust. Brake system cleaner should be used to flush the dust into a drain pan. After the brake components are wiped with a damp rag, dispose of the contaminated rag(s) and brake cleaner in a covered and labeled container. Try to use non-asbestos replacement parts whenever possible.*
Note: *In addition to the specified intervals, the brake system should inspected each time the wheels are removed or a malfunction is indicated. Because of the obvious safety considerations, the following brake system checks are some of the most important maintenance procedures you can perform on your vehicle.*

Symptoms of brake system problems

1   The disc brakes have built-in wear indicators which should make a high-pitched squealing or scraping noise when they're worn to the replacement point. When you hear this noise, replace the pads immediately

or expensive damage to the brake discs could result.
2   Any of the following symptoms could indicate a potential brake system defect. The vehicle pulls to one side when the brake pedal is depressed, the brakes make squealing or dragging noises when applied, brake travel is excessive, the pedal pulsates and brake fluid leaks are noted (usually on the inner side of the tire or wheel). If any of these conditions are noted, inspect the brake system immediately.

Brake lines and hoses

Refer to illustrations 26.7a and 26.7b
Note: *Steel tubing is used throughout the brake system, with the exception of flexible, reinforced hoses at the front wheels and as connectors at the rear axle. Periodic inspection of these lines is very important.*
3   Park the vehicle on level ground and turn the engine off.
4   Remove the wheel covers. Loosen, but do not remove, the lug nuts on all four wheels.
5   Raise the vehicle and support it securely on jackstands. **Caution:** *If the vehicle is equipped with Automatic Ride Control (ARC), make sure the air suspension switch is turned to the OFF position before the vehicle is raised to prevent damage to the system components (see Chapter 10).*
6   Remove the wheels (see *Jacking and towing* at the front of this manual, or refer to your owner's manual, if necessary).
7   Check all brake lines and hoses for cracks, chafing of the outer cover, leaks, blisters and distortion. Check the brake hoses at front and rear of the vehicle for softening, cracks, bulging, or wear from rubbing on other components **(see illustrations)**. Check all threaded fittings for leaks and make sure the brake hose mounting bolts and clips are secure.
8   If leaks or damage are discovered, they

must be fixed immediately. Refer to Chapter 9 for detailed brake system repair procedures.

Disc brakes

Refer to illustration 26.11
9   If it hasn't already been done, raise the vehicle and support it securely on jackstands. **Caution:** *If the vehicle is equipped with Automatic Ride Control (ARC), make sure the air suspension switch is turned to the OFF position before the vehicle is raised to prevent damage to the system components (see Chapter 10).* Remove the front wheels.
10   The disc brake calipers, which contain the pads, are now visible. Each caliper has an outer and an inner pad - all pads should be checked.
11   Note the pad thickness by looking through the inspection hole in the caliper **(see illustration)**. If the lining material is 1/8-inch thick or less, or if it is tapered from end-to-end, the pads should be replaced (see Chapter 9). Keep in mind that the lining material is riveted or bonded to a metal plate or shoe - the metal portion is not included in this measurement.
12   Check the condition of the brake disc. Look for score marks, deep scratches and overheated areas (they will appear blue or discolored). If damage or wear is noted, the disc can be removed and resurfaced by an automotive machine shop or replaced with a new one. Refer to Chapter 9 for more detailed inspection and repair procedures.
13   Remove the calipers without disconnecting the brake hoses (see Chapter 9). Lubricate the caliper slide rails and the inner pad slots on the steering knuckles with the special caliper slide grease listed in this Chapter's Specifications.

Drum brakes

Refer to illustrations 26.15 and 26.17
14   Refer to Chapter 9 and remove the rear brake drums.
15   Note the thickness of the lining material

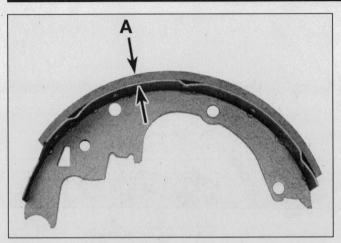

26.15 If the lining is bonded to the brake shoe, measure the lining thickness from the outer surface to the metal shoe, as shown here; if the lining is riveted to the shoe, measure from the lining outer surface to the rivet head

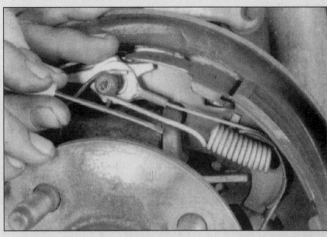

26.17 Carefully peel back the rubber boot on each end of the wheel cylinder - if the exposed area is covered with brake fluid, or if fluid runs out, the wheel cylinder must be overhauled or replaced

1

on the rear brake shoes and look for signs of contamination by brake fluid or grease **(see illustration)**. If the lining material is within 1/16-inch of the recessed rivets or metal shoes, replace the brake shoes with new ones. The shoes should also be replaced if they are cracked, glazed (shiny lining surfaces), or contaminated with brake fluid or grease. See Chapter 9 for the replacement procedure.

16   Check the shoe return and hold-down springs and the adjusting mechanism to make sure they are installed correctly and in good condition. Deteriorated or distorted springs, if not replaced, could allow the linings to drag and wear prematurely.

17   Check the wheel cylinders for leakage by carefully peeling back the rubber boots **(see illustration)**. Slight moisture behind the boots is acceptable. If brake fluid is noted behind the boots or if it runs out of the wheel cylinder, the wheel cylinders must be overhauled or replaced (see Chapter 9).

18   Check the drums for cracks, score marks, deep scratches and hard spots, which will appear as small discolored areas. If imperfections cannot be removed with emery cloth, the drums must be resurfaced by an automotive machine shop (see Chapter 9 for more detailed information).

19   Refer to Chapter 9 and install the brake drums.

20   Install the wheels, but don't lower the vehicle yet.

Parking brake

Note: *The parking brake cable and linkage should be periodically lubricated (see Section 15). This maintenance procedure helps prevent the parking brake cable adjuster or the linkage from binding and adversely affecting the operation or adjustment of the parking brake.*

21   The easiest, and perhaps most obvious, method of checking the parking brake is to park the vehicle on a steep hill with the parking brake set and the transmission in Neutral. If the parking brake doesn't prevent the vehicle from rolling, refer to Chapter 9 and adjust it.

27   Front wheel bearing check, repack and adjustment (2WD models)

Refer to illustrations 27.1, 27.7, 27.9, 27.15, 27.16 and 27.23

1   In most cases the front wheel bearings will not need servicing until the brake pads are changed. However, the bearings should be checked whenever the front of the vehicle is raised for any reason. Several items, including a torque wrench and special grease, are required for this procedure **(see illustration)**.

2   With the vehicle securely supported on jackstands, spin each wheel and check for noise, rolling resistance and freeplay. **Caution:** *If the vehicle is equipped with Automatic Ride Control (ARC), make sure the air suspension switch is turned to the OFF position before the vehicle is raised to prevent damage to the system components (see Chapter 10).*

3   Move the wheel in-and-out on the spindle **(see illustration 25.11a)**. If there's any noticeable movement, the bearings should be checked and then repacked with grease or replaced if necessary.

4   Remove the wheel.

5   Remove the brake caliper (see Chapter 9) and hang it out of the way on a piece of wire. **Warning:** *DO NOT allow the brake caliper to hang by the rubber hose!* On 1995 and later models, remove the caliper mounting bracket.

6   Pry the grease cap out of the hub with a screwdriver or hammer and chisel.

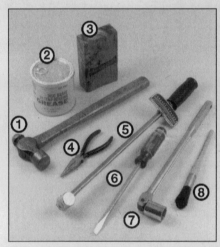

27.1 Tools and materials needed for front wheel bearing maintenance

1   *Hammer* - A common hammer will do just fine

2   *Grease* - High-temperature grease that is formulated specially for front wheel bearings should be used

3   *Wood block* - If you have a scrap piece of 2 x4, it can be used to drive the new seal into the hub

4   *Needle-nose pliers* - Used to straighten and remove the cotter pin in the spindle

5   *Torque wrench* - This is very important in this procedure; if the bearing is too tight, the wheel won't turn freely - if it's too loose, the wheel will "wobble" on the spindle. Either way, it could mean extensive damage.

6   *Screwdriver* - Used to remove the seal from the hub (a long screwdriver would be preferred)

7   *Socket/breaker bar* - Needed to loosen the nut on the spindle if it's extremely tight

8   *Brush* - Together with some clean solvent, this will be used to remove old grease from the hub and spindle

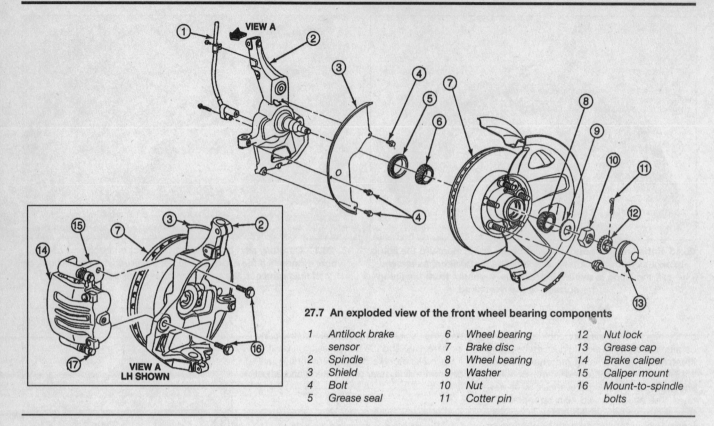

27.7 An exploded view of the front wheel bearing components

1	Antilock brake	6	Wheel bearing	12	Nut lock
	sensor	7	Brake disc	13	Grease cap
2	Spindle	8	Wheel bearing	14	Brake caliper
3	Shield	9	Washer	15	Caliper mount
4	Bolt	10	Nut	16	Mount-to-spindle
5	Grease seal	11	Cotter pin		bolts

7   Straighten the bent ends of the cotter pin, then pull the cotter pin out of the retaining nut and spindle **(see illustration)**. Discard the cotter pin and use a new one during reassembly.

8   Remove the retainer, the adjusting nut and flat washer from the end of the spindle.

9   Pull the hub assembly out slightly, then push it back into its original position. This should force the outer bearing off the spindle enough so it can be removed **(see illustration)**.

10   Pull the hub off the spindle.

11   Use a screwdriver to pry the grease seal out of the rear of the hub. As this is done, note how the seal is installed.

12   Remove the inner wheel bearing from the hub.

13   Use solvent to remove all traces of old grease from the bearings, hub and spindle. A small brush may prove helpful; however make sure no bristles from the brush embed themselves inside the bearing rollers. Allow the parts to air dry.

14   Carefully inspect the bearings for cracks, heat discoloration, worn rollers, etc. Check the bearing races inside the hub for wear and damage. If the bearing races are defective, the hubs should be taken to a machine shop with the facilities to remove the old races and press new ones in. Note that the bearings and races come as matched sets and new bearings should never be installed on old races.

15   Use high-temperature front wheel bearing grease to pack the bearings. Work the grease completely into the bearings, forcing it between the rollers, cone and cage from the back side **(see illustration)**.

16   Apply a thin coat of grease to the spindle at the outer bearing seat, inner bearing seat, shoulder and seal seat **(see illustration)**.

17   Put a small quantity of grease inboard of each bearing race inside the hub. Using your finger, form a dam at these points to provide extra grease availability and to keep thinned grease from flowing out of the bearing.

18   Place the grease-packed inner bearing into the rear of the hub and put a little more grease outward of the bearing.

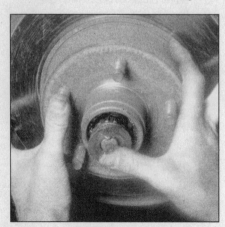

27.9 Pull out on the hub to dislodge the outer bearing

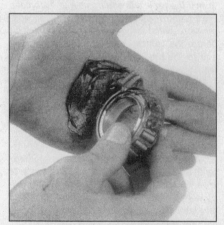

27.15 Pack each wheel bearing by working the grease into the rollers from the back side

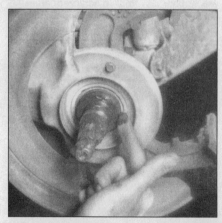

27.16 Apply a thin coat of grease to the spindle, particularly where the seal rides

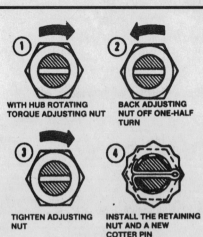

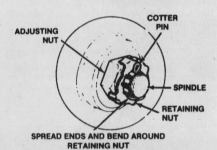

WITH HUB ROTATING TORQUE ADJUSTING NUT

BACK ADJUSTING NUT OFF ONE-HALF TURN

TIGHTEN ADJUSTING NUT

INSTALL THE RETAINING NUT AND A NEW COTTER PIN

COTTER PIN

ADJUSTING NUT

SPINDLE

RETAINING NUT

SPREAD ENDS AND BEND AROUND RETAINING NUT

27.23 Wheel bearing adjustment procedure (2WD models)

19   Place a new seal over the inner bearing and tap the seal evenly into place with a hammer and block of wood until it's flush with the hub.
20   Carefully place the hub assembly onto the spindle and push the grease-packed outer bearing into position.
21   Install the flat washer and adjusting nut. Tighten the nut only slightly.
22   Spin the hub in a forward direction to seat the bearings and remove any grease or burrs which could cause excessive bearing play later.
23   While spinning the wheel, tighten the adjusting nut to the specified torque (Step 1 in this Chapter's Specifications) **(see illustration)**.

24   Loosen the nut 1/2-turn, no more.
25   Tighten the nut to the specified torque (Step 3 in this Chapter's Specifications). Install a new cotter pin through the hole in the spindle and retainer nut. If the holes don't line up, don't turn the nut. Instead, remove the retainer and try it in a different position. The notches in the retainer are offset for this purpose. Keep trying the retainer in different positions until the holes line up.
26   Bend the ends of the cotter pin until they're flat against the nut. Cut off any extra length which could interfere with the grease cap.
27   Install the grease cap, tapping it into place with a hammer.
28   Install the caliper (see Chapter 9).
29   Install the wheel and lug nuts. Tighten the lug nuts to the torque listed in this Chapter's Specifications.
30   Check the bearings in the manner described earlier in this Section.
31   Lower the vehicle.

28 Front hub lock, spindle bearing and wheel bearing maintenance (1994 and earlier 4WD models)

Note: *On 1995 and later 4WD models, the wheel bearings are sealed bearings, incorporated into the hub and bearing assembly (see Chapter 10) - no maintenance is possible.*

Removal (manual and automatic hubs)
Refer to illustrations 28.1, 28.7a and 28.7b
1   In most cases the front wheel bearings will not need servicing until the brake pads are changed. However, the bearings should be checked whenever the front of the vehicle is raised for any reason. Several items, including a torque wrench and special grease, are required for this procedure **(see illustration 27.1)**. In addition to these tools you'll need a four-pronged spindle nut spanner wrench for manual locking hubs or a 2-3/8 inch hex locknut wrench for automatic locking hubs **(see illustration)**. These may be available at four-

28.1 The tool on the left is used for automatic locking hubs; the tool on the right is used for manual locking hubs

wheel drive shops or auto parts stores. You'll also need a pair of snap-ring pliers.
2   With the vehicle securely supported on jackstands, spin each wheel and check for noise, rolling resistance and freeplay. **Caution:** *If the vehicle is equipped with Automatic Ride Control (ARC), make sure the air suspension switch is turned to the OFF position before the vehicle is raised to prevent damage to the system components (see Chapter 10).*
3   Move the wheel in-and-out on the spindle **(see illustration 25.11a)**. If there's any noticeable movement, the bearings should be checked and then repacked with grease or replaced if necessary.
4   The lubrication of hub locks, as well as spindle needle and thrust bearings on vehicles so equipped, should be checked at the intervals specified in the maintenance schedule.
5   Remove the wheel.
6   Remove the brake caliper (see Chapter 9) and hang it out of the way on a piece of wire. **Warning:** *DO NOT allow the brake caliper to hang by the hose!*
7   Remove the retaining washers from the wheel studs **(see illustration)**. Make alignment marks on the locking hub and wheel hub, then remove the locking hub **(see illustration)**.

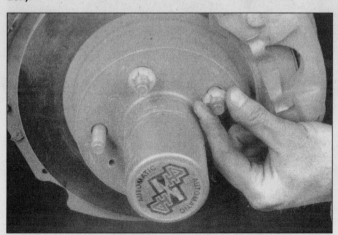

28.7a Pry the retaining washers up slightly to engage the wheel stud threads, then unscrew them from the studs

28.7b Make alignment marks on the locking hub and wheel hub, then pull the locking hub off the wheel studs

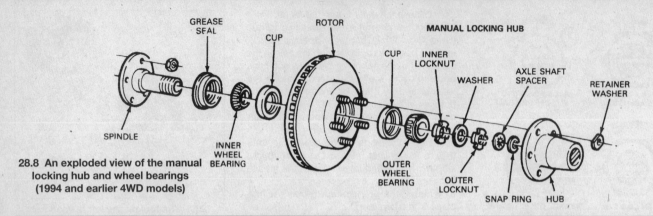

28.8 An exploded view of the manual locking hub and wheel bearings (1994 and earlier 4WD models)

Manual locking hubs

Refer to illustrations 28.8, 28.9, 28.10, 28.16 and 28.21

8.   Using snap-ring pliers, carefully expand the snap-ring just enough to remove it from the end of the spindle shaft **(see illustration)**.

9   Remove the axleshaft spacer **(see illustration)**.

10   Remove the outer wheel bearing locknut with a four-prong spanner wrench **(see illustration)**. **Note:** *This nut is very tight. Don't try to remove it with a makeshift tool.*

11   Remove the inner wheel bearing locknut with the spanner wrench. Be sure the notch in the wrench is positioned over the locknut pin.

12   Perform Steps 9 through 20 of Section 27 to repack the bearings. Be sure to use the type of grease listed in this Chapter's Specifications.

13   Install the inner locknut on the spindle and tighten it to the specified torque (Step 1 in this Chapter's Specifications).

14   Spin the brake disc several turns in each direction to seat the bearings.

15   Loosen the inner locknut 1/4-turn, then retighten it to the specified torque (Step 3 in this Chapter's Specifications).

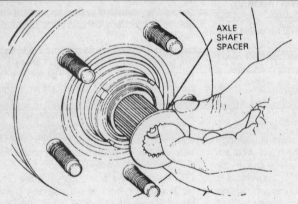

28.9   Remove the snap-ring, then the axleshaft spacer

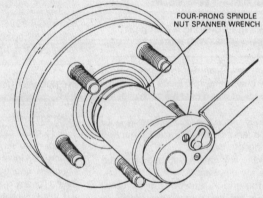

28.10   A four-pronged socket is required to remove the outer locknut; the nut is very tight, so don't use makeshift tools

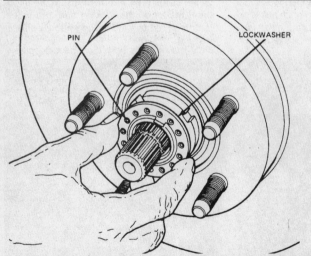

28.16   Align the pin in the locknut with one of the holes in the lockwasher; if necessary, adjust the locknut position slightly to align the pin with a hole

28.21   To remove the internal components from a manual locking hub, insert a small screwdriver behind the retaining ring and work it gently out of its groove - DO NOT remove the screw from the plastic dial

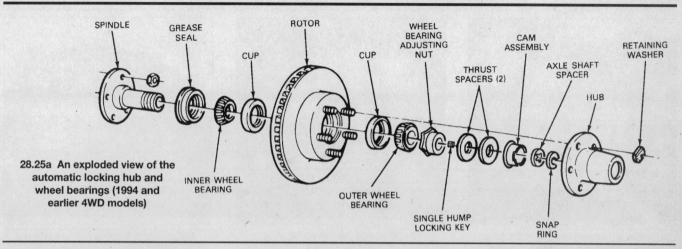

28.25a An exploded view of the automatic locking hub and wheel bearings (1994 and earlier 4WD models)

1

16   Install the lockwasher and align the lock-nut pin with one of the holes in the lockwasher **(see illustration)**. If necessary, turn the inner locknut slightly to align the hole and pin.

17   Install the outer locknut and tighten it to the specified torque (Step 4 in this Chapter's Specifications).

18   Lubricate the needle bearing spacer and needle bearing (if equipped) with the same grease used for the wheel bearings. Install

them on the spindle.

19   Install the axleshaft spacer.

20   Install the snap-ring on the spindle.

21   Remove the lock ring that secures the inner components in the locking hub **(see illustration)**. **Caution:** *Don't remove the screw from the plastic dial.*

22   Remove the internal assembly, spring and clutch gear. Lubricate the components with the specified grease.

23   Reassemble the hub and install the lock ring.

24   Install the locking hub on the wheel studs and secure it with the retainer washers. Proceed to Step 35.

Automatic locking hubs

Refer to illustrations 28.25a, 28.25b, 28.26, 28.27a, 28.27b, 28.27c, 28.28, 28.30, 28.33a, 28.33b and 28.34

25   Using snap-ring pliers, carefully expand the snap-ring just enough to remove it from the end of the spindle shaft **(see illustrations)**. **Note:** *If you don't have snap-ring pliers, carefully pry the snap-ring off with a screwdriver. Hold a finger against the snap-ring as shown in the illustration so the snap-ring doesn't fly off.*

26   Remove the axleshaft spacer **(see illustration)**.

27   Carefully pull the plastic cam assembly from the wheel bearing adjusting nut **(see illustration)**. **Caution:** *Don't pry the cam off or you may damage it. If it's hard to remove, try turning it as you pull. Pull off the two plastic thrust spacers and remove the locking key with a magnet* **(see illustrations)**. **Note:** *The thrust spacers are thin and flexible, with a tendency to cock sideways and jam on the spindle. Hook your fingernails behind the spacers and pull evenly at two or more points. If necessary, rotate the adjusting nut slightly to relieve pressure on the locking key.*

28.25b The snap-ring can be lifted off with a screwdriver if snap-ring pliers aren't available - hold the center of the snap-ring with a finger as shown so it doesn't fly off

28.26 Remove the axleshaft spacer

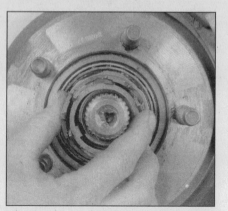

28.27a Remove the cam from the spindle

28.27b Remove the two plastic thrust spacers

28.27c Remove the locking key (arrow) with a pencil magnet - don't try to unscrew the nut before removing the key

28.28 On automatic locking hubs, loosen the wheel bearing adjusting nut with a special 2-3/8 inch hex locknut wrench

28.30 Install the wheel bearing adjusting nut

28.33a The adjusting nut has internal slots - align one of them with the keyway slot

28   Remove the wheel bearing adjusting nut from the spindle. It may be loose enough to turn with fingers. If not, use a 2-3/8 inch hex socket **(see illustration)**. **Caution:** *Be sure to remove the locking key before you remove the nut or the spindle threads will be damaged.*

29   Perform Steps 9 through 20 of Section 27 to repack the wheel bearings. Be sure to use the correct grease (listed in this Chapter's Specifications).

30   Install the wheel bearing adjusting nut **(see illustration)**. While spinning the brake disc, tighten the nut to the specified torque (Step 1 in this Chapter's Specifications).

31   Loosen the nut 1/4-turn.

32   Retighten the nut to the final torque (Step 3 in this Chapter's Specifications).

33   Align the center of the spindle keyway slot with the closest slot in the wheel bearing adjusting nut **(see illustration)**. If necessary, tighten the adjusting nut so the next slot aligns with the keyway slot. Be sure the slots line up exactly.

a) *Install the locking key in the keyway slot, under the adjusting nut. Don't force the key in or it will be damaged. If it is difficult to insert, make sure the slots are lined up exactly.*

b) *Install the two thrust spacers **(see illustration 28.27b)**.*

c) *Line up the key in the fixed cam with the keyway slot in the spindle, then push the cam on over the adjusting nut **(see illustration)**.*

d) *Install the axleshaft spacer.*

e) *Install the snap-ring on the end of the spindle.*

34   Align the three legs on the automatic locking hub with the pockets in the cam **(see illustration)**, then install the locking hub and secure it with the retainer washers.

All models

35   Install the wheel and lug nuts. Lower the vehicle and tighten the lug nuts to the torque listed in this Chapter's Specifications.

36   Check the endplay of the wheel on the spindle and measure the amount of torque required to turn the hub. Compare your find-

ings with this Chapter's Specifications. If the measurements are incorrect, readjust the wheel bearings.

---

29   Windshield wiper blade check and replacement

---

1   Road film can build up on the wiper blades and affect their efficiency, so they should be washed regularly with a mild detergent solution.

Check

2   The windshield wiper and blade assembly should be inspected periodically. Even if you don't use your wipers, the sun and elements will dry out the rubber portions, causing them to crack and break apart. If inspection reveals hardened or cracked rubber, replace the wiper blades. If inspection reveals nothing unusual, wet the windshield, turn the wipers on, allow them to cycle several times, then shut them off. An uneven wiper pattern across the glass or streaks over clean glass indicate that the blades should be replaced.

3   The operation of the wiper mechanism

can loosen the fasteners, so they should be checked and tightened, as necessary, at the same time the wiper blades are checked (see Chapter 12 for further information regarding the wiper mechanism).

Front wiper blade replacement

Refer to illustrations 29.5a and 29.5b

4   Park the wiper blades in a convenient position to be worked on. To do this, run the wipers, then turn the ignition key to Off when the wiper blades reach the desired position.

5   Lift the blade slightly from the windshield. Squeeze the retaining lever to release the blade **(see illustration)**, unhook the wiper arm from the blade **(see illustration)** and take the blade off. **Caution:** *Do not press too hard on the spring lock or it will be distorted.*

6   Slide the new blade onto the wiper arm hook until the blade locks. Make sure the spring lock secures the blade to the pin.

Rear wiper blade replacement

Refer to illustrations 29.8a and 29.8b

7   Park the wiper blades in a convenient position to be worked on. To do this, run the

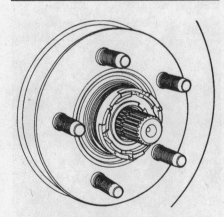

28.33b Position the cam over the wheel bearing adjusting nut - use extreme care to align the key accurately with the slot in the spindle

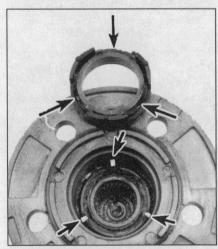

28.34 Align the legs in the automatic locking hub with the pockets on the cam when installing the hub

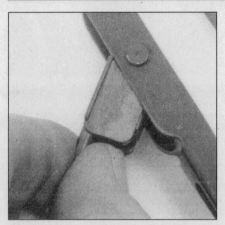

29.5a Squeeze the retaining lever . . .

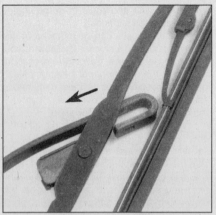

29.5b . . . push the wiper blade down the arm to disengage the hook and separate the blade from the arm

wipers, then turn the ignition key to Off when the wiper blades reach the desired position.

8   Press straight in on the retaining lever with a small screwdriver **(see illustration)**. Lift the blade off the stud on the wiper arm **(see illustration)**.

9   To install, push the blade onto the wiper arm stud until it locks.

Front or rear wiper element replacement

Refer to illustrations 29.10a and 29.10b

10   Insert a screwdriver blade between the wiper blade and element **(see illustration)**. Twist the screwdriver clockwise while press-

ing in and down to separate the element from the end retaining claw. **Note:** *On later models, simply squeeze the end of the element with needlenose pliers and withdraw the blade* **(see illustration)**.

11   Slide the element out of the remaining retaining claws.

12   Starting at either end of the blade, slide a new element into the second retaining claw (not the one closest to the end of the blade). Slide it through the other retaining claws until it reaches the end of the blade.

13   Bend the element and slide it back into the claw at the end of the blade.

30 Cooling system servicing (draining, flushing and refilling)

1

Warning: *Do not allow antifreeze to come in contact with your skin or painted surfaces of the vehicle. Rinse off spills immediately with plenty of water. Antifreeze is highly toxic if ingested. Never leave antifreeze lying around in an open container or in puddles on the floor; children and pets are attracted by it's sweet smell and may drink it. Check with local authorities about disposing of used anti-*

29.8a To remove a rear wiper blade, press on the retaining lever with a small screwdriver . . .

29.8b . . . and pull the blade off the stud on the wiper arm

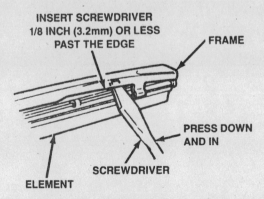

29.10a To remove a wiper element on early models, pry it out of the retaining claw at either end of the blade

INSERT SCREWDRIVER 1/8 INCH (3.2mm) OR LESS PAST THE EDGE — FRAME

PRESS DOWN AND IN

SCREWDRIVER

ELEMENT

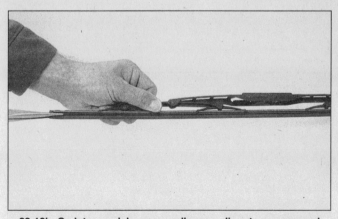

29.10b On later models, use needlenose pliers to remove and install the blade

freeze. Many communities have collection centers which will see that antifreeze is disposed of safely. Never dump used antifreeze on the ground or pour it into drains.

1   Periodically, the cooling system should be drained, flushed and refilled to replenish the antifreeze mixture and prevent formation of rust and corrosion, which can impair the performance of the cooling system and cause engine damage. When the cooling system is serviced, all hoses and the radiator cap should be checked and replaced if necessary.

Draining

Refer to illustration 30.4

2   Apply the parking brake and block the wheels. If the vehicle has just been driven, wait several hours to allow the engine to cool down before beginning this procedure.

3   Once the engine is completely cool, remove the radiator cap.

4   Move a large container under the radiator drain to catch the coolant **(see illustration)**. Attach a 3/8-inch diameter hose to the drain fitting to direct the coolant into the container, then open the drain fitting (a pair of pliers may be required to turn it). **Note:** *On later models, there may be a plastic splash shield below the radiator. Release one side of it (its held on with plastic pushpins) to access the drain fitting.*

5   While the coolant is draining, check the condition of the radiator hoses, heater hoses and clamps (see Section 23 if necessary).

6   Replace any damaged clamps or hoses (see Chapter 3 for detailed replacement procedures).

Flushing

7   Once the system is completely drained, flush the radiator with fresh water from a garden hose until the water runs clear at the drain. The flushing action of the water will remove sediments from the radiator but will not remove rust and scale from the engine and cooling tube surfaces.

8   These deposits can be removed by the chemical action of a cleaner available at your local auto parts store. Follow the procedure outlined in the manufacturer's instructions. If the radiator is severely corroded, damaged or leaking, it should be removed (see Chapter 3) and taken to a radiator repair shop.

9   The heater core should be backflushed whenever the cooling system is flushed. To do this, disconnect the heater return hose from the thermostat housing or engine. Slide a female garden hose fitting into the heater hose and secure it with a clamp. This will allow you to attach a garden hose securely.

10   Attach the end of a garden hose to the fitting you installed in the heater hose.

11   Disconnect the heater inlet hose and position it to act as a drain.

12   Turn the water on and off several times to create a surging action through the heater core. Then turn the water on full force and allow it to run for approximately five minutes.

13   Turn off the water and disconnect the garden hose from the female fitting. Remove the fitting from the heater return hose, then reconnect the hoses to the engine.

14   Remove the overflow hose from the coolant recovery reservoir. Drain the reservoir and flush it with clean water, then reconnect the hose.

Refilling

15   Close and tighten the radiator drain. Install and tighten the block drain plug(s).

16   Place the heater temperature control in the maximum heat position.

17   Slowly add new coolant (a 50/50 mixture of water and antifreeze) to the radiator until it is full. Add coolant to the reservoir up to the lower mark.

18   Leave the radiator cap off and run the engine in a well-ventilated area until the thermostat opens (coolant will begin flowing

30.4 The radiator drain (arrow) is located on the bottom-left of the radiator - connect a length of rubber hose to the drain and let it hang into the drain pan

through the radiator and the upper radiator hose will become hot).

19   Turn the engine off and let it cool. Add more coolant mixture to bring the coolant level back up to the lip on the radiator filler neck.

20   Squeeze the upper radiator hose to expel air, then add more coolant mixture if necessary. Replace the radiator cap.

21   Start the engine, allow it to reach normal operating temperature and check for leaks.

31   Driveshaft and driveaxle yoke lubrication (4WD models)

At the specified intervals, the slip yokes on the driveshafts and (on 1994 and earlier models) the right front driveaxle should be lubricated (see this Chapter's Specifications for the correct lubricant). This requires removal of the driveshafts and driveaxle (see Chapter 8 for the procedures).

Chapter 2 Part A
4.0L (pushrod) V6 engine

Contents

2A

Specifications

General

Displacement	4.0 liters (244 cubic inches)
Cylinder numbers (front-to-rear)	
Left (driver's) side	4-5-6
Right side	1-2-3
Firing order	1-4-2-5-3-6

Camshaft

Lobe lift (intake and exhaust)	
1995 and earlier	0.2756 inch
1996 and later	0.272 inch
Allowable lobe lift loss	0.005 inch
Endplay	0.0025 to 0.0064 inch
Journal-to-bearing (oil) clearance	0.001 to 0.003 inch
Bearing inside diameter (standard)	
No. 1	1.954 to 1.955 inch
No. 2	1.939 to 1.940 inch
No. 3	1.919 to 1.920 inch
No. 4	1.924 to 1.925 inch

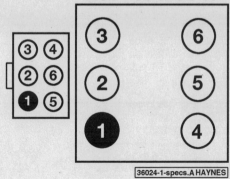

4.0L PUSHROD V6

Cylinder locations and coil terminal identification

36024-1-specs.A HAYNES

Camshaft

Journal diameter (standard)

No. 1	1.951 to 1.952 inch
No. 2	1.937 to 1.938 inch
No. 3	1.922 to 1.923 inch
No. 4	1.907 to 1.908 inch

Front bearing location

1998 and earlier models	0.040 to 0.060 inch below face of block
1999 and later models	0.020 to 0.035 inch below face of block

Oil pan-to-transmission spacer thickness

Coded yellow	0.010 inch
Coded blue	0.020 inch
Coded pink	0.030 inch

Crankshaft rear seal

Dimension from rear face of block

1995 and earlier	0.413 to 0.425 inch
1996 and later	flush with block
Square to crankshaft centerline within	0.015 inch

Torque specifications

Ft-lbs (unless otherwise indicated)

Camshaft sprocket bolt	44 to 50
Camshaft thrust plate bolts	84 to 120 in-lbs

Crankshaft pulley bolt*

Step 1	30 to 37
Step 2	Tighten an additional 90 degrees

Cylinder head bolts*

1995 and earlier

Step 1	44
Step 2	59
Step 3	Tighten an additional 90 degrees

1996 and later

Step 1	22 to 26
Step 2	52 to 56
Step 3	Tighten an additional 90 degrees

Flywheel/driveplate bolts

1995 and earlier

Step 1	108 to 132 in-lbs
Step 2	50 to 55

1996 to 1998

Step 1	22
Step 2	Tighten an additional 90 degrees

1999 and later

Step 1	10
Step 2	52
Engine mount insulator-to-frame nuts	65 to 97
Engine mount insulator-to-bracket nuts	65 to 97
Engine mount bracket-to-block bolts	45 to 60
Exhaust manifold bolts	15 to 18
Exhaust pipe-to-manifold nuts	25 to 33
Timing chain cover bolts	13 to 15

Intake manifold bolts/nuts

1995 and earlier

Step 1	72 in-lbs
Step 2	144 in-lbs
Step 3	192 in-lbs

1996 to 1998

Step 1	36 in-lbs
Step 2	72 in-lbs
Step 3	144 in-lbs

1999 and later

Step 1	27 in-lbs
Step 2	89 in-lbs
Step 3	10
Step 4	12
Intake manifold studs-to-block	72 to 84 in-lbs
Oil pump drive gear bolt	156 to 180 in-lbs
Oil pump pick-up tube-to-pump bolts	84 to 120 in-lbs
Oil pump-to-block bolts	156 to 180 in-lbs

Oil pan-to-block bolts..	60 to 84 in-lbs
Timing chain tensioner bolts..	84 to 96 in-lbs
Timing chain guide bolts..	84 to 108 in-lbs

Torque specifications   **Ft-lbs** (unless otherwise indicated)

Valve cover bolts	
1995 and earlier..	53 to 70 in-lbs
1996 to 1998	71 to 89 in-lbs
1999 and later	62 in-lbs
Rocker arm shaft support bolts	
1994 and earlier, 1999 and later	
Step 1..	24
Step 2..	Tighten an additional 90 degrees
1995 to 1998	46 to 52

*Use new bolt(s) on reassembly.

1 General information

This Part of Chapter 2 is devoted to in-vehicle repair procedures for the 4.0L pushrod V6 engine, as well as procedures such as timing chain and sprocket and oil pan removal which require removal of the engine from the vehicle. All information concerning engine removal and installation and engine block and cylinder head overhaul can be found in Part D of this Chapter.

The following repair procedures are based on the assumption that the engine is installed in the vehicle. If the engine has been removed from the vehicle and mounted on a stand, many of the steps outlined in this Part of Chapter 2 will not apply.

The Specifications included in this Part of Chapter 2 apply only to the procedures contained in this Part. Part D of Chapter 2 contains the Specifications necessary for cylinder head and engine block rebuilding.

2 Repair operations possible with the engine in the vehicle

Many major repair operations can be accomplished without removing the engine from the vehicle.

Clean the engine compartment and the exterior of the engine with some type of degreaser before any work is done. It will make the job easier and help keep dirt out of the internal areas of the engine.

Depending on the components involved, it may be helpful to remove the hood to improve access to the engine as repairs are performed (refer to Chapter 11 if necessary). Cover the fenders to prevent damage to the paint. Special pads are available, but an old bedspread or blanket will also work.

If vacuum, exhaust, oil or coolant leaks develop, indicating a need for gasket or seal replacement, the repairs can generally be made with the engine in the vehicle. The intake and exhaust manifold gaskets and cylinder head gaskets are all accessible with the engine in place. **Note:** *Removing the oil pan on a 4.0L engine requires removing the engine from the vehicle.*

Exterior engine components, such as the intake and exhaust manifolds, the water pump, the starter motor, the alternator, the distributor and the fuel system components can be removed for repair with the engine in place.

Since the cylinder heads can be removed without pulling the engine, valve component servicing can also be accomplished with the engine in the vehicle. Replacement of the timing chain and sprockets requires removal of the oil pan, so it is not possible with the engine in the vehicle.

3 Top Dead Center (TDC) for number one piston - locating

Refer to illustrations 3.5a and 3.5b
Note: *The 4.0L engine is not equipped with a distributor. Piston position must be determined by feeling for compression at the number one spark plug hole, then aligning the ignition timing marks as described in Step 5.*

1   Top Dead Center (TDC) is the highest point in the cylinder that each piston reaches as it travels up-and-down when the crankshaft turns. Each piston reaches TDC on the compression stroke and again on the exhaust stroke, but TDC generally refers to piston position on the compression stroke.

2   Positioning the piston(s) at TDC is an essential part of many other repair procedures discussed in this manual.

3   Before beginning this procedure, be sure to place the transmission in Neutral and apply the parking brake or block the rear wheels. Remove the spark plugs (see Chapter 1). Disable the ignition system by disconnecting the wiring harness connector from the ignition coil pack, located above the left valve cover.

4   In order to bring any piston to TDC, the crankshaft must be turned using one of the methods outlined below. When looking at the front of the engine, normal crankshaft rotation is clockwise.

a) *The preferred method is to turn the crankshaft with a socket and ratchet attached to the bolt threaded into the front of the crankshaft.*

b) *A remote starter switch, which may save some time, can also be used. Follow the instructions included with the switch. Once the piston is close to TDC, use a socket and ratchet as described in the previous paragraph.*

c) *If an assistant is available to turn the ignition switch to the Start position in short bursts, you can get the piston close to TDC without a remote starter switch. Make sure your assistant is out of the vehicle, away from the ignition switch, then use a socket and ratchet as described in Paragraph a) to complete the procedure.*

5   The crankshaft pulley has 35 teeth, evenly spaced every 10-degrees around the pulley, and a gap where a 36th tooth would be. The gap is located at 60-degrees Before Top Dead Center (BTDC) **(see illustration)**. Turn the crankshaft (see Paragraph 4 above)

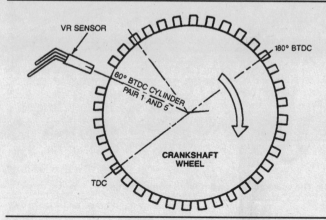

3.5a The crankshaft pulley has a gap at 60-degrees BTDC - TDC is located at the sixth tooth from the gap

2A

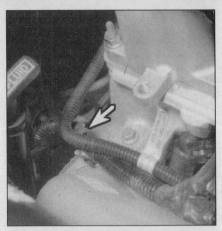

3.5b The pointer on the front of the engine lines up with a notch in the crankshaft pulley to indicate TDC

4.7 On the passenger's side, disconnect the vacuum hose at the coupling above the valve cover

4.8 Lift the wiring harness off the stud with a thumbnail - don't pry it

until you feel compression at the number one spark plug hole, then turn it slowly until the sixth tooth from the missing tooth is aligned with the Variable Reluctance (VR) sensor and the TDC notch is aligned with the pointer (located at the front of the engine) **(see illustration)**.

6   After the number one piston has been positioned at TDC on the compression stroke, TDC for any of the remaining pistons can be located by rotating the crankshaft in 120-degree increments and following the firing order. Divide the crankshaft pulley into three equal sections with chalk marks at each point, each indicating 120-degrees of crankshaft rotation. Rotating the engine past TDC for number 1 cylinder to the next mark will place the engine at TDC for cylinder number 4.

4   Valve covers - removal and installation

Removal

1   Disconnect the negative cable from the battery.
2   Remove the fresh air intake shield and tube (see Chapter 1). If necessary, disconnect the fuel supply and return lines to provide removal access for the valve covers (see Chapter 4). **Warning:** *Relieve fuel system pressure as described in Chapter 4 before disconnecting any fuel lines.*

Right valve cover
Models through 1998
Refer to illustrations 4.7, 4.8, 4.9a, 4.9b and 4.9c
3   Remove the alternator and ignition coil

pack (refer to Chapter 5). **Note:** *This Step may not be necessary on 1997 and later models.*
4   Remove the bolt that secures the air conditioning refrigerant line above the upper intake manifold (if not already done when removing the coil pack).
5   Detach the spark plug wires from the valve cover clips.
6   Carefully pry the two wiring harnesses loose from the valve cover with a tool such as a door panel clip remover (available at auto parts stores).
7   Disconnect the vacuum hose at the coupling above the valve cover **(see illustration)**. **Note:** *On 1997 and later models, there are three vacuum hoses to disconnect. Label them with masking tape before disconnecting.*
8   Carefully lift the engine wiring harness clip with your thumb at the point shown to separate it from the valve cover **(see illustration)**. Don't pull on the harness.

1999 and later
9   Label and then disconnect the spark plug wires.
10   Drain the cooling system (see to Chapter 1). Disconnect the upper radiator hose from the intake manifold.

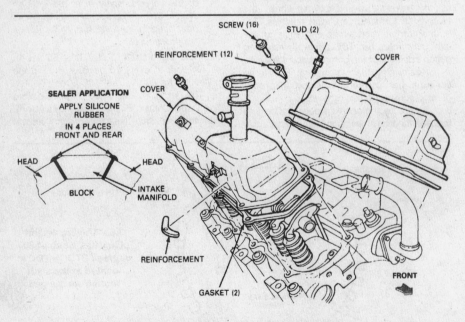

4.9a The valve covers and related components

SEALER APPLICATION
APPLY SILICONE RUBBER
IN 4 PLACES
FRONT AND REAR

SCREW (16)   STUD (2)
REINFORCEMENT (12)   COVER
COVER
HEAD   HEAD
BLOCK   INTAKE MANIFOLD
REINFORCEMENT
GASKET (2)   FRONT

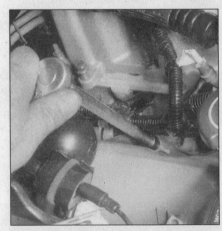

4.9b Remove the valve cover bolts with a ratchet and extension . . .

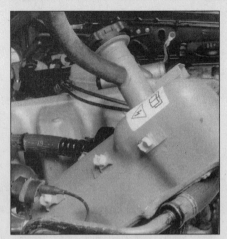

4.9c . . . and lift the cover off - you may have to angle it out, as shown here - tap it gently with a soft-face hammer if necessary to break the gasket seal

4.16 Disconnect the compressor electrical connector from the back of the compressor, then remove the mounting bolts and shift the compressor forward as shown - DO NOT disconnect any refrigerant lines

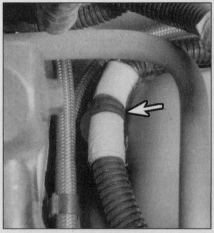

4.20 Wiring harnesses are secured to the valve cover by clips (arrow)

2A

11   Disconnect the PCV tube from the valve cover.
12   Disconnect any interfering electrical wiring harnesses, cables and heater or vacuum hoses.

All models
13   Remove the valve cover bolts and reinforcing plates **(see illustrations)**. Lift the valve cover off **(see illustration)**. Tap it gently with a soft-face hammer if necessary to break the gasket seal. **Note:** *On 1997 and later models, there are no reinforcing plates, and the valve cover bolts are "captive" on the covers. They do not come off.*

Left valve cover
Models through 1998
Refer to illustrations 4.16, 4.20 and 4.27
14   If you haven't already done so, remove the bolt from the air conditioning refrigerant line above the upper intake manifold.
15   Disconnect the electrical connector for the air conditioning compressor clutch. Carefully pry the wiring harness from the back of the compressor with a door panel clip remover or similar tool.
16   Remove the mounting bolts from the air conditioning compressor (see Chapter 3). Lift the air conditioning compressor, then position it out of the way **(see illustration)**. **Warning:** *DO NOT disconnect any refrigerant lines!*
17   Disconnect the brake booster vacuum hose.
18   Label and disconnect the vacuum hoses from the fitting on the plenum.
19   Detach the PCV hose. Remove the PCV hose and valve (see Chapter 1).
20   Carefully pry the wiring harness away from the valve cover with a door trim panel remover or similar tool **(see illustration)**. Place the harnesses out of the way.
21   Disconnect the driver's side spark plug wires from the plugs and detach them from the clips on the valve cover. Position the wires out of the way.

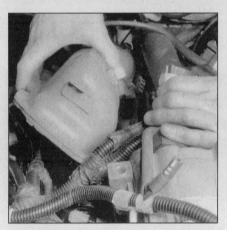

4.27 Move the wiring harness out of the way and lift the valve cover off

22   Carefully lift the engine wiring harness clip with your thumb to separate it from the valve cover. Don't pull on the harness.
23   Remove the bolt that secures the fuel line clip to the front of the engine. Move the fuel line just enough to provide access to the front valve cover bolt. DO NOT disconnect any fuel lines without first relieving fuel pressure (see Chapter 4).

1999 and later
24   Remove the upper portion of the intake manifold (refer to Chapter 4).
25   Remove the EGR valve, tube and brackets.
26   Remove the dipstick tube.

All models
27   Remove the valve cover bolts and reinforcing plates **(see illustration 4.9a)**. Lift the valve cover off **(see illustration)**. Tap it gently with a soft-face hammer if necessary to break the gasket seal. **Note:** *On 1997 and later models, there are no reinforcing plates, and the valve cover bolts are "captive" on the covers. They do not come off.*

4.29 Just before installing the new gasket, peel the plastic film off the self-sticking sealant on the valve-cover side of the gasket

Installation
Refer to illustration 4.29
28   Clean the gasket surfaces on the intake manifold, cylinder head and valve cover. Use a scraper to remove the pieces of old gasket material, then wipe off all residue with lacquer thinner or acetone.
29   Most valve cover gaskets are equipped with self-sticking sealant on the valve cover side. Pull the plastic film off the gasket **(see illustration)** and stick the gasket to the valve cover.
30   Apply silicone sealant to the seam where the cylinder head joins the intake manifold **(see illustration 4.9a)**. Apply a 1/8-inch ball of sealant to the valve cover bolt holes on the outer (exhaust) side of the cylinder head (not necessary on 1997 and later models). **Note:** *Apply silicone sealant to one side of the engine at a time (if both valve covers were removed), then install the valve cover.*
31   The remainder of installation is the reverse of removal. Tighten the valve cover bolts evenly, starting with the center bolts and working out, to the torque listed in this Chapter's Specifications.

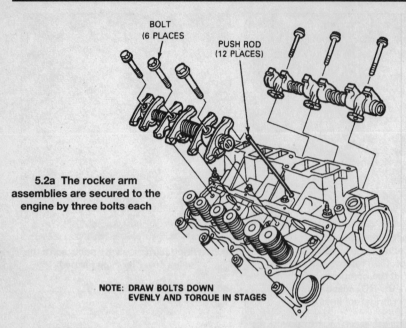

BOLT
(6 PLACES)

PUSH ROD
(12 PLACES)

5.2a The rocker arm assemblies are secured to the engine by three bolts each

NOTE: DRAW BOLTS DOWN
EVENLY AND TORQUE IN STAGES

5.2b Loosen the bolts evenly to prevent the shafts from being bent by valve spring pressure

5   Rocker arms and pushrods - removal, inspection and installation

Removal

Refer to illustrations 5.2a, 5.2b, 5.3, 5.4a and 5.4b

1   Refer to Section 4 and remove the valve cover(s).

2   Loosen the rocker arm shaft support bolts two turns at a time, starting with the center bolt and working out, until the bolts can be removed by hand **(see illustrations)**.

3   Lift the rocker arm shaft assembly off the cylinder head **(see illustration)**. The pins will hold the components together. Mark each shaft assembly so it can be returned to the same side of the engine.

4   Lift the pushrods out of the engine **(see illustration)**. Place the pushrods in order in a holder **(see illustration)** so they can be returned to their original positions. Be sure to store them so you can reinstall them with the same end facing up.

5.3 Once the bolts are loose, lift the rocker assembly off the engine

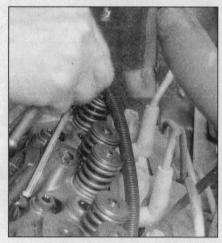

5.4a Remove the pushrods . . .

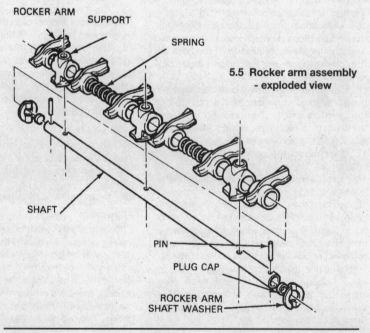

ROCKER ARM

SUPPORT

SPRING

5.5 Rocker arm assembly - exploded view

SHAFT

PIN

PLUG CAP

ROCKER ARM
SHAFT WASHER

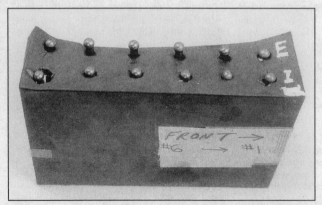

5.4b . . . and place them in a labeled holder so they can be returned to their original positions

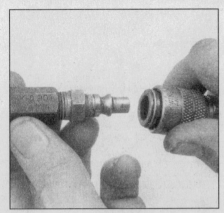

6.4 This is what the air hose adapter that threads into the spark plug hole looks like - they're commonly available from auto parts stores

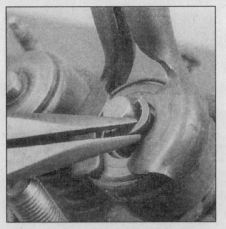

6.9a Compress the valve spring, then remove the valve stem locks

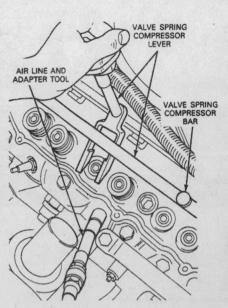

6.9b Here's the lever-type tool and bar used to compress the valve spring - these are special Ford tools, but equivalents may be available from automotive tool companies

Inspection

Refer to illustration 5.5

5   Remove the pins and disassemble the rocker arm assembly **(see illustration)**. Place the parts in order on a clean workbench. Be sure you don't mix up the parts - they must be reassembled in exactly the same order they were before disassembling.

6   Check each rocker arm for wear, cracks and other damage, especially where the pushrods and valve stems contact the rocker arm faces.

7   Make sure all oil holes are open and not plugged. Plugging can be cleared with a piece of wire.

8   Check each rocker arm bore, and its corresponding position on the rocker shaft, for wear, cracks and galling. If the rocker arms or shaft are damaged, replace them with new ones.

9   Inspect the pushrods for cracks and excessive wear at the ends. Roll each pushrod across a piece of plate glass to see if it's bent (if it wobbles, it's bent).

10   If necessary, remove the plug from each end of the rocker shaft. Drill into one plug and insert a long steel rod through it to knock out the other plug. Knock out the first plug in the same manner.

11   If the rocker arm shaft plugs were removed, install new ones with a hammer and suitable drift.

12   Assemble the rocker assembly **(see illustration 5.5)**. Lubricate at friction points (pushrod ends, rocker arm bores and ends) with engine assembly lube.

13   Install new cotter pins in the ends of the rocker shaft. Be sure the rocker shaft oil holes will face down when the shaft is installed. The position of the oil holes is indicated by a notch on the front of each shaft.

Installation

14   Coat each end of each pushrod with engine assembly lube, then install them in the engine. If you are reinstalling the original pushrods, be sure to return them to their original positions.

15   Coat the rocker arm pads with engine assembly lube.

16   Install the rocker arm assembly on the engine. The notch on the front end of each shaft should face down.

17   Position the rocker arm ball ends in the pushrods.

18   Tighten the rocker shaft support bolts two turns at a time, working from the center bolt out, to the torque listed in this Chapter's Specifications.

19   The remainder of installation is the reverse of the removal Steps.

6   Valve springs, retainers and seals - replacement

Refer to illustrations 6.4, 6.9a, 6.9b and 6.10
Note: *Broken valve springs and defective valve stem seals can be replaced without removing the cylinder heads. Two special tools and a compressed air source are normally required to perform this operation, so read through this Section carefully and rent or buy the tools before beginning the job. If compressed air isn't available, a length of nylon rope can be used to keep the valves from falling into the cylinder during this procedure.*

1   Refer to Section 4 and remove the valve cover from the affected cylinder head. If all of the valve stem seals are being replaced, remove both valve covers.

2   Remove the spark plug from the cylinder which has the defective component. If all of the valve stem seals are being replaced, all of the spark plugs should be removed.

3   Turn the crankshaft until the piston in the affected cylinder is at top dead center on the compression stroke (refer to Section 3 for instructions). If you're replacing all of the valve stem seals, begin with cylinder number one and work on the valves for one cylinder at a time. Move from cylinder-to-cylinder following the firing order sequence (see this Chapter's Specifications).

4   Thread an adapter into the spark plug

hole **(see illustration)** and connect an air hose from a compressed air source to it. Most auto parts stores can supply the air hose adapter. **Note:** *Many cylinder compression gauges utilize a screw-in fitting that may work with your air hose quick-disconnect fitting.*

5   Remove the rocker assembly on the affected side of the engine (see Section 5). If all of the valve stem seals are being replaced, remove both rocker assemblies.

6   Apply compressed air to the cylinder. **Warning:** *The piston may be forced down by compressed air, causing the crankshaft to turn suddenly. If the wrench used when positioning the number one piston at TDC is still attached to the bolt in the crankshaft nose, it could cause damage or injury when the crankshaft moves.*

7   The valves should be held in place by the air pressure.

8   Stuff shop rags into the cylinder head holes above and below the valves to prevent parts and tools from falling into the engine.

9   Using an appropriate valve spring compressor, compress the valve spring and remove the valve stem locks with small needle-nose pliers or a magnet **(see illustration)**. **Note:** *A couple of different types of tools are available for compressing the valve springs with the cylinder head in place. One type grips the lower spring coils and presses on the retainer as the knob is turned, while the other type, shown here **(see illustration)**, utilizes a bar installed in place of the rocker arm shaft for leverage. Both types work very well, although the knob type is more readily available.*

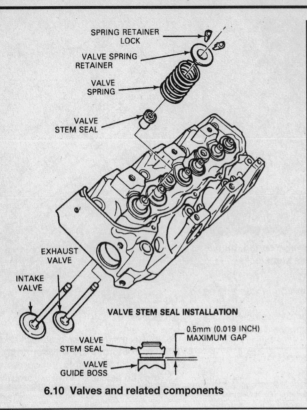

6.10 Valves and related components

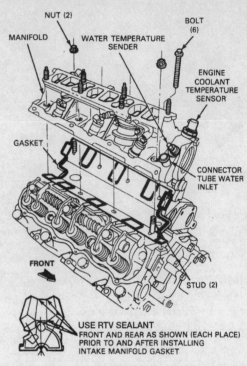

7.6a Intake manifold installation details

10   Remove the spring retainer and valve spring, then remove the valve stem seal **(see illustration)**. **Note:** *If air pressure fails to hold the valve in the closed position during this operation, the valve face or seat is probably damaged. If so, the cylinder head will have to be removed for additional repair operations.*

11   Wrap a rubber band or tape around the top of the valve stem so the valve won't fall into the combustion chamber, then release the air pressure.

12   Inspect the valve stem for damage. Rotate the valve in the guide and check the end for eccentric movement, which would indicate that the valve is bent.

13   Move the valve up-and-down in the guide and make sure it doesn't bind. If the valve stem binds, either the valve is bent or the guide is damaged. In either case, the cylinder head will have to be removed for repair.

14   Reapply air pressure to the cylinder to retain the valve in the closed position, then remove the tape or rubber band from the valve stem.

15   Lubricate the valve stem with engine oil and install a new stem seal. A special seal installer tool is recommended to install the seal, your local auto parts store should have an appropriate tool.

16   Install the spring in position over the valve.

17   Install the valve spring retainer. Compress the valve spring and carefully position the valve stem locks in the groove. Apply a small dab of grease to the inside of each lock to hold it in place.

18   Remove the pressure from the spring tool and make sure the valve stem locks are seated.

19   Disconnect the air hose and remove the adapter from the spark plug hole.

20   Refer to Section 5 and install the rocker arm assembly.

21   Install the spark plug(s) and hook up the wire(s).

22   Refer to Section 4 and install the valve cover(s).

23   Start and run the engine, then check for oil leaks and unusual sounds coming from the valve cover area.

7   Intake manifold - removal and installation

Removal

Refer to illustrations 7.6a and 7.6b

1   Disconnect the negative cable from the battery. Refer to Chapter 1 and remove the drivebelt.

2   Remove the upper intake manifold (see Chapter 4).

3   Remove the valve covers (see Section 4). Refer to Chapter 3 and remove the cooling fan and shroud.

4   Disconnect the electrical connectors from the coolant temperature sender (see Chapter 3) and coolant temperature sensor (see Chapter 6) at the front of the intake manifold.

5   Disconnect the heater hose from the front of the intake manifold. Either remove the thermostat and housing (see Chapter 3) or disconnect the upper radiator hose from the outlet fitting.

6   Remove the intake manifold nuts and bolts and lift the manifold off **(see illustrations)**. If it's stuck, tap it lightly with a soft-face hammer to break the gasket seal. If necessary, pry the manifold off, but pry between a casting protrusion and the engine - don't pry against gasket surfaces.

Installation

Refer to illustrations 7.8, 7.9 and 7.11

7   Clean away all traces of old gasket material. Remove oil and dirt with a cloth and solvent, such as acetone or lacquer thinner.

8   Apply RTV sealant around the water jacket ports (the smaller ports at the ends of the cylinder heads) and to the front sealing surface. Also apply a bead of sealant to the four corners where the intake manifold joins the engine block **(see the accompanying illustration and illustration 7.6a)**.

9   Install the intake manifold gasket **(see illustration)**, then reapply sealant to the four corners.

10   Install the intake manifold over the studs. Install the nuts and bolts and tighten them hand-tight. **Note:** *Install the intake manifold within five minutes after applying the sealant. If allowed to set up (approximately 15 minutes), the sealant won't seal properly.*

11   Tighten the nuts and bolts to the torque listed in this Chapter's Specifications, following the recommended tightening sequence **(see illustration)**.

12   The remainder of installation is the reverse of removal.

13   Run the engine and check for oil, coolant and vacuum leaks.

7.6b Remove the eight intake manifold bolts and nuts

7.8 Before installing the intake manifold gasket, apply a film of silicone sealant around the water jacket ports as well as the front sealing surface - be sure to apply a bead to the corners (arrows) at the front and rear of the manifold

7.9 As soon as the sealant is applied, install the gasket and place a bead of sealant in the corners

8 Exhaust manifolds - removal and installation

Removal

Refer to illustration 8.5

1   If removing the left exhaust manifold, remove the oil dipstick tube bracket from the engine. On 1997 and later models, disconnect the EGR tube from the exhaust manifold, then remove the EGR valve (see Chapter 6).

2   If the power steering pump hoses obstruct manifold removal, disconnect them from the pump (see Chapter 10). Cap the hoses and fittings to keep out dirt and place the hose ends out of the way.

3   If removing the right exhaust manifold, detach the heater hose bracket and disconnect the heater hoses (see Chapter 3).

4   Detach the exhaust pipe from the manifold(s). For access to the exhaust pipe nuts, it may be necessary to raise the vehicle and support it securely on jackstands. **Caution:** *If the vehicle is equipped with Automatic Ride Control (ARC), make sure the air suspension switch is turned to the OFF position before the vehicle is raised to prevent damage to the*

system components *(see Chapter 10).*

5   Remove the bolts and the exhaust manifold from the cylinder head **(see illustration)**.

Installation

6   Using a scraper, thoroughly clean the mating surfaces on cylinder head, exhaust manifold and exhaust pipe. Remove residue with a solvent such as acetone or lacquer thinner.

7   Apply some graphite grease to the cylinder head mating surface and place the manifold on the cylinder head. Tighten the bolts evenly to the torque listed in this Chapter's Specifications. **Note:** *These engines were originally assembled without exhaust manifold gaskets and can be reassembled the same way so long as the mating surfaces are perfectly flat and not damaged in any way. Warped or damaged manifolds may require a gasket or machining. Gaskets are available from aftermarket sources at your local auto parts store.*

8   Connect the exhaust pipe to the exhaust

manifold and tighten the nuts evenly to the torque listed in this Chapter's Specifications.

9   The remainder of installation is the reverse of the removal steps. If the heater hoses were disconnected, fill the cooling system (see Chapter 1).

10   Run the engine and check for exhaust leaks.

9 Cylinder heads - removal and installation

Removal

Note: *Head bolt removal requires a Torx bit. Obtain the necessary tool before starting. DO NOT try to use an Allen wrench since it may round out the bolt head.*

1   Disconnect the negative cable from the battery.

2   Drain the cooling system (see Chapter 3).

3   Remove the valve covers (see Section 4).

4   Remove the rocker arms and pushrods (see Section 5).

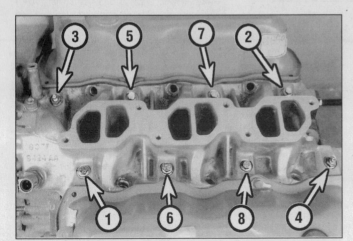

7.11 Intake manifold tightening sequence

8.5 Remove the exhaust manifold bolts (arrows)

9.15a Remove the cylinder head bolts with a T55 Torx bit - don't use an Allen wrench, since it may round out the bolt heads

5   Remove the intake manifold (see Section 7).
6   Remove the drivebelt (see Chapter 1).

Left-side

7   If not already done, detach the air conditioning compressor from the engine and place it out of the way **(see illustration 4.12).** DO NOT disconnect any refrigerant lines!
8   Remove the compressor mounting bracket on 1997 and later models.
9   Detach the power steering pump and bracket and set them out of the way (see Chapter 10). Don't disconnect the power steering hoses.
10   There's a wiring harness attached to the rear of the head. The harness is secured to the plastic retainer with tape. Unwrap or cut the tape to detach the harness from the retainer.

Right-side

11   Remove the alternator and bracket (see Chapter 5).
12   Remove the ignition coil pack and bracket (see Chapter 5).

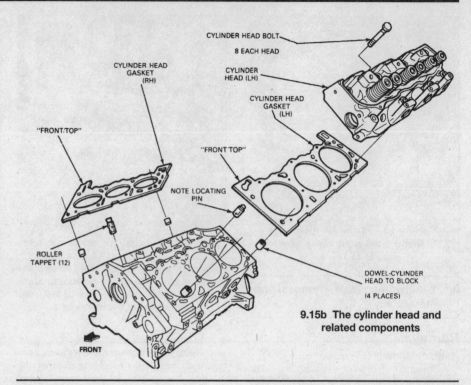

9.15b The cylinder head and related components

Both sides

Refer to illustrations 9.15a, 9.15b and 9.16
13   Remove the spark plugs (see Chapter 1).
14   Remove the exhaust manifolds (see Section 8).
15   Remove and discard the cylinder head bolts **(see illustrations).** The bolts must be replaced with new ones whenever they are removed. Loosen the bolts in several stages.
16   Lift the cylinder head off the engine. If it's difficult to remove, carefully pry it off. Pry against a casting protrusion, not against gasket surfaces **(see illustration).**

Installation

Refer to illustrations 9.17, 9.20 and 9.23
17   Thoroughly remove all traces of gasket

material with a gasket scraper and clean all parts with solvent **(see illustration).** Use a rag and acetone or lacquer thinner to remove any traces of oil from the gasket mating surfaces. See Chapter 2 Part B for cylinder head inspection procedures.
18   Use a tap of the correct size to chase the threads in the cylinder head bolt holes.
19   Recheck all cylinder head bolt holes and cylinder bores for any traces of coolant, oil or other foreign matter. Remove as needed.
20   Position the new gaskets over the dowel pins on the block. Don't use sealant on the gaskets. Be sure the FRONT TOP marks are positioned correctly **(see illustration).**
21   Install the cylinder heads and new cylinder head bolts. Tighten the cylinder head bolts hand-tight only at this time.
22   Referring to Section 7, apply the sealer

9.16 If necessary, pry the cylinder head loose; pry against a casting protrusion so the gasket surfaces won't be damaged

9.17 Remove all traces of the old gasket with a scraper, taking care not to gouge the sealing surfaces

9.20 Be sure the FRONT TOP mark on the cylinder head gasket is positioned correctly

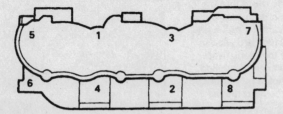

9.23 Cylinder head tightening sequence

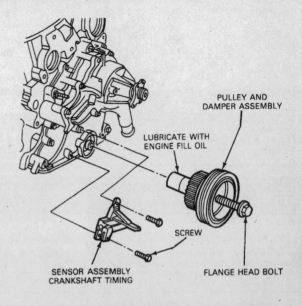

PULLEY AND DAMPER ASSEMBLY

LUBRICATE WITH ENGINE FILL OIL

SCREW

SENSOR ASSEMBLY CRANKSHAFT TIMING

FLANGE HEAD BOLT

10.2 The crankshaft pulley/damper assembly and timing sensor

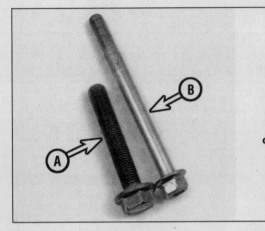

10.3a To remove the crankshaft pulley, remove the pulley bolt (A) and install a longer bolt that's smaller in diameter (B) into the crankshaft . . .

and position the intake manifold on the engine. Install the intake manifold bolts/nuts and tighten them hand-tight.

23 Alternately tighten the cylinder head bolts, then the intake manifold bolts in sequence to the torque listed in this Chapter's Specifications. Follow the tightening sequence shown for the cylinder head bolts (see illustration); refer to Section 7 for the intake manifold tightening sequence. **Caution:** *The cylinder head bolts and intake manifold bolts must be tightened in alternate stages and in sequence to ensure correct alignment and gasket seal. Failure to follow the proper procedure may result in gasket failure.*

24 The remainder of installation is the reverse of the removal steps.

25 Run the engine and check for oil, coolant and vacuum leaks.

10 Crankshaft pulley, front oil seal and timing chain cover - removal and installation

Crankshaft pulley removal and installation

Refer to illustrations 10.2, 10.3a and 10.3b

1 Remove the drivebelt (see Chapter 1).

2 Remove the crankshaft pulley bolt **(see illustration)**. **Note:** *Ford recommends replacing this bolt whenever it is removed.*

3 If necessary, insert a longer, smaller-diameter bolt into the crankshaft so the puller will have something to seat against. Remove the crankshaft pulley with a puller that bolts the pulley hub **(see illustrations)**. **Caution:** *DO NOT use a jaw-type puller that grips the outer edge of the pulley or the crankshaft*

damper will be damaged.

4 Using clean engine oil, lubricate the surface of the pulley where the front seal rides. Install the crankshaft pulley using an installation tool (available at most auto parts stores). Don't hammer the pulley on. Tighten the pulley bolt to the torque listed in this Chapter's Specifications.

Front oil seal replacement

Refer to illustration 10.5

5 Remove the crankshaft pulley (see Steps 1 through 3 above). Carefully pry the oil seal out with a screwdriver or seal removal tool **(see illustration)**.

6 Clean the seal bore and check it for nicks or gouges.

7 Coat the lip of the new seal with clean engine oil and drive it into the bore with a socket or large piece of pipe slightly smaller in diameter than the seal. The open side of

the seal faces into the engine. Install the crankshaft pulley (see Step 4 above).

Timing chain cover

Removal

Refer to illustration 10.13a, 10.13b and 10.14

8 Remove the crankshaft pulley (see Steps 1 through 3 above). Also remove the crankshaft timing sensor. **Note:** *Removal of the timing chain cover requires removal of the oil pan, which involves removing the engine from the vehicle, see Part D of this Chapter for engine removal.*

9 Drain the cooling system (see Chapter 1).

10 Remove the engine cooling fan and radiator (see Chapter 3).

11 Detach the air conditioning compressor and bracket (if equipped) and set them aside (see Chapter 3). DO NOT disconnect any refrigerant lines! Remove the oil pan (see

10.3b . . . so the puller screw (arrow) will have something to push against

10.5 To remove the front cover seal with the cover on the engine, pry it out with a screwdriver or seal removal tool, taking care not to gouge the cover

2A

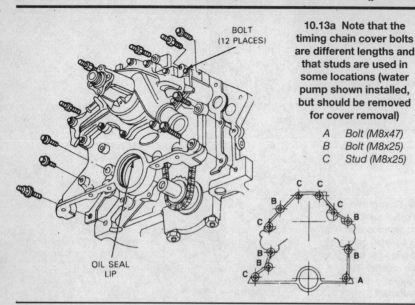

BOLT
(12 PLACES)

OIL SEAL
LIP

10.13a Note that the timing chain cover bolts are different lengths and that studs are used in some locations (water pump shown installed, but should be removed for cover removal)

A   Bolt (M8x47)
B   Bolt (M8x25)
C   Stud (M8x25)

10.13b Label the studs and brackets so they can be returned to their original locations

10.14 Remove the cover from the engine - if it's stuck, recheck to make sure all bolts and studs have been removed

Section 14).

12   Remove the water pump (see Chapter 3). Remove the heater and radiator hoses as needed to provide removal access for the timing chain cover.

13   Remove the brackets, studs and cover bolts **(see illustrations)**. Note: *The bolts are different lengths, and bolts and studs go in different locations. Label them so they can be installed in the correct location.*

14   Take the cover off **(see illustration)**. If it's stuck, tap it lightly with a soft-face hammer or pry it carefully to break the gasket seal. Don't use excessive force or you'll crack the cover. If it's difficult to remove, check to make sure you've removed all the bolts.

Installation

15   Thoroughly clean and inspect all parts

and use a scraper to remove all traces of gasket material. Remove oil film with a solvent such as lacquer thinner or acetone.

16   Apply RTV sealant to the gasket mating surfaces. Install the guide sleeves (if removed). Install the front cover and start the studs bolts two or three turns by hand. Note that the bolts are different lengths; be sure to install them in the correct location **(see illustration 10.13a)**.

17   Tighten the cover bolts evenly to the torque listed in this Chapter's Specifications.

18   Install the crankshaft timing sensor.

19   Install the crankshaft pulley (see Step 4 above).

20   The remainder of installation is the reverse of the removal steps.

21   Run the engine and check for oil or coolant leaks.

11   Timing chain and sprockets - inspection, removal and installation

Camshaft endplay check

Refer to illustration 11.4

1   Remove the timing chain cover (see Section 10).

2   Remove the rocker arm shafts (see Section 5).

3   Push the camshaft to the rear as far as it will go.

4   Install a dial indicator with its pointer on the camshaft sprocket bolt **(see illustration)**. Set the indicator to zero.

5   Pry the camshaft forward with a large screwdriver or prybar between the camshaft sprocket and the block. Note the dial indicator reading and compare with the endplay listed in this Chapter's Specifications. Replace the camshaft thrust plate (see Section 13) if endplay is excessive.

Timing chain and tensioner inspection

Refer to illustration 11.7

6   Remove the timing chain cover (see Section 10).

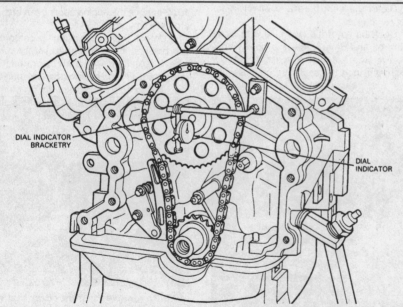

DIAL INDICATOR
BRACKETRY

DIAL
INDICATOR

11.4 To check camshaft endplay, position the dial indicator with its tip against the camshaft sprocket bolt

11.7 Unbolt the chain tensioner from the block - when checking timing chain deflection, mark a reference point on the block (arrow)

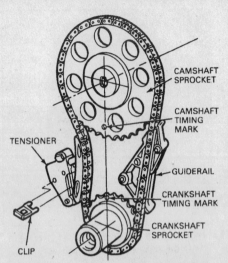

11.19 Timing chain installation details

11.20a Remove the timing chain and sprockets together

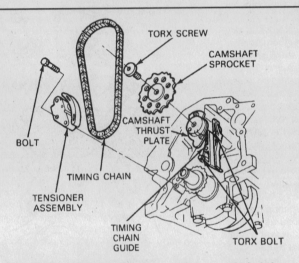

11.20b Timing chain and sprockets - exploded view

11.28 To retract the timing chain tensioner so it can be installed, squeeze the tensioner pad and use a sharp-tipped probe to push the ratchet mechanism down, then in to release the latch and let the tensioner retract

7   Remove the timing chain tensioner (see illustration).

8   Rotate the crankshaft counterclockwise (as viewed from the front of the engine) to take up the slack in the right side of the chain.

9   Mark a reference point on the block approximately halfway along the chain and measure from that point to the right side of the chain (see illustration 11.7).

10   Turn the crankshaft clockwise to take up the slack on the left side of the chain.

11   Push the chain out and measure from the reference point to the chain.

12   The difference between the two measurements is deflection. If deflection is excessive, replace the timing chain (see below).

13   Check the tensioner for wear and damage. If tensioner face wear is excessive, replace the tensioner. Before reinstalling the tensioner, it must be retracted (see Step 28). Tighten the tensioner bolts to the torque listed in this Chapter's Specifications.

Chain and sprocket

Removal

Refer to illustrations 11.19, 11.20a and 11.20b

14   Drain the cooling system and engine oil (see Chapter 1).

15   Remove the oil pan (see Section 14).

16   Replace the oil filter (see Chapter 1).

17   Remove the drivebelt (see Chapter 1).

18   Remove the timing chain cover (see Section 10).

19   Turn the crankshaft (see Section 3, Step 4) until the timing marks are aligned (see illustration). Remove the timing chain tensioner (see illustration 11.7). Remove the camshaft sprocket bolt.

20   Remove the sprockets together with the chain (see illustrations). Do not disturb the crankshaft or camshaft while the timing chain and sprockets are removed.

21   If necessary, remove the bolts that secure the chain guide and remove it from the block.

Installation

Refer to illustration 11.28

22   Be sure the crankshaft and camshaft are positioned so the sprocket timing marks will align correctly after the sprockets are installed (see illustration 11.19).

23   If the chain guide was removed, install it. Be sure its pin is inserted into the block oil hole, then tighten the guide bolts to the torque listed in this Chapter's Specifications.

24   Place the sprockets in the chain with their timing marks aligned (see illustration 11.19).

25   Install the sprockets and chain together on the crankshaft and camshaft.

26   Make sure the sprocket timing marks are still aligned (see illustration 11.19). The guide side of the chain must be straight, without any slack, for the marks to align accurately.

27   Install the camshaft sprocket bolt and tighten to the torque listed in this Chapter's Specifications.

28   Squeeze the tensioner pad with your fingers. At the same time, use a sharp-tipped

12.4 Store the lifters in a marked box so they can be returned to their original positions

12.8a Check the pushrod seat (arrow) in the top of each lifter for wear

12.8b The roller on roller lifters must turn freely - check for wear and excessive play as well

probe to push the ratchet mechanism down, then in. This will release the latch and let the tensioner retract **(see illustration)**. Retain the tensioner pad in the retracted position by holding it or installing a clip **(see illustration 11.19)** or similar device.

29   Install the timing chain tensioner and tighten its bolts to the torque listed in this Chapter's Specifications.

30   The remainder of installation is the reverse of removal.

31   Run the engine and check for oil or coolant leaks.

12   Valve lifters - removal, inspection and installation

Removal

Refer to illustration 12.4

1   Remove the intake manifold (see Section 7).

2   Remove the cylinder heads (see Section 9). Not all of the lifters can be removed with the cylinder heads in place.

3   The lifters protrude from the bores, so if there isn't a lot of varnish buildup, simply pull them out of their bores with your fingers. On engines with a lot of sludge and varnish, work the lifters up and down, using carburetor spray cleaner to loosen the deposits. If the lifters are particularly stubborn, special tools designed to grip and remove lifters are manufactured by many tool companies and are widely available.

4   Before removing the lifters, arrange to store them in a clearly labeled box to ensure that they're installed in their original locations **(see illustration)**.

5   Remove the lifters and store them where they won't get dirty.

Inspection

Refer to illustrations 12.4, 12.8a and 12.8b

6   Parts for hydraulic valve lifters are not available separately. The work required to

remove them from the engine again if cleaning is unsuccessful outweighs any potential savings from repairing them.

7   Clean the lifters with solvent and dry them thoroughly without mixing them up.

8   Check each lifter wall, pushrod seat and roller for scuffing, score marks and uneven wear. Replace any lifter that shows these conditions. If the lifter walls are worn (which isn't very likely), inspect the lifter bores in the engine block as well. If the pushrod seats are worn **(see illustration)**, check the pushrod ends. Make sure each roller turns freely **(see illustration)**.

9   If new lifters are being installed, a new camshaft must also be installed. If the camshaft is replaced, then use new lifters as well. The manufacturer recommends that the lifters and camshaft be replaced as a set if any lifter needs replacement. Never install used lifters unless the original camshaft is used and the lifters can be installed in their original locations!

10   The original lifters, if they're being reinstalled, must be returned to their original locations.

Installation

Refer to illustration 12.11

11   Install the lifters in the bores. Coat them with moly-based engine assembly lube. Note that when the lifters are installed, the alignment tab must fit in the groove in the lifter bore **(see illustration)**.

12   The remainder of installation is the reverse of removal.

13   Camshaft - removal, inspection and installation

Endplay check

1   Refer to Section 11 for this procedure.

Lobe lift check

2   In order to determine the extent of cam lobe wear, the lobe lift should be checked

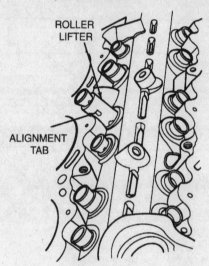

12.11 The alignment tab on each lifter must be positioned in its groove in the lifter bore

prior to camshaft removal. Refer to Section 4 and remove the valve covers. The rocker arm assembly must also be removed (see Section 5), but leave the pushrods in place.

3   Position the number one piston to TDC on the compression stroke (see Section 3).

4   Beginning with the number one cylinder, mount a dial indicator on the engine and position the plunger in-line with and resting on the first pushrod.

5   Zero the dial indicator, then very slowly turn the crankshaft in the normal direction of rotation until the indicator needle stops and begins to move in the opposite direction. The point at which it stops indicates maximum cam lobe lift.

6   Record this figure for future reference, then reposition the piston at TDC on the compression stroke.

7   Move the dial indicator to the remaining number one cylinder pushrod and repeat the check. Be sure to record the results for each valve.

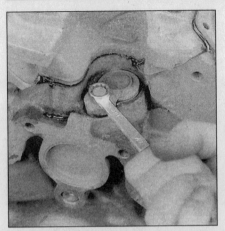

13.17a Remove the oil pump drive gear bolt . . .

13.17b . . . and lift the gear out of the block

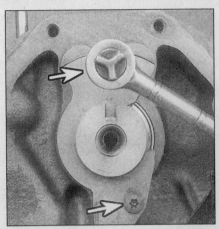

13.25 Remove the camshaft thrust plate bolts with a Torx bit and take the thrust plate off - note which end of the thrust plate is up so it can be installed correctly

13.26 As you're withdrawing it, support the camshaft near the block with both hands - be careful not to let the camshaft nick the bearings

8   Repeat the check for the remaining valves. Since each piston must be at TDC on the compression stroke for this procedure, work from cylinder-to-cylinder, following the firing order sequence.

9   After the check is complete, compare the results to this Chapter's Specifications. If camshaft lobe lift is less than specified for any of the lobes, cam lobe wear has occurred and a new camshaft should be installed.

Removal

Refer to illustrations 13.17a, 13.17b, 13.25 and 13.26.

Note: *This procedure requires engine removal.*

10   Disconnect the negative cable from the battery.

11   Drain the cooling system and engine oil (see Chapter 1).

12   Remove the radiator and fan (see Chapter 3).

13   Remove the drivebelt (see Chapter 1).

14   Remove the spark plug wires (see Chapter 1). Remove the ignition coil pack and bracket (see Chapter 5).

15   Remove the alternator (see Chapter 5).

16   Remove the engine (see Chapter 2, Part B).

17   On all 1993 models and 1994 Federal emissions models, remove the oil pump drive gear from the block (see illustrations). On

1994 California emissions models and all 1995 and later models, mark the exact orientation of the camshaft position sensor in relation to the block. The camshaft position sensor is installed in the same location as the oil pump drive gear on earlier models. Remove the camshaft position sensor and synchronizer assembly (see Chapter 6).

18   Remove the intake manifold (see Section 7). The upper intake manifold can be left attached to the lower manifold.

19   Remove the valve covers (see Section 4).

20   Remove the rocker arms and pushrods (see Section 5).

21   Remove the valve lifters (see Section 12).

22   Remove the oil pan (see Section 14).

23   Remove the timing chain cover (see Section 10).

24   Remove the timing chain and sprockets (see Section 11).

25   Remove the camshaft thrust plate bolts with a Torx bit **(see illustration)** and take the thrust plate off the block.

26   Carefully pull the camshaft out of the block, rotating it as you pull **(see illustration)**. Don't let the camshaft lobes or journals nick the camshaft bearings.

Inspection

27   After the camshaft has been removed

from the engine, cleaned with solvent and dried, inspect the bearing journals for uneven wear, pits and galling. If the journals are damaged, the bearing inserts in the engine are probably damaged as well. Both the camshaft and bearings will have to be replaced with new ones. Using a telescoping gauge, measure the inside diameter of each camshaft bearing and record the results (take two measurements, 90-degrees apart, at each bearing).

28   Measure the camshaft bearing journals with a micrometer to determine if they're excessively worn or out-of-round. If they're out-of-round, the camshaft should be replaced with a new one. Subtract the bearing journal diameters from the corresponding bearing inside diameter measurements to obtain the oil clearance. If it's excessive, new bearings must be installed. **Note:** *Camshaft bearing replacement requires special tools and expertise that place it outside the scope of the home mechanic. Take the block to an automotive machine shop to ensure that the job is done correctly.*

29   Check the camshaft lobes for heat discoloration, score marks, chopped areas, pitting, flaking and uneven wear. If the lobes are in good condition the camshaft can be reused.

30   Make sure the camshaft oil passages are clear and clean.

Installation

31   Lubricate the camshaft bearing journals and cam lobes with camshaft assembly lube.

32   Slide the camshaft into the engine. Support the cam near the block and be careful not to scrape or nick the bearings.

33   Coat both sides of the thrust plate with camshaft assembly lube, then install it. Tighten the bolts to the torque listed in this Chapter's Specifications.

34   The remainder of installation is the reverse of the removal Steps.

On models with a camshaft position sensor, install it as described in Chapter 6.

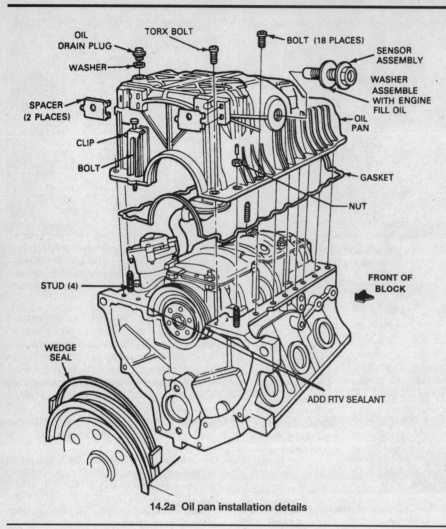

OIL
DRAIN PLUG
TORX BOLT
WASHER
BOLT (18 PLACES)
SENSOR
ASSEMBLY
WASHER
ASSEMBLE
WITH ENGINE
FILL OIL
SPACER
(2 PLACES)
OIL
PAN
CLIP
GASKET
BOLT
NUT
STUD (4)
FRONT OF
BLOCK
WEDGE
SEAL
ADD RTV SEALANT

14.2a Oil pan installation details

14.2b Use a Torx driver to remove the two bolts (arrows) adjacent to the crankshaft rear oil seal

14.3 The oil baffle is secured to the main bearing caps

14 Oil pan and baffle - removal and installation

Removal

Refer to illustrations 14.2a, 14.2b and 14.3

1   Remove the engine from the vehicle (see Part B of this Chapter).

2   Remove the oil pan fasteners **(see illustration)** and remove the oil pan from the engine. Use a Torx bit to remove the bolt on each side of the crankshaft rear oil seal **(see illustration)**.

3   If necessary, remove the baffle fasteners and remove the baffle **(see illustration)**.

Installation

Refer to illustrations 14.5, 14.9a, 14.9b, and 14.10

Caution: *Since two of the oil pan bolts are attached to the transmission, spacers are used between the oil pan and transmission. Failure to select the spacers correctly can cause oil pan damage or oil leaks when the pan is installed.*

4   Using a gasket scraper, thoroughly clean all old gasket material from the oil pan and its mounting surface. Remove residue

and oil film with a solvent such as acetone or lacquer thinner.

5   Place the oil pan on the engine block. Place a straightedge across the transmission

mounting surface on the engine block and one of the mounting pads for the oil pan-to-transmission bolts **(see illustration)**.

6   Measure the gap between the mounting

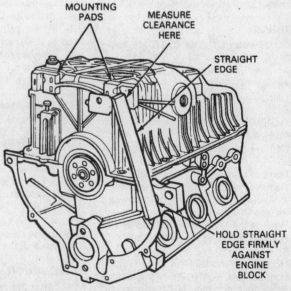

MOUNTING
PADS
MEASURE
CLEARANCE
HERE
STRAIGHT
EDGE
HOLD STRAIGHT
EDGE FIRMLY
AGAINST
ENGINE
BLOCK

14.5 To select oil pan-to-transmission spacers, place a straightedge on the transmission mounting surface and each of the oil pan-to-transmission mounting pads in turn and measure the gap between the straightedge and the mounting pad

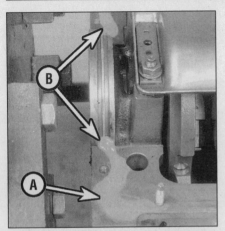

14.9a At the rear of the engine, apply thin beads of sealant to the outside edges of the gasket surface (A) and also to the crankshaft rear oil seal corners (B)

14.9b Apply beads of sealant to the corners of the front seal cover (arrows) and at the seams where the timing chain cover meets the engine block

14.10 The long Torx screw next to the crankshaft rear oil seal passes through a boss on the pan, then through the pan flange and into the block

15.2 Remove the oil pump bolts and lift the pump off the engine

15.3 As you remove the intermediate driveshaft, note the different shapes of its ends and the location of the retainer ring on the shaft

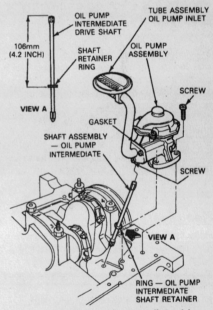

15.4 Oil pump and intermediate drive shaft installation details

pad and straightedge with a feeler gauge.

7 Move the straightedge to the other mounting pad and measure its gap.

8 Select spacers for the mounting pads to compensate for the gap. Spacers are available in three thickness (see this Chapter's Specifications).

9 At the rear of the engine, apply a thin bead of RTV sealant to the outside edge of the gasket surface and another bead to the area around the crankshaft rear oil seal **(see illustration)**. At the front of the engine, apply a bead of sealant to each of the engine block-to-timing-chain cover seams and to the corners of the front seal cover **(see illustration)**.

10 Install a new oil pan gasket on the engine. Install the oil pan fasteners and tighten evenly to the torque listed in this Chapter's Specifications. The long screw next to the crankshaft rear oil seal passes through a boss on the oil pan, then through the oil pan flange and into the engine block **(see illustration)**.

11 Install the spacers on the mounting pads before bolting the engine to the transmission.

15 Oil pump - removal and installation

Removal

Refer to illustrations 15.2, 15.3 and 15.4

1 Remove the oil pan and baffle (see Section 14).

2 Remove the oil pump bolts and the oil pump from engine **(see illustration)**.

3 Withdraw the oil pump intermediate driveshaft **(see illustration)**.

4 Unbolt the inlet tube and screen from the oil pump **(see illustration)**.

Installation

5 Fill one of the oil pump ports with clean engine oil and rotate the oil pump drive by hand to prime it.

6 Install the oil pump pick-up tube on the oil pump, using a new gasket.

7 Insert the intermediate drive shaft into the engine, pointed end first. The pressed-on retaining ring on the shaft should be positioned as shown in illustration 15.4.

8 Install the oil pump on the engine block, using a new gasket. Install the oil pump bolts and tighten them to the torque listed in this Chapter's Specifications.

9 The remainder of installation is the reverse of removal.

10 Run the engine and make sure oil pressure comes up to normal quickly. If it doesn't, stop the engine and find out the cause. Severe engine damage can result from running an engine with insufficient oil pressure!

16 Crankshaft oil seals - replacement

Front seal - timing chain cover in place

1 Refer to Section 10 for this procedure.

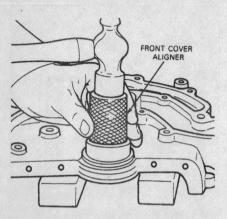

16.5 Drive the seal in squarely and evenly - a large socket or piece of pipe can be used if a seal driver isn't available

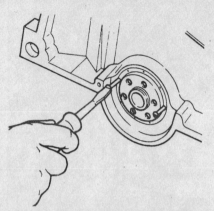

16.11 Thread two sheet metal screws into the crankshaft rear oil seal to pry it out

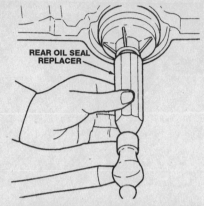

16.15 Install the crankshaft rear oil seal with the open end facing into the engine - use a socket or large piece of pipe if a seal driver isn't available

Front seal - timing chain cover removed

Refer to illustration 16.5

2   Support the timing chain cover on wooden blocks, then drive out the seal with a punch and hammer.

3   Check the seal bore in the timing chain cover for nicks or burrs.

4   Coat the seal lip and outer circumference with engine oil.

5   Position the new seal with its lip facing IN. Drive in the new seal with a seal driver **(see illustration)**. If a seal driver isn't available, use a socket or piece of pipe the same diameter as the seal.

Rear seal

Refer to illustrations 16.11, 16.15 and 16.16

6   Remove the transmission (see Chapter 7).

7   Remove the clutch (if equipped) (see Chapter 8).

8   Remove the flywheel or driveplate (see Section 17).

9   Remove the engine rear plate.

10   With a sharp awl or similar tool, punch two holes in the seal on opposite sides just above the point where the main bearing cap meets the cylinder block.

11   Thread a sheet metal screw into each hole **(see illustration)**. Pry against the sheet metal screws with two large screwdrivers or small prybars to remove the seal. **Caution:** *Be very careful not to scratch or gouge the crankshaft seal surface.*

12   Clean the oil seal bore in the block and main bearing cap. Check the seal bore and crankshaft sealing surface for nicks or burrs.

13   Apply a thin coat of clean engine oil to the outer diameter of the seal. Apply a thin coat of multi-purpose grease or engine oil to the contact surfaces of the seal and crankshaft.

14   Position the seal in the bore with its open end facing IN.

15   Drive the seal in with a seal driver until it is securely seated **(see illustration)**. Use a

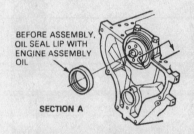

16.16 The crankshaft rear oil seal must be square with the crankshaft centerline and within the specified dimension from the rear face of the cylinder block

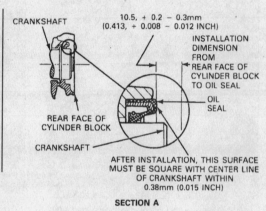

SECTION A

socket or piece of pipe the same diameter as the seal if a seal driver isn't available.

16   After installation, squareness of the rear seal with the crankshaft centerline and the dimension from the seal to the rear face of the block must be within the tolerances listed in this Chapter's Specifications **(see illustration).**

17   Flywheel/driveplate - removal and installation

Refer to illustrations 17.3 and 17.4

1   Raise the vehicle and support it securely on jackstands, then refer to Chapter 7 and remove the transmission. **Caution:** *If the vehicle is equipped with Automatic Ride Control (ARC), make sure the air suspension switch is turned to the OFF position before the vehicle is raised to prevent damage to the system components (see Chapter 10).*

2   On manual transmission equipped vehicles, remove the pressure plate and clutch disc (see Chapter 8).

3   Apply alignment marks on the flywheel/driveplate and crankshaft to ensure

correct alignment during reinstallation **(see illustration).**

4   Remove the bolts that secure the flywheel/driveplate to the crankshaft **(see illustration)**. If the crankshaft turns, wedge a screwdriver through the starter opening to jam the flywheel. **Note:** *1997 and later mod-*

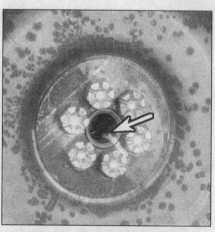

17.3 Apply alignment marks on the flywheel/driveplate and crankshaft

17.4 Remove the mounting bolts and the flywheel/driveplate from the crankshaft - the manual transmission flywheel is heavy, so be sure to support it during removal

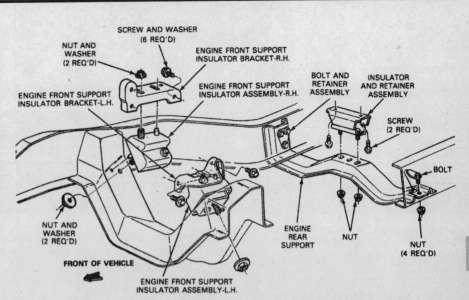

18.8 Typical engine and transmission mount details

els have eight flywheel bolts, while earlier models have six.

5   Remove the flywheel/driveplate from the crankshaft. Since the flywheel is fairly heavy, be sure to support it while removing the last bolt.

6   Clean the flywheel to remove grease and oil. Inspect the surface for cracks, rivet grooves, burned areas and score marks. Light scoring can be removed with emery cloth. Check for cracked and broken ring gear teeth. Lay the flywheel on a flat surface and use a straightedge to check for warpage.

7   Clean and inspect the mating surfaces of the flywheel/driveplate and the crankshaft. If the crankshaft rear seal is leaking, replace it before reinstalling the flywheel/driveplate.

8   Position the flywheel/driveplate against the crankshaft. Be sure to align the marks made during removal. Note that some engines have an alignment dowel or staggered bolt holes to ensure correct installation. Before installing the bolts, apply thread locking compound to the threads.

9   Wedge a screwdriver through the starter motor opening to keep the flywheel/driveplate from turning as you tighten the bolts to the torque listed in this Chapter's Specifications.

10   The remainder of installation is the reverse of the removal procedure.

18   Engine mounts - check and replacement

1   Engine mounts seldom require attention, but broken or deteriorated mounts should be replaced immediately or the added strain placed on the driveline components may cause damage or wear.

Check

2   During the check, the engine must be raised slightly to remove the weight from the mounts.

3   Raise the vehicle and support it securely on jackstands, then position a jack under the engine oil pan. **Caution:** *If the vehicle is equipped with Automatic Ride Control (ARC), make sure the air suspension switch is turned to the OFF position before the vehicle is raised to prevent damage to the system components (see Chapter 10).* Place a large block of wood between the jack head and the oil pan, then carefully raise the engine just enough to take the weight off the mounts. **Warning:** *DO NOT place any part of your body under the engine when it's supported only by a jack!*

4   Check the mounts to see if the rubber is cracked, hardened or separated from the metal plates. Sometimes the rubber will split right down the center.

5   Check for relative movement between the mount plates and the engine or frame (use a large screwdriver or pry bar to attempt to move the mounts). If movement is noted, lower the engine and tighten the mount fasteners.

6   Rubber preservative should be applied to the mounts to slow deterioration.

Replacement

Refer to illustration 18.8

7   Disconnect the negative battery cable from the battery, then raise the vehicle and support it securely on jackstands (if not already done).

8   Remove the fasteners and detach the mount from the frame bracket **(see illustration)**.

9   Raise the engine slightly with a jack or hoist (make sure the fan doesn't hit the radiator or shroud). Remove the mount-to-block bolts and detach the mount.

10   Installation is the reverse of removal. Use thread locking compound on the mount bolts and be sure to tighten them securely.

Notes

Chapter 2 Part B 4.0L SOHC V6 engine

Contents

2B

Specifications

General

Displacement	4.0 liters (244 cubic inches)
Bore and stroke	3.953 x 3.31 inches
Cylinder numbers (front-to-rear)	
Left (driver's) side	4-5-6
Right side	1-2-3
Firing order	1-4-2-5-3-6

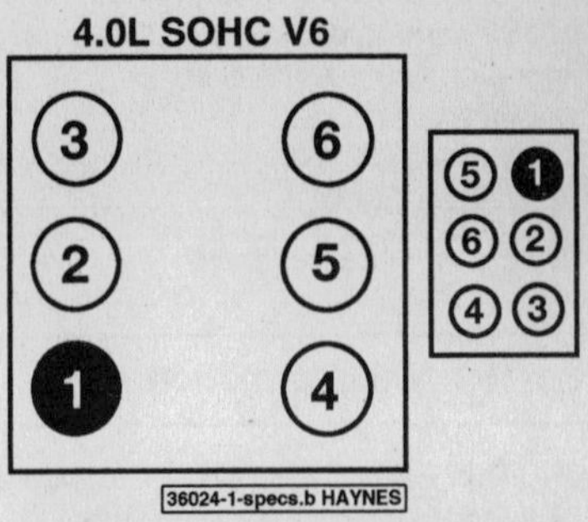

4.0L SOHC V6

```
┌─────────────────┐
│  ③        ⑥     │      ┌──────┐
│                 │      │ ⑤  ① │
│  ②        ⑤     │      │ ⑥  ② │
│                 │      │ ④  ③ │
│  ●①       ④     │      └──────┘
└─────────────────┘
```

36024-1-specs.b HAYNES

Cylinder locations and coil terminal identification

Camshafts

Lobe lift (intake and exhaust)	0.259 inch
Allowable lobe lift loss	0.005 inch
Endplay	0.0003 to 0.007 inch
Journal diameter (all)	1.100 to 1.104 inches
Bearing inside diameter (all)	1.102 to 1.104 inches
Journal-to-bearing (oil) clearance	
Standard	0.002 to 0.004 inch
Service limit	0.006 inch

Torque specifications

	Ft-lbs (unless otherwise indicated)
Camshaft sprocket bolt	62
Camshaft bearing cap bolts	
Step 1	53.5 in-lbs
Step 2	132 to 150 in-lbs
Crankshaft damper bolt*	
Step 1	44
Step 2	Tighten an additional 90 degrees
Crankshaft pulley bolts	20 to 28
Cylinder head bolts (new)	
8 mm bolts	23 to 25
12 mm bolts	
Step 1	26
Step 2	Tighten an additional 90 degrees
Step 3	Tighten an additional 90 degrees
Flywheel/driveplate bolts	
1999 and earlier	
Step 1	19 to 25
Step 2	Tighten an additional 90 degrees
2000 models	
Step 1	10
Step 2	52

Torque specifications (continued)

	Ft-lbs (unless otherwise indicated)
Engine mount insulator-to-frame nuts	65 to 97
Engine mount insulator-throughbolts/nuts	51 to 67
Exhaust manifold nuts	15 to 18
Exhaust pipe-to-manifold nuts	25 to 33
Engine front cover bolts	13 to 15
Intake manifold bolts/nuts	
Lower manifold (all except 2001 models)	107 to 123 in-lbs
Manifold (2001 models)	89 in-lbs
Jackshaft sprocket bolts	
Front	
Step 1	31 to 34
Step 2	Tighten an additional 75 degrees
Rear	
Step 1	168 in-lbs
Step 2	Tighten an additional 37 degrees
Oil pump drive gear bolt	156 to 180 in-lbs
Oil pump pick-up tube bolts	80 to 115 in-lbs
Oil pump-to-block bolts	156 to 180 in-lbs
Oil pan (sheet metal)-to-crankcase reinforcement section bolts	71 to 88 in-lbs
Crankcase reinforcement section-to-block	
Perimeter bolts/nuts	62 to 88 in-lbs
Threaded inserts	12
Inside bolts	
Step 1	10 to 12 in-lbs
Step 2	23 to 26
Timing chain tensioner	
Through 1998	
Left	35 to 39
Right	31 to 33
1999 and later	49
Timing chain cassette bolts	
Left	89 to 123 in-lbs
Right	80 to 97 in-lbs
Valve cover bolts	71 to 88 in-lbs

* Bolt(s) must be replaced.

1   General information

This Part of Chapter 2 is devoted to in-vehicle repair procedures for the 4.0L SOHC (Singe Overhead Camshaft) V6 engine installed in 1997 and later models, as well as procedures such as timing chain and sprocket and oil pan (crankcase reinforcement section) removal which require removal of the engine from the vehicle. All information concerning engine removal and installation and engine block and cylinder head overhaul can be found in Part D of this Chapter.

The following repair procedures are based on the assumption that the engine is installed in the vehicle. If the engine has been removed from the vehicle and mounted on a stand, many of the steps outlined in this Part of Chapter 2 will not apply.

The Specifications included in this Part of Chapter 2 apply only to the procedures contained in this Part. Part D of Chapter 2 contains the Specifications necessary for cylinder head and engine block rebuilding.

2   Repair operations possible with the engine in the vehicle

Many major repair operations can be accomplished without removing the engine

from the vehicle.

Clean the engine compartment and the exterior of the engine with some type of degreaser before any work is done. It will make the job easier and help keep dirt out of the internal areas of the engine.

Depending on the components involved, it may be helpful to remove the hood to improve access to the engine as repairs are performed (refer to Chapter 11 if necessary). Cover the fenders to prevent damage to the paint. Special pads are available, but an old bedspread or blanket will also work.

If vacuum, exhaust, oil or coolant leaks develop, indicating a need for gasket or seal replacement, the repairs can generally be made with the engine in the vehicle. The intake and exhaust manifold gaskets and cylinder head gaskets are all accessible with the engine in place.

Exterior engine components, such as the intake and exhaust manifolds, the water pump, the starter motor, the alternator and the fuel system components can be removed for repair with the engine in place.

Since the cylinder heads can be removed without pulling the engine, valve component servicing can also be accomplished with the engine in the vehicle. Since the camshafts are located above the cylinder heads, camshaft replacement can be done with the engine in-vehicle. Replacement of

the timing chain and sprockets requires removal of the engine oil pan and reinforcement section, so it is not possible with the engine in the vehicle.

3   Top Dead Center (TDC) for number one piston - locating

Refer to illustration 3.5

Note: *The 4.0L SOHC engine is not equipped with a distributor. Piston position must be determined by feeling for compression at the number one spark plug hole, then aligning the ignition timing marks as described in Step 5.*

1   Top Dead Center (TDC) is the highest point in the cylinder that each piston reaches as it travels up-and-down during crankshaft rotation. Each piston reaches TDC on the compression stroke and again on the exhaust stroke, but TDC generally refers to piston position on the compression stroke.

2   Positioning the piston(s) at TDC is an essential part of many other repair procedures discussed in this manual.

3   Before beginning this procedure, be sure to place the transmission in Neutral and apply the parking brake or block the rear wheels. Remove the spark plugs (see Chapter 1). Disable the ignition system by disconnecting the wiring harness connector from the ignition coil pack, located above the left valve cover.

3.5 Timing marks on 4.0L SOHC V6 - align the pointer on the crankshaft position sensor (A) with the mark in the crankshaft pulley (B)

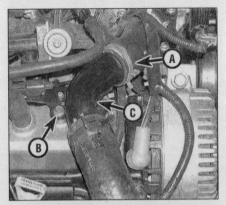

4.3 On the passenger's side, disconnect the upper radiator hose (A), remove the bolt (B) and remove the hose with the pipe (C)

4   In order to bring any piston to TDC, the crankshaft must be turned using one of the methods outlined below. When looking at the front of the engine, normal crankshaft rotation is clockwise.

  a) *The preferred method is to turn the crankshaft with a socket and ratchet attached to the bolt threaded into the front of the crankshaft.*

  b) *A remote starter switch, which may save some time, can also be used. Follow the instructions included with the switch. Once the piston is close to TDC, use a socket and ratchet as described in the previous paragraph.*

  c) *If an assistant is available to turn the ignition switch to the Start position in short bursts, you can get the piston close to TDC without a remote starter switch. Make sure your assistant is out of the vehicle, away from the ignition switch, then use a socket and ratchet as described in Paragraph a) to complete the procedure.*

5   Turn the crankshaft (see Paragraph 4 above) until you feel compression at the number one spark plug hole, then turn it slowly until the TDC notch is aligned with the pointer on the crankshaft position sensor **(see illustration)**. **Note:** *There are marks for TDC and*

for 10 degrees BTDC. Make sure you are on the TDC mark.

6   After the number one piston has been positioned at TDC on the compression stroke, TDC for any of the remaining pistons can be located by turning the crankshaft and following the firing order. Divide the crankshaft pulley into three equal sections with chalk marks at each point, each indicating 120 degrees of crankshaft rotation. Rotating the engine past TDC no. 1 to the next mark will place the engine at TDC for cylinder no. 4.

4   Valve covers - removal and installation

Removal

1   Disconnect the negative cable from the battery.
2   On models through 2000, refer to Chapter 4 and remove the upper intake manifold. **Warning:** *On all models, relieve the fuel system pressure as described in Chapter 4 before disconnecting any fuel lines.*

Right valve cover
Models through 2000
Refer to illustrations 4.3, 4.4 and 4.6
3   Drain the cooling system (refer to Chap-

ter 1) and disconnect the upper radiator hose and pipe **(see illustration)**.
4   Remove the bolts that secure the heater hose bracket and disconnect the transmission dipstick tube **(see illustration)**.
5   Detach the spark plug wires from the valve cover clips.
6   Remove the valve cover bolts **(see illustration)**. Lift the valve cover off. Tap gently with a soft-face hammer if necessary to break the gasket seal.

2001 models
7   Detach the mass air flow sensor wiring connector move the wiring aside.
8   Loosen the clamp screws and remove the air filter outlet pipe.
9   Drain the cooling system (refer to Chapter 1) and disconnect the upper radiator hose and pipe.
10   Label and disconnect all hoses attached to the valve cover.
11   Remove the bolts that secure the heater hose bracket and disconnect the transmission dipstick tube.
12   Detach the heater tube and bracket assembly.
13   Detach the spark plug wires and remove them from the valve cover clips.
14   Detach the electrical wires from the fuel injectors.
15   Remove the valve cover bolts. Lift the valve cover off. Tap gently with a soft-face hammer if necessary to break the gasket seal.

Left valve cover
Models through 1998
Refer to illustration 4.18
16   Disconnect the electrical connector at the camshaft position sensor (see Chapter 6).
17   Disconnect the electrical connector and vacuum connections at the differential pressure feedback transducer (See Chapter 6).
18   Remove the bolt and cover from the main electrical connect at the valve cover, then push in the clip to detach the other side of the connector from the valve cover **(see illustration)**.

1999 to 2000 models
19   Disconnect the fuel line (refer to Chapter 4).

2B

4.4 Remove the two heater hose bracket bolts (A) and disconnect the transmission dipstick tube (B)

4.6 Right-hand valve cover bolts (arrows) - left-hand bolts similar

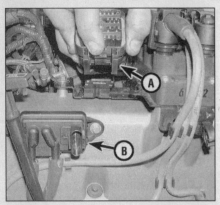

4.18 Loosen the bolt and disconnect the chassis side of the engine electrical connector, then push in the clip (A) to disconnect the other side from the valve cover - (B) is the differential pressure feedback transducer

20   Disconnect the wiring from the ignition coil, the camshaft position sensor and any other interfering components. Disconnect the vacuum hose from the EGR transducer.
21   Disconnect the large electrical connector from the valve cover.

2001 models

22   Detach the accelerator and cruise control (if equipped) cables from the throttle body. Detach the cables from the bracket and move the cables aside.
23   Detach the spark plug wires and remove them from the valve cover clips.
24   Detach the exhaust gas recirculation (EGR) valve vacuum hose, the solenoid hose and the electrical connector.
25   Detach the camshaft position sensor and the differential pressure feedback electrical connectors.
26   Detach the spark plug wire retainer.
27   Detach the fuel vapor hose and the vacuum hose.
28   Remove the bolt and the electrical connector and bracket. Move them aside.
29   Remove the bolt and the EGR valve vacuum regulator solenoid.
30   Detach the differential pressure feedback with the hoses from the exhaust manifold-to-EGR valve tube.
31   Disconnect the fuel supply line from the fuel rail. **Warning:** *Relieve the fuel system pressure as described in Chapter 4 before disconnecting any fuel lines.*
32   Remove the bolt and move the fuel supply line aside.
33   Detach the fitting and disconnect the exhaust manifold-to-EGR valve tube.
34   Detach the hose and unscrew the hose fitting from the valve cover.

All models

35   Disconnect the radio interference electrical connector at the valve cover.
36   Refer to Chapter 5 and label and remove the spark plug wires and electrical connector from the ignition coil-pack.
37   Remove the valve cover bolts and lift the

5.2a The special valve spring compressor hooks under the camshaft at (A), pushes on the valve spring retainer at (B), and is operated by a 1/2-inch-drive breaker bar placed at (C)

valve cover off. Tap it gently with a soft-face hammer if necessary to break the gasket seal.

Installation

38   Clean the gasket surfaces on the intake manifold, cylinder head and valve cover. Use a scraper to remove the pieces of old gasket material, then wipe off all residue with lacquer thinner or acetone.
39   Most valve cover gaskets are equipped with self-sticking sealant on the valve cover side. Pull the plastic film off the gasket and stick the gasket to the valve cover.
40   The remainder of installation is the reverse of the removal Steps. Tighten the valve cover bolts evenly, starting with the center bolts and working out, to the torque listed in this Chapter's Specifications.

---

5   Rocker arms and lash adjusters- removal, inspection and installation

Note: *There are two methods of removing the rocker arms and lash adjusters on this model engine. The method recommended by the manufacturer accomplishes the removal of the rocker arms without the removal of the camshaft(s), using a special valve spring compressor made specifically for the SOHC engine. The valve spring compressor uses the camshaft as a pivot point and, with a ratchet or breaker bar attached, pushes down on the valve spring to release tension on the rocker arm. The alternative method requires the removal of the camshaft (see Section 12). Either method will achieve the same results, but it is much easier using the manufacturers special tool, if it can be located.*

Removal

Refer to illustrations 5.2a, 5.2b and 5.3
1   Remove the valve cover(s) (see Section 4).
2   Install the special valve spring compressor and compress the spring just enough to remove the rocker arm **(see illustrations)**.

5.2b Compress the valve spring just enough to allow the rocker arm to be removed

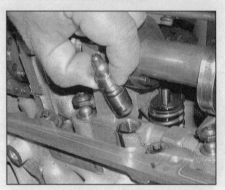

5.3 Remove the lash adjuster and store it and the rocker arm in organized manner so the components will be returned to their original locations

Caution: *Camshaft rocker arms and hydraulic lash adjusters MUST be reinstalled in the same location they were removed from. Label and store all components to avoid confusion during reassembly.*

3   Remove the hydraulic lash adjuster **(see illustration)**. If there are many miles on the vehicle, the adjusters may have become varnished and difficult to remove. Apply a little penetrating oil around the lash adjuster to help loosen the varnish.

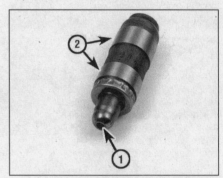

5.4 Inspect the lash adjuster for signs of excessive wear or damage, such as pitting, scoring or signs of overheating (bluing or discoloration) - the areas of wear are the rocker arm pivot point (1) and the side surfaces where the lifter body contacts the cylinder head bore (2)

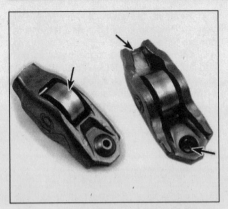

5.6 Check the rocker arm roller, the valve stem contact point and lash adjuster contact point (arrows)

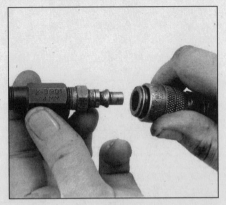

6.4 This is what the air hose adapter that threads into the spark plug hole looks like - they're commonly available from auto parts stores

6.9a Compress the valve spring, then remove the valve stem locks with a magnetic tool

6.9b Remove the valve seal with pliers

Inspection

Refer to illustrations 5.4 and 5.6

4   Inspect each adjuster carefully for signs of wear or damage **(see illustration)**. Since the lash adjusters frequently become clogged as mileage increases, we recommend replacing them if you're concerned about their condition or if the engine is exhibiting valve noise.

5   A thin wire or paper clip can be placed in the oil hole to move the plunger and make sure it's not stuck. **Note:** *The lash adjuster must have no more than 1.5 mm of total plunger travel.* It's recommended that if replacement of any of the adjusters is necessary, that the entire set be replaced. This will avoid the need to repeat the repair procedure as the others require replacement in the future.

6   Inspect the rocker arms for signs of wear or damage **(see illustration)**.

Installation

7   Before installing the lash adjusters, bleed them of air. Stand the adjusters upright in a container of oil. Use a thin wire or paper clip to work the plunger up and down. This "primes" the adjuster and removes the air. Leave the adjusters in the oil until ready to install.

8   Lubricate the valve stem tip, rocker arm, and lash adjuster bore with clean engine oil.

9   Install the lash adjusters and, with the valve spring depressed as in Step 2, install each rocker arm.

10   The remainder of installation is the reverse of the removal procedure.

11   When starting the engine after replacing the adjusters, there will normally be some noise until all the air is bled from the lash adjusters. After the engine is warmed-up, raise the speed from idle to 3,000 rpm for one minute. Stop the engine and let it cool down. All of the noise should be gone when it is restarted.

6   Valve springs, retainers and seals - replacement

Refer to illustrations 6.4, 6.9a, 6.9b, 6.14 and 6.16

Note: *Broken valve springs and/or defective*

valve stem seals can be replaced without removing the cylinder heads. The method recommended by the manufacturer accomplishes the removal of the valve springs and seals without the removal of the camshafts, although it does require the use of a special tool specified by the manufacturer; a valve spring compressor which is made specifically for the SOHC engine **(see illustration 5.2a)**. The valve spring compressor uses the camshaft as a pivot point and, with a ratchet attached, pushes down on the spring to release tension on the valve stem locks. The alternative method uses a more commonly available tool, but will require the removal of the cylinder head. The valve springs are recessed in this cylinder head design, so most aftermarket clamp-type spring compressors can't be used. With the cylinder head off the vehicle, a standard C-clamp-type compressor can be used with an adapter made for recessed spring pockets. In either repair procedure, a compressed air source is normally required to perform this operation, so read through this Section carefully and rent or buy the tools before beginning the job.

1   Remove the valve cover (see Section 4).

2   Remove the spark plug from the cylinder with the defective component. If all of the valve stem seals are being replaced, remove all the spark plugs.

3   Turn the crankshaft until the piston in the affected cylinder is at Top Dead Center (TDC) on the compression stroke (see Section 3). If you're replacing all of the valve stem seals, begin with cylinder number one and work on the valves for one cylinder at a time. Move from cylinder-to-cylinder following the firing order sequence (see this Chapter's Specifications).

4   Thread an air hose adapter into the spark plug hole **(see illustration)** and connect an air hose from a compressed air source to it. Most auto parts stores can supply the air hose adapter. **Note:** *Many cylinder compression gauges utilize a screw-in fitting that may work with your air hose quick-disconnect fitting.*

5   Apply compressed air to the cylinder. **Warning:** *The piston may be forced down by*

compressed air, causing the crankshaft to turn suddenly. If the wrench used when positioning the number one piston at TDC is still attached to the bolt in the crankshaft nose, it could cause damage or injury when the crankshaft moves.

6   The valves should be held in place by the air pressure.

7   Stuff shop rags into the cylinder head holes above and below the valves to prevent parts and tools from falling into the engine.

8   Compress the spring and remove the rocker arm (see Section 5). Rocker arms and hydraulic lash adjusters MUST be reinstalled with the same camshaft lobe that they were removed from. Label and store all components to avoid confusion during reassembly.

9   Keeping the spring compressed, remove the valve stem locks with small needle-nose pliers or a magnet **(see illustration)**. Remove the spring retainer and valve spring. Remove the valve stem seal **(see illustration)**. If air pressure fails to hold the valve in the closed position during this operation, the valve face and/or seat is probably damaged. If so, the cylinder head will have to be removed for additional repair operations.

10   Wrap a rubber band or tape around the top of the valve stem so the valve won't fall into the combustion chamber, then release the air pressure.

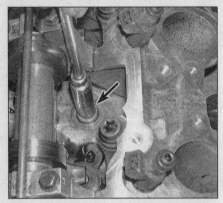

6.14 Install the seal by tapping it in place with a deep socket

6.16 Apply a small dab of grease to each valve stem lock as shown here before installation - it'll hold them in place on the valve stem as the spring is released

11 Inspect the valve stem for damage. Rotate the valve in the guide and check the end for eccentric movement, which would indicate that the valve is bent.

12 Move the valve up-and-down in the guide and make sure it doesn't bind. If the valve stem binds, either the valve is bent or the guide is damaged. In either case, the head will have to be removed for repair.

13 Reapply air pressure to the cylinder to retain the valve in the closed position, then remove the tape or rubber band from the valve stem.

14 Lubricate the valve stem with engine oil and install a new seal. A socket that will fit over the seal and is deep enough to make contact with the seat can be used to carefully tap the new seal into place (see illustration). **Caution:** *The valve seal used on the 4.0L SOHC engines is a combination seal and spring seat. Never place a valve spring directly against an aluminum cylinder head (without a seal/spring seat); the hardened spring would damage the cylinder head.*

15 Install the spring in position over the valve.

16 Install the valve spring retainer. Compress the valve spring and carefully position the valve stem lock in the groove. Apply a small dab of grease to the inside of each valve stem lock to hold it in place if necessary (see illustration).

17 Remove the pressure from the spring tool and make sure the valve stem lock is seated.

18 Disconnect the air hose and remove the adapter from the spark plug hole.

19 The remaining installation steps are the reverse of removal.

20 Start and run the engine, then check for oil leaks and unusual sounds coming from the valve cover area.

7   Intake manifold - removal and installation

Removal (1997 through 2000 models)

Refer to illustration 7.4

1 Disconnect the negative cable from the battery. Refer to Chapter 1 and remove the drivebelt.

2 Remove the upper intake manifold (see Chapter 4). **Warning:** *Follow the procedure for relieving the fuel system pressure before disconnecting any fuel components.*

3 Remove the valve covers (see Section 4). Refer to Chapter 3 and remove the cooling fan and shroud.

4 Remove the lower intake manifold bolts and lift the manifold off (see illustration). If it's stuck, tap it lightly with a soft-face hammer to break the gasket seal. **Caution:** *The manifold is plastic, don't pry against the gasket surfaces.*

Installation (1997 through 2000 models)

Refer to illustration 7.6

5 Clean away all traces of old gasket material. Remove oil and dirt with a cloth and solvent, such as lacquer thinner.

6 Install new O-ring gaskets around each of the six intake runners (see illustration). **Note:** *No sealant is required.*

7 Install the manifold to the cylinder heads, making sure the bolt holes are aligned. Install the bolts and tighten them finger-tight. **Note:** *Do not move the manifold back and forth, or the O-ring seals could move out of position, causing vacuum leaks.*

8 Tighten the bolts in the stages listed in this Chapter's Specifications, starting with the center bolts and working towards the ends.

9 The remainder of installation is the reverse of the removal steps.

10 Run the engine and check for oil, coolant and vacuum leaks.

Removal/Installation (2001 models)

This procedure is covered in Chapter 4, Section 13.

8   Exhaust manifolds - removal and installation

Removal

Refer to illustration 8.5

1 Disconnect the negative battery cable.

2 Disconnect the EGR tube from the left manifold and the EGR valve (see Chapter 6).

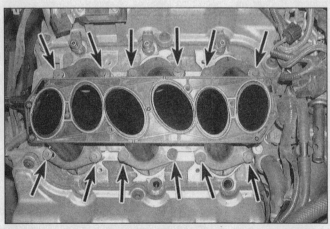

7.4 Remove the bolts (arrows) from the lower intake manifold - 1997 through 2000 models

7.6 When replacing the lower intake manifold, position new O-ring gaskets (arrows) around each runner - 1997 through 2000 models

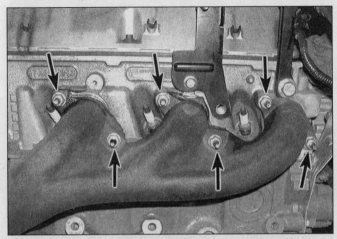

8.5 Exhaust manifold nuts (arrows) - right side shown,
left side similar

8.7 Place the new gasket over the studs on the cylinder head

3   Disconnect the two hoses at the differential pressure feedback EGR system **(see illustration 4.9)**.
4   Detach the exhaust pipe from the manifold(s). For access to the exhaust pipe nuts, it may be necessary to raise the vehicle and support it securely on jackstands. *Caution: If the vehicle is equipped with Automatic Ride Control (ARC), make sure the air suspension switch is turned to the OFF position before the vehicle is raised to prevent damage to the system components (see Chapter 10).*
5   Unbolt the manifold from the cylinder head and take it off **(see illustration)**. **Note:** *On the left side manifold, remove the nut holding the engine oil dipstick tube.*

Installation

Refer to illustration 8.7

6   Using a scraper, thoroughly clean the mating surfaces on cylinder head, manifold and exhaust pipe. Remove residue with a solvent such as acetone or lacquer thinner.
7   Check that the mating surfaces are perfectly flat and not damaged in any way. Warped or damaged manifolds may require

machining. Install the new gasket to the cylinder head studs and place the manifold on the cylinder head **(see illustration)**. Tighten the bolts evenly to the torque listed in this Chapter's Specifications.
8   Connect the exhaust pipe to the manifold and tighten the nuts evenly to the torque listed in this Chapter's Specifications.
9   The remainder of installation is the reverse of the removal steps.
10   Run the engine and check for exhaust leaks.

9   Cylinder heads - removal and installation

Caution: *The engine must be completely cool when the heads are removed. Failure to allow the engine to cool off could result in head warpage.*
Note: *Cylinder head removal on this engine is a difficult, time-consuming job, requiring a number of special tools. Read through the entire section and obtain the necessary tools before beginning the procedure.*

Removal

Refer to illustrations 9.3, 9.7, 9.8, 9.14, 9.15, 9.16 and 9.17

1   Disconnect the negative battery cable and refer to Chapter 1 to drain the cooling system.
2   Remove the upper intake manifold and label and disconnect the electrical connectors from the fuel injectors (see Chapter 4). Remove the lower intake manifold (see Section 7).
3   Refer to Section 8 and remove the exhaust manifolds. On the left side, remove the bolt and the engine oil dipstick tube **(see illustration)**.
4   Refer to Chapter 3 and remove the engine cooling fan/shroud assembly.
5   Refer to Chapter 1 and remove the drivebelt.
6   Refer to Chapter 5 and remove the alternator.
7   Remove the bolts on the accessory bracket and set the bracket aside **(see illustration)**.
8   Referring to Chapter 6, disconnect the electrical connectors at the camshaft position

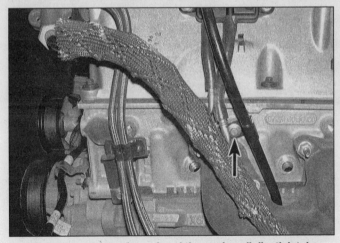

9.3 Remove the nut (arrow) and the engine oil dipstick tube

9.7 Remove the bolts (A) from the accessory bracket and set it aside - (B) indicates the bolt securing the wiring harness to the left cylinder head

9.8 Remove the engine ground strap (arrow) before removing the right cylinder head

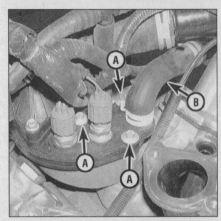

9.14 Remove the three bolts (A) and the thermostat housing (not the thermostat cover) - remove the bypass hose (B)

9.15 Remove the hydraulic chain tensioner (arrow) - left side shown, right side similar

sensor and crankshaft position sensor, then remove the EGR valve. Remove the engine ground strap from the rear of the right cylinder head **(see illustration)**.

9   Remove the valve covers (see Section 4).

10   The wiring harness is retained at the front of the left cylinder head; remove the bolt and position the harness aside **(see illustration 9.7)**.

11   Disconnect the electrical connectors at the engine coolant temperature sensor and coolant temperature gauge sending units (at the thermostat housing). These wires are part of the harness that must be pulled back (along with the injector harness) to allow intake manifold removal.

12   Separate the transmission dipstick tube from the heater hose bracket, then remove the bracket **(see illustration 4.4)**.

13   Disconnect the heater hoses from the engine, and remove the bypass hose **(see illustration 9.14)**.

14   Remove the bolts and the thermostat housing **(see illustration)**.

15   On the left cylinder head, remove the hydraulic chain tensioner **(see illustration)**.

16   Remove the bolt from the camshaft sprocket, then remove the Torx bolt below it in the head, which secures the camshaft chain "cassette" to the head **(see illustration)**.

17   Remove the bolt holding the chain cassette to the head and pull up on the chain while slipping the sprocket out. Keep the slack out of the chain (to prevent dropping it below) and tie it to the cassette with a large rubber band **(see illustration)**.

18   The cylinder heads can be removed with the camshafts, rocker arms and lash adjusters in place. Remove the cylinder head bolts, following the reverse of the tightening sequence **(see illustration 9.28)**. Loosen the bolts in sequence 1/4-turn at a time. **Note:** *There are two 8 mm external Torx bolts (one on each side of the chain opening in the head) and eight 12 mm internal Torx bolts.*

19   If the head is to be completely over-hauled, refer to Section 5 and remove the rocker arms, and Section 12 for removal of

the camshafts. Repeat Steps 15 through 18 for the right-hand cylinder head. The sprocket is at the rear of the right cylinder head, and a special, offset tool is used to remove the sprocket bolt, but it is possible (though difficult) to remove it with a conventional wrench. Tie the chain for the rear sprocket up with a rubber band as in Step 17. **Caution:** *The sprocket bolt for the right cylinder head is a left-hand thread.*

20   Use a pry bar at the corners of the head-to-block mating surface to break the gasket seal. Do not pry between the cylinder head and engine block in the gasket sealing area.

21   Lift the cylinder head(s) off the engine. If resistance is felt, place a wood block against the end and strike the wood block with a hammer. Store the cylinder heads on wood blocks to prevent damage to the gasket sealing surfaces.

22   Remove the old cylinder head gasket(s). Before removing, note which gasket goes on which side, they are different and cannot be interchanged.

23   Cylinder head disassembly and inspec-

tion procedures are covered in detail in Chapter 2, Part D.

Installation

Refer to illustrations 9.27, 9.28, 9.32, 9.33 and 9.35

24   The mating surfaces of the cylinder heads and block must be perfectly clean when the heads are installed. Use a gasket scraper to remove all traces of carbon and old gasket material, then clean the mating surfaces with lacquer thinner or acetone. If there's oil on the mating surfaces when the cylinder heads are installed, the gaskets may not seal correctly and leaks may develop. When working on the engine block, cover the open areas of the engine with shop rags to keep debris out during repair and reassembly. Use a vacuum cleaner to remove any debris that falls into the cylinders.

25   Check the engine block and cylinder head mating surfaces for nicks, deep scratches and other damage.

26   Use a tap of the correct size to chase the threads in the cylinder head bolt holes. Dirt, corrosion, sealant and damaged threads

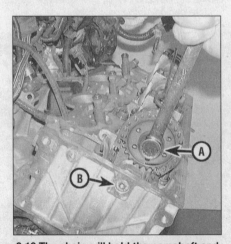

9.16 The chain will hold the camshaft and sprocket in position while you remove the sprocket bolt (A) - then remove the upper cassette bolt (B)

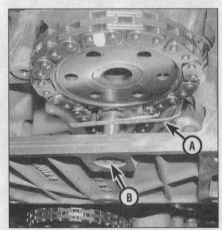

9.17 Slip a heavy-duty rubber band (A) over the chain, just below the sprocket, then remove the cassette bolt (B) - slip the sprocket out of the chain and the rubber band will prevent the slack from falling

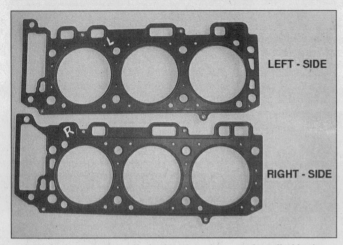

9.27 Make sure the cylinder head gaskets are installed in the correct location - they are not interchangeable

9.28 Cylinder head bolt tightening sequence - bolts (A) are 8 mm bolts, the rest are 12 mm bolts

2B

will affect torque readings.

27 Make sure the new gaskets are on the correct cylinder banks, and located on the dowels in the block. They are not interchangeable **(see illustration)**.

28 Carefully position the cylinder heads on the engine block without disturbing the gaskets. Install the cylinder head bolts and following the recommended sequence **(see illustration)**, tighten the bolts to the torque listed in this Chapter's Specifications. All bolts are tightened in the first Step, then only the eight 12 mm bolts are tightened in Steps 2 and 3. Mark a stripe on each of the 12 mm head bolts to help keep track of the bolts that have been tightened the additional 90 degrees. **Note:** *The method used for the head bolt tightening procedure is referred to as "torque-angle" or "torque-to-yield" method. A special torque angle gauge (available at most auto parts stores) is available to attach to a breaker bar and socket for better accuracy during the tightening procedure.*

29 Pull up the slack chain and insert the camshaft sprocket, align it with the camshaft

and install the sprocket bolt. **Caution:** *Do not tighten the sprocket bolt at this time, the sprocket must rotate freely on the camshaft!* Do this on both cylinder heads and install the bolts that hold the chain cassettes to the heads.

30 Install the camshaft chain tensioning tool into the chain tensioner location on the left cylinder head.

31 Rotate the engine at least one revolution and position cylinder number 1 at TDC (see Section 3).

32 Clamp the special crankshaft holding tool over the crankshaft damper (flush with the rear edge of the damper) and with the straight end against the bottom of the block **(see illustration)**. Tighten the bolt on the holding tool.

33 Install the camshaft holding tool on the rear of the left camshaft **(see illustration 9.35)**. Rotate the camshaft with a wrench to align the off-center slot on the rear of the camshaft with the tool. The slot is positioned down when aligned with the tool. Tighten the tool bolts securely. Install the camshaft

sprocket holding tool on the left camshaft sprocket **(see illustration)**. Tighten the holding tool bolts to hold the camshaft sprocket stationary, then tighten the camshaft sprocket bolt to the torque listed in this Chapter's Specifications.

34 Remove the tensioning tool from the left cylinder head and install the hydraulic tensioner. Install the tensioning tool into the right cylinder head tensioner location. **Note:** *To reach the tensioner on the right head, remove the inner fenderwell liner (see Chapter 11).* Remove the camshaft and camshaft sprocket holding tools.

35 Install the camshaft holding tool onto the front of the right camshaft **(see illustration)**. Rotate the camshaft to align the slots in the end of the camshaft with the tool. Tighten the holding tool bolts securely. Install camshaft sprocket holding tool onto the rear of the camshaft **(see illustration 9.33)** and tighten the sprocket bolt to the torque listed in this Chapter's Specifications. **Caution:** *The right camshaft sprocket bolt is left-hand thread.*

9.32 Clamp the special crankshaft holding tool around the damper as shown, then tighten the bolt (arrow) - this locks the crankshaft at TDC

9.33 Install the camshaft sprocket holding tool to the cylinder head inserting the pins on the aligning tool (A) into the holes in the sprocket - tighten the holder bolts (B), then tighten the sprocket bolt (C)

9.35 Camshaft positioning tool at the front of the right camshaft - align the tool's projection with the slot in the end of the camshaft

36   Remove the chain tensioning tool and install the hydraulic tensioner. Remove the camshaft and crankshaft holding tools.

37   The remaining installation steps are the reverse of removal.

38   Change the engine oil and filter (Chapter 1), then start the engine and check carefully for oil and coolant leaks.

10   Crankshaft pulley and front oil seal - removal and installation

Removal

Refer to illustrations 10.4, 10.5a, 10.5b and 10.6

1   Disconnect the negative battery cable and remove the drivebelt (see Chapter 1).

2   Remove the engine cooling fan/shroud assembly (see Chapter 3).

3   Remove the crankshaft pulley bolts, if equipped. **Note:** *Some early 1997 models may have a pulley that is separate from the damper, later models have a one-piece pulley/damper.*

4   Use a breaker bar and socket to remove the crankshaft pulley center bolt **(see illustration)**. Discard the bolt and obtain a new

10.4 To remove the crankshaft pulley/damper, remove the center bolt (arrow)

one for installation.

5   Using a bolt-type puller, pull the pulley/damper from the crankshaft **(see illustration)**. **Note:** *Because the pulley is recessed, an adapter may be needed between the puller bolt and the crankshaft.*

6   Use a seal puller to remove the crankshaft front oil seal **(see illustration)**. A screwdriver may be used instead, if the tip is wrapped with tape to avoid scratching the crankshaft.

7   Clean the seal bore and check it for nicks or gouges. Also examine the area of the hub that rides in the seal for signs of abnormal wear or scoring. For many popular engines, a repair sleeve is available to restore a smooth finish to the sealing surface. Check with your auto parts store.

Installation

Refer to illustration 10.8

8   Coat the lip of the new seal with clean engine oil and drive it into the bore with a socket or section of pipe slightly smaller in diameter than the seal **(see illustration)**. The open side of the seal faces into the engine.

9   Using clean engine oil, lubricate the sealing surface of the hub. Install the crankshaft pulley/damper with a special installation tool, available at most auto parts stores. Do not use a hammer to install the pulley/damper. Install a new center bolt and tighten it to the torque listed in this Chapter's Specifications. **Note:** *You must use a new pulley bolt.*

10   The remainder of the installation is the reverse of the removal procedure.

11   Timing chain and sprockets - inspection, removal and installation

Note: *This is a difficult procedure, involving special tools and the removal of the engine from the vehicle. Read through the entire Section and obtain the necessary tools before beginning the procedure.*

Removal

Refer to illustrations 11.4, 11.5, 11.6, 11.7, 11.8, 11.9, 11.10a and 11.10b

1   Refer to Chapter 2 Part D and remove the engine from the vehicle. The balance of this procedure is written assuming you have the engine out and mounted on an engine stand.

2   If you're disassembling the engine for major repair or overhaul, refer to Section 9 and remove the cylinder heads.

3   Refer to Section 13 and remove the lower oil pan cover, oil pump pickup tube and the crankcase reinforcement section.

4   Remove the bolts/nuts and the front cover from the engine block **(see illustration)**.

5   This engine uses a jackshaft (in place of a camshaft on a conventional pushrod engine) to drive the camshaft timing chains and the oil pump **(see illustration)**. The left cylinder bank camshaft is driven by a chain at the front of the engine and the right cylinder bank camshaft is driven by a chain from the rear of the jackshaft. On four-wheel drive models, the balance shaft assembly is driven by a chain from the crankshaft sprocket.

6   Remove the bolts retaining the jackshaft chain tensioner and the jackshaft chain guide **(see illustration)**.

7   Remove the bolt in the center of the jackshaft sprocket, then remove the sprocket with the jackshaft chain **(see illustration)**.

8   If not previously removed, remove the top cassette bolt **(see illustration 9.16)** and the camshaft sprocket **(see Section 9)**, remove the lower cassette-to-block bolt. Then remove the cassette, chain and jackshaft sprocket **(see illustration)**.

9   To remove the right camshaft chain, remove the large plug covering the rear of the jackshaft **(see illustration)**. **Note:** *Obtain a new plug for reassembly, the plug is not reusable.*

10   Remove the Torx bolt and spacer through the opening, then remove the cassette bolt and the chain and cassette **(see illustrations)**.

10.5 Remove the pulley/damper with a puller that bolts to the pulley hub; an adapter may be needed between the puller bolt and the crankshaft snout

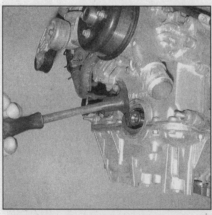

10.6 Use a seal puller to remove the old crankshaft seal, taking care not to damage the crankshaft or the seal bore in the cover

10.8 Drive the new seal in with a large socket or short section of appropriate-size pipe

11.4 Remove the engine front cover bolts (arrows) - the water pump may remain attached to the cover, if desired

11.5 Timing chain components - 2WD model show (4WD models are equipped with a balance shaft chain attached to the crankshaft sprocket)

1   Jackshaft sprocket
2   Jackshaft chain
3   Chain/cassette for left camshaft
4   Crankshaft sprocket

11.6 Remove the bolts (arrows) retaining the jackshaft chain tensioner (A) and the chain guide (B)

11.7 Remove the jackshaft sprocket bolt (arrow), then remove the sprocket and chain

11.8 Remove the bolt (arrow) and remove the left camshaft chain cassette and chain

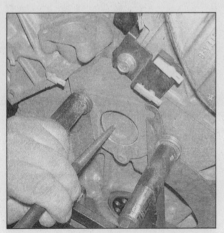

11.9 Remove the rear jackshaft plug by tapping it sideways (similar to a core plug), then use pliers to extract it

11.10a Remove this bolt and spacer (arrow) . . .

11.10b . . . then remove this bolt (arrow) and the chain and cassette

2B

11.16 Install the new front cover gasket

12.7a Areas to look for excessive wear or damage on the camshafts are the bearing surfaces and camshaft lobes (arrows)

Installation

Refer to illustration 11.16

11   If the crankshaft has been rotated during this procedure, make sure the number one piston is at the top of it's stroke (TDC). The crankshaft keyway should point straight up.

12   Install the left camshaft timing chain cassette to the block. Secure the excess chain at the top of the cassette with a rubber band **(see illustration 9.17)**. Install the rear jackshaft sprocket bolt and tighten it to the torque listed in this Chapter's specifications. Install the rear jackshaft plug.

13   Install the right camshaft chain and cassette and tighten the lower bolt to the torque listed in this Chapter's Specifications.

14   Drape the jackshaft chain over the jackshaft sprocket and engage the chain onto the crankshaft sprocket. Install jackshaft sprocket onto the jackshaft and tighten the jackshaft sprocket bolt to the torque listed in this Chapter's Specifications.

15   Install the jackshaft chain guide and tensioner **(see illustration 11.6)**. Tighten the bolts to the torque listed in this Chapter's Specifications.

16   Clean the front surface of the engine block and front cover with lacquer thinner and install a new gasket **(see illustration)**. Install the front cover, tightening the bolts to

the torque listed in this Chapter's Specifications.

16   The remainder of installation is the reverse of the removal Steps. The timing procedure for each individual camshaft, using the special tools is described in Section 9.

17   Run the engine and check for oil or coolant leaks.

12 Camshafts - removal, inspection and installation

Removal

1   Remove the valve covers (see Section 4).

2   Follow the procedure in Section 5 for removing the rocker arms and see Section 9 for the procedure to disconnect the camshaft sprockets and timing chains from the camshafts.

3   Mount a dial indicator to the front of the cylinder head and measure the camshaft endplay of each camshaft. If the clearance is greater than the value listed in this Chapter's Specifications, replace the camshaft and/or the cylinder head.

4   Loosen the bearing cap bolts in 1/4-turn increments, following the reverse sequence

of the tightening procedure **(see illustration 12.13)**, until they can be removed by hand.

5   Remove the bearing caps and lift the camshaft off the cylinder head. Don't mix up the camshafts or any of the components. They must all go back to their original locations, and on the same cylinder head they were removed from.

6   Repeat this procedure for removal of the other camshaft.

Inspection

Refer to illustrations 12.7a, 12.7b, 12.9a and 12.9b

7   Visually examine the cam lobes and bearing journals for score marks, pitting, galling and evidence of overheating (blue, discolored areas). Look for flaking of the hardened surface of each lobe **(see illustrations)**.

8   Using a micrometer, measure the diameter of each camshaft journal and the lift of each camshaft lobe **(see Chapter 2, Part D)**. Compare your measurements with the Specifications listed at the front of this Chapter, and if the diameter of any one of these is less than specified, replace the camshaft.

9   Check the oil clearance for each camshaft bearing as follows:

a) Clean the bearing surfaces and the camshaft journals with lacquer thinner or acetone.

b) Carefully lay the camshaft(s) in place in the cylinder head. Don't install the rocker arms or lash adjusters and don't use any lubrication.

c) Lay a strip of Plastigage on each journal **(see illustration)**.

d) Install the camshaft caps.

e) Tighten the caps, a little at a time, to the torque listed in this Chapter's Specifications. **Note:** Don't turn the camshaft while the Plastigage is in place.

f) Remove the bolts and detach the caps.

g) Compare the width of the crushed Plastigage (at its widest point) to the scale on the Plastigage envelope **(see illustration)**.

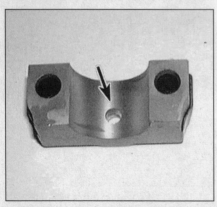

12.7b Also inspect the bearing surface of each camshaft cap (arrow)

12.9a Lay a strip of Plastigage on each camshaft journal

h) If the clearance is greater than specified, and the diameter of any journal is less than specified, replace the camshaft. If the journal diameters are within specifications but the oil clearance is too great, the cylinder head is worn and must be replaced.

10   Scrape off the Plastigage with your fingernail or the edge of a credit card - don't scratch or nick the journals or bearing surfaces.

Installation

Refer to illustration 12.13

11   If the lash adjusters and/or rocker arms have been removed, install them in their original locations **(see Section 5)**.

12   Apply moly-based engine assembly lubricant to the camshaft lobes and bearing journals, then install the camshaft(s).

13   Install the camshaft caps in the correct locations, and following the correct bolt tightening sequence **(see illustration)**, tighten the bolts to the torque listed in this Chapter's Specifications.

14   Refer to Section 9 for the timing procedure for each camshaft, using the special tools.

15   The remainder of installation is the reverse of the removal procedure.

13   Oil pan - removal and installation

Note: *The complete oil pan assembly is comprised of two sections. The lower section is a sheet metal oil pan, while the upper section is a large aluminum casting that serves as a structural reinforcement for the lower part of the crankcase. The lower oil pan and the oil pump pick-up/screen may be removed with the engine in-vehicle, but the engine must be removed from the vehicle to remove the reinforcement section for to access the oil pump and crankshaft.*

Removal

Refer to illustrations 13.2, 13.3 and 13.5

1   Remove the engine from the vehicle (see Part D of this Chapter).

2   Remove the lower oil pan fasteners **(see illustration)** and remove the sheet metal oil pan.

3   Remove the two bolts and the oil pump pickup/screen **(see illustration)**.

4   To remove the reinforcement section,

12.9b Compare the width of the crushed Plastigage to the scale on the envelope to determine the oil clearance

begin by removing the eight bolts in the center of the section **(see illustration 13.16)**.

5   Remove the two rear reinforcement section-to-block bolts **(see illustration)**, then the outside bolts on each side.

6   Carefully remove the reinforcement section from the block. **Caution:** *Do not pry between the reinforcement section and the block on any gasket sealing surface.*

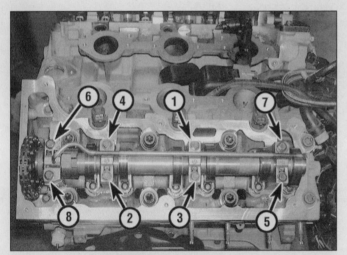

12.13 Camshaft bearing cap tightening sequence

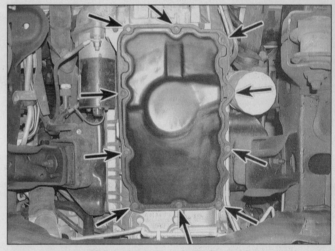

13.2 Lower oil pan bolts (arrows)

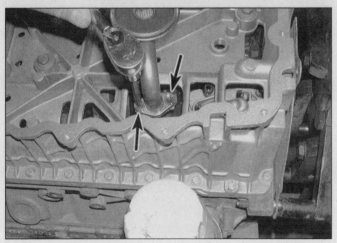

13.3 Remove the oil pump pickup tube bolts (arrows)

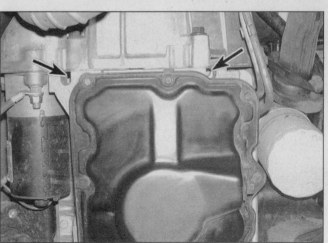

13.5 Remove the two rearmost bolts (arrows)

2B

Installation

Refer to illustrations 13.8, 13.9a, 13.9b and 13.16

Caution: *Since two of the oil pan bolts are attached to the transmission, spacers are used between the oil pan and transmission. Failure to check the clearance and select the shims correctly can cause oil pan damage or oil leaks when the pan is installed. The engine must be removed from the engine stand for this check.*

7   Using a gasket scraper, thoroughly clean all old gasket material from the engine block, reinforcement section and oil pan. Remove residue and oil film with a solvent such as acetone or lacquer thinner.

8   The eight center bolts of the reinforcement section are threaded into the main cap bolt heads. Threaded inserts are installed in the reinforcement section to create support for the bolts without stressing the reinforcement section. Use an Allen wrench to back the inserts out until they are in the "index" position before installing the reinforcement section to the engine block **(see illustration)**.

9   Apply RTV sealant to the rear main cap and the bottom of the front cover **(see illustrations)**. On the front cover, there are four spots to seal, where the cover meets the block, and on either side of the front crankshaft seal, where the rubber end gasket will meet the side gaskets.

10   Position the new gaskets on the engine block. Install the two side gaskets, then the rubber end seals, making sure that the ends of the side gaskets fit into the notches on the rubber end seals. Add a small amount of RTV sealant over the joints between the side gaskets and end seals. Install the reinforcement section.

11   Install, but do not tighten, the reinforcement section side rail bolts, including the two rear bolts.

12   Place a straightedge across the transmission mounting surface on the cylinder block and one of the mounting pads for the reinforcement section-to-transmission bolts. Measure the clearance between the mounting pad and straightedge with a feeler gauge (see Chapter 2A, **illustration 14.5**).

13   Move the straightedge to the other mounting pad and measure the clearance. The clearance should not exceed 0.010-inch. If the reinforcement section protrudes over the rear edge of the engine block, tap it toward the front of the engine with a soft-faced hammer until it's flush (it may protrude a maximum of 0.002-inch). If the clearance is excessive, tap it to the rear of the engine. If the correct clearance can not be obtained, shims are available to install on the reinforcement section mounting pad to correct the situation.

14   Tighten the perimeter bolts, including the two rear bolts, to the torque listed in this Chapter's Specifications.

15   Tighten the reinforcement section threaded inserts to the torque listed in this Chapter's Specifications.

16   Install the interior reinforcement section bolts, using new seals on the two front bolts (the front two bolts are distinguished by their silver color). Tighten them in sequence to the torque listed in this Chapter's Specifications **(see illustration)**.

17   Repeat the reinforcement section-to-engine block clearance check (see Step 13).

18   Install the oil pump pickup tube, place a new oil pan gasket in position and install the oil pan and bolts **(see illustration 13.2)**. Tighten the bolts to the torque listed in this Chapter's Specifications.

---

14   Oil pump - removal and installation

Removal

Refer to illustration 14.2

1   Remove the lower and upper oil pan sections (see Section 13). **Note:** *This involves removing the engine from the vehicle* **(see Part D of this Chapter)**.

2   Remove the oil pump bolts and take the pump off the engine **(see illustration)**.

3   Pull the oil pump intermediate driveshaft out of the engine.

13.8 Turn the threaded reinforcement section inserts out (counterclockwise) several turns each with an Allen wrench to ensure they do not contact the main cap bolt head when the section is initially installed

Installation

Refer to illustrations 14.5

4   Fill one of the pump ports with clean engine oil and rotate the pump by hand to prime it.

5   Insert the intermediate drive shaft into the engine, pointed end first, until it engages the oil pump drive **(see illustration)**. Make sure the travel limit clip is in place, and closer to the camshaft position sensor end. It fits tightly on a non-machined section of the driveshaft.

6   Install the oil pump on the block, using a new gasket. Make sure the pump engages the driveshaft. Install the oil pump bolts and tighten to the torque listed in this Chapter's Specifications.

7   The remainder of installation is the reverse of removal.

8   Run the engine and make sure oil pressure comes up to normal quickly. If it doesn't, stop the engine and find out the cause. Severe engine damage can result from running an engine with insufficient oil pressure!

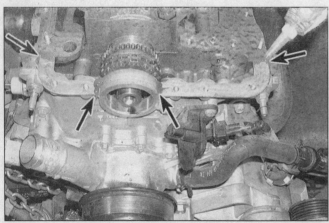

13.9a Apply RTV sealant to the four places indicated (arrows) on the front cover . . .

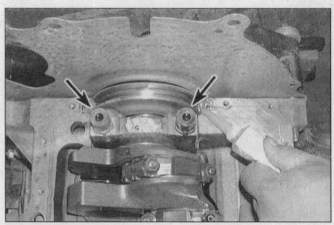

13.9b . . . and the two places indicated on the rear main cap (arrows)

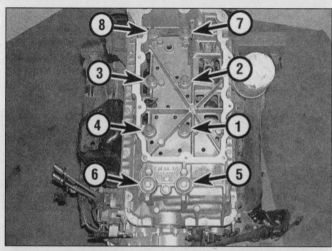

13.16 Crankcase reinforcement section bolt tightening sequence

14.2 Remove the two oil pump bolts (arrows)

15 Flywheel/driveplate - removal and installation

This procedure is identical to that for the 4.0L pushrod V6. Refer to Section 17 in Part A of this Chapter.

16 Rear main oil seal - replacement

This procedure is identical to that for the 4.0L pushrod V6. Refer to Section 16 in Part A of this Chapter.

17 Engine mounts - check and replacement

Check

1   Engine mounts seldom require attention, but broken or deteriorated mounts should be replaced immediately or the added strain placed on the driveline components may cause damage or wear.

2   During the check, the engine must be raised slightly to remove the weight from the mounts.

3   Raise the vehicle and support it securely on jackstands, then position a jack under the engine oil pan. **Caution:** *If the vehicle is equipped with Automatic Ride Control (ARC), make sure the air suspension switch is turned to the OFF position before the vehicle is raised to prevent damage to the system components (see Chapter 10).* Place a large block of wood between the jack head and the oil pan, then carefully raise the engine just enough to take the weight off the mounts. **Warning:** *DO NOT place any part of your body under the engine when it's supported only by a jack!*

4   Check the mounts to see if the rubber is cracked, hardened or separated from the metal plates. Sometimes the rubber will split right down the center.

5   Check for relative movement between the mount plates and the engine or frame (use a large screwdriver or pry bar to attempt to move the mounts). If movement is noted, lower the engine and tighten the mount fasteners.

6   Rubber preservative should be applied to the mounts to slow deterioration.

Replacement

Refer to illustration 17.9

7   Disconnect the negative battery cable from the battery, then raise the vehicle and support it securely on jackstands. **Caution:** *If the vehicle is equipped with Automatic Ride Control (ARC), make sure the air suspension switch is turned to the OFF position before the vehicle is raised to prevent damage to the system components (see Chapter 10).*

8   Remove the engine electric cooling fan/shroud assembly (see Chapter 3).

9   Remove the nuts holding the mount to the engine bracket **(see illustration)**. These nuts are reached from above, inside the engine compartment.

10   Raise the engine slightly with a jack or hoist. Remove the mount-throughbolt/nut and detach the mount from the chassis bracket.

11   Installation is the reverse of removal. Use thread-locking compound on the mount bolts and be sure to tighten them securely.

14.5 Insert the oil pump driveshaft into the pump, then guide the driveshaft into the engine, making sure the travel limit clip (arrow) is in place

17.9 Remove the engine mount nuts (arrows) from above

Notes

Chapter 2 Part C
5.0L V8 engine

Contents

2C

Specifications

General

Displacement	5.0 liters (302 cubic inches)
Cylinder numbers (front to rear)	
Right side	1-2-3-4
Left (driver's) side	5-6-7-8
Firing order	1-3-7-2-6-5-4-8

Camshaft

Lobe lift	
Intake	0.2637 inch
Exhaust	0.2801 inch
Endplay	
Standard	0.001 to 0.007 inch
Service limit	0.009 inch
Journal diameter	
No. 1	2.0815 inches
No. 2	2.0665 inches
No. 3	2.0515 inches
No. 4	2.0365 inches
No. 5	2.0215 inches
Runout limit	0.005 inch
Bearing inside diameter	
No. 1	2.0835 inches
No. 2	2.0685 inches
No. 3	2.0535 inches
No. 4	2.0385 inches
No. 5	2.0235 inches
Journal-to-bearing (oil) clearance	
Standard	0.001 to 0.003 inch
Service limit	0.006 inch
Front bearing location	0.005 to 0.020 inch below front face of block

5.0L V8

Cylinder locations and coil terminal identification

Timing chain

Timing chain deflection.. 1/2 inch maximum

Torque specifications
Ft-lbs (unless otherwise indicated)

Camshaft sprocket bolt .... 40 to 45
Camshaft thrust plate-to-engine block bolts.. 108 to 144 in-lbs
Cylinder head bolts
    Step 1 30
    Step 2 50
    Step 3 .... Tighten an additional 90 degrees
Engine mount-to-frame nuts .... 95 to 125
Exhaust manifold bolts 30
Flywheel/driveplate mounting bolts... 75 to 85
Lower intake manifold-to-cylinder head bolts
    Step 1 .... 89 in-lbs
    Step 2 24
Oil cooler-to-engine block adapter bolt .... 40 to 65
Oil pan mounting bolts
    Side bolts .... 110 to 144 in-lbs
    Corner bolts.. 144 to 216 in-lbs
Oil pick-up tube-to-main bearing cap nut .... 23 to 31
Oil pick-up tube-to-oil pump bolts .... 12 to 18
Oil pump mounting bolts .... 23 to 31
Rocker arm fulcrum bolts .... 18 to 25
Timing chain cover bolts .... 12 to 18
Valve cover bolts .... 144 to 180 in-lbs
Crankshaft pulley-to-crankshaft bolt.. 110 to 130

1   General information

This Part of Chapter 2 is devoted to in-vehicle repair procedures for the 5.0L V8 engine. All information concerning engine removal and installation and engine block and cylinder head overhaul can be found in Part D of this Chapter.

The following repair procedures are based on the assumption that the engine is installed in the vehicle. If the engine has been removed from the vehicle and mounted on a stand, many of the steps outlined in this Part of Chapter 2 will not apply.

The Specifications included in this Part of Chapter 2 apply only to the procedures contained in this Part. Part D of Chapter 2 contains the Specifications necessary for cylinder head and engine block rebuilding.

2   Repair operations possible with the engine in the vehicle

Many major repair operations can be accomplished without removing the engine from the vehicle.

Clean the engine compartment and the exterior of the engine with some type of pressure washer before any work is done. It will make the job easier and help keep dirt out of the internal areas of the engine.

It may help to remove the hood to improve access to the engine as repairs are performed (see Chapter 11 if necessary).

If vacuum, exhaust, oil or coolant leaks develop, indicating a need for gasket or seal replacement, the repairs can generally be made with the engine in the vehicle. The intake and exhaust manifold gaskets, crankshaft oil seals and cylinder head gaskets are all accessible with the engine in place.

Exterior engine components, such as the intake and exhaust manifolds, the water pump, the starter motor, the alternator, the ignition and fuel system components can be removed for repair with the engine in place. The oil pan requires engine removal to replace.

Since the cylinder heads can be removed without pulling the engine, valve component servicing can also be accomplished with the engine in the vehicle. Replacement of the timing chain and sprockets, since the oil pan must be removed, should be done with the engine out of the vehicle.

In extreme cases caused by a lack of necessary equipment, repair or replacement of piston rings, pistons, connecting rods and rod bearings is possible with the engine in the vehicle. However, this practice is not recommended because of the cleaning and preparation work that must be done to the components involved.

3   Top Dead Center (TDC) for number one piston - locating

Refer to illustration 3.4
Note: *The 5.0L engine is not equipped with a distributor. Piston position must be determined by feeling for compression at the number one spark plug hole, then aligning the ignition timing marks.*

1   Top Dead Center (TDC) is the highest point in the cylinder that each piston reaches as it travels up-and-down when the crankshaft turns. Each piston reaches TDC on the compression stroke and again on the exhaust stroke, but TDC generally refers to piston position on the compression stroke. The timing marks on the crankshaft pulley installed on the front of the crankshaft are referenced to the number one piston at TDC on the compression stroke.

2   Positioning the piston(s) at TDC is an essential part of many procedures such as rocker arm removal, valve adjustment, timing chain and sprocket replacement and distributor removal.

3   In order to bring any piston to TDC, the crankshaft must be turned using one of the methods outlined below. When looking at the front of the engine, normal crankshaft rotation is clockwise. **Warning:** *Before beginning this procedure, be sure to place the transmission in Neutral and disable the ignition system by disconnecting the primary electrical connectors at the ignition coil pack/modules (see Chapter 5).*

a) *The preferred method is to turn the crankshaft with a large socket and breaker bar attached to the crankshaft pulley bolt threaded into the front of the crankshaft.*

b) *A remote starter switch, which may save some time, can also be used. Attach the switch leads to the S (switch) and B (battery) terminals on the starter relay. Once the piston is close to TDC, use a socket and breaker bar as described in the previous paragraph.*

c) *If an assistant is available to turn the ignition switch to the Start position in short bursts, you can get the piston close to TDC without a remote starter*

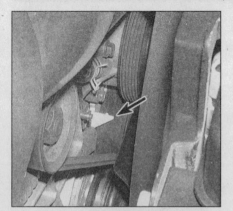

3.4 With the no. 1 cylinder on the compression stroke, align the mark on the crankshaft pulley with the timing indicator (arrow)

4.6 Location of the right-hand valve cover bolts (arrows indicate two)

4.8 Detach the fuel lines from the clips (arrows) on the left-hand valve cover

2C

switch. Use a socket and breaker bar as described in Paragraph a) to complete the procedure.

d) Turn the crankshaft clockwise with a socket and breaker bar while you hold your finger over the number one spark plug hole.

4   When the piston approaches TDC, air pressure will be felt escaping at the spark plug hole. Continue turning the crankshaft until the notch in the crankshaft pulley is aligned with the TDC mark on the front cover **(see illustration)** . At this point number one cylinder is at TDC on the compression stroke.

5   After the number one piston has been positioned at TDC on the compression stroke, TDC for any of the remaining cylinders can be located by turning the crankshaft in 90-degree increments and following the firing order listed in this Chapter's Specifications (i.e. rotating the crankshaft 90-degrees from the no. 1 TDC position will bring no. 3 cylinder to TDC).

---

4   Valve covers - removal and installation

Removal

1   Disconnect the negative cable from the battery.

Right side cover

Refer to illustration 4.6

2   Remove the air intake duct.

3   Disconnect the electrical connector on the idle air control valve (IAC) (see Chapter 6).

4   Refer to Chapter 4 and disconnect the throttle and cruise control cables, remove the throttle cable bracket, and disconnect the electrical connector at the throttle position switch.

5   Refer to Chapter 6 and disconnect both ends of the EGR tube, then remove the EGR valve and spacer.

6   Remove the valve cover bolts **(see illustration)**, then detach the cover from the head. **Note:** *If the cover is stuck to the head,*

bump one end with a block of wood and a hammer to jar it loose. If that doesn't work, try to slip a flexible putty knife between the head and cover to break the gasket seal. Don't pry at the cover-to-head joint or damage to the sealing surfaces may occur (leading to oil leaks in the future). Some valve covers are made of plastic - be extra careful when tapping or pulling on them.

Left side cover

Refer to illustration 4.8

7   Refer to Chapter 4 and remove the upper intake plenum assembly. **Warning:** *Do not disconnect any fuel system components until the fuel system pressure is relieved as described in Chapter 4.*

8   Detach the fuel lines from the valve cover clips **(see illustration)**.

9   Release the spark plug wires from the clip on the valve cover.

10   Remove the valve cover bolts.

Installation

11   The mating surfaces of each cylinder head and valve cover must be perfectly clean when the covers are installed. Use a gasket scraper to remove all traces of sealant and old gasket material, then clean the mating surfaces with lacquer thinner or acetone. If there's sealant or oil on the mating surfaces when the cover is installed, oil leaks may develop.

12   Clean the mounting bolt threads with a die to remove any corrosion and restore damaged threads. Make sure the threaded holes in the head are clean - run a tap into them to remove corrosion and restore damaged threads. Apply a small amount of light oil to the bolt threads.

13   The gaskets should be mated to the covers before the covers are installed. The gaskets on the covered models have a steel reinforcing section. Make sure the chamfered edges of the metal portion of the gasket are facing toward the gasket surface of the cover, not the cylinder head.

14   Carefully position the cover on the head and install the bolts/nuts. Tighten the bolts in two steps to the torque listed in this Chapter's Specifications. Wait two minutes

between the first and the second round of tightening. **Caution:** *Be careful with plastic valve covers, if equipped, they are easily damaged, so don't over tighten the bolts!*

15   The remaining installation steps are the reverse of removal.

16   Start the engine and check carefully for oil leaks as the engine warms up.

---

5   Rocker arms and pushrods - removal, inspection and installation

Removal

Refer to illustrations 5.2 and 5.4

1   Remove the valve cover(s) from the cylinder head(s) (see Section 4).

2   Beginning at the front of one cylinder head, remove the rocker arm fulcrum bolts **(see illustration)**. Store them separately in marked containers to ensure that they will be reinstalled in their original locations. **Note:** *If*

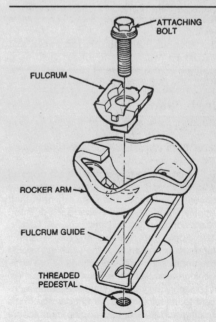

5.2 Rocker arm and related components

5.4 A perforated cardboard box provides ideal pushrod storage to ensure that they are reinstalled in their original locations

6.4 This is what the air hose adapter that threads into the spark plug hole looks like - they're commonly available from auto parts stores

6.8 Once the spring is depressed, the valve stem locks can be removed with a small magnet or needle-nose pliers (a magnet is preferred to prevent dropping the keepers)

the pushrods are the only items being removed, loosen each bolt just enough to allow the rocker arms to be rotated to the side so the pushrods can be lifted out.

3   Lift off the rocker arms, fulcrums and fulcrum guides **(see illustration 5.2)**. Store them in the marked containers with the bolts (they must be reinstalled in their original locations).

4   Remove the pushrods and store them separately to make sure they don't get mixed up during installation **(see illustration)**.

Inspection

5   Check each rocker arm for wear, cracks and other damage, especially where the pushrods and valve stems contact the rocker arm faces.

6   Make sure the hole at the pushrod end of each rocker arm is open.

7   Check each rocker arm pivot area and fulcrum for wear, cracks and galling. If the rocker arms are worn or damaged, replace them with new ones and use new fulcrums as well.

8   Inspect the pushrods for cracks and excessive wear at the ends. Roll each pushrod across a piece of plate glass to see if it's bent (if it wobbles, it's bent).

Installation

Caution: *Make sure that both lifters for each cylinder are on the base circle of the cam lobe (both valves closed) before tightening the rocker arm bolts.*

9   Lubricate the lower end of each pushrod with clean engine oil or moly-based engine assembly lubricant and install them in their original locations. Make sure each pushrod seats completely in the lifter.

10   Apply moly-based engine assembly lubricant to the ends of the valve stems and the upper ends of the pushrods before positioning the rocker arms, fulcrums and guides.

11   Apply moly-based engine assembly lubricant to the fulcrums to prevent damage to the mating surfaces before engine oil pressure builds up. Set the rocker arms and guides in place, then install the fulcrums and bolts.

6   Valve springs, retainers and seals - replacement

Refer to illustrations 6.4, 6.8, 6.9a and 6.9b
Note: *Broken valve springs and defective valve stem seals can be replaced without removing the cylinder heads. Two special tools and a compressed air source are normally required to perform this operation, so read through this Section carefully and rent or buy the tools before beginning the job.*

1   Remove the valve cover from the cylinder head(s) (see Section 4). If all of the valve stem seals are being replaced, remove both valve covers.

2   Remove the spark plug from the cylinder which has the defective component. If all of the valve stem seals are being replaced, all of the spark plugs should be removed.

3   Turn the crankshaft until the piston in the affected cylinder is at Top Dead Center on the compression stroke (see Section 3). If you're replacing all of the valve stem seals, begin with cylinder number one and work on the valves for one cylinder at a time. Move from cylinder-to-cylinder following the firing order sequence (see this Chapter's Specifications).

4   Thread an adapter into the spark plug hole **(see illustration)** and connect an air hose from a compressed air source to it. Most auto parts stores can supply the air hose adapter. **Note:** *Many cylinder compression gauges utilize a screw-in fitting that may work with your air hose quick-disconnect fitting.*

5   Remove the bolt, fulcrum and rocker arm for the valve with the defective part and pull out the pushrod. If all of the valve stem seals are being replaced, all of the rocker arms and pushrods should be removed (see Section 5).

6   Apply compressed air to the cylinder. **Warning:** *The piston may be forced down by compressed air, causing the crankshaft to turn suddenly. If the wrench used when positioning the number one piston at TDC is still*

attached to the bolt in the crankshaft nose, it could cause damage or injury when the crankshaft moves.

7   The valves should be held in place by the air pressure.

8   Stuff shop rags into the cylinder head holes above and below the valves to prevent parts and tools from falling into the engine, then use a valve spring compressor to compress the spring. Remove the valve stem locks with small needle-nose pliers or a magnet **(see illustration)**. **Note:** *A couple of different types of tools are available for compressing the valve springs with the head in place. One type, shown here, grips the lower spring coils and presses on the retainer as the knob is turned, while the other type utilizes the rocker arm bolt for leverage. Both types work very well, although the lever type is usually less expensive.*

9   Remove the spring retainer and valve spring assembly, then remove the valve guide seal **(see illustrations)**. **Note:** *If air pressure fails to hold the valve in the closed position during this operation, the valve face or seat is probably damaged. If so, the cylinder head will have to be removed for additional repair operations.*

10   Wrap a rubber band or tape around the top of the valve stem so the valve won't fall into the combustion chamber, then release the air pressure.

11   Inspect the valve stem for damage. Rotate the valve in the guide and check the end for eccentric movement, which would indicate that the valve is bent.

12   Move the valve up-and-down in the guide and make sure it doesn't bind. If the valve stem binds, either the valve is bent or the guide is damaged. In either case, the head will have to be removed for repair.

13   Reapply air pressure to the cylinder to retain the valve in the closed position, then remove the tape or rubber band from the valve stem.

14   Lubricate the valve stem with engine oil and the valve stem tip with engine assembly lubricant, then install a new guide seal.

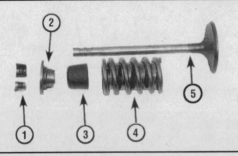

6.9a Typical valve and related components

1   Keepers
2   Retainer
3   Valve stem seal
4   Valve spring
5   Valve

6.9b Remove the seal from the valve guide with a pair of pliers

15   Install the spring in position over the valve.

16   Install the valve spring retainer. Compress the valve spring and carefully position the valve stem locks in the groove. Apply a small dab of grease to the inside of each lock to hold it in place.

17   Remove the pressure from the spring tool and make sure the valve stem locks are seated.

18   Disconnect the air hose and remove the adapter from the spark plug hole.

19   Install the rocker arm(s) and pushrod(s) (see Section 5).

20   Install the spark plug(s) and connect the spark plug wire(s) (see Chapter 1).

21   Install the valve cover(s) (see Section 4).

22   Start and run the engine, then check for oil leaks and unusual sounds coming from the valve cover area.

7   Intake manifold - removal and installation

Removal

Refer to illustration 7.8

1   Drain the cooling system (see Chapter 1).

2   Refer to Chapter 4 and remove the upper intake manifold, then disconnect the EFI wiring, labeling all connectors. **Warning:** *Do not disconnect any fuel system components until the fuel system pressure is relieved as described in Chapter 4.*

3   Remove the nut retaining the wiring harness to the intake manifold, and remove the ground strap at the rear of the manifold.

4   Disconnect the radiator and water pump bypass hoses from the water outlet (see Chapter 3). Disconnect the throttle body cooler hoses.

5   Disconnect the electrical connectors from the coolant temperature sending unit, air charge temperature sensor, throttle position sensor, idle speed control solenoid, EGR sensors, fuel injectors and fuel charging assembly (see Chapter 4).

6   Disconnect the heater water hoses and PCV heater hoses from the manifold.

7   Refer to Chapter 6 and remove the camshaft position sensor and camshaft synchronizer. **Caution:** *The engine must be set to TDC for number 1 cylinder (see Section 3) before removing the synchronizer. Do not move the crankshaft after removing the syn-*

chronizer or the engine fuel system will be out of time.

8   Loosen the lower intake manifold bolts and nuts in 1/4-turn increments until they can be removed by hand **(see illustration)**. Keep track of the location of stud/bolts.

9   The manifold will probably be stuck to the cylinder heads and force may be required to break the gasket seal. A prybar can be used to pry up the manifold, but make sure all bolts and nuts have been removed first! **Caution:** *Don't pry between the block and manifold or the heads and manifold or damage to the gasket sealing surfaces may occur, leading to vacuum and oil leaks. Pry only at a manifold casting protrusion.*

Installation

Refer to illustrations 7.12 and 7.17

Caution: *The mating surfaces of the cylinder heads, block and manifold must be perfectly clean when the manifold is installed. Gasket removal solvents in aerosol cans are available at most auto parts stores and may be helpful when removing old gasket material that's stuck to the heads and manifold (since the manifold is made of aluminum, aggressive scraping can cause damage!) Be sure to follow directions printed on the container.*

Note: *The manufacturer recommends the use of guide pins when installing the manifold. To*

make these, buy four extra manifold bolts. Cut the heads off the bolts, then grind a taper and cut a screwdriver slot in the cut ends.

10   Use a gasket scraper to remove all traces of sealant and old gasket material, then clean the mating surfaces with lacquer thinner or acetone. If there's old sealant or oil on the mating surfaces when the manifold is installed, oil or vacuum leaks may develop. When working on the heads and block, cover the lifter valley with shop rags to keep debris out of the engine. Use a vacuum cleaner to remove any gasket material that falls into the intake ports in the heads.

11   Use a tap of the correct size to chase the threads in the bolt holes, then use compressed air (if available) to remove the debris from the holes. **Warning:** *Wear safety glasses or a face shield to protect your eyes when using compressed air!* Remove excessive carbon deposits and corrosion from the exhaust and coolant passages in the heads and manifold.

12   Apply a 1/8-inch wide bead of RTV

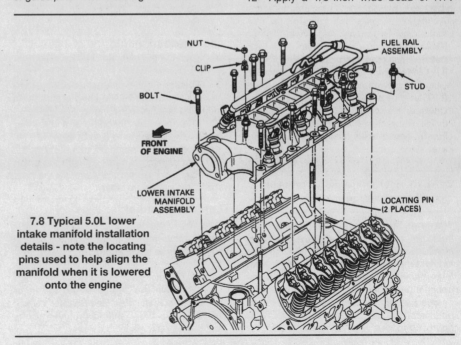

7.8 Typical 5.0L lower intake manifold installation details - note the locating pins used to help align the manifold when it is lowered onto the engine

NUT
CLIP
BOLT
FRONT OF ENGINE
LOWER INTAKE MANIFOLD ASSEMBLY
FUEL RAIL ASSEMBLY
STUD
LOCATING PIN (2 PLACES)

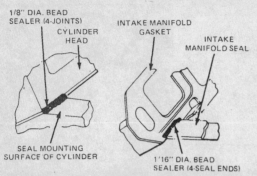

7.12 Apply a bead of RTV sealant to the corners where the block, heads and manifold meet (left), then position the gaskets and seals and apply an additional bead of sealant (right)

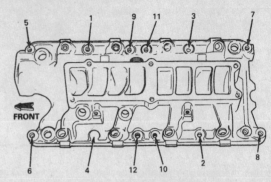

7.17 Lower intake manifold bolt tightening sequence

sealant to the four corners where the manifold, block and heads converge **(see illustration)**. **Note:** *This sealant sets up in 10 minutes. Do not take longer to install and tighten the manifold once the sealant is applied, or leaks may occur.*

13 Apply a small dab of contact adhesive to the manifold gasket mating surface on each cylinder head. Position the gaskets on the cylinder heads. The upper side of each gasket will have a TOP or THIS SIDE UP label stamped into it to ensure correct installation.

14 Position the end seals on the block, then apply a 1/8-inch wide bead of RTV sealant to the four points where the end seals meet the heads **(see illustration 7.12)**. **Note:** *The end seals have tabs that fit into notches in the side gaskets.*

15 Make sure all intake port openings, coolant passage holes and bolt holes are aligned correctly.

16 Carefully set the manifold in place while the sealant is still wet. **Caution:** *Don't disturb the gaskets and don't move the manifold fore-and-aft after it contacts the seals on the block. Make sure the end seals haven't been disturbed.*

17 Install the bolts and following the recommended tightening sequence, tighten them to the torque listed in this Chapter's Specifications **(see illustration)**.

18 The remaining installation steps are the reverse of removal.

19 Change the engine oil and refill the cooling system (see Chapter 1).

20 Start the engine and check carefully for oil and coolant leaks at the intake manifold joints.

8   Exhaust manifold(s) - removal and installation

Warning: *The air conditioning compressor must be removed before the left exhaust manifold can be removed. The air conditioning system is under high pressure. Do not loosen any hose fittings or remove any components until after the system has been discharged. Air conditioning refrigerant should be properly discharged into an EPA-approved recovery/recycling unit at a dealer* service department or an automotive air conditioning repair facility. Always wear eye protection when disconnecting air conditioning system fittings.

Removal

1 If removing the left exhaust manifold, have the air-conditioning refrigerant discharged and recovered. Disconnect the cable from the negative terminal of the battery.

2 Remove the spark plug wires and their mounting brackets.

3 Raise the vehicle and support it securely on jackstands. **Caution:** *If the vehicle is equipped with Automatic Ride Control (ARC), make sure the air suspension switch is turned to the OFF position before the vehicle is raised to prevent damage to the system components (see Chapter 10).*

5 Working under the vehicle, apply penetrating oil to the exhaust pipe-to-manifold studs and nuts. **Caution:** *Have patience - these can easily be broken off in the manifold, causing a great deal of extra work and frustration. Let the penetrating fluid sit as long as possible to loosen the rusty parts.*

6 Remove the air intake duct.

7 Remove the drivebelt.

Right side manifold

8 Remove the bolts and the drivebelt tensioner.

9 Disconnect the alternator electrical connectors, and remove the alternator with its mounting bracket (see Chapter 5).

10 Remove the pushpins and the inner fenderwell splash shield (see Chapter 11).

11 Remove the manifold mounting bolts. The heat shield will come off with the manifold. **Caution:** *Be careful handling the heat shield, the edges are sharp.*

Left side manifold

Refer to illustration 8.12

12 Remove the nut and the engine oil dipstick tube **(see illustration)**.

13 Remove the air-conditioning compressor (see Chapter 3).

14 Remove the pushpins and the inner fenderwell splash shield (see Chapter 11).

15 Remove the manifold mounting bolts and the exhaust manifold.

8.12 Remove the nut (arrow) and the engine oil dipstick tube

Installation

16 Check the manifold for cracks and make sure the bolt threads are clean and undamaged. The manifold and cylinder head mating surfaces must be clean before the manifolds are reinstalled - use a gasket scraper to remove all carbon deposits and old gasket material.

17 Position the manifold and gasket (if used) on the head and install the mounting bolts.

18 When tightening the mounting bolts, work from the center to the ends and be sure to use a torque wrench. Tighten the bolts in three equal steps until the torque listed in this Chapter's Specifications is reached.

19 The remaining installation steps are the reverse of removal.

20 Start the engine and check for exhaust leaks. If the air-conditioning compressor was removed, have the system evacuated, charged and leak tested.

9   Cylinder head(s) - removal and installation

Warning: *The cylinder heads are heavy. It is highly recommended to have an assistant help you lift them off the engine to avoid injury.*

Caution: *The engine must be completely cool when the heads are removed. Failure to*

9.13 Be certain that the cylinder head gaskets are positioned with the correct side up (note the Front mark on the type used here) and located over the dowels (arrow indicates front dowel)

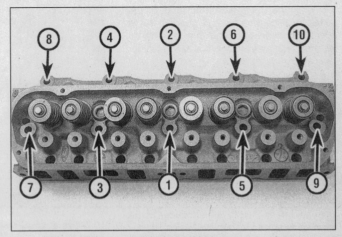

9.16 Cylinder head bolt tightening sequence - always use NEW cylinder head bolts

2C

10.4 Use the recommended puller to remove the crankshaft pulley - if a puller that applies force to the outer edge is used, the pulley will be damaged

allow the engine to cool off could result in head warpage.

Removal

1   If removing the left cylinder head, have the air-conditioning refrigerant discharged and recovered (see the **Warning** at the beginning of Section 8). Disconnect the cable from the negative terminal of the battery.
2   Remove the valve cover(s) (see Section 4). Remove the pushrods and rocker arms (see Section 5).
3   Remove the intake manifold (see Section 7). **Note:** *The cooling system must be drained first.*
4   Remove the drivebelt (see Chapter 1) and drivebelt tensioner.
5   Remove the exhaust manifold(s) (see Section 8).
6   At the rear of the right cylinder head, remove the bolt holding the transmission dipstick tube.
7   Loosen the cylinder head bolts in 1/4-turn increments until they can be removed by hand. Work from bolt-to-bolt in a pattern that's the reverse of the tightening sequence

(see illustration 9.16). Discard the cylinder head bolts.
8   Lift the cylinder head(s) off the engine. If resistance is felt, DO NOT pry between the cylinder head and engine block as damage to the mating surfaces will result. To dislodge the cylinder head, place a wood block against the end of it and strike the wood block with a hammer. Store the cylinder heads on wood blocks to prevent damage to the gasket-sealing surfaces.
9   Cylinder head disassembly and inspection procedures are covered in detail in Chapter 2, Part D.

Installation

Refer to illustrations 9.13 and 9.16

10   The mating surfaces of the cylinder heads and engine block must be perfectly clean when the cylinder heads are installed. Use a gasket scraper to remove all traces of carbon and old gasket material, then clean the mating surfaces with lacquer thinner or acetone. If there's oil on the mating surfaces when the cylinder heads are installed, the gaskets may not seal correctly and leaks may develop. When working on the engine block, cover the lifter valley with shop rags to keep debris out of the engine. Use a vacuum cleaner to remove any debris that falls into the cylinders.
11   Check the engine block and cylinder head mating surfaces for nicks, deep scratches and other damage. If damage is slight, it can be removed with a file - if it's excessive, machining may be the only alternative.
12   Use a tap of the correct size to chase the threads in the cylinder head bolt holes. Mount each bolt in a vise and run a die down the threads to remove corrosion and restore the threads. Dirt, corrosion, sealant and damaged threads will affect torque readings.
13   Position the new gasket(s) over the locating dowels in the engine block. Make sure it's facing the correct direction and that all bolt and coolant passage holes are

aligned. **Note:** *Most gaskets will either be marked FRONT or TOP to be sure the gasket is positioned correctly* **(see illustration)**.
14   Carefully position the cylinder head(s) on the engine block without disturbing the gasket(s).
15   Before installing the NEW cylinder head bolts, lightly oil the threads on all of the bolts.
16   Install the bolts finger tight. Follow the recommended sequence and tighten the bolts, in steps, to the torque listed in this Chapter's Specifications **(see illustration)**.
17   The remaining installation steps are the reverse of removal.
18   Change the engine oil and filter (see Chapter 1), then start the engine and check carefully for oil and coolant leaks.

10   Crankshaft front oil seal - replacement

Refer to illustrations 10.4, 10.6 and 10.8

1   Disconnect the negative battery cable.
2   Remove the drivebelt (see Chapter 1).
3   Remove the bolt from the center of the crankshaft pulley. You'll most likely have to prevent the crankshaft from turning. Remove the torque converter access cover and wedge a large screwdriver between the teeth of the starter ring gear, allowing the screwdriver to rest against the transmission housing. Two other methods include holding an extension through one of the holes in the crankshaft pulley and against the block, or using a strap-type wrench around the circumference of the pulley.
4   Use a puller to detach the crankshaft pulley **(see illustration)**. **Caution:** *Don't use a puller with jaws that grip the outer edge of the pulley. The puller must be the type shown in the illustration that utilizes bolts to apply force to the pulley hub only.* Clean the crankshaft nose and the seal contact surface on the crankshaft pulley with lacquer thinner or acetone. Leave the Woodruff key in place in the crankshaft keyway.

5   Examine the seal contact area of the pulley for excessive wear or scoring. If it is scored or undersize, you may be able to restore it with a repair sleeve available at your local auto parts store.

6   The crankshaft front oil seal can be removed with the use of a special tool **(see illustration)** or with the careful use of a screwdriver with the end taped. Be careful not to damage the cover or scratch the wall of the seal bore. If the engine has accumulated a lot of miles, apply penetrating oil to the seal-to-cover joint and allow it to soak in before attempting to remove the seal.

7   Position the new seal in the bore with the open end of the seal facing IN. A small amount of oil applied to the outer edge of the new seal will make installation easier.

8   The seal can be reinstalled in either of two ways. Either use a special tool or the seal can be driven into the bore with a large socket and hammer until it's completely seated and square to the cover **(see illustration)**. Select a socket that's the same outside diameter as the seal (a section of pipe can be used if a socket isn't available).

9   Lubricate the oil seal contact surface of the crankshaft pulley hub with multi-purpose grease or clean engine oil, then install the pulley on the end of the crankshaft. The keyway in the pulley must be aligned with the Woodruff key in the crankshaft nose. **Note:** *Apply a small amount of RTV sealant to the Woodruff key groove in the pulley before installation.*

10   Use a special pulley installation tool (available at most auto parts stores) to install the pulley. As an alternative method, slip the large washer over the bolt, install the bolt and tighten it to press the pulley into place. **Caution:** *Never use a hammer to drive the pulley on.* Tighten the bolt to the torque listed in this Chapter's Specifications.

11   The remaining installation steps are the reverse of removal.

12   Add coolant and check the oil level. Run the engine and check for oil and coolant leaks.

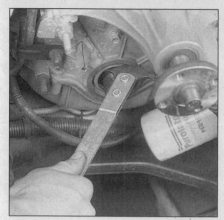

10.6 Use a screwdriver or oil seal removal tool (shown) to work the seal out of the timing chain cover - be very careful not to damage to cover or nick the crankshaft!

10.8 Drive the seal in place with a seal-installation tool or a large socket and hammer

---

11 Timing chain cover - removal and installation

Note: *On this model, the manufacturer recommends removing the engine from the vehicle to perform timing chain cover removal, since the oil pan must be removed to properly seal the bottom of the timing cover. Refer to Part D of this Chapter for engine removal. The following procedure is an alternative method, which allows the oil pan to remain attached, making for an in-vehicle procedure. Read the procedure carefully before beginning.*

Removal

Refer to illustration 11.8

1   Refer to Chapter 3 and remove the engine cooling fan/shroud and water pump, and disconnect the radiator and heater hoses.

2   Drain the engine oil and remove the oil filter (Chapter 1).

3   Remove the crankshaft pulley (see Section 10). Disconnect the electrical connector from the crankshaft position sensor, and

remove the sensor (see Chapter 6).

4   Unbolt and remove all accessory brackets attached to the timing chain cover. When unbolting the power steering pump (see Chapter 10), tie it aside with the hoses still connected. On models equipped with air conditioning, remove the compressor front support bracket, leaving the compressor in place (see Chapter 3).

5   Position the number one piston at TDC on the compression stroke (Section 3).

6   Remove the oil pan-to-timing chain cover bolts.

7   Use a razor knife (thin blade) or razor blade to cut the oil pan gasket flush with the engine block face, by inserting the blade behind the bottom corners of the front cover-to-block mating surfaces. The idea is to make a clean cut of the oil pan gasket so that cover removal doesn't tear the original gasket.

8   Remove the bolts and separate the timing chain cover from the engine block **(see illustration)**. If it's stuck, tap it gently with a soft-face hammer. **Caution:** *DO NOT use excessive force or you may crack the cover. If the cover is difficult to remove, double check to make sure all of the bolts are out.* Don't overlook the two cover bolts at the top.

11.8 With the water pump and oil pan-to-cover bolts removed, remove these four bolts (arrows)

11.13 Before installing the timing chain cover and gasket, apply a bead of RTV sealant at the junction of the oil pan and block as shown

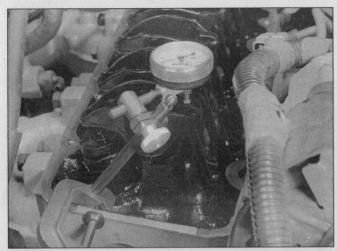

12.5 A dial indicator installed to measure timing chain deflection (this setup can also be used to check camshaft lobe lift)

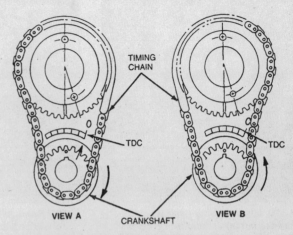

12.6 Timing chain deflection check - dial indicator measurement procedure

2C

Installation

Refer to illustration 11.13

9   Stuff a shop rag into the oil pan opening to keep debris out of the engine.

10   Use a gasket scraper to remove all traces of old gasket material and sealant from the cover, oil pan and engine block, then clean them with lacquer thinner or acetone.

11   Cut the front section from a new oil pan gasket to fit the timing chain cover.

12   Attach the gasket section to the timing chain cover with contact adhesive. Allow the adhesive to dry thoroughly.

13   Apply a 1/8-inch bead of RTV sealant to the oil pan-to-block joints (see illustration).

14   Lubricate the front crankshaft oil seal lip with engine oil.

15   Apply a thin coat of RTV sealant to the block side of the new cover gasket, then position it on the engine. The dowel pins will hold it in place as the cover is installed.

16   Apply a thin coat of RTV sealant to the gasket surface of the cover and attach it to the engine. Be careful not to dislodge the oil pan gasket as you install the cover.

12.12 With the timing chain cover removed, establish a reference point on the engine block and measure from that point to the chain

17   It may be necessary to compress the rubber seal by forcing the cover down before installing the cover bolts to the block.

18   Install the four cover-to-engine block bolts then the oil pan-to-cover bolts. Tighten the cover-to-block bolts to the torque listed in this Chapter's Specifications, then tighten the oil pan-to-cover bolts to the torque listed in this Chapter's Specifications. Make sure the gasket remains in place.

19   Install the remaining components in the reverse order of removal. **Note:** *When reinstalling the pulley (see Section 10), apply a dab of RTV sealant to the keyway in the pulley first.*

20   Add engine oil and coolant (Chapter 1).

21   Run the engine and check for leaks.

12   Timing chain and sprockets - inspection, removal and installation

Timing chain inspection

Cover on engine

Refer to illustrations 12.5 and 12.6

1   Disconnect the negative battery cable from the battery.

2   Place the engine at Top Dead Center (TDC) compression for the number one cylinder (see Section 3).

3   Rotate the crankshaft almost 360-degrees. Place the number one piston several degrees before TDC (BTDC).

4   Remove the right valve cover (see Section 4).

5   Attach a dial indicator to the cylinder head with the plunger in-line with and resting on the number one rocker arm at the end contacting the pushrod (see illustration).

6   Turn the crankshaft clockwise until the number one piston is at TDC. This will take up the slack on the right side of the timing chain (see illustration - view A).

7   Zero the dial indicator.

8   Slowly turn the crankshaft counterclockwise until the slightest movement is seen on the dial indicator. Stop and note how far the number one piston has moved away from the TDC mark by looking at the ignition marks (see illustration 12.6 - view B).

9   If the mark has moved more than 10-degrees, install a new timing chain and sprockets.

Cover removed from engine

Refer to illustration 12.12

Note: *Refer to Section 11 to remove the timing chain cover.*

10   Position the number one piston at TDC on the compression stroke (see Section 3).

11   Rotate the crankshaft in a counterclockwise direction to take up the slack in the left side of the chain.

12   Establish a reference point on the block and measure from that point to the chain (see illustration).

13   Reinstall the crankshaft pulley bolt. Using this bolt, turn the crankshaft clockwise with a wrench until the slack is taken up on the right side of the chain.

14   Force the left side of the chain out with your fingers and measure the distance between the reference point and the chain. The difference between the two measurements is the deflection.

15   If the deflection exceeds 1/2-inch, install a new timing chain and sprockets. **Note:** *Whenever a new timing chain is required, the entire set (chain, camshaft and crankshaft sprockets) must be replaced as an assembly.*

Gear Inspection

16   Inspect the camshaft gear for damage or wear, the teeth can be grooved or worn enough to cause a poor meshing of the gear and the chain and cause timing chain failure.

17   Inspect the crankshaft gear for damage or wear. The crankshaft gear is a steel gear, but the teeth can be grooved or worn enough to also cause a poor meshing of the gear and the chain which can lead to chain failure.

12.20 Align the timing marks on the crankshaft and camshaft sprockets (arrows) as shown here before removing the sprockets from the shafts

12.22 Remove both timing gears and the chain as a unit (be sure to align the timing gear marks first)

12.27 Position the crankshaft with the key facing up (12 o'clock)

Timing chain and sprocket removal

Refer to illustrations 12.20 and 12.22

18   Make sure the number one piston is at TDC (see Section 3).
19   Remove the timing chain cover (see Section 11). Try to avoid turning the crankshaft during crankshaft pulley removal.
20   Make sure the crankshaft and camshaft sprocket timing marks are aligned **(see illustration)**. If they aren't, install the crankshaft pulley bolt and use it to turn the crankshaft clockwise until the two marks are aligned.
21   Remove the camshaft sprocket mounting bolt.
22   Pull the sprocket/chain off the camshaft and detach the chain from the crankshaft sprocket **(see illustration)**. Don't lose the pin in the end of the camshaft.
23   The crankshaft sprocket can be levered off with two large screwdrivers or a prybar.

Timing chain and sprocket installation

Refer to illustration 12.27

24   Use a gasket scraper to remove all traces of old gasket material and sealant from the cover and engine block. Stuff a shop rag into the opening at the front of the oil pan to keep debris out of the engine. Wipe the cover and block sealing surfaces with a cloth saturated with lacquer thinner or acetone.
25   Check the cover flange for distortion, particularly around the bolt holes. **Note:** *If the timing chain cover oil seal has been leaking, refer to Section 10 and install a new one.*
26   Align the keyway in the crankshaft sprocket with the Woodruff key in the end of the crankshaft. Press the sprocket onto the crankshaft with the crankshaft pulley bolt, a large socket and some washers or tap it gently into place until it's completely seated. **Caution:** *If resistance is encountered, DO NOT hammer the sprocket onto the shaft. It may eventually move into place, but it may be cracked in the process and fail later, causing*

extensive engine damage.
27   Turn the crankshaft until the key is facing up (12 o'clock position) **(see illustration)**.
28   Drape the chain over the camshaft sprocket and turn the sprocket until the timing mark faces down (6 o'clock position). Mesh the chain with the crankshaft sprocket and position the camshaft sprocket on the end of the camshaft. If necessary, turn the camshaft so the dowel pin fits into the sprocket hole.
29   When correctly installed, a straight line should pass through the center of the camshaft, the camshaft timing mark (in the 6 o'clock position), the crankshaft timing mark (in the 12 o'clock position) and the center of the crankshaft **(see illustration 12.20)**. DO NOT proceed until the valve timing is correct!
30   Apply thread locking compound to the threads and install the camshaft sprocket bolt. Tighten the bolt to the torque listed in this Chapter's Specifications.

31   Reinstall the timing chain cover (see Section 11).
32   Refer to Section 10 and reinstall the crankshaft pulley.
33   Reinstall the remaining parts in the reverse order of removal.
34   Add coolant and check the oil level. Run the engine and check for oil and coolant leaks.

13 Valve lifters - removal, inspection and installation

Removal

Refer to illustrations 13.4 and 13.5

1   Remove the lower intake manifold (see Section 7).
2   Remove the rocker arms and pushrods (see Section 5).

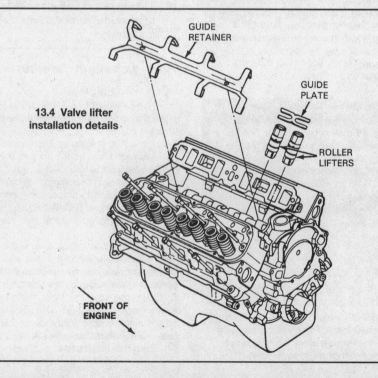

GUIDE RETAINER

GUIDE PLATE

ROLLER LIFTERS

13.4 Valve lifter installation details

FRONT OF ENGINE

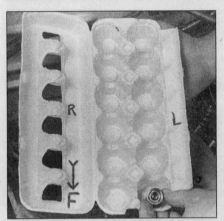

13.5 Be sure to store the lifters in an organized manner to make sure they're reinstalled in their original locations

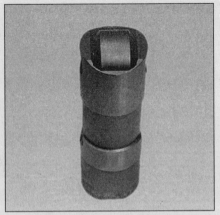

13.6 The roller on roller lifters must turn freely - check for wear and excessive play as well

14.12a Thread a long bolt into the camshaft sprocket bolt hole to use as a handle when removing the camshaft

2C

3 There are several ways to extract the lifters from the bores. Special tools designed to grip and remove lifters are manufactured by many tool companies and are widely available, but may not be needed in every case. On newer engines without a lot of varnish buildup, the lifters can often be removed with a small magnet or even with your fingers. A machinist's scribe with a bent end can be used to pull the lifters out by positioning the point under the retainer ring in the top of each lifter. **Caution:** *Don't use pliers to remove the lifters unless you intend to replace them with new ones (along with the camshaft).* The pliers may damage the precision-machined and hardened lifters, rendering them useless. On engines with a lot of sludge and varnish, work the lifters up and down, using carburetor cleaner spray to loosen the deposits.

4 The guide retainer and guide plates must be removed before the lifters are withdrawn **(see illustration)**.

5 Remove the lifters and store them in a clearly labeled box to ensure that they're reinstalled in their original locations **(see illustration)**.

Inspection

Refer to illustration 13.6

6 Check the rollers carefully for wear and damage and make sure they turn freely without excessive play **(see illustration)**.

7 Check each lifter wall and pushrod seat for scuffing, score marks and uneven wear.

8 Unlike conventional lifters, used roller lifters can be reinstalled with a new camshaft and the original camshaft can be used if new lifters are installed, provided the used parts are in good condition.

Installation

9 Before installing the lifter(s) the air should be bled out of them as much as possible. Stand the lifter(s) upright in a container with enough oil to cover them. Use one of the pushrods to work the plunger to prime the lifter and remove all the air.

10 The original lifters, if they're being

reinstalled, must be returned to their original locations. Coat them with moly-based engine assembly lubricant.

11 Install the lifters in the bores.

12 Install the guide plates and retainer.

13 Install the pushrods and rocker arms.

14 Install the intake manifold and valve covers.

15 Reinstall the remaining parts in the reverse order of removal.

14 Camshaft - removal, inspection and installation

Camshaft lobe lift check

1 In order to determine the extent of cam lobe wear, the lobe lift should be checked prior to camshaft removal. Remove the valve covers (see Section 4).

2 Position the number one piston at TDC on the compression stroke (see Section 3).

3 Beginning with the number one cylinder, mount a dial indicator on the engine and position the plunger in-line with and resting on the first rocker arm **(see illustration 12.5)**.

4 Zero the dial indicator, then very slowly turn the crankshaft in the normal direction of rotation until the indicator needle stops and begins to move in the opposite direction. The point at which it stops indicates maximum cam lobe lift.

5 Record this figure for future reference, then reposition the piston at TDC on the compression stroke.

6 Move the dial indicator to the remaining number one cylinder pushrod and repeat the check. Be sure to record the results for each valve.

7 Repeat the check for the remaining valves. Since each piston must be at TDC on the compression stroke for this procedure, work from cylinder-to-cylinder following the firing order sequence.

8 After the check is complete, compare the results to this Chapter's Specifications. If camshaft lobe lift is less than specified, cam

14.12b Remove the camshaft carefully to avoid damaging the bearings

lobe wear has occurred and a new camshaft should be installed.

Removal

Refer to illustrations 14.12a and 14.12b

9 Refer to the appropriate Sections and remove the pushrods, the valve lifters and the timing chain and camshaft sprocket. The radiator should be removed as well (see Chapter 3). You also may have to remove the air conditioning condenser and the grille, but wait and see if the camshaft can be pulled out of the engine.

10 Check the camshaft endplay with a dial indicator. If it's greater than specified, replace the thrust plate with a new one when the camshaft is reinstalled.

11 Remove the camshaft thrust plate bolts.

12 Thread a long bolt into the camshaft that can be used to pull the camshaft from the engine block and also provide leverage to support the camshaft so the lobes don't get nicked or gouged on the bearings as it's withdrawn **(see illustrations)**.

Inspection

13 After the camshaft has been removed from the engine, cleaned with solvent and

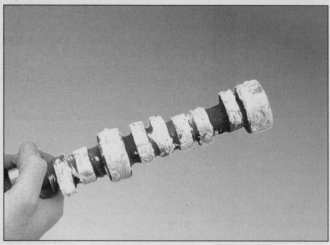

14.16 Apply moly-based engine assembly lubricant to the camshaft lobes and journals prior to installation

16.2 Remove the oil pump pick-up tube bracket nut (arrow) . . .

dried, inspect the bearing journals for uneven wear, pitting and evidence of seizure. If the journals are damaged, the bearing inserts in the block are probably damaged as well. Both the camshaft and bearings will have to be replaced. **Note:** *Camshaft bearing replacement requires special tools and expertise that make it a difficult job to do at home. However, the special tool required for bearing removal/installation is available at many stores that carry automotive tools, possibly even found at a tool rental company. It is advisable though, if the bearings are bad, that the engine be removed and the block taken to an automotive machine shop to ensure that the job is done correctly.*
14   Measure the bearing journals with a micrometer to determine if they are excessively worn or out-of-round (see Part D of this Chapter).
15   Check the camshaft lobes for heat discoloration, score marks, chipped areas, pitting and uneven wear. If the lobes are in good condition and if the lobe lift measurements are as specified in this Chapter, the camshaft can be reused.

Installation

Refer to illustration 14.16

16   Lubricate the camshaft bearing journals and cam lobes with moly-based engine assembly lubricant **(see illustration)**.
17   Slide the camshaft into the engine. Support the cam near the block and be careful not to scrape or nick the bearings.
18   Apply moly-based engine assembly lubricant to both sides of the thrust plate, then position it on the block. Install the bolts and tighten them to the torque listed in this Chapter's Specifications.
19   Refer to the appropriate Sections and install the lifters, pushrods, rocker arms, timing chain/sprocket, timing chain cover and valve covers.
20   The remaining installation steps are the reverse of removal.
21   Before starting and running the engine,

change the oil and install a new oil filter (see Chapter 1).

15   Oil pan - removal and installation

Removal

1   This procedure requires engine removal in the covered models. Refer to Part D of this Chapter for engine removal.
2   Remove the oil pan mounting bolts.
3   Carefully separate the pan from the block. Don't pry between the block and pan or damage to the sealing surfaces may result and oil leaks could develop. Instead, dislodge the pan with a large rubber mallet or a block of wood and a hammer. **Note:** *These models have a cast-aluminum oil pan, do not gouge the soft aluminum on the gasket sealing surfaces.*

Installation

Note: *Follow the installation instructions included with an OEM or aftermarket gasket set - they supersede the information included here.*
4   Remove the gasket(s) using a gasket scraper or putty knife, if necessary. Remove all traces of old gasket material and sealant from the pan and block.
5   Clean the mating surfaces with lacquer thinner or acetone. Make sure the bolt holes in the block are clean.
6   Install the new gasket on the bottom of the block, using a small dab of RTV sealant at either side of the rear main cap and the juncture of the front cover with the block. Then apply another dab over each of these points once the gasket is on the block. Install the pan within 10 minutes of applying the RTV.
7   Carefully position the pan against the block and install the bolts finger tight. Make sure the gasket hasn't shifted, then tighten the bolts in three steps to the torque listed in this Chapter's Specifications. Start at the center of the pan and work out toward the

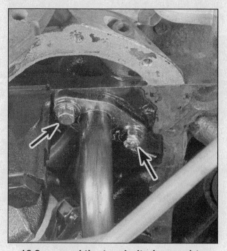

16.3 . . . and the two bolts (arrows) to detach the oil pick-up tube

ends in a spiral pattern. **Note:** *The four larger bolts (two at each end) have a different torque Specification than the remaining bolts.*
8   The remaining steps are the reverse of removal. **Caution:** *Don't forget to refill the engine with oil and replace the filter before starting it* (see Chapter 1).
9   Start the engine and check carefully for oil leaks at the oil pan.

16   Oil pump - removal and installation

Removal

Refer to illustrations 16.2, 16.3 and 16.5

1   Remove the oil pan (see Section 15).
2   Remove the oil pick-up tube-to-main bearing cap nut **(see illustration)**.
3   Remove the oil pump pick-up tube mounting bolts **(see illustration)**.
4   Remove the oil pump mounting bolts and lower the oil pump assembly. If the pump is

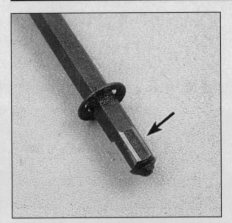

16.5 Replace the oil pump driveshaft if either end is worn (arrow)

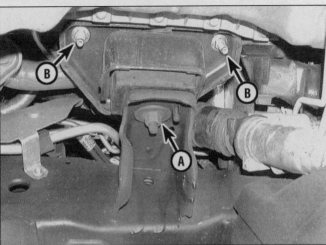

19.1 Once the weight is taken off the mounts, remove the lower mount-to-frame nut (A) - then raise the engine and remove the mount-to-block bolts (B)

faulty, or you suspect that it's faulty, install a new one - do not attempt to repair the original.

5　The oil pump hex driveshaft will come out with the pump. Examine the ends for wear **(see illustration)**. **Caution:** *If this shaft breaks, serious engine damage can result. If it looks worn, replace it.*

Installation

6　Prime the oil pump prior to installation. Pour clean oil into the pick-up and turn the pump shaft by hand.

7　If you separate the pump from the pick-up tube, use a new gasket and tighten the bolts securely.

8　As the pump is reinstalled, fit the oil pump driveshaft into the pump. It must seat all the way. DO NOT try to force it. If it doesn't align with the bottom of the camshaft synchronizer, turn the pump slightly and try again. **Note:** *The end with the stop-clip goes*

up into the engine block.

9　Install the mounting bolts/nut and tighten them to the torque listed in this Chapter's Specifications.

17　Driveplate - removal and installation

The driveplate procedure is virtually the same for the 5.0L V8 as for the other engines in the covered models. Refer to Part A of this Chapter for the removal and installation procedures. **Note:** *The original flywheel bolts are coated with Teflon thread sealant. If you are reusing the original bolts, clean the bolt threads and the crankshaft bolt holes and apply Teflon thread sealant to the bolt threads. New bolts come already coated with sealant, so don't apply sealant to new bolts.*

18　Crankshaft rear main seal - replacement

The rear main seal replacement procedure is virtually the same for the 5.0L V8 as for the other engines in the covered models. Refer to Part A of this Chapter for the procedure.

19　Engine mounts - check and replacement

Refer to illustration 19.1

This procedure is virtually the same for the 5.0L V8 as for the other engines in the covered models. Refer to Part A of this Chapter for the check and replacement procedures, but refer to the illustration in this Section for the fastener locations on the V8 models.

2C

Notes

Chapter 2 Part D
General engine overhaul procedures

Contents

Specifications

General
Oil pressure, all engines (engine hot at 2000 rpm) .... 40 to 60 psi
Cylinder head warpage limit .... 0.003 (in any 6 inches), 0.006 inch overall
Compression pressure .... Lowest reading cylinder must be within 75 psi of highest reading cylinder (100 psi minimum)

4.0L pushrod V6 engine

Cylinder bore
Diameter .... 3.9527 to 3.9543 inches
Out-of-round limit .... 0.005 inch
Taper limit... 0.001 inch

Valves and related components
Valve face angle.. 44 degrees
Seat angle.. 45 degrees
Seat width .... 0.060 to 0.080 inch
Minimum valve margin width................................ 1/32 inch
Stem diameter
  Intake.. 0.3159 to 0.3167 inch
  Exhaust.. 0.3149 to 0.3156 inch
Stem-to-guide clearance
  Intake.. 0.0008 to 0.0025 inch
  Exhaust.. 0.0018 to 0.0035 inch
Valve spring
  Free length .... 1.91 inches
  Installed height .... 1-37/64 to 1-39/64 inches
Valve lifter
  Diameter .... 0.8742 to 0.8755 inch
  Lifter-to-bore clearance
    Standard .... 0.0005 to 0.0022 inch
    Service limit.. 0.005 inch

4.0L pushrod V6 engine (continued)

Crankshaft and connecting rods

Connecting rod journal
Diameter	2.1252 to 2.1260 inches
Taper and out-of-round limit	0.0003 inch

Bearing oil clearance
Desired	0.0008 to 0.0020 inch
Allowable	0.0005 to 0.0022 inch

Connecting rod side clearance (endplay)
Standard	0.0036 to 0.0106 inch
Service limit	0.014 inch

Main journal
Diameter*	2.2433 to 2.2441 inches
Taper and out of round limit	0.0003 inch

Main bearing oil clearance
Desired	0.0005 to 0.0022 inch
Allowable	0.0005 to 0.0019 inch
Crankshaft endplay	0.002 to 0.0125 inch

*Note: The crankshaft journals can't be machined more than 0.010 inch under the standard dimension.

Pistons and rings

Piston diameter	3.9527 to 3.9543 inches

Piston ring end gap
Compression rings	0.015 to 0.023 inch
Oil ring	0.015 to 0.055 inch
Piston ring side clearance (compression rings)	0.0020 to 0.0033 inch

Torque specifications*

Ft-lbs (unless otherwise indicated)

Main bearing cap bolts
Step 1	25
Step 2	72

Connecting rod cap
1996 and earlier	18 to 24

1997 and later
Step 1	15
Step 2	Tighten an additional 90 degrees

*Note: Refer to Chapter 2, Part A for additional torque specifications.

4.0L SOHC V6 engine

Cylinder bore

Diameter	3.953 inches
Out-of-round limit	0.001 inch
Taper	0.001 inch maximum

Valves and related components

Intake valve
Seat angle	45 degrees
Seat width	0.06 to 0.094 inch
Seat runout limit	0.002 inch maximum

Stem diameter
Standard	0.274 to 0.275 inch
Valve stem-to-guide clearance	0.001 to 0.002 inch
Valve face angle	45 degrees
Valve face runout limit	0.001 inch maximum

Exhaust valve
Seat angle	45 degrees
Seat width	0.050 to 0.083 inch
Seat runout limit	0.002 inch maximum

Stem diameter
Standard	0.0275 inch
Valve stem-to-guide clearance	0.001 to 0.003 inch
Valve face angle	45 degrees
Valve face runout limit	0.001 inch maximum

Valve spring
Free length, intake and exhaust	1.70 inches
Out-of-square limit	2 degrees maximum
Installed height, intake and exhaust	1.569 to 1.601 inches

Crankshaft and connecting rods

Crankshaft endplay	0.002 to 0.0126 inch
Connecting rods	
Connecting rod journal	
Diameter	2.125 to 2.126 inches
Out-of-round and taper limits	0.0003 inch
Bearing oil clearance	
Desired	0.0003 to 0.0024 inch
Allowable	0.0005 to 0.002 inch
Connecting rod side clearance (endplay)	0.0036 to 0.0106 inch
Main bearing journal	
Diameter	2.243 to 2.244 inches
Out-of-round and taper limits	0.0003 inch
Bearing oil clearance	
Desired	0.0008 to 0.0015 inch
Allowable	0.0005 to 0.002 inch

Pistons and rings

Piston diameter	
Standard	3.9520 to 3.9528 inches
Coded 0.5 mm	3.9716 to 3.9724 inches
Coded 1.0 mm	3.990 to 3.991 inches
Piston-to-bore clearance limit	0.0012 to 0.0020 inch
Piston ring end gap	
Compression ring (top)	0.008 to 0.018 inch
Compression ring (bottom)	0.018 to 0.028 inch
Oil ring	Snug fit
Piston ring side clearance	
Compression rings	0.002 to 0.003 inch
Oil ring	Snug fit

Torque specifications*

	Ft-lbs (unless otherwise indicated)
Main bearing cap bolts	72
Connecting rod cap nuts/bolts**	
Step 1	162 in-lbs
Step 2	Tighten an additional 90-degrees
Jackshaft retainer plate bolts	80 to 115 in-lbs
Balance shaft assembly bolts	19 to 21
Balance shaft chain guide bolts	80 to 97 in-lbs
Balance shaft tensioner bolts	21 to 22

*Note: Refer to Chapter 2, Part B for additional torque specifications.

**Note: Use new nuts and bolts.

5.0L V8 engine

Cylinder bore

Diameter	4.000 to 4.0012 inches
Out-of-round	
Standard	0.0015 inch
Service limit	0.005 inch
Taper	0.010 inch maximum

Valves and related components

Intake valve	
Seat angle	45 degrees
Seat width	0.060 to 0.080 inch
Seat runout limit	0.002 inch maximum
Stem diameter	
Standard	0.3415 to 0.3423 inch
0.015 oversize	0.3565 to 0.3573 inch
0.030 oversize	0.3715 to 0.3723 inch
Valve stem-to-guide clearance	
Standard	0.0010 to 0.0027 inch
Service limit	0.0055 inch maximum
Valve face angle	44 degrees
Valve face runout limit	0.002 inch maximum
Exhaust valve	
Seat angle	45-degrees
Seat width	0.060 to 0.080 inch
Seat runout limit	0.002 inch maximum

2D

5.0L V8 engine (continued)

Stem diameter	
Standard	0.3410 to 0.3418 inch
0.015 oversize	0.3561 to 0.3568 inch
0.030 oversize	0.3711 to 0.3718 inch
Valve stem-to-guide clearance	
Standard	0.0015 to 0.0032 inch
Service limit	0.0055 inch maximum
Valve face angle	44-degrees
Valve face runout limit	0.002 inch maximum
Valve spring	
Free length	
Intake	2.06 inches
Exhaust	1.88 inch
Installed height	
Intake	1-3/4 to 1-13/16 inches
Exhaust	1-37/64 to 1-41/64 inches
Out-of-square limit	5/64 inch
Hydraulic lifter	
Diameter (standard)	0.8740 to 0.8745
Lifter-to-bore clearance	
Standard	0.0007 to 0.0027 inch
Service limit	0.005 inch maximum
Collapsed tappet gap	
Desired	0.091 to 0.151 inch
Allowable	0.071 to 0.0171 inch

Crankshaft and connecting rods

Crankshaft	
Endplay	
Standard	0.004 to 0.008 inch
Service limit	0.012 inch maximum
Runout to rear face of block	0.005 inch maximum
Connecting rods	
Connecting rod journal	
Diameter	2.1228 to 2.1236 inches
Out-of-round/taper limit	0.0006 inch per inch maximum
Bearing oil clearance	
Desired	0.0008 to 0.0015 inch
Allowable	0.0007 to 0.0024 inch
Connecting rod side clearance (endplay)	
Standard	0.010 to 0.020 inch
Service limit	0.023 inch maximum
Main bearing journal	
Diameter	2.2482 to 2.2490 inches
Out-of-round limit	0.0006 inch
Taper limit	0.0004 inch per inch
Bearing oil clearance	0.0008 to 0.0015 inch

Pistons and rings

Piston diameter	
Coded red	3.9987 to 3.9993 inch
Coded blue	3.9999 to 4.0005 inch
Coded yellow	4.0011 to 4.0017 inch
Piston-to-bore clearance limit	0.0012 to 0.0020 inch
Piston ring end gap	
Top compression ring	0.010 to 0.020 inch
Second compression ring	0.018 to 0.028 inch
Oil ring	0.010 to 0.040 inch
Piston ring side clearance	
Compression rings	0.0013 to 0.0033 inch
Service limit	0.006 inch
Oil ring	Snug fit

Torque specifications*

	Ft-lbs (unless otherwise indicated)
Main bearing cap bolts	61 to 68
Connecting rod cap nuts	19 to 24

*Note: *Refer to Chapter 2, Part C for additional torque specifications.*

1 General information - engine overhaul

Included in this portion of Chapter 2 are the general overhaul procedures for the cylinder head and internal engine components.

The information ranges from advice concerning preparation for an overhaul and the purchase of replacement parts to detailed, step-by-step procedures covering Removal and installation of internal engine components and the inspection of parts.

The following Sections have been written based on the assumption that the engine has been removed from the vehicle. For information concerning in-vehicle engine repair, as well as removal and installation of the external components necessary for the overhaul, see Part A (4.0L pushrod V6), Part B (4.0L SOHC V6) or Part C (5.0L V8) of this Chapter.

The Specifications included in this Part are only those necessary for the inspection and overhaul procedures which follow. Refer to Chapter 2, Part A, B or C for additional Specifications.

It's not always easy to determine when, or if, an engine should be completely overhauled, as a number of factors must be considered.

High mileage is not necessarily an indication that an overhaul is needed, while low mileage doesn't preclude the need for an overhaul. Frequency of servicing is probably the most important consideration. An engine that's had regular and frequent oil and filter changes, as well as other required maintenance, will most likely give many thousands of miles of reliable service. Conversely, a neglected engine may require an overhaul very early in its life.

Excessive oil consumption is an indication that piston rings, valve seals and/or valve guides are in need of attention. Make sure that oil leaks aren't responsible before deciding that the rings and/or guides are bad. Perform a cylinder compression check to determine the extent of the work required (see Section 4). Also check the vacuum readings under various conditions (see Section 3).

Loss of power, rough running, knocking or metallic engine noises, excessive valve train noise and high fuel consumption rates may also point to the need for an overhaul, especially if they're all present at the same time. If a complete tune-up doesn't remedy the situation, major mechanical work is the only solution.

An engine overhaul involves restoring the internal parts to the specifications of a new engine. During an overhaul, the piston rings are replaced and the cylinder walls are reconditioned (re-bored and/or honed). If a re-bore is done by an automotive machine shop, new oversize pistons will also be installed. The main bearings, connecting rod bearings and camshaft bearings are generally replaced with new ones and, if necessary, the crankshaft may be reground to restore the journals. Generally, the valves are serviced as well, since they're usually in less-than-perfect condition at this point. While the engine is being overhauled, other components, such as the distributor, starter and alternator, can be rebuilt as well. The end result should be a like new engine that will give many trouble free miles. **Note:** *Critical cooling system components such as the hoses, drivebelts, thermostat and water pump should be replaced with new parts when an engine is overhauled. The radiator should be checked carefully to ensure that it isn't clogged or leaking (see Chapter 3). If you purchase a rebuilt engine or short block, some rebuilders will not warranty their engines unless the radiator has been professionally flushed. Also, we don't recommend overhauling the oil pump - always install a new one when an engine is rebuilt.*

Before beginning the engine overhaul, read through the entire procedure to familiarize yourself with the scope and requirements of the job. Overhauling an engine isn't difficult, but it is time-consuming. Plan on the vehicle being tied up for a minimum of two weeks, especially if parts must be taken to an automotive machine shop for repair or reconditioning. Check on availability of parts and make sure that any necessary special tools and equipment are obtained in advance.

Most work can be done with typical hand tools, although a number of precision measuring tools are required for inspecting parts to determine if they must be replaced. Often an automotive machine shop will handle the inspection of parts and offer advice concerning reconditioning and replacement. **Note:** *Always wait until the engine has been completely disassembled and all components, especially the engine block, have been inspected before deciding what service and repair operations must be performed by an automotive machine shop. Since the block's condition will be the major factor to consider when determining whether to overhaul the original engine or buy a rebuilt one, never purchase parts or have machine work done on other components until the block has been thoroughly inspected. As a general rule, time is the primary cost of an overhaul, so it doesn't pay to install worn or substandard parts.*

As a final note, to ensure maximum life and minimum trouble from a rebuilt engine, everything must be assembled with care in a spotlessly-clean environment.

2 Oil Pressure check

Refer to illustrations 2.2a, 2.2b, 2.2c and 2.3

1 Low engine oil pressure can be a sign of an engine in need of rebuilding. A "low oil pressure" indicator (often called an "idiot light") is not a test of the oiling system. Such indicators only come on when the oil pressure is dangerously low. Even a factory original oil pressure gauge in the instrument panel is only a relative indication, although much better for driver information than a warning light. A better test is with a mechanical (not electrical) oil pressure gauge. When used in conjunction with an accurate tachometer, an engine's oil pressure performance can be compared to the Specifications for that year and model.

2 Locate the oil pressure indicator sending unit **(see illustrations)**.

3 Remove the oil pressure sending unit

2D

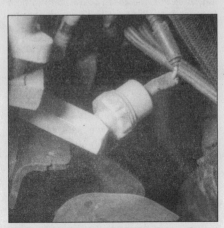

2.2a Oil pressure sending unit location - 4.0L pushrod V6

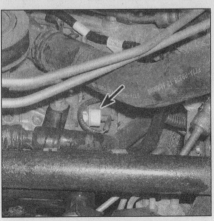

2.2b Oil pressure sending unit location - 4.0L SOHC V6

2.2c Oil pressure sending unit location - 5.0L V8

and install a fitting which will allow you to directly connect your hand-held, mechanical oil pressure gauge **(see illustration)**. Use Teflon tape or sealant on the threads of the adapter and the fitting on the end of your gauge hose.

4   Connect an accurate tachometer to the engine, according to the tachometer manufacturer's instructions.

5   Check the oil pressure with the engine running (full operating temperature) at the specified engine speed, and compare it to this Chapter's Specifications. If it's extremely low, the bearings and/or oil pump are probably worn out.

3   Cylinder compression check

Refer to illustration 3.6

1   A compression check will give you an indication of the mechanical condition of the engine (pistons, rings, valves, head gaskets). Specifically, it can tell you if the compression is down due to leakage caused by worn piston rings, defective valves and seats or a blown head gasket. **Note:** *The engine must be at normal operating temperature for this check and the battery must be fully charged.*

2   Begin by cleaning the area around the spark plugs before you remove them (compressed air works best for this). This will prevent dirt from getting into the cylinders as the compression check is being done.

3   Remove all of the spark plugs from the engine (see Chapter 1).

4   Block the throttle wide open.

5   Disconnect the primary wires from the coil(s).

6   With the compression gauge in the number one spark plug hole, crank the engine over at least four compression strokes and watch the gauge **(see illustration)**. The compression should build up quickly in a healthy engine. Low compression on the first stroke, followed by gradually increasing pressure on successive strokes, indicates worn piston rings. A low compression reading on the first stroke, which does not build up during successive strokes, indicates leaking valves or a blown head gasket (a cracked head could also be the cause). Record the highest gauge reading obtained.

7   Repeat the procedure for the remaining cylinders and compare the results to the Specifications.

8   Add some engine oil (about three squirts from a plunger-type oil can) to each cylinder, through the spark plug hole, and repeat the test.

9   If the compression increases after the oil is added, the piston rings are definitely worn. If the compression does not increase significantly, the leakage is occurring at the valves or head gasket. Leakage past the valves may be caused by burned valve seats and/or faces or warped, cracked or bent valves.

10   If two adjacent cylinders have equally low compression, there is a strong possibility that the head gasket between them is blown.

2.3 Remove the oil pressure sending unit and attach an oil pressure gauge - be sure the fittings you use have the same thread as the sending unit

The appearance of coolant in the combustion chambers or the crankcase would verify this condition.

11   If the compression is unusually high, the combustion chambers are probably coated with carbon deposits. If that is the case, the cylinder heads should be removed and decarbonized.

12   If compression is way down or varies greatly between cylinders, it would be a good idea to have a leak-down test performed by an automotive repair shop. This test will pinpoint exactly where the leakage is occurring and how severe it is.

4   Vacuum gauge diagnostic checks

Refer to illustration 4.6

A vacuum gauge provides valuable information about what is going on in the engine at a low-cost. You can check for worn rings or cylinder walls, leaking head or intake manifold gaskets, incorrect carburetor adjustments, restricted exhaust, stuck or burned valves, weak valve springs, improper ignition or valve timing and ignition problems.

Unfortunately, vacuum gauge readings are easy to misinterpret, so they should be used in conjunction with other tests to confirm the diagnosis.

Both the absolute readings and the rate of needle movement are important for accurate interpretation. Most gauges measure vacuum in inches of mercury (in-Hg). The following references to vacuum assume the diagnosis is being performed at sea level. As elevation increases (or atmospheric pressure decreases), the reading will decrease. For every 1,000 foot increase in elevation above approximately 2000 feet, the gauge readings will decrease about one inch of mercury.

Connect the vacuum gauge directly to intake manifold vacuum, not to ported (throttle body) vacuum. Be sure no hoses are left disconnected during the test or false readings will result.

3.6 A compression gauge with a threaded fitting for the spark plug hole is preferred over the type that requires hand pressure to maintain the seal - be sure to open the throttle valve as far as possible during the compression check

Before you begin the test, allow the engine to warm up completely. Block the wheels and set the parking brake. With the transmission in Park, start the engine and allow it to run at normal idle speed. **Warning:** *Carefully inspect the fan blades for cracks or damage before starting the engine. Keep your hands and the vacuum gauge clear of the fan and do not stand in front of the vehicle or in line with the fan when the engine is running.*

Read the vacuum gauge; an average, healthy engine should normally produce about 17 to 22 inches of vacuum with a fairly steady needle. Refer to the following vacuum gauge readings and what they indicate about the engine's condition **(see illustration)**:

A **low steady reading** usually indicates a leaking gasket between the intake manifold and cylinder head(s) or throttle body, a leaky vacuum hose, late ignition timing or incorrect camshaft timing. Check ignition timing with a timing light and eliminate all other possible causes, utilizing the tests provided in this Chapter before you remove the timing chain cover to check the timing marks.

If the reading is a **low, fluctuating reading** (three to eight inches below normal and it fluctuates at that low reading), suspect an intake manifold gasket leak at an intake port or a faulty fuel injector.

If the needle has **regular drops** of about two-to-four inches at a steady rate, the valves are probably leaking. Perform a compression check or leak-down test to confirm this.

An **irregular drop** or down-flick of the needle can be caused by a sticking valve or an ignition misfire. Perform a compression check or leak-down test and read the spark plugs.

A **rapid vibration** of about four in.-Hg vibration at idle combined with exhaust smoke indicates worn valve guides. Perform a leak-down test to confirm this. If the rapid vibration occurs with an increase in engine speed, check for a leaking intake manifold

Low , steady reading   Low, fluctuating needle   regular drops

Irregular drops   Rapid vibration

Large fluctuation   Slow Fluctuation

STD-O-OBR HAYNES

4.6 Typical vacuum gauge readings

gasket or head gasket, weak valve springs, burned valves or ignition misfire.

A **slight fluctuation**, say one inch up and down, may mean ignition problems. Check all the usual tune-up items and, if necessary, run the engine on an ignition analyzer.

If there is a **large fluctuation**, perform a compression or leak-down test to look for a weak or dead cylinder or a blown head gasket.

If the needle moves slowly through a wide range, check for a clogged PCV system, incorrect idle fuel mixture, carburetor/throttle body or intake manifold gasket leaks.

Check for a **slow fluctuation** by quickly opening the throttle until the engine reaches about 2,500 rpm and let it snap shut. Normally the reading should drop to near zero, rise above normal idle reading (about 5 in.-Hg over) and then return to the previous idle reading. If the vacuum returns slowly and doesn't peak when the throttle is snapped shut, the rings may be worn. If there is a long delay, look for a restricted exhaust system (often the muffler or catalytic converter). An easy way to check this is to temporarily disconnect the exhaust ahead of the suspected part and redo the test.

5 Engine removal - methods and precautions

If you have decided that an engine must be removed for overhaul or major repair work, several preliminary steps should be taken.

Locating a suitable work area is extremely important. A shop is, of course, the most desirable place to work. Adequate work space, along with storage space for the vehicle, will be needed. If a shop or garage is not available, at the very least a flat, level, clean work surface made of concrete or asphalt is required.

Cleaning the engine compartment and engine before beginning the removal procedure will help keep tools clean and organized.

An engine hoist or A-frame will be needed. Make sure that the equipment is rated in excess of the combined weight of the engine and its accessories. Safety is of primary importance, considering the potential hazards involved in lifting the engine out of the vehicle.

If the engine is being removed by a novice, a helper should be available. Advice and aid from someone more experienced would also be helpful. There are many instances when one person cannot simultaneously perform all of the operations required when lifting the engine out of the vehicle.

Plan the operation ahead of time. Arrange for or obtain all of the tools and equipment you will need prior to beginning the job. Some of the equipment necessary to perform engine removal and installation safely and with relative ease are (in addition to an engine hoist) a heavy duty floor jack, complete sets of wrenches and sockets as described in the front of this manual, wooden blocks and plenty of rags and cleaning solvent for mopping up spilled oil, coolant and gasoline. If the hoist is to be rented, make sure that you arrange for it in advance and perform beforehand all of the operations possible without it. This will save you money and time.

Plan for the vehicle to be out of use for a considerable amount of time. A machine shop will be required to perform some of the work which the do-it-yourselfer cannot accomplish due to a lack of special equipment. These shops often have a busy schedule, so it would be wise to consult them

2D

before removing the engine in order to accurately estimate the amount of time required to rebuild or repair components that may need work.

Always use extreme caution when removing and installing the engine. Serious injury can result from careless actions. Plan ahead, take your time and a job of this nature, although major, can be accomplished successfully.

6   Engine - removal and installation

Warning 1: *The air conditioning system is under high pressure. DO NOT loosen any fittings or remove any components until after the system has been discharged. Air conditioning refrigerant should be properly discharged into an EPA-approved container at a dealer service department or an automotive air conditioning repair facility. Always wear eye protection when disconnecting air-conditioning system fittings.*

Warning 2: *Gasoline is extremely flammable, so take extra precautions when you work on any part of the fuel system. Don't smoke or allow open flames or bare light bulbs near the work area, and don't work in a garage where*

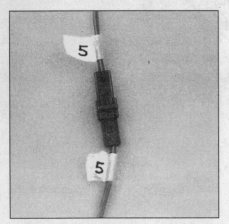

6.6 Label each wire before unplugging the connector

a natural gas-type appliance (such as a water heater or a clothes dryer) with a pilot light is present. Since gasoline is carcinogenic, wear latex gloves when there's a possibility of being exposed to fuel, and, if you spill any fuel on your skin, rinse it off immediately with soap and water. Mop up any spills immediately and do not store fuel-soaked rags where they could ignite. The fuel system is under constant pressure, so, if any fuel lines are to be disconnected, the fuel pressure in the system must be relieved first (see Chapter 4 for more information). When you perform any kind of work on the fuel system, wear safety glasses and have a Class B type fire extinguisher on hand.*

Removal

Refer to illustration 6.6, 6.14a, 6.14b, 6.28 and 6.29

Note: *Have the air-conditioning refrigerant recovered before beginning the procedure.*

1   Relieve the fuel system pressure (see Chapter 4).

2   Disconnect the negative cable, then the positive cable from the battery and remove the battery.

3   Cover the fenders and cowl and remove the hood (see Chapter 11). Special pads are available to protect the fenders, but an old bedspread or blanket will also work.

4   Remove the air cleaner assembly.

5   Drain the cooling system (see Chapter 1).

6   Label the vacuum lines, emissions system hoses, electrical connectors, ground straps and fuel lines that would interfere with engine removal, to ensure correct reinstallation, then detach them. Pieces of masking tape with numbers or letters written on them work well **(see illustration)** . If there's any possibility of confusion, make a sketch of the engine compartment and clearly label the lines, hoses and wires.

7   Label and detach all coolant hoses from the engine, including the heater hoses at the firewall.

8   Remove the cooling fan/shroud and radiator (see Chapter 3).

9   Remove the drivebelt(s) (see Chapter 1).

10   Disconnect the accelerator cable, and speed control cable from the engine (see Chapter 4).

11   Unbolt the power steering pump (see Chapter 10). Leave the lines/hoses attached and make sure the pump is kept in an upright position in the engine compartment (use wire or rope to restrain it out of the way).

12   On air conditioned models, disconnect the refrigerant lines and remove the compressor (see Chapter 3). On 5.0L V8 models, remove the air conditioning condenser.

13   Remove the air inlet tube.

14   Disconnect the ground wires and electrical connectors at the firewall and valve cover **(see illustrations)**.

15   Raise and suitably support the vehicle on sturdy jackstands. Drain the engine oil (see Chapter 1) and remove the filter.

16   Remove the starter motor (see Chapter 5).

17   Remove the alternator (see Chapter 5).

18   Unbolt the exhaust system from the engine (see Chapter 2, Part A, B or C).

19   If equipped with an automatic transmission, remove the torque converter access cover and remove the torque converter-to-driveplate fasteners (see Chapter 7B).

20   Support the transmission with a jack. Position a block of wood between them to prevent damage to the transmission. Special transmission jacks with safety chains are available - use one if possible.

21   Attach an engine sling or a length of sturdy chain to the engine and then to the engine hoist. **Note:** *Some engines have lifting brackets already installed on them, usually on the exhaust manifold studs. Other engine models do not have them, but they are available separately from a dealer parts department, or you can attach a chain to exhaust studs/bolts on either side of the engine, one near the front and one at the rear.*

22   Roll the hoist into position and connect the sling to it. Take up the slack in the sling or chain, but don't lift the engine. **Warning:** *DO NOT place any part of your body under the engine when it's supported only by a hoist or*

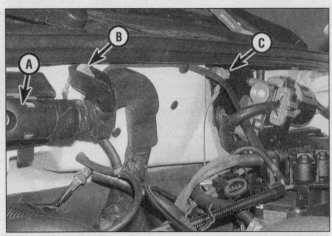

6.14a Disconnect the PCM electrical connector (A), PCM ground (B), and engine ground strap (C) (4.0L SOHC V6 shown)

6.14b Loosen the bolt (arrow) and disconnect the main engine harness electrical connector (4.0L SOHC V6 shown)

other lifting device.

23 Remove the transmission-to-engine block bolts (see Chapter 7).

24 Remove the engine mount through-bolts or mounting stud nuts from both sides (see Chapter 2, Part A, B or C).

25 Recheck to be sure nothing is still connecting the engine to the transmission or vehicle. Disconnect anything still remaining.

26 Raise the engine slightly. Carefully work it forward to separate it from the transmission. If you're working on a vehicle with an automatic transmission, be sure the torque converter stays in the transmission (use a C-clamp or locking pliers clamped to the transmission housing to keep the converter from sliding out). Slowly raise the engine out of the engine compartment.

27 Remove the flywheel or driveplate (see Chapter 2, Part A, B or C).

28 Mount the engine on an engine stand **(see illustration)**.

29 Once the engine is removed, support the transmission with a chain or pipe that crosses from side to side to hold the transmission as the floor jack is removed **(see illustration)**.

Installation

30 Check the engine and transmission mounts. If they're worn or damaged, replace them.

31 Carefully lower the engine into the engine compartment - make sure the engine mounts line up.

32 If you're working on an automatic transmission equipped vehicle, remove the vise grips and guide the torque converter into the crankshaft following the procedure outlined in Chapter 7, Part B.

33 If you're working on a manual transmission equipped vehicle, apply a dab of high-temperature grease to the input shaft and guide it into the crankshaft pilot bearing until the bellhousing is flush with the engine block.

34 Install the transmission-to-engine bolts and tighten them securely. **Caution:** *DO NOT use the bolts to force the transmission and*

engine together!

35 Reinstall the remaining components in the reverse order of removal.

36 Add coolant and oil as needed. Run the engine and check for leaks and proper operation of all accessories, then install the hood and test drive the vehicle.

7   Engine rebuilding alternatives

The do-it-yourselfer is faced with a number of options when performing an engine overhaul. The decision to replace the engine block, piston/connecting rod assemblies and crankshaft depends on a number of factors, with the number one consideration being the condition of the block. Other considerations are cost, access to machine shop facilities, parts availability, time required to complete the project and the extent of prior mechanical experience on the part of the do-it-yourselfer.

Some of the rebuilding alternatives include:

Individual parts - If the inspection procedures reveal that the engine block and most engine components are in reusable condition, purchasing individual parts may be the most economical alternative. The block, crankshaft and piston/connecting rod assemblies should all be inspected carefully. Even if the block shows little wear, the cylinder bores should be surface honed.

Crankshaft kit - This rebuild package consists of a reground crankshaft and a matched set of pistons and connecting rods. The pistons will already be installed on the connecting rods. Piston rings and the necessary bearings will be included in the kit. These kits are commonly available for standard cylinder bores, as well as for engine blocks which have been bored to a regular oversize.

Short block - A short block consists of an engine block with renewed crankshaft and piston/connecting rod assemblies already installed. All new bearings are incorporated and all clearances will be correct. The exist-

ing cylinder head(s), camshaft, valve train components and external parts can be bolted to the short block with little or no machine shop work necessary.

Long block - A long block consists of a short block plus an oil pump, oil pan, cylinder heads, valve covers, camshaft and valve train components, timing sprockets, timing chain and timing cover. All components are installed with new bearings, seals and gaskets incorporated throughout. The installation of manifolds and external parts is all that is necessary.

Used engine assembly - While overhaul provides the best assurance of a like-new engine, used engines available from wrecking yards and importers are often a very simple and economical solution. Many used engines come with warranties, but always give any engine a thorough diagnostic check-out before purchase. Check compression, vacuum and also for signs of oil leakage. If possible, have the seller run the engine, ether in the vehicle or on a test stand so you can be sure it runs smoothly with no knocking or other noises.

Give careful thought to which alternative is best for you and discuss the situation with local automotive machine shops, auto parts dealers or parts store countermen before ordering or purchasing replacement parts.

8   Engine overhaul - disassembly sequence

Caution 1: *The cylinder head bolts on the 4.0L pushrod V6 and 5.0L V8 engines are "torque-to-yield" bolts and are NOT reusable. A predetermined stretch of the bolt gives the even clamping load needed to seal the cylinders properly. Once removed they must be replaced.*

Caution 2: *The connecting rod bolts/nuts on V6 engines (pushrod and SOHC) are "torque-to-yield" design and are NOT reusable. A predetermined stretch of the bolt, calculated by the manufacturer, gives the added rigidity*

6.28 Use long, high-strength bolts (arrows) to hold the engine block on the engine stand - make sure they are tight before lowering the hoist and placing the entire weight of the engine on the stand

6.29 Use a piece of pipe or chain to support the transmission once the engine has been removed, then remove the floor jack that supported the transmission during engine removal

required with this cylinder block. Once removed they must be replaced with new bolts/nuts.

1   It's much easier to disassemble and work on the engine if it's mounted on a portable engine stand. A stand can often be rented quite cheaply from an equipment rental yard. Before the engine is mounted on a stand, the flywheel/driveplate should be removed from the engine.

2   If a stand isn't available, it's possible to disassemble the engine with it blocked up on the floor. Be extra careful not to tip or drop the engine when working without a stand.

3   If you're going to obtain a rebuilt engine, all external components must come off first, to be transferred to the replacement engine, just as they will if you're doing a complete engine overhaul yourself. These include:

> Alternator and brackets
> Emissions control components
> Camshaft position sensor/synchronizer
> Spark plug wires and spark plugs
> Thermostat and housing cover
> Water pump
> EFI components
> Intake/exhaust manifolds
> Oil filter (replace)
> Engine mounts
> Driveplate
> Engine rear plate
> Crankshaft damper

Note: *When removing the external components from the engine, pay close attention to details that may be helpful or important during installation. Note the installed position of gaskets, seals, spacers, pins, brackets, washers, bolts and other small items.*

4   If you're obtaining a short block, which consists of the engine block, crankshaft, pistons and connecting rods all assembled, then the cylinder head(s), oil pan and oil pump will have to be removed as well. See *Engine rebuilding alternatives* for additional information regarding the different possibilities to be considered.

5   If you're planning a complete overhaul, the engine must be disassembled and the components removed in the following order:

> Driveplate
> Valve covers
> Intake manifold
> Exhaust manifolds
> Rocker arms
> Pushrods (4.0L pushrod V6 and 5.0L V8 engines)
> Camshafts and hydraulic lash adjusters (4.0L SOHC V6 engines)
> Valve lifters (4.0L pushrod V6 and 5.0L V8 engines)
> Vibration damper
> Timing chain cover
> Timing chain(s), sprockets, guides and tensioners
> Camshaft (4.0L pushrod V6 and 5.0L V8 engines)
> Cylinder heads
> Oil pan
> Oil pump (replace)

9.1 A small plastic bag, with an appropriate label, can be used to store the valve train components so they can be kept together and reinstalled in the original position

> Jackshaft and Balance shaft assembly (4.0L SOHC V6)
> Piston/connecting rod assemblies
> Crankshaft and main bearings ((replace)

6   Before beginning the disassembly and overhaul procedures, make sure the following items are available. Also, refer to *Engine overhaul - reassembly sequence* for a list of tools and materials needed for engine reassembly.

> Common hand tools
> Small cardboard boxes and plastic bags for storing parts
> Gasket scraper
> Ridge reamer
> Vibration damper puller
> Micrometers
> Telescoping gauges
> Dial indicator set
> Valve spring compressor
> Cylinder surfacing hone
> Piston ring groove cleaning tool
> Electric drill motor
> Tap and die set
> Wire brushes
> Oil gallery brushes
> Cleaning solvent

9   Cylinder head - disassembly

Refer to illustrations 9.1, 9.2 and 9.3
Note: *New and rebuilt cylinder heads are commonly available for most engines at dealerships and auto parts stores. Due to the fact that some specialized tools are necessary for the disassembly and inspection procedures, and replacement parts may not be readily available, it may be more practical and economical for the home mechanic to purchase replacement head(s) rather than taking the time to disassemble, inspect and recondition the original(s).*

1   Cylinder head disassembly involves removal of the intake and exhaust valves and related components. If they're still in place,

9.2 Use a valve spring compressor to compress the spring, then remove the valve stem locks from the valve stem with needle-nose pliers or a magnet

remove the rocker arm nuts, pivot balls and rocker arms from the cylinder head studs. Label the parts or store them separately **(see illustration)** so they can be reinstalled in their original locations and in the same valve guides they are removed from.

2   Compress the springs on the first valve with a spring compressor and remove the valve stem locks **(see illustration)**. Carefully release the valve spring compressor and remove the retainer, sleeve (if used), the spring and the spring seat (if used).

3   Pull the valve out of the head, then remove the oil seal from the guide. If the valve binds in the guide (won't pull through), push it back into the head and deburr the area around the stem lock groove with a fine file or whetstone **(see illustration)**.

4   Repeat the procedure for the remaining valves. Remember to keep all the parts for each valve together so they can be reinstalled in the same locations.

5   Once the valves and related components have been removed and stored in an organized manner, the head should be thoroughly cleaned and inspected. If a complete

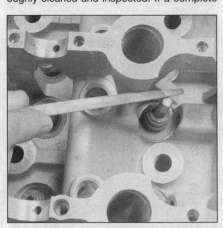

9.3 If the valve won't pull through the guide, deburr the edge of the stem end and the area around the top of the valve stem lock groove with a file or whetstone

10.12 Check the cylinder head gasket surface for warpage by trying to slip a feeler gauge under the straightedge (see this Chapter's Specifications for the maximum warpage allowed and use a feeler gauge of that thickness)

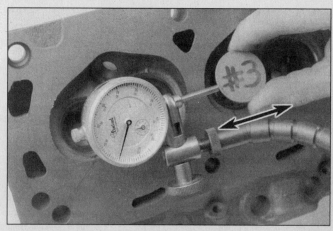

10.14 Lay the head on its edge, pull each valve out about 1/8 inch, set up a dial indicator with the probe touching the valve stem, wiggle the valve and measure its movement

engine overhaul is being done, finish the engine disassembly procedures before beginning the cylinder head cleaning and inspection process.

10 Cylinder head - cleaning and inspection

1   Thorough cleaning of the cylinder head(s) and related valve train components, followed by a detailed inspection, will enable you to decide how much valve service work must be done during the engine overhaul. **Note:** *If the engine was severely overheated, the cylinder head is probably warped (see Step 12).*

Cleaning

2   Scrape all traces of old gasket material and sealing compound off the head gasket, intake manifold and exhaust manifold sealing surfaces. Be very careful not to gouge the cylinder head. Special gasket removal solvents that soften gaskets and make removal much easier are available at auto parts stores.

3   Remove all built up scale from the coolant passages.

4   Run a stiff wire brush through the various holes to remove deposits that may have formed in them.

5   Run an appropriate-size tap into each of the threaded holes to remove corrosion and thread sealant that may be present. If compressed air is available, use it to clear the holes of debris produced by this operation. **Warning:** *Wear eye protection when using compressed air!*

6   Clean the exhaust manifold stud threads, if equipped.

7   Clean the cylinder head with solvent and dry it thoroughly. Compressed air will speed the drying process and ensure that all holes and recessed areas are clean. **Note:** *Decarbonizing chemicals are available and may prove very useful when cleaning cylinder*

heads and valve train components. They are very caustic and should be used with caution. Be sure to follow the instructions on the container.

8   Clean the valvetrain components with solvent and dry them thoroughly (don't mix them up during the cleaning process). **Note:** *Compressed air will speed the drying process and can be used to clean out the oil passages.*

9   Clean all the valve springs, spring seats, stem locks and retainers with solvent and dry them thoroughly. Work the components from one valve at a time to avoid mixing up the parts.

10   Scrape off any heavy deposits that may have formed on the valves, then use a motorized wire brush to remove deposits from the valve heads and stems. Again, make sure the valves don't get mixed up.

Inspection

Note: *Be sure to perform all of the following inspection procedures before concluding that machine shop work is required. Make a list of the items that need attention.*

Cylinder head

Refer to illustration 10.12 and 10.14

11   Inspect the head very carefully for cracks, evidence of coolant leakage and other damage. If cracks are found, check with an automotive machine shop concerning repair. If repair isn't possible, a new cylinder head should be obtained.

12   Using a straightedge and feeler gauge, check the head gasket mating surface for warpage **(see illustration)**. Check the head both straight across and corner-to-corner. If the warpage exceeds the limit listed in this Chapter's Specifications, it can be resurfaced at an automotive machine shop. **Note:** *If the 5.0L cylinder heads are resurfaced, the intake manifold flanges will also require machining.*

13   Examine the valve seats in each of the combustion chambers. If they're pitted, cracked or burned, the head will require valve

service that's beyond the scope of the home mechanic.

14   Check the valve stem-to-guide clearance by measuring the lateral movement of the valve stem with a dial indicator attached securely to the head **(see illustration)**. The valve must be in the guide and approximately 1/16-inch off the seat. The total valve stem movement indicated by the gauge needle must be divided by two to obtain the actual clearance. After this is done, if there's still some doubt regarding the condition of the valve guides they should be checked by an automotive machine shop (the cost should be minimal).

Valves

Refer to illustrations 10.15 and 10.16

15   Carefully inspect each valve face for uneven wear, deformation, cracks, pits and burned areas **(see illustration)** . Check the valve stem for scuffing and galling and the neck for cracks. Rotate the valve and check for any obvious indication that it's bent. Look for pits and excessive wear on the end of the

2D

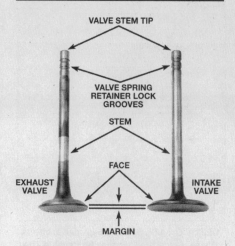

10.15 Check for valve wear at the points shown here

VALVE STEM TIP

VALVE SPRING RETAINER LOCK GROOVES

STEM

FACE

EXHAUST VALVE

INTAKE VALVE

MARGIN

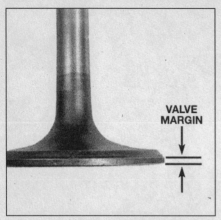

10.16 The margin width on each valve must be as specified (if no margin exists, the valve cannot be reused)

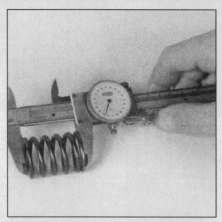

10.17 Measure the free length of each valve spring with a dial or vernier caliper

10.18 Check each valve spring for out-of-square, if it is bent it should be replaced

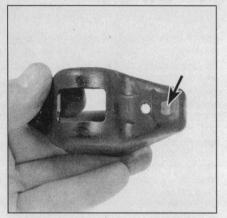

10.22a On 4.0L pushrod and 5.0L V8 engines, check the rocker arms where the valve stem rides (arrow) . . .

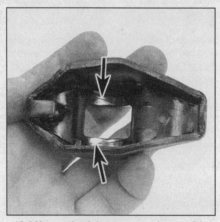

10.22b . . . the fulcrum seats (arrows) in the top of the rocker arm . . .

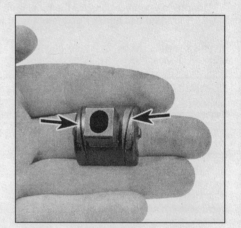

10.22c . . . and the fulcrums themselves for wear and galling (arrows)

stem. The presence of any of these conditions indicates the need for valve service by an automotive machine shop.

16   Measure the margin width on each valve **(see illustration)**. Any valve with a margin narrower than 1/32-inch will have to be replaced with a new one.

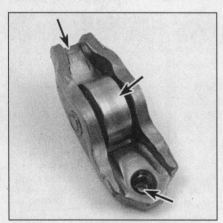

10.22d On 4.0L SOHC V6 engines, check the rocker arms for wear (arrows) at the valve stem end, roller and the pocket that contacts the lash adjuster

Valve components

Refer to illustrations 10.17 and 10.18

17   Check each valve spring for wear (on the ends) and pits. Measure the free length and compare it to the Specifications **(see illustration)**. Any springs that are shorter than specified have sagged and should not be reused. The tension of all springs should be checked with a special fixture before deciding that they're suitable for use in a rebuilt engine (take the springs to an automotive machine shop for this check).

18   Stand each spring on a flat surface and check it for out-of-square **(see illustration)**. If any of the springs are distorted or sagged, replace all of them with new parts.

19   Check the spring retainers and stem locks for obvious wear and cracks. Any questionable parts should be replaced with new ones, as extensive damage will occur if they fail during engine operation.

Rocker arm components

Refer to illustration 10.22a, 10.22b, 10.22c and 10.22d

20   Clean all the parts thoroughly. Make sure all oil passages are open.

21   Check the rocker arm faces (the areas that contact the pushrod ends and valve stems) for pits, wear, galling, score marks and rough spots.

22   Check the rocker arm pivot contact areas and fulcrums. Look for cracks in each rocker arm and bolt **(see illustrations)**.

23   Inspect the pushrod ends for scuffing and excessive wear. Roll each pushrod on a flat surface, like a piece of plate glass, to determine if it's bent.

24   Check the rocker arm studs in the cylinder heads (4.0L pushrod V6 and 5.0L V8) for damaged threads and secure installation.

25   Any damaged or excessively worn parts must be replaced with new ones.

26   If the inspection process indicates that the valve components are in generally poor condition and worn beyond the limits specified, which is usually the case in an engine that's being overhauled, reassemble the valves in the cylinder head and refer to Section 11 for valve servicing recommendations.

27   Clean all the parts thoroughly. Make sure all oil passages are open.

28   Any damaged or excessively worn parts must be replaced with new ones.

29   If the inspection process indicates that the valve components are in generally poor

12.3a On pushrod models with the type of seal shown, use a hammer and a seal installer (or a deep socket, as shown here) to drive the seal onto the valve guide/head casting boss

12.3b Installing a valve stem seal on an 4.0L SOHC V6 engine - the socket must contact the flange (spring seat) of the seal

12.5 4.0L SOHC V6 engines use valve stem seals that are a valve spring seat and seal combined into one component - make sure replacement parts are the same as the originals

condition and worn beyond the limits specified, which is usually the case in an engine that's being overhauled, reassemble the valves in the cylinder head and refer to Section 11 for valve servicing recommendations.

11 Valves - servicing

1   Because of the complex nature of the job and the special tools and equipment needed, servicing of the valves, the valve seats and the valve guides, commonly known as a valve job, should be done by a professional.

2   The home mechanic can remove and disassemble the head, do the initial cleaning and inspection, then reassemble and deliver it to a dealer service department or an automotive machine shop for the actual service work. Doing the inspection will enable you to see what condition the head and valve train components are in and will ensure that you know what work and new parts are required when dealing with an automotive machine shop.

3   The dealer service department or automotive machine shop will remove the valves and springs, recondition or replace the valves and valve seats, recondition the valve guides, check and replace the valve springs, spring retainers or rotators and keepers (as necessary), replace the valve seals with new ones, reassemble the valve components and make sure the installed spring height is correct. The cylinder head gasket surface will also be resurfaced if it's warped.

4   After the valve job has been performed by a professional, the head will be in like-new condition. When the head is returned, be sure to clean it again before installation on the engine to remove any metal particles and abrasive grit that may still be present from the valve service or head resurfacing operations. Use compressed air, if available, to blow out all the oil holes and passages. **Warning:** *Always wear eye protection when using compressed air!*

12.6a Apply a small dab of grease to each valve stem lock as shown here before installation - it'll hold them in place on the valve stem as the spring is released

12 Cylinder head - reassembly

Refer to illustrations 12.3a, 12.3b, 12.5, 12.6a, 12.6b and 12.8

1   Regardless of whether or not the head was sent to an automotive repair shop for valve servicing, make sure it's clean before beginning reassembly.

2   If the head was sent out for valve servicing, the valves and related components will already be in place. Begin the reassembly procedure with Step 9.

3   On all engines, lubricate and install the valves, then install new seals on each of the valve guides. Using a hammer and deep socket, gently tap each seal into place until it's seated on the guide **(see illustrations)**. Don't twist or cock the seals during installation or they will not seat properly on the valve stems.

4   Beginning at one end of the head, lubricate and install the first valve. Apply clean engine oil to the valve stem.

5   Place the spring seat or shim (if used)

12.6b Compress the springs with a valve spring compressor and position the valve stem locks in the upper groove, then slowly release the compressor and make sure the locks seat properly

over the valve guide and set the valve spring and retainer in place. **Note:** *On 4.0L SOHC V6 engines, the valve seal has the spring seat/shim incorporated into one piece* **(see illustration)**. *A valve spring should never sit directly against an aluminum cylinder head.*

6   Apply a small dab of grease to each valve stem lock to hold it in place **(see illustration)** . Compress the springs with a valve spring compressor and carefully install the locks in the upper groove **(see illustration)**, then slowly release the compressor and make sure the locks seat properly.

7   Repeat the procedure for the remaining valves. Be sure to return the components to their original locations - don't mix them up!

8   Check the installed valve spring height with a ruler graduated in 1/64-inch increments or a dial caliper. If the head was sent out for service work, the installed height should be correct (but don't automatically assume that it is). The measurement is taken from the top of each spring seat or shim to

2D

the bottom of the retainer **(see illustration)**. If the height is greater than the figure listed in this Chapter's Specifications, a shim can be added under the spring to correct it. **Caution:** *Don't, under any circumstances, shim the springs to the point where the installed height is less than specified.*

9   Apply moly-base engine assembly lubricant to the rocker arm faces and the fulcrums, then install the rocker arms and fulcrums on the cylinder head studs.

13 Jackshaft, balance shaft and bearings (4.0L SOHC V6 only) - removal and inspection

Removal

Refer to illustrations 13.3 and 13.5

1   The camshafts, timing chains and sprockets and the jackshaft chain and sprocket must be removed before extracting the jackshaft (see Part B of this Chapter).

2   Check the jackshaft endplay by attaching a dial-indicator to the front of the block, with the indicator on the jackshaft. Pry the jackshaft back and forth and compare the endplay to Specifications.

3   From above, at the top-rear of the block, remove the bolt holding the oil pump drive and pull out the drive assembly **(see illustration)**.

4   Remove the two jackshaft thrust plate bolts, and any spacer that may be in place. Installing a long bolt in the front of the jackshaft to use as a handle, guide the jackshaft gently out of the block, without nicking the journals on the jackshaft bearings in the block.

5   Compress the balance shaft tensioner by hand and insert a drill bit or pin to hold it in place, then remove the tensioner bolts and the tensioner **(see illustration)**.

6   Slide the balance shaft chain and crankshaft sprocket from the crankshaft and balance shaft sprocket. **Note:** *Do not remove the bolt from the balance shaft sprocket.*

12.8 Valve spring installed height is the distance from the spring seat on the head to the bottom of the spring retainer

7   Rotate the engine on the engine stand so that the bottom of the block is up. Remove the two rear bolts and the balance shaft assembly. **Note:** *Two of the balance shaft assembly mounting bolts are the ones removed to take off the tensioner.*

Inspection

8   The jackshaft rides in a bearing at the front of the block and a bushing at the rear. Check the jackshaft bearings in the block for wear and damage. Look for galling, pitting and discolored areas.

9   Jackshaft bearing replacement requires special tools and expertise that place it outside the scope of the home mechanic. Take the block to an automotive machine shop to ensure the job is done correctly.

14 Pistons/connecting rods - removal

Refer to illustrations 14.1, 14.3, 14.4 and 14.7
Note: *Prior to removing the piston/connecting rod assemblies, remove the cylinder head(s), the oil pan and the oil pump by refer-*

13.3 Remove the oil pump drive assembly retaining bolt and clamp (A) and pull the drive assembly (B) straight up and out of the engine block

ring to the appropriate Sections in Chapter 2, Part A, B or C.

1   Use your fingernail to feel if a ridge has formed at the upper limit of ring travel (about 1/4-inch down from the top of each cylinder). If carbon deposits or cylinder wear have produced ridges, they must be completely removed with a special tool **(see illustration)**. Follow the manufacturer's instructions provided with the tool. Failure to remove the ridges before attempting to remove the piston/connecting rod assemblies may result in piston breakage. **Note:** *Do not let the tool cut into the ring travel area more than 1/32-inch.*

2   After the cylinder ridges have been removed, turn the engine upside-down so the crankshaft is facing up.

3   Before the connecting rods are removed, check the endplay with a dial indicator or with feeler gauges **(see illustration)**. Slide them between the first connecting rod and the crankshaft throw until the play is removed. The endplay is equal to the thickness of the feeler gauge(s). If the endplay exceeds the service limit, new connecting rods will be required. If new rods (or a new crankshaft) are installed, the endplay may fall

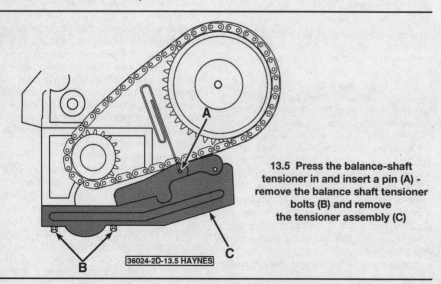

13.5 Press the balance-shaft tensioner in and insert a pin (A) - remove the balance shaft tensioner bolts (B) and remove the tensioner assembly (C)

`36024-2D-13.5 HAYNES`

14.1 A ridge reamer is required to remove the ridge from the top of each cylinder - do this before removing the pistons!

14.3 Check the connecting rod side clearance (endplay) with a dial indicator or with a feeler gauge as shown

14.4 Mark the rod bearing caps in order from the front of the engine to the rear

14.7 To prevent damage to the crankshaft journals and cylinder walls, slip sections of hose over the rod bolts (if assembled with this type of rod bolt) before removing the piston/rod assemblies

under the specified minimum (if it does, the rods will have to be machined to restore it - consult an automotive machine shop for advice if necessary). Repeat the procedure for the remaining connecting rods.

4   Check the connecting rods and caps for identification marks. If they aren't plainly marked, use a small center-punch, number stamping die **(see illustration)**, or scribe, to make the appropriate number of indentations, or marks, on each rod and cap (1, 2, 3, etc., depending on the engine type and cylinder they're associated with).

5   Loosen each of the connecting rod cap nuts 1/2-turn at a time until they can be removed by hand.

6   Remove the connecting rod cap and bearing insert. Don't drop the bearing insert out of the cap.

7   Slip a short length of plastic or rubber hose over each connecting rod cap bolt to protect the crankshaft journal and cylinder wall as the piston is removed **(see illustration)**.

8   Remove the bearing insert and push the connecting rod/piston assembly out through the top of the engine. Use a wooden or plastic hammer handle to push on the upper

bearing surface in the connecting rod. If resistance is felt, double-check to make sure that all of the ridge was removed from the cylinder.

9   Repeat the procedure for the remaining cylinders.

10   After removal, reassemble the connecting rod caps and bearing inserts in their respective connecting rods and install the cap nuts finger tight. Leaving the old bearing inserts in place until reassembly will help prevent the connecting rod bearing surfaces from being accidentally nicked or gouged.

11   Don't separate the pistons from the connecting rods (see Section 18 for additional information).

15   Crankshaft - removal

Refer to illustrations 15.1, 15.3 and 15.4
Note: The crankshaft can be removed only after the engine has been removed from the vehicle. It's assumed that the flywheel or driveplate, vibration damper, timing chain(s), oil pan, oil pump and piston/connecting rod assemblies have been removed.

1   Before the crankshaft removal procedure is started, check the endplay. Mount a dial indicator with the stem in line with the crankshaft and just touching the end of the crankshaft **(see illustration)**.

2   Push the crankshaft all the way to the rear and zero the dial indicator. Next, pry the crankshaft to the front as far as possible and check the reading on the dial indicator. The distance that it moves is the endplay. If it's greater than limit listed in this Chapter's Specifications, check the crankshaft thrust surfaces for wear. If no wear is evident, new main bearings should correct the endplay.

3   If a dial indicator isn't available, feeler gauges can be used. Gently pry or push the crankshaft all the way to the front of the engine. Slip feeler gauges between the crankshaft and the front face of the thrust main bearing to determine the clearance **(see illustration)**.

4   Check the main bearing caps to see if they're marked to indicate their locations. They should be numbered consecutively from the front of the engine to the rear. If they

2D

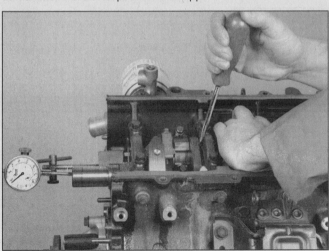

15.1 Checking crankshaft endplay with a dial indicator

15.3 Checking crankshaft endplay with a feeler gauge

15.4 The main bearing caps are usually marked to indicate their position and installed direction (arrows) - they should be numbered consecutively from the front of the engine to the rear (number 1 at the front)

aren't, mark them with number stamping dies or a center-punch **(see illustration)**. Main bearing caps generally have a cast-in arrow, which points to the front of the engine.

5   Loosen the main bearing cap bolts 1/4-turn at a time each, until they can be removed

16.1a A hammer and a large punch can be used to knock the core plugs sideways in their bores

16.8 All bolt holes in the block - particularly the main bearing cap and head bolt holes - should be cleaned and restored with a tap (be sure to remove debris from the holes after this is done)

by hand. Note if any stud bolts are used and make sure they're returned to their original locations when the crankshaft is reinstalled.

6   Gently tap the caps with a soft-face hammer, then separate them from the engine block. If necessary, use the bolts as levers to remove the caps. Try not to drop the bearing inserts if they come out with the caps.

7   Carefully lift the crankshaft out of the engine. It may be a good idea to have an assistant available, since the crankshaft is quite heavy. With the bearing inserts in place in the engine block and main bearing caps, return the caps to their respective locations on the engine block and tighten the bolts finger tight.

---

16   Engine block - cleaning

Refer to illustrations 16.1a, 16.1b, 16.8 and 16.10

Caution: *The core plugs (also known as freeze plugs or soft plugs) may be difficult or impossible to retrieve if they're driven into the block coolant passages.*

1   Using the wide end of a punch **(see illustration)** tap in on the outer edge of the core plug to turn the plug sideways in the bore.

16.1b Pull the core plugs from the block with pliers

16.10 A large socket on an extension can be used to drive the new core plugs into the bores

Then, using a pair of pliers, pull the core plug from the engine block **(see illustration)**. Don't worry about the condition of the old core plugs as they are being removed because they will be replaced on reassembly with new plugs.

2   Using a gasket scraper, remove all traces of gasket material from the engine block. Be very careful not to nick or gouge the gasket sealing surfaces.

3   Remove the main bearing caps and separate the bearing inserts from the caps and the engine block (see Section 15). Tag the bearings, indicating which cylinder they were removed from and whether they were in the cap or the block, then set them aside.

4   Remove all of the threaded oil gallery plugs from the block. The plugs are usually very tight - they may have to be drilled out and the holes re-tapped. Use new plugs when the engine is reassembled.

5   If the engine is extremely dirty it should be taken to an automotive machine shop to be steam cleaned or hot tanked.

6   After the block is returned, clean all oil holes and oil galleries one more time. Brushes specifically designed for this purpose are available at most auto parts stores. Flush the passages with warm water until the water runs clear, dry the block thoroughly and wipe all machined surfaces with a light, rust preventive oil. If you have access to compressed air, use it to speed the drying process and to blow out all the oil holes and galleries. **Warning:** *Wear eye protection when using compressed air!*

7   If the block isn't extremely dirty or sludged up, you can do an adequate cleaning job with hot soapy water and a stiff brush. Take plenty of time and do a thorough job. Regardless of the cleaning method used, be sure to clean all oil holes and galleries very thoroughly, dry the block completely and coat all machined surfaces with light oil.

8   The threaded holes in the block must be clean to ensure accurate torque readings during reassembly. Run the proper size tap into each of the holes to remove rust, corrosion, thread sealant or sludge and restore damaged threads **(see illustration)**. If possible, use compressed air to clear the holes of debris produced by this operation. Now is a good time to clean the threads on the head bolts and the main bearing cap bolts as well.

9   Reinstall the main bearing caps and tighten all bolts finger tight.

10   After coating the sealing surfaces of the new core plugs with core plug sealant (such as Permatex no. 1 sealant or Loctite 540 retaining compound), install them in the engine block **(see illustration)**. Make sure they're driven in straight and seated properly or leakage could result. Special tools are available for this purpose, but a large socket, with an outside diameter that will just slip into the core plug, a 1/2-inch drive extension and a hammer will work just as well.

11   Apply a thread sealant (such as Teflon pipe thread sealant) to the new oil gallery plugs and thread them into the holes in the block. Make sure they're tightened securely.

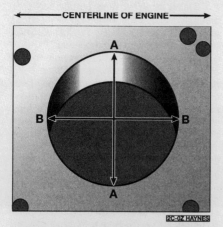

17.4a Measure the diameter of each cylinder at a right angle to the engine centerline (A), and parallel to engine centerline (B) - out-of-round is the difference between A and B; taper is the difference between A and B at the top of the cylinder and A and B at the bottom of the cylinder

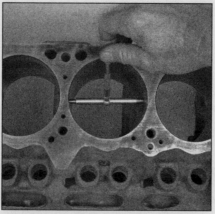

17.4b The ability to "feel" when the telescoping gauge is at the correct point will be developed over time, so work slowly and repeat the check until you're satisfied the bore measurement is accurate

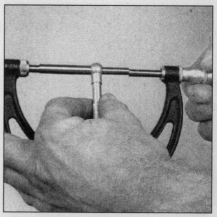

17.4c The gauge is then measured with a micrometer to determine the bore size

2D

12   If the engine isn't going to be reassembled right away, cover it with a large plastic trash bag to keep it clean.

17   Engine block - inspection

Refer to illustrations 17.4a, 17.4b and 17.4c

1   Before the block is inspected, it should be cleaned as described in Section 16.
2   Visually check the block for cracks, rust and corrosion. Look for stripped threads in the threaded holes. It's also a good idea to have the block checked for hidden cracks by an automotive machine shop that has the special equipment to do this type of work. If defects are found, have the block repaired, if possible, or replaced.
3   Check the cylinder bores for scuffing and scoring.
4   Check the cylinders for taper and out-of-round conditions as follows **(see illustrations)**:
5   Measure the diameter of each cylinder at the top (just under the ridge area), center and bottom of the cylinder bore, parallel to the crankshaft axis.
6   Next measure each cylinder's diameter at the same three locations perpendicular to the crankshaft axis.
7   The taper of the cylinder is the difference between the bore diameter at the top of the cylinder and the diameter at the bottom. The out-of-round specification of the cylinder bore is the difference between the parallel and perpendicular readings. Compare your results to those listed in this Chapter's Specifications.
8   Repeat the procedure for the remaining pistons and cylinders.
9   If the cylinder walls are badly scuffed or scored, or if they're out-of-round or tapered beyond the limits given in this Chapter's

18.3a A "bottle brush" hone will produce better results if you've never honed cylinders before

Specifications, have the engine block rebored and honed at an automotive machine shop. If a rebore is done, oversize pistons and rings will be required.
10   If the cylinders are in reasonably good condition and not worn to the outside of the limits, and if the piston-to-cylinder clearances can be maintained properly, then they don't have to be rebored. Honing is all that's necessary (see Section 18).

18   Cylinder honing

Refer to illustrations 18.3a and 18.3b

1   Prior to engine reassembly, the cylinder bores must be honed so the new piston rings will seat correctly and provide the best possible combustion chamber seal. **Note:** *If you don't have the tools or don't want to tackle the honing operation, most automotive machine shops will do it for a reasonable fee.*
2   Before honing the cylinders, install the main bearing caps and tighten the bolts to

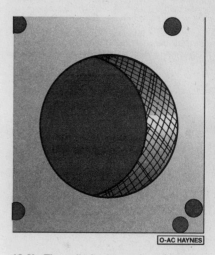

18.3b The cylinder hone should leave a smooth, crosshatch pattern with the lines intersecting at approximately a 60- degree angle

the torque listed in this Chapter's Specifications. Make sure you use only the original main cap bolts, not the new ones for final assembly.
3   Two types of cylinder hones are commonly available - the flex hone or "bottle brush" type and the more traditional surfacing hone with spring-loaded stones. Both will do the job, but for the less experienced mechanic the "bottle brush" hone will probably be easier to use. You'll also need some kerosene or honing oil, rags and an electric drill motor. Proceed as follows:

a) *Mount the hone in the drill motor, compress the stones and slip it into the first cylinder* **(see illustration)**. *Be sure to wear safety goggles or a face shield!*
b) *Lubricate the cylinder with plenty of honing oil, turn on the drill and move the hone up-and-down in the cylinder at a pace that will produce a fine crosshatch pattern on the cylinder walls. Ideally, the crosshatch lines should intersect at approximately a 60-degree angle* **(see illustration)**. *Be sure to use plenty of*

19.4a The piston ring grooves can be cleaned with a special tool, as shown here . . .

19.4b . . . or a section of a broken ring

lubricant and don't take off any more material than is absolutely necessary to produce the desired finish. **Note:** *Piston ring manufacturers may specify a smaller crosshatch angle than the traditional 60-degrees - read and follow any instructions included with the new rings.*

c) *Don't withdraw the hone from the cylinder while it's running. Instead, shut off the drill and continue moving the hone up-and-down in the cylinder until it comes to a complete stop, then compress the stones and withdraw the hone. If you're using a "bottle brush" type hone, stop the drill motor, then turn the chuck in the normal direction of rotation while withdrawing the hone from the cylinder.*

d) *Wipe the oil out of the cylinder and repeat the procedure for the remaining cylinders.*

4   After the honing job is complete, chamfer the top edges of the cylinder bores with a small file so the rings won't catch when the pistons are installed. Be very careful not to nick the cylinder walls with the end of the file.

5   The entire engine block must be washed again very thoroughly with warm, soapy water to remove all traces of the abrasive grit produced during the honing operation. **Note:** *The bores can be considered clean when a lint-free white cloth - dampened with clean engine oil - used to wipe them out doesn't pick-up any more honing residue, which will show up as gray areas on the cloth. Be sure to run a brush through all oil holes and galleries and flush them with running water.*

6   After rinsing, dry the block and apply a coat of light rust preventive oil to all machined surfaces. Wrap the block in a plastic trash bag to keep it clean and set it aside until reassembly.

19   Pistons/connecting rods - inspection

Refer to illustrations 19.4a, 19.4b, 19.10 and 19.11

1   Before the inspection process can be carried out, the piston/connecting rod assemblies must be cleaned and the original piston rings removed from the pistons. **Note:** *Always use new piston rings when the engine is reassembled.*

2   Using a piston ring installation tool, carefully remove the rings from the pistons **(see illustration 23.11)**. Be careful not to nick or gouge the pistons in the process.

3   Scrape all traces of carbon from the top of the piston. A hand-held wire brush or a piece of fine emery cloth can be used once the majority of the deposits have been scraped away. Do not, under any circumstances, use a wire brush mounted in a drill motor to remove deposits from the pistons. The piston material is soft and may be eroded away by the wire brush.

4   Use a piston ring groove cleaning tool to remove carbon deposits from the ring grooves. If a tool isn't available, a piece broken off the old ring will do the job **(see illustrations)**. Be very careful to remove only the carbon deposits - don't remove any metal and do not nick or scratch the sides of the ring grooves.

5   Once the deposits have been removed, clean the piston/rod assemblies with solvent and dry them with compressed air (if available). Make sure the oil return holes in the back sides of the ring grooves are clear.

6   If the pistons and cylinder walls aren't damaged or worn excessively, and if the engine block is not rebored, new pistons won't be necessary. Normal piston wear appears as even vertical wear on the piston thrust surfaces and slight looseness of the top ring in its groove. New piston rings, however, should always be used when an engine is rebuilt.

7   Carefully inspect each piston for cracks around the skirt, at the pin bosses and at the ring lands.

8   Look for scoring and scuffing on the thrust faces of the skirt, holes in the piston crown and burned areas at the edge of the crown. If the skirt is scored or scuffed, the engine may have been suffering from overheating and/or abnormal combustion, which caused excessively high operating temperatures. The cooling and lubrication systems should be checked thoroughly. A hole in the piston crown is an indication that abnormal

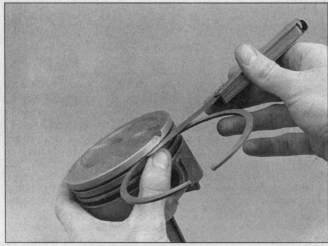

19.10 Check the ring side clearance with a feeler gauge at several points around the groove

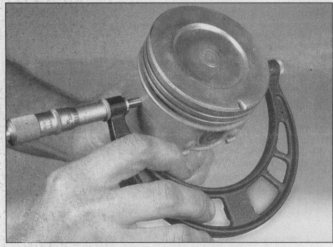

19.11 Measure the piston diameter at a 90-degree angle to the piston pin and in line with it

20.1 The oil holes should be chamfered so sharp edges don't gouge or scratch the new bearings

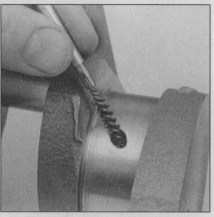

20.2 Use a wire or stiff plastic bristle brush to clean the oil passages in the crankshaft

the rod and cap bearing surfaces clean and inspect them for nicks, gouges and scratches. After checking the rods, replace the old bearings, slip the caps into place and tighten the nuts finger tight. **Note:** *If the engine is being rebuilt because of a connecting rod knock, be sure to install new rods.*

20   Crankshaft - inspection

Refer to illustrations 20.1, 20.2, 20.5 and 20.7

1   Remove all burrs from the crankshaft oil holes with a stone, file or scraper **(see illustration)**.

2   Clean the crankshaft with solvent and dry it with compressed air (if available). Be sure to clean the oil holes with a stiff brush **(see illustration)** and flush them with solvent.

3   Check the main and connecting rod bearing journals for uneven wear, scoring, pits and cracks.

4   Check the rest of the crankshaft for cracks and other damage. It should be magnafluxed to reveal hidden cracks - an automotive machine shop will handle the procedure.

5   Using a micrometer, measure the diameter of the main and connecting rod journals and compare the results to this Chapter's Specifications **(see illustration)**. By measuring the diameter at a number of points around each journal's circumference, you'll be able to determine whether or not the journal is out-of-round. Take the measurement at each end of the journal, near the crank throws, to determine if the journal is tapered.

6   If the crankshaft journals are damaged, tapered, out-of-round or worn beyond the limits given in the Specifications, have the crankshaft reground by an automotive machine shop. Be sure to use the correct size bearing inserts if the crankshaft is reconditioned.

7   Check the oil seal journals at each end of the crankshaft for wear and damage. If the seal has worn a groove in the journal, or if it's nicked or scratched **(see illustration)**, the new seal may leak when the engine is reassembled. In some cases, an automotive machine shop may be able to repair the journal by pressing on a thin sleeve. If repair isn't feasible, a new or different crankshaft should be installed.

8   Refer to Section 21 and examine the main and rod bearing inserts.

combustion (pre-ignition) was occurring. Burned areas at the edge of the piston crown are usually evidence of spark knock (detonation). If any of the above problems exist, the causes must be corrected or the damage will occur again.

piston across the skirt, at a 90-degree angle to, and in line with, the piston pin **(see illustration)**. Subtract the piston diameter from the bore diameter to obtain the clearance. If it's greater than specified, the block will have to be rebored and new pistons and rings installed.

12   Check the piston-to-rod clearance by twisting the piston and rod in opposite directions. Any noticeable play indicates excessive wear, which must be corrected. The piston/connecting rod assemblies should be taken to an automotive machine shop to have the pistons and rods resized and new pins installed.

13   If the pistons must be removed from the connecting rods for any reason, they should be taken to an automotive machine shop. While they are there have the connecting rods checked for bend and twist, since automotive machine shops have special equipment for this purpose. **Note:** *Unless new pistons and/or connecting rods must be installed, do not disassemble the pistons and connecting rods.*

14   Check the connecting rods for cracks and other damage. Temporarily remove the rod caps, lift out the old bearing inserts, wipe

10   ... have to be used.

11   Check the piston-to-bore clearance by measuring the bore (see Section 17) and the piston diameter. Make sure the pistons and bores are correctly matched. Measure the

20.5 Measure the diameter of each crankshaft journal at several points to detect taper and out-of-round conditions

20.7 If the seals have worn grooves in the crankshaft journals, or if the seal contact surfaces are nicked or scratched, the new seals will leak

21   Main and connecting rod bearings - inspection

Refer to illustration 21.1

1   Even though the main and connecting rod bearings should be replaced with new ones during the engine overhaul, the old bearings should be retained for close examination, as they may reveal valuable information about the condition of the engine **(see illustration)**.

2D

2   Bearing failure occurs because of lack of lubrication, the presence of dirt or other foreign particles, overloading the engine and corrosion. Regardless of the cause of bearing failure, it must be corrected before the engine is reassembled to prevent it from happening again.

3   When examining the bearings, remove them from the engine block, the main bearing caps, the connecting rods and the rod caps and lay them out on a clean surface in the same general position as their location in the engine. This will enable you to match any bearing problems with the corresponding crankshaft journal.

4   Dirt and other foreign particles get into the engine in a variety of ways. It may be left in the engine during assembly, or it may pass through filters or the PCV system. It may get into the oil, and from there into the bearings. Metal chips from machining operations and normal engine wear are often present. Abrasives are sometimes left in engine components after reconditioning, especially when parts are not thoroughly cleaned using the proper cleaning methods. Whatever the source, these foreign objects often end up embedded in the soft bearing material and are easily recognized. Large particles will not embed in the bearing and will score or gouge the bearing and journal. The best prevention for this cause of bearing failure is to clean all parts thoroughly and keep everything spotlessly clean during engine assembly. Frequent and regular engine oil and filter changes are also recommended.

5   Lack of lubrication (or lubrication breakdown) has a number of interrelated causes. Excessive heat (which thins the oil), overloading (which squeezes the oil from the bearing face) and oil leakage or throw off (from excessive bearing clearances, worn oil pump or high engine speeds) all contribute to lubrication breakdown. Blocked oil passages, which usually are the result of misaligned oil holes in a bearing shell, will also oil starve a bearing and destroy it. When lack of lubrication is the cause of bearing failure, the bearing material is wiped or extruded from the steel backing of the bearing. Temperatures may increase to the point where the steel backing turns blue from overheating.

6   Driving habits can have a definite effect on bearing life. Low speed operation in too high a gear (lugging the engine) puts very high loads on bearings, which tends to squeeze out the oil film. These loads cause the bearings to flex, which produces fine cracks in the bearing face (fatigue failure). Eventually the bearing material will loosen in pieces and tear away from the steel backing. Short trip driving leads to corrosion of bearings because insufficient engine heat is produced to drive off the condensed water and corrosive gases. These products collect in the engine oil, forming acid and sludge. As the oil is carried to the engine bearings, the acid attacks and corrodes the bearing material.

7   Incorrect bearing installation during engine assembly will lead to bearing failure

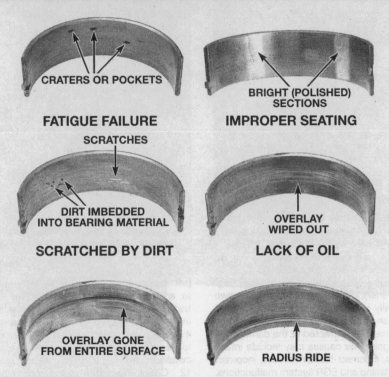

21.1 Typical bearing failures

as well. Tight-fitting bearings leave insufficient bearing oil clearance and will result in oil starvation. Dirt or foreign particles trapped behind a bearing insert result in high spots on the bearing which lead to failure.

22   Engine overhaul - reassembly sequence

1   Before beginning engine reassembly, make sure you have all the necessary new parts, gaskets and seals as well as the following items on hand:

   Common hand tools
   A 1/2-inch drive torque wrench
   A 3/8-inch drive torque wrench (in-lb measurement)
   Piston ring installation tool
   Piston ring compressor
   Camshaft/crankshaft holding/alignment tools (4.0L SOHC V6 only)
   Vibration damper installation tool
   Short lengths of rubber or plastic hose to fit over connecting rod bolts (pushrod engines)
   Plastigage
   Feeler gauges
   A fine-tooth file
   New engine oil
   Engine assembly lube or moly-base grease
   Gasket sealant
   Thread locking compound

2   In order to save time and avoid problems, engine reassembly must be done in the following general order:

4.0L pushrod V6 and 5.0L V8 engines

   New camshaft bearings (recommended to be done by an automotive machine shop)
   Piston rings
   Crankshaft and main bearings
   Piston/connecting rod assemblies
   Oil pump
   Oil pan
   Camshaft
   Valve lifters
   Timing chain and sprockets
   Timing chain cover
   Cylinder heads
   Rocker arms and pushrods
   Intake and exhaust manifolds
   Valve covers
   Driveplate

4.0L SOHC V6 engine

   Piston rings
   Crankshaft and main bearings
   Piston/connecting rod assemblies
   Cylinder heads
   Valve lifters
   Rocker arms
   Camshafts
   Camshaft caps
   Balance shaft and chain
   Jackshaft and chain
   Timing chains and sprockets
   Timing chain guides and tensioners
   Timing chain cover
   Oil pump
   Crankcase reinforcement section
   Oil pan (lower)
   Valve covers
   Intake and exhaust manifolds
   Driveplate

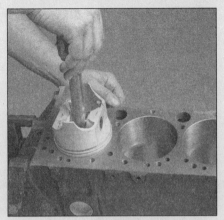

23.3 When checking piston ring end gap, the ring must be square in the cylinder bore (this is done by pushing the ring down with the top of a piston as shown)

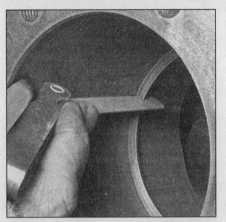

23.4 With the ring square in the cylinder, measure the end gap with a feeler gauge

23.8a Installing the spacer/expander in the oil control ring groove . . .

23 Piston rings - installation

Refer to illustrations 23.3, 23.4, 23.8a, 23.8b and 23.11

1   Before installing the new piston rings, the ring end gaps must be checked. It's assumed that the piston ring side clearance has been checked and verified correct (see Section 19).

2   Lay out the piston/connecting rod assemblies and the new ring sets so the ring sets will be matched with the same piston and cylinder during the end gap measurement and engine assembly.

3   Insert the top (number one) ring into the first cylinder and square it up with the cylinder walls by pushing it in with the top of the piston (see illustration). The ring should be near the bottom of the cylinder, at the lower limit of ring travel.

4   To measure the end gap, slip a feeler gauge between the ends of the ring until a gauge equal to the gap width is found (see illustration). The feeler gauge should slide between the ring ends with a slight amount of drag. Compare the measurement to this Chapter's Specifications. If the gap is larger or smaller than specified, double-check to make sure you have the correct rings before proceeding. If there is any doubt contact the parts store where the rings were purchased, to verify that the correct ring set is being used.

5   Excess end gap isn't critical unless it's greater than the specified limit. Again, double-check to make sure you have the correct rings for your engine.

6   Repeat the procedure for each ring that will be installed in the first cylinder and for each ring in the remaining cylinders. Remember to keep rings, pistons and cylinders matched up.

7   Once the ring end gaps have been checked/corrected, the rings can be installed on the pistons.

8   The oil control ring (lowest one on the piston) is usually installed first. It's composed of three separate components. Slip the spacer/expander into the groove (see illustration). Next, install the lower side rail. Don't use a piston ring installation tool on the oil ring side rails, as they may be damaged. Instead, place one end of the side rail into the groove between the spacer/expander and the ring land, hold it firmly in place and slide a finger around the piston while pushing the rail into the groove (see illustration). Next, install the upper side rail in the same manner.

9   After the three oil ring components have been installed, check to make sure that both the upper and lower side rails can be turned smoothly in the ring groove.

10   The number two (middle) ring is installed next. It's usually stamped with a mark which must face up, toward the top of the piston. **Note:** *Always follow the instructions printed on the ring package or box - different manufacturers may require different approaches. Do not mix up the top and middle rings, as they have different cross sections.*

11   Use a piston ring installation tool and make sure the identification mark is facing the top of the piston, then slip the ring into the middle groove on the piston (see illustration). Don't expand the ring any more than necessary to slide it over the piston.

12   Install the number one (top) ring in the same manner. Make sure the mark is facing up. Be careful not to confuse the number one and number two rings.

13   Repeat the procedure for the remaining pistons and rings.

2D

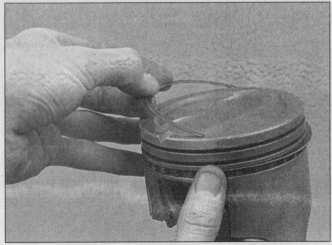

23.8b . . . followed by the side rails - DO NOT use a piston ring installation tool when installing the oil ring side rails

23.11 Installing the compressor rings with a ring expander - the mark (arrow) must face up

24.11 Lay the Plastigage strips (arrow) on the main bearing journals, parallel to the crankshaft centerline

24.15 Compare the width of the crushed Plastigage to the scale on the envelope to determine the main bearing oil clearance (always take the measurement at the widest point of the Plastigage); be sure to use the correct scale - standard and metric ones are included

24.22 Before installing the rear main cap, apply a small amount of sealant to the cap mating surfaces and in the corners where the cap meets the block

24 Crankshaft - installation and main bearing oil clearance check

Caution: *The connecting rod bolts/nuts on V6 engines (pushrod and SOHC) are "torque-to-yield" design and are NOT reusable. A pre-determined stretch of the bolt, calculated by the manufacturer, gives the added rigidity required with this cylinder block. Once removed they must be replaced with new bolts/nuts.*

1   Crankshaft installation is the first step in engine reassembly. It's assumed at this point that the engine block and crankshaft have been cleaned, inspected and repaired or reconditioned.
2   Position the engine with the bottom facing up.
3   Remove the main bearing cap bolts and lift out the caps. Lay them out in the proper order to ensure correct installation.
4   If they're still in place, remove the original bearing inserts from the block and the main bearing caps. Wipe the bearing surfaces of the block and caps with a clean, lint-free cloth. They must be kept spotlessly clean.

Main bearing oil clearance check

Refer to illustrations 24.11 and 24.15

5   Clean the back sides of the new main bearing inserts and lay one in each main bearing saddle in the block. If one of the bearing inserts from each set has a large groove in it, make sure the grooved insert is installed in the block. Lay the other bearing from each set in the corresponding main bearing cap. Make sure the tab on the bearing insert fits into the recess in the block or cap. **Caution:** *The oil holes in the block must line up with the oil holes in the bearing insert. Do not hammer the bearing into place and don't nick or gouge the bearing faces. No lubrication should be used at this time.*
6   The flanged thrust bearing must be

installed in the third cap and saddle on all engines.
7   Clean the faces of the bearings in the block and the crankshaft main bearing journals with a clean, lint-free cloth.
8   Check or clean the oil holes in the crankshaft, as any dirt here can go only one way - straight through the new bearings.
9   Once you're certain the crankshaft is clean, carefully lay it in position in the main bearings.
10   Before the crankshaft can be permanently installed, the main bearing oil clearance must be checked.
11   Cut several pieces of the appropriate-size Plastigage (they must be slightly shorter than the width of the main bearings) and place one piece on each crankshaft main bearing journal, parallel with the journal axis **(see illustration)**.
12   Clean the faces of the bearings in the caps and install the caps in their respective positions (don't mix them up) with the arrows pointing toward the front of the engine (see Section 15). Don't disturb the Plastigage.
13   Starting with the center main and working out toward the ends, tighten the main bearing cap bolts, in three steps, to the torque listed in this Chapter's Specifications. **Note:** *Don't rotate the crankshaft at any time during this operation.*
14   Remove the bolts and carefully lift off the main bearing caps. Keep them in order. Don't disturb the Plastigage or rotate the crankshaft. If any of the main bearing caps are difficult to remove, tap them gently from side-to-side with a soft-face hammer to loosen them.
15   Compare the width of the crushed Plastigage on each journal to the scale printed on the Plastigage envelope to obtain the main bearing oil clearance **(see illustration)**. Check the Specifications to make sure it's correct.

16   If the clearance is not as specified, the bearing inserts may be the wrong size (which means different ones will be required). Before deciding that different inserts are needed, make sure that no dirt or oil was between the bearing inserts and the caps or block when the clearance was measured. If the Plastigage was wider at one end than the other, the journal may be tapered (refer to Section 20).
17   Carefully scrape all traces of the Plastigage material off the main bearing journals and/or the bearing faces. Use your fingernail or the edge of a credit card - don't nick or scratch the bearing faces.

Final crankshaft installation

Refer to illustration 24.22

18   Carefully lift the crankshaft out of the engine.
19   Clean the bearing faces in the block, then apply a thin, uniform layer of moly-based engine assembly lubricant to each of the bearing surfaces. Be sure to coat the thrust faces as well as the journal face of the thrust bearing.
20   Make sure the crankshaft journals are clean, then lay the crankshaft back in place in the block.
21   Clean the faces of the bearings in the caps, then apply lubricant to them.
22   Install the caps in their respective positions with the arrows pointing toward the front of the engine. Before installing the rear main cap, apply a small amount of RTV sealant on either side of the rear main recess in the block **(see illustration)**. **Note:** *The rear main cap should be installed within 4 minutes of applying the sealant.*
23   Install the main cap bolts.
24   Tighten all, except the thrust bearing cap bolts (number 3) to the torque listed in this Chapter's Specifications (work from the center out and approach the final torque in three steps).
25   Tighten the thrust bearing cap bolts finger tight.

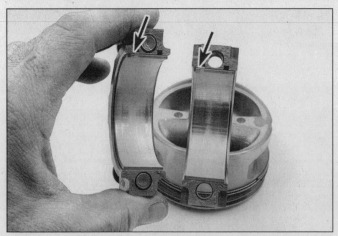

25.3 Insert the connecting rod bearing halves, making sure the bearing tab (arrows) are in the notches in the rod and cap

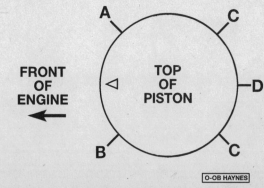

25.5 Ring end gap positions - align the oil ring spacer gap at (D), the oil ring side rails at (C) (one inch either side of the pin centerline) and the compression rings at (A) and (B) (one inch either side of the pin centerline)

26   Pry the crankshaft forward and while holding pressure on the crankshaft, pry the thrust bearing cap backward. Forcing these two in opposite directions, against each other, will align the thrust bearing surfaces.

27   While keeping forward pressure on the crankshaft, re-tighten ALL main bearing cap bolts to the torque listed in this Chapter's Specifications.

28   Rotate the crankshaft a number of times by hand to check for any obvious binding.

29   The final step is to check the crankshaft endplay with a feeler gauge or a dial indicator (see Section 14) The endplay should be correct if the crankshaft thrust faces aren't worn or damaged and new bearings have been installed.

30   Install the rear main oil seal (see Chapter 2, Part A).

25   Pistons/connecting rods - installation and rod bearing oil clearance check

Caution: *The connecting rod bolts/nuts on V6 engines (pushrod and SOHC) are "torque-to-yield" design and are NOT reusable. A pre-determined stretch of the bolt, calculated by the manufacturer, gives the added rigidity required with this cylinder block. Once removed they must be replaced with new bolts/nuts. During clearance checks using plastigage, use the old bolts/nuts and torque to Specifications, but use only new bolts/nuts for final assembly.*

1   Before installing the piston/connecting rod assemblies, the cylinder walls must be perfectly clean, the top edge of each cylinder must be chamfered, and the crankshaft must be in place.

2   Remove the cap from the end of the number one connecting rod (refer to the marks made during removal). Remove the original bearing inserts and wipe the bearing surfaces of the connecting rod and cap with a clean, lint-free cloth. They must be kept spotlessly clean.

Connecting rod bearing oil clearance check

Refer to illustrations 25.3, 25.5, 25.9, 25.11, 25.13 and 25.17

3   Clean the back side of the new upper bearing insert, then lay it in place in the connecting rod **(see illustration)**. Make sure the tab on the bearing fits into the recess in the rod. Don't hammer the bearing insert into place and be very careful not to nick or gouge the bearing face. Don't lubricate the bearing at this time.

4   Clean the back side of the other bearing insert and install it in the rod cap. Again, make sure the tab on the bearing fits into the recess in the cap, and don't apply any lubricant. It's critically important that the mating surfaces of the bearing and connecting rod are perfectly clean and oil free when they're assembled.

5   Position the piston ring gaps at intervals around the piston **(see illustration)**.

6   Slip a section of plastic or rubber hose over each connecting rod bolt **(see illustration 14.7)**.

7   Lubricate the piston and rings with clean engine oil and attach a piston ring compres-

sor to the piston. Leave the skirt protruding about 1/4-inch to guide the piston into the cylinder. The rings must be compressed until they're flush with the piston.

8   Rotate the crankshaft until the number one connecting rod journal is at BDC (bottom dead center) and apply a coat of engine oil to the cylinder walls.

9   With the arrow or notches on top of the piston facing the front of the engine **(see illustration)**, gently insert the piston/connecting rod assembly into the number one cylinder bore and rest the bottom edge of the ring compressor on the engine block.

10   Tap the top edge of the ring compressor to make sure it's contacting the block around its entire circumference.

11   Gently tap on the top of the piston with the end of a wooden hammer handle while guiding the end of the connecting rod into place on the crankshaft journal **(see illustration)**. The piston rings may try to pop out of the ring compressor just before entering the cylinder bore, so keep some downward pressure on the ring compressor. Work slowly, and if any resistance is felt as the piston enters the cylinder, stop immediately and fix the problem before proceeding. Do not, for

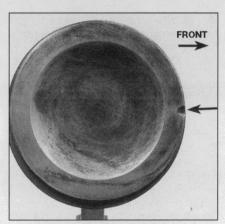

25.9 Install the piston with the mark/notch in the top of the piston facing the front of the engine

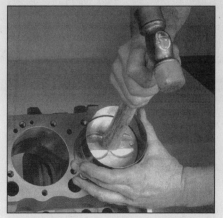

25.11 The piston can be driven gently into the cylinder bore with the end of a wooden or plastic hammer handle

2D

25.13 Lay the Plastigage strips on each rod bearing journal, parallel to the crankshaft centerline

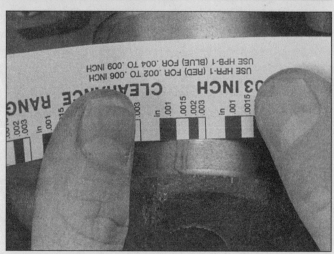

25.17 Measuring the width of the crushed Plastigage to determine the rod bearing oil clearance (be sure to use the correct scale - standard and metric ones are included)

any reason, force the piston into the cylinder - you might break a ring and/or the piston.

12   Once the piston/connecting rod assembly is installed, the connecting rod bearing oil clearance must be checked before the rod cap is permanently bolted in place.

13   Cut a piece of the appropriate-size Plastigage slightly shorter than the width of the connecting rod bearing and lay it in place on the number one connecting rod journal, parallel with the journal axis **(see illustration)**.

14   Clean the connecting rod cap bearing face, remove the protective hoses from the connecting rod bolts and install the rod cap. Make sure the mating mark on the cap is on the same side as the mark on the connecting rod.

15   Install the nuts or bolts and tighten them to the torque listed in this Chapter's Specifications, working up to it in three steps. **Note:** *Use a thin-wall socket to avoid erroneous torque readings that can result if the socket is wedged between the rod cap and nut/bolt. If the socket tends to wedge itself between the nut and the cap, lift up on it slightly until it no longer contacts the cap. Do not rotate the crankshaft at any time during this operation.*

16   Remove the nuts and detach the rod cap, being very careful not to disturb the Plastigage.

17   Compare the width of the crushed Plastigage to the scale printed on the Plastigage envelope to obtain the oil clearance **(see illustration)**. Compare it to the Specifications to make sure the clearance is correct.

18   If the clearance is not as specified, the bearing inserts may be the wrong size (which means different ones will be required). Before deciding that different inserts are needed, make sure that no dirt or oil was between the bearing inserts and the connecting rod or cap when the clearance was measured. Also, recheck the journal diameter. If the Plastigage was wider at one end than the other, the journal may be tapered (refer to Section 20).

Final connecting rod installation

Refer to illustration 25.20

19   Carefully scrape all traces of the Plastigage material off the rod journal and/or bearing face. Be very careful not to scratch the bearing - use your fingernail or the edge of a credit card.

20   On SOHC V6 engines, install new rod bolts/nuts at this time. Loosen the nuts until there is 2 mm of clearance between the nut and the rod cap, then tap the bolt out of the rod by tapping on the nut. Once the bolt comes loose, remove the nut and bolt. Insert the new bolts in the rod, aligning the bolt head with the recess in the connecting rod **(see illustration)**.

21   Make sure the bearing faces are perfectly clean, then apply a uniform layer of clean moly-base engine assembly lube to both of them. You'll have to push the piston into the cylinder to expose the face of the bearing insert in the connecting rod - be sure to slip the protective hoses over the rod bolts first.

22   Slide the connecting rod back into place on the journal, remove the protective hoses from the rod cap bolts, install the rod cap and tighten the new nuts to the torque listed in this Chapter's Specifications. Again, work up to the torque in three steps, and make sure the heads of the new rod bolts are aligned with the recess in the rod.

23   Repeat the entire procedure for the remaining pistons/connecting rods.

24   The important points to remember are:
a) *Keep the back sides of the bearing inserts and the insides of the connecting rods and caps perfectly clean when assembling them.*
b) *Make sure you have the correct piston/rod assembly for each cylinder.*
c) *The notches or mark on the piston must face the FRONT of the engine.*

d) *Lubricate the cylinder walls with clean oil.*
e) *Lubricate the bearing faces when installing the rod caps after the oil clearance has been checked.*

25   After all the piston/connecting rod assemblies have been properly installed, rotate the crankshaft a number of times by hand to check for any obvious binding.

26   As a final step, the connecting rod endplay must be checked. Refer to Section 14 for this procedure.

27   Compare the measured endplay to the Specifications to make sure it's correct. If it was correct before disassembly and the original crankshaft and rods were reinstalled, it should still be right. If new rods or a new crankshaft were installed, the endplay may be inadequate. If so, the rods will have to be removed and taken to an automotive machine shop for resizing.

25.20 Align the head of the new rod bolts squarely to the recess in the end of the rod (arrow)

26 Jackshaft and balance shaft (4.0L SOHC V6 only) - installation

Jackshaft

1   Lubricate the jackshaft bearing journals with engine assembly lubricant.

2   Slide the jackshaft into the engine, using a long bolt screwed into the front of the jackshaft as a handle. Support the shaft near the block and be careful not to scrape or nick the bearings. Install the jackshaft retainer plate and tighten the bolts to the torque listed in this Chapter's Specifications.

Balance shaft

Refer to illustration 26.5

3   Lubricate the front bearing and rear journal of the balance shaft with engine oil and insert the balance shaft assembly carefully onto the block. The block should be turned bottom-end up on the engine stand, and the crankshaft should be positioned with cylinder number 1 at TDC.

4   Install and tighten the balance shaft bolts to the torque listed in this Chapter's Specifications.

5   Turn the balance shaft sprocket until the two dots on the sprocket align over the hole in the front of the assembly, and the timing mark on the front of the lower shaft aligns with the balance shaft-to-block mating line **(see illustration)**. Install a 4 mm pin or drill bit into the hole to hold it in this position. **Note:** *It make take quite a few turns of the balance shaft sprocket to get the shaft and sprocket timing marks sets to align, because of the reduction gearing in the assembly.*

6   Install the balance shaft chain and the crankshaft sprocket. Install the balance shaft chain guide and tensioner (with the tensioner pinned) **(see illustration 13.5)**. Tighten the chain guide and tensioner bolts to the torque listed in this Chapter's Specifications. Remove the pin from the tensioner and remove the alignment pin from the balance shaft sprocket.

7   Refer to Chapter 2B and install the

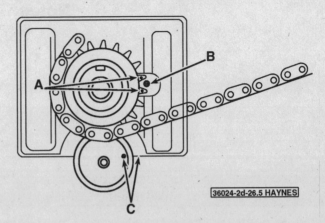

26.5 To align the balance shaft for chain installation, turn the sprocket until the marks on the sprocket (A) align with the hole in the assembly (B) and the lower shaft mark (C) aligns with the mating line of the assembly and engine block

camshaft timing chain cassettes, jackshaft chain and jackshaft sprocket.

27 Initial start-up and break-in after overhaul

Warning: *Have a fire extinguisher handy when starting the engine for the first time.*

1   Once the engine has been installed in the vehicle, double-check the engine oil and coolant levels.

2   With the spark plugs out of the engine and the ignition system disabled, crank the engine until oil pressure registers on the gauge or the light goes out.

3   Install the spark plugs, hook up the plug wires and restore the ignition system functions.

4   Start the engine. It may take a few moments for the fuel system to build up pressure, but the engine should start without a great deal of effort. **Note:** *If backfiring occurs through the throttle body, recheck the valve timing and ignition timing.*

5   After the engine starts, it should be

allowed to warm up to normal operating temperature. While the engine is warming up, make a thorough check for fuel, oil and coolant leaks. During the initial break-in period, the engine rpm should be kept above 1500 rpm, to provide sufficient oil pressure.

6   Shut the engine off and recheck the engine oil and coolant levels.

7   Drive the vehicle to an area with no traffic, accelerate from 30 to 50 mph, then allow the vehicle to slow to 30 mph with the throttle closed. Repeat the procedure 10 or 12 times. This will load the piston rings and cause them to seat properly against the cylinder walls. Check again for oil and coolant leaks.

8   Drive the vehicle gently for the first 500 miles (no sustained high speeds) and keep a constant check on the oil level. It is not unusual for an engine to use oil during the break-in period.

9   At approximately 500 to 600 miles, change the oil and filter.

10   For the next few hundred miles, drive the vehicle normally. Do not pamper it or abuse it.

11   After 2000 miles, change the oil and filter again and consider the engine broken in.

Notes

Chapter 3
Cooling, heating and air conditioning systems

Contents

3

Specifications

General

Thermostat
Type	Wax pellet
Opening temperature	192-degrees to 199-degrees F
Fully open	212-degrees

Air conditioning system refrigerant type
1993 and earlier	R-12
1994 and later	R-134a

Air conditioning system refrigerant capacity
1993 and earlier	32 ounces
1994 through 1996	26 ounces
1997 and later	22 ounces

Torque specifications

	Ft-lbs (unless otherwise indicated)
Cooling fan-to-clutch bolt	14 to 19
Fan clutch-to-water pump nut	34 to 46
Oil cooler mounting bolt (5.0L V8)	40 to 65
Thermostat housing bolts	
4.0L pushrod V6 and 5.0L V8	12 to 18
4.0L SOHC V6	78 to 92 in-lbs
Water pump bolts	
V6	72 to 108 in-lbs
V8	15 to 21
Accumulator pressure switch	
Metal base	60 to 120 in-lbs
Plastic base	Hand tighten only

1 General information

Engine cooling system

All vehicles covered by this manual employ a pressurized engine cooling system with thermostatically-controlled coolant circulation. An impeller-type water pump mounted on the front of the block pumps coolant through the engine. The coolant flows around each cylinder and toward the rear of the engine. Cast-in coolant passages direct coolant around the intake and exhaust ports, near the spark plug areas and in close proximity to the exhaust valve guides.

A wax pellet type thermostat is located in a housing near the front of the engine. During warm up, the closed thermostat prevents coolant from circulating through the radiator. As the engine nears normal operating temperature, the thermostat opens and allows hot coolant to travel through the radiator, where it's cooled before returning to the engine.

The cooling system is sealed by a pressure type radiator cap, which raises the boiling point of the coolant and increases the cooling efficiency of the radiator. If the system pressure exceeds the cap pressure relief value, the excess pressure in the system forces the spring-loaded valve inside the cap off its seat and allows the coolant to escape through the overflow tube into a coolant reservoir. When the system cools the excess coolant is automatically drawn from the reservoir back into the radiator.

The coolant reservoir serves as both the point at which fresh coolant is added to the cooling system to maintain the proper fluid level and as a holding tank for coolant.

This type of cooling system is known as a closed design because coolant that escapes past the pressure cap is saved and reused.

Heating system

The heating system consists of a blower fan and heater core located in the heater box, the hoses connecting the heater core to the engine cooling system and the heater/air conditioning control head on the dashboard. Hot engine coolant is circulated through the heater core. When the heater mode is activated, a flap door opens to expose the heater box to the passenger compartment. A fan switch on the control head activates the blower motor, which forces air through the core, heating the air.

Air conditioning system

The air conditioning system consists of a condenser mounted in front of the radiator, an evaporator mounted adjacent to the heater core, a compressor mounted on the engine, a filter-drier (accumulator) which contains a high pressure relief valve and the plumbing connecting all of the above components.

A blower fan forces the warmer air of the passenger compartment through the evaporator core (sort of a radiator-in-reverse), transferring the heat from the air to the refrigerant. The liquid refrigerant boils off into low pressure vapor, taking the heat with it when it leaves the evaporator.

2 Antifreeze - general information

Refer to illustration 2.4

Warning: *Do not allow antifreeze to come in contact with your skin or painted surfaces of the vehicle. Rinse off spills immediately with plenty of water. Antifreeze is highly toxic if ingested. Never leave antifreeze lying around in an open container or in puddles on the floor; children and pets are attracted by it's sweet smell and may drink it. Check with local authorities about disposing of used antifreeze. Many communities have collection centers which will see that antifreeze is disposed of safely. Never dump used antifreeze on the ground or pour it into drains.*

The cooling system should be filled with a water/ethylene glycol based antifreeze solution which will prevent freezing down to at least -20-degrees F (even lower in cold climates). It also provides protection against corrosion and increases the coolant boiling point. The engines in the covered vehicles have aluminum heads. The manufacturer recommends that only coolant designated as safe for aluminum engine components be used.

The cooling system should be drained, flushed and refilled at least every other year (see Chapter 1). The use of antifreeze solutions for periods of longer than two years is likely to cause damage and encourage the formation of rust and scale in the system.

Before adding antifreeze to the system, check all hose connections. Antifreeze can leak through very minute openings.

The exact mixture of antifreeze to water which you should use depends on the relative weather conditions. The mixture should contain at least 50-percent antifreeze, but should never contain more than 70-percent antifreeze. Consult the mixture ratio chart on the container before adding coolant. Hydrometers are available at most auto parts stores to test the coolant **(see illustration)**. Use antifreeze which meets The manufacturers specifications for engines with aluminum heads.

3 Thermostat - check and replacement

Warning: *Do not remove the radiator cap, drain the coolant or replace the thermostat until the engine has cooled completely.*

Check

1   Before assuming the thermostat is to blame for a cooling system problem, check the coolant level, drivebelt tension (see

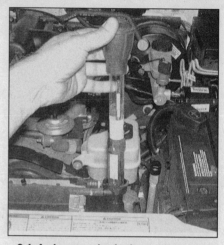

2.4 An inexpensive hydrometer can be used to test the level of antifreeze protection in your coolant

Chapter 1) and temperature gauge (or light) operation.

2   If the engine seems to be taking a long time to warm up (based on heater output or temperature gauge operation), the thermostat is probably stuck open. Replace the thermostat with a new one.

3   If the engine runs hot, use your hand to check the temperature of the upper radiator hose. If the hose isn't hot, but the engine is, the thermostat is probably stuck closed, preventing the coolant inside the engine from escaping to the radiator. Replace the thermostat. **Caution:** *Don't drive the vehicle without a thermostat. The computer may stay in open loop and emissions and fuel economy will suffer.*

4   If the upper radiator hose is hot, it means that the coolant is flowing and the thermostat is open. Consult the Troubleshooting section at the front of this manual for cooling system diagnosis.

Replacement

Refer to illustrations 3.8a, 3.8b, 3.8c and 3.12

5   Disconnect the negative battery cable from the battery.

6   Remove the air cleaner air duct from the throttle body and air cleaner,

7   Drain the cooling system (see Chapter 1). If the coolant is relatively new or in good condition (see Chapter 1), save it and reuse it.

8   Follow the upper radiator hose to the engine to locate the thermostat housing. The housing is located at the front of the intake manifold **(see illustrations)**.

9   Loosen the hose clamp, then detach the hose from the fitting. If it's stuck, grasp it near the end with a pair of adjustable pliers and twist it to break the seal, then pull it off. If the hose is old or deteriorated, cut it off and install a new one.

10   If the outer surface of the large fitting that mates with the hose is deteriorated (corroded, pitted, etc.) it may be damaged further by hose removal. If it is, the thermostat hous-

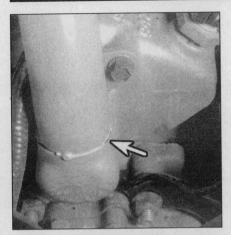

3.8a Thermostat housing location (arrow) - 4.0L pushrod V6

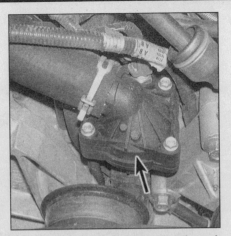

3.8b Thermostat housing location (arrow) - 4.0L SOHC V6

3.8c Thermostat housing location (arrow) - 5.0L V8

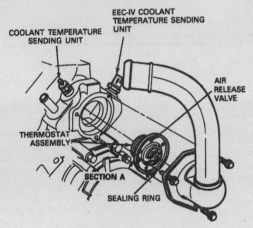

3.12 Thermostat and related components (typical)

4.3 Use a socket or screwdriver to loosen the hose clamp (arrow)

ing cover will have to be replaced.

11   Remove the bolts and detach the housing cover. If the cover is stuck, tap it with a soft-face hammer to jar it loose. Be prepared for some coolant to spill out as the gasket seal is broken.

12   Note how it's installed (which end is facing toward the engine), then remove the thermostat **(see illustration)**. It may be necessary to rotate the thermostat to free it.

13   Stuff a rag into the engine opening, then remove all traces of corrosion from the housing and cover with a gasket scraper. Remove the rag from the opening and clean the mating surfaces with lacquer thinner or acetone.

14   Make sure the sealing ring is positioned correctly on the thermostat **(see illustration 3.12)**.

15   Install the new thermostat in the housing with the bridge section toward the outlet housing cover. Make sure the air release valve is in the up position **(see illustration 3.12)**.

16   Install the housing cover and bolts. Tighten the bolts to the torque listed in this Chapter's Specifications.

17   Reattach the hose to the fitting and tighten the hose clamp securely.

18   Refill the cooling system (see Chap-

ter 1).

19   Install the air cleaner air duct.

20   Start the engine and allow it to reach normal operating temperature, then check for leaks and proper thermostat operation (as described in Steps 2 through 4).

4   Radiator - removal and installation

Warning: *Wait until the engine is completely cool before beginning this procedure.*

Removal

Refer to illustrations 4.3, 4.5, 4.6 and 4.8

1   Disconnect the negative battery cable from the battery.

2   Drain the cooling system (see Chapter 1). If the coolant is relatively new or in good condition, save it and reuse it.

3   Loosen the hose clamps and slide them back on the hoses, then detach the radiator hoses from the fittings **(see illustration)**. If they're stuck, grasp each hose near the end with a pair of adjustable pliers and twist it to break the seal, then pull it off - be careful not to distort the radiator fittings! If the hoses are

4.5 Remove the shroud mounting bolts and place the shroud over the fan

old or deteriorated, cut them off and install new ones. **Note:** *On spring-type hose clamps, squeeze the clamp with pliers and slide it back along the hose.*

4   Disconnect the reservoir hose from the radiator filler neck.

5   Remove the screws that attach the fan shroud to the radiator and slide the shroud toward the engine **(see illustration)**. Let the shroud rest on the fan.

3

4.6 On automatic transmission models, hold the cooler line fitting with a backup wrench and loosen the line with another wrench

6 If the vehicle is equipped with an automatic transmission, hold the cooler fittings with a backup wrench and disconnect the lines from the radiator **(see illustration)**. Use a drip pan to catch spilled fluid.

7 Plug the lines and fittings to avoid spillage and contamination.

8 Remove the radiator mounting bolts **(see illustration)**. On 5.0L V8 models, detach the plastic clips at the bottom of the radiator that hold the condenser to the radiator.

9 Carefully lift the radiator up and out of its lower mounting pads or grommets. Remove the radiator and be careful not to spill coolant on the vehicle or scratch the paint.

10 With the radiator removed, it can be inspected for leaks and damage. If it needs repair, have a radiator shop or dealer service department perform the work as special techniques are required.

11 Bugs and dirt can be removed from the radiator with compressed air and a soft brush. Don't bend the cooling fins as this is done.

Installation

12 Check the radiator mounts for deterioration and make sure there's nothing in them when the radiator is installed.

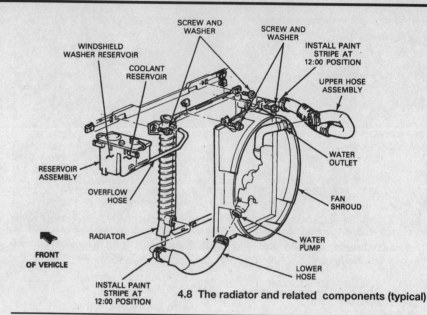

4.8 The radiator and related components (typical)

13 Installation is the reverse of the removal procedure with the following additions:

a) After installation, fill the cooling system with the proper mixture of antifreeze and water (see Chapter 1).

b) Start the engine and check for leaks. Allow the engine to reach normal operating temperature, indicated by the upper radiator hose becoming hot. Recheck the coolant level and add more if required.

c) If you're working on an automatic transmission equipped vehicle, check and add transmission fluid as needed (see Chapter 1).

5 Cooling fan and viscous clutch - inspection, removal and installation

Warning: *To avoid possible injury or damage, DO NOT operate the engine with a damaged fan. Do not attempt to repair fan blades - replace a damaged fan with a new one.*

Viscous clutch inspection

1 Disconnect the negative cable from the battery.

2 Rock the fan back and forth by hand to check for excessive bearing play.

3 With the engine cold, turn the fan blades by hand. The fan should turn freely.

4 Visually inspect for substantial fluid leakage from the clutch assembly. If problems are noted, replace the clutch assembly.

5 Reconnect the battery cable and warm up the engine. With the engine completely warmed up, turn off the ignition switch and disconnect the negative battery cable from the battery. Turn the fan by hand. Some drag should be evident. If the fan turns easily, replace the fan clutch.

Removal and installation

Refer to illustrations 5.9a, 5.9b and 5.11

6 Disconnect the negative cable from the battery.

7 Disconnect the reservoir hose from the radiator filler neck.

8 Remove the screws securing the shroud

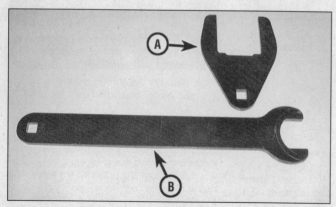

5.9a To loosen the large nut that holds the fan clutch to the water pump, use the two tools shown - tool (A) is used with a breaker bar

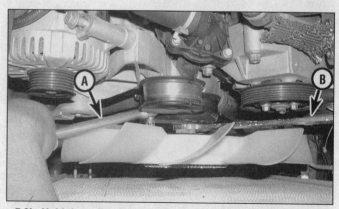

5.9b Hold the pulley from turning (with tool A) and turn the nut counterclockwise (with tool B) to loosen the fan from the water pump hub

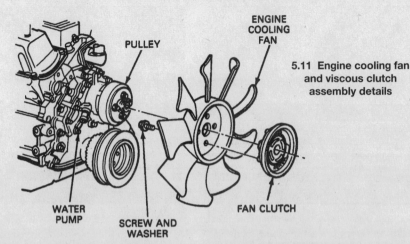

PULLEY

ENGINE
COOLING
FAN

5.11 Engine cooling fan
and viscous clutch
assembly details

WATER
PUMP

SCREW AND
WASHER

FAN CLUTCH

6.4 The coolant reservoir is secured to
the inner fender panel by screws - early
model shown

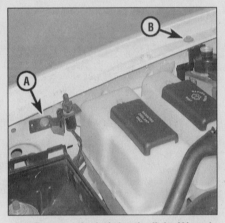

6.8 Remove the bolt for the light (A) and
the speed control servo (B), then set them
both aside

6.9 Disconnect the washer pump
electrical connector (arrow)

6.10a Remove the two reservoir mounting
bolts from the inner fender (arrows) . . .

6.10b . . . and the two nuts (arrow) - the
other nut is located at the other end of the
reservoir under the air cleaner housing

to the radiator and slide the shroud toward the engine (see Section 4).

9   Loosen the clutch hub nut as follows:

a) **Note:** *This nut has standard right-hand threads and must be rotated counter-clockwise for removal. Remove the large nut securing the clutch hub assembly to the water pump drivebelt hub.*

b) *Use the two special tools, available at most auto parts stores* **(see illustrations).**

c) *If the special tools are not available, use a strap wrench to hold the assembly from rotating and use a large adjustable wrench to remove the large nut.*

10   Lift the shroud and the fan/clutch assembly out of the engine compartment together. **Note:** *The shroud must come out with the fan assembly.*

11   Inspect the fan blades for damage and defects. Replace it if necessary by removing the bolts securing the fan to the clutch assembly **(see illustration).** If the fan clutch is stored, position it with the radiator side facing down.

12   Installation is the reverse of the removal steps. Be sure to tighten the fan and clutch nut evenly and to the torque listed in this Chapter's Specifications.

6   Coolant reservoir - removal and installation

Note: *On these models, the coolant recovery reservoir and windshield washer reservoir are combined in a single mounting.*

1   Disconnect the radiator overflow hose from the base of the coolant reservoir (see Chapter 1).

2   Connect a hose to the fitting on the coolant reservoir and drain the coolant from the reservoir.

1994 and earlier models

Refer to illustrations 6.4

3   Disconnect the electrical connector for the windshield washer pump. Disconnect the windshield washer hoses and drain the windshield washer reservoir.

4   Remove the reservoir attachment to the inner fender panel **(see illustration).**

5   Lift the reservoir out of the engine compartment.

6   Installation is the reverse of removal.

1995 and later models

Refer to illustrations 6.8, 6.9, 6.10a and 6.10b

7   Refer to Chapter 4 and remove the air cleaner assembly and inlet tube.

8   Detach the underhood light switch and the speed control servo and set them aside **(see illustration).**

9   Disconnect the windshield washer hose (behind the wiper motor), and at the windshield washer reservoir, disconnect the electrical connector for the windshield washer pump **(see illustration).**

10   Remove the bolts and nuts and remove the reservoir from the engine compartment **(see illustrations).**

11   Installation is the reverse of the removal procedure.

3

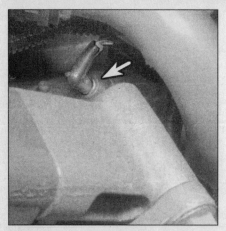

8.2a The 4.0L pushrod V6 temperature sending unit (arrow) is located on the left side of the intake manifold

8.2b Temperature sending unit location (arrow) - 5.0L V8 models

8.2c Remove the upper intake manifold to access the temperature sending unit (arrow) on the 4.0L SOHC V6

7 Engine oil cooler (5.0L V8 only) - removal and installation

Removal

Warning: *The engine should be completely cool for this procedure.*

1   Some 5.0L V8 models equipped with a towing package have an engine oil cooler. Coolant flows through the cooler from two hoses attached to pipes on the cooler body.

2   To replace the oil cooler, remove the oil filter and drain the cooling system (see Chapter 1). **Note:** *If the cooler is not being removed from the vehicle, do not drain the cooling system or disconnect the coolant hoses at the oil cooler.*

3   Disconnect the coolant hoses from the oil cooler.

4   Remove the long bolt holding the oil cooler housing to the oil filter adapter housing.

5   Pull the oil cooler from the filter adapter.

Installation

6   Clean the oil filter adapter and the back of the oil cooler of any old gasket material.

7   Using a new O-ring against the oil filter adapter, install the oil cooler, aligning its locator pins over the locating boss on the filter adapter body. Install a new O-ring on the cooler-to-adapter bolt and tighten it the to the torque listed in this Chapter's Specifications.

8   The remainder of installation is the reverse of removal. Reconnect the coolant hoses if they were removed, then install a new oil filter, refill and bleed the cooling system (see Chapter 1), and run the engine to check for oil or coolant leaks.

8 Coolant temperature sending unit - removal and installation

Refer to illustrations 8.2a, 8.2b and 8.2c

1   Allow the cooling system to completely

cool down. Drain coolant from the radiator to approximately four inches below the upper radiator hose. This will minimize coolant loss during this procedure.

2   Disconnect the electrical connector from the top of the sending unit **(see illustrations)**. **Note:** *On SOHC V6 models, the upper intake manifold must be removed to access the sending unit. Refer to Chapter 2, Part B for the procedure. It may be possible to access the sender with the thermostat housing removed.*

3   Wrap the threads of the new sending unit with Teflon tape.

4   Remove the old sending unit from the engine.

5   Install the new sending unit and tighten it securely.

6   Connect the sending unit electrical connector.

7   Refill the cooling system (see Chapter 1).

8   Start the engine and check for leaks.

9 Water pump - check and replacement

Check

Refer to illustration 9.4

1   A failure in the water pump can cause serious engine damage due to overheating.

2   There are three ways to check the operation of the water pump while it's installed on the engine. If the pump is defective, it should be replaced with a new or rebuilt unit.

3   With the engine running at normal operating temperature, squeeze the upper radiator hose. If the water pump is working properly, a pressure surge should be felt as the hose is released. **Warning:** *Keep your hands away from the fan blades!*

4   Water pumps are equipped with weep or vent holes. If a failure occurs in the pump seal, coolant will leak from the hole **(see illustration)**. In most cases you'll need a flashlight to find the hole on the water pump

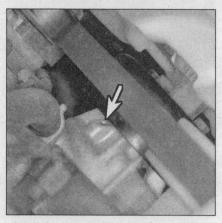

9.4 If the water pump seal fails, coolant will leak from the weep hole (arrow)

from underneath to check for leaks.

5   If the water pump shaft bearings fail there may be a howling sound at the front of the engine while it's running. Shaft wear can be felt if the water pump pulley is rocked up and down. Don't mistake drivebelt slippage, which causes a squealing sound, for water pump bearing failure.

Replacement

Refer to illustrations 9.9, 9.12a and 9.12b

Warning: *Wait until the engine is completely cool before beginning this procedure.*

6   Disconnect the negative battery cable from the battery.

7   Drain the cooling system (see Chapter 1). If the coolant is relatively new or in good condition, save it and reuse it.

8   Remove the cooling fan and shroud (see Section 5). **Note:** *It may also be necessary on V6 models to remove the radiator for clearance* (see Section 4).

9   Loosen the bolts on the water pump pulley, then remove the drivebelt (see Chapter 1) and remove the pulley. On SOHC V6 models, also remove the drivebelt idler pulley **(see illustration)**.

9.9 Remove the drivebelt idler pulley on SOHC V6 models

9.12a Water pump bolt locations (arrows) - V6 models

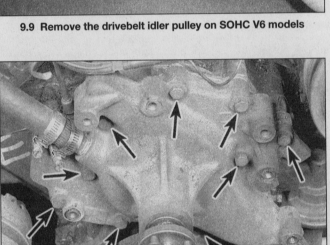

9.12b 5.0L V8 water pump bolt locations (arrows) - remove the nuts and set aside the two wiring harnesses

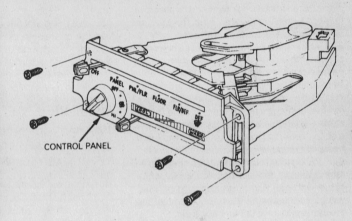

CONTROL PANEL

10.3 Heater control assembly mounting screws - early models with cable controls

10 Loosen the clamps and detach the hoses from the water pump. If they're stuck, grasp each hose near the end with a pair of adjustable pliers and twist it to break the seal, then pull it off. If the hoses are deteriorated, cut them off and install new ones. **Note:** *On the 5.0L V8, it may be easier to disconnect the hose from the oil cooler, rather than at the water pump.*

11 If necessary, remove the alternator and its mounting bracket from the water pump (see Chapter 5).

12 Remove the bolts and detach the water pump from the engine **(see illustrations)**. Note the locations of the various lengths and different types of bolts/studs as they're removed to ensure correct installation. If the pump is stuck, strike it with a soft-faced hammer, do not pry between the pump and the engine, or you could damage the gasket sealing surfaces.

13 Clean the bolt threads and the threaded holes in the engine to remove corrosion and sealant.

14 Compare the new pump to the old one to make sure they're identical.

15 Remove all traces of old gasket material from the engine gasket surface with a gasket scraper.

16 Clean the engine and new water pump gasket mating surfaces with lacquer thinner or acetone.

17 Apply a thin coat of RTV sealant to the engine side of the new gasket.

18 Apply a thin layer of RTV sealant to the gasket mating surface of the new pump, then carefully mate the gasket and the pump. Slip a couple of bolts through the pump mounting holes to hold the gasket in place.

19 Carefully attach the pump and gasket to the engine and thread the bolts into the holes finger tight.

20 Install the remaining bolts (if they also hold the alternator bracket in place, be sure to reposition the bracket at this time). Tighten them evenly to the torque listed in this Chapter's Specifications in 1/4-turn increments. Don't overtighten them or the pump may be distorted.

21 Reinstall all parts removed for access to the pump.

22 Refill the cooling system and check the drivebelt tension (see Chapter 1). Run the engine and check for leaks.

10 Heater and air-conditioning control assembly - removal and installation

Warning: *On models equipped with airbags, always disable the airbag system before working in the vicinity of the impact sensors, steering column or instrument panel to avoid the possibility of accidental deployment of the airbag, which could cause personal injury (see Chapter 12).*

Removal

1994 and earlier models

Refer to illustration 10.3

1 Disconnect the negative cable from the battery.

2 Remove the instrument cluster trim panel (see Chapter 12).

3 Remove the four screws securing the control panel to the instrument panel **(see illustration)**.

4 Pull the control assembly through the opening in the instrument panel far enough to

3

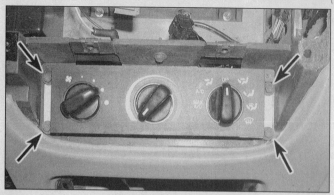

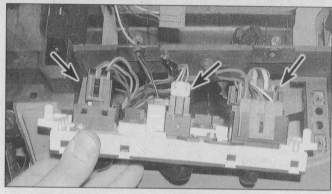

10.11 Control assembly mounting screws (arrows) - later models

10.12 Disconnect the electrical connectors (arrows)

allow removal of the electrical connections. Carefully spread the clips on the electrical connectors and disconnect the connectors for the blower switch and illumination light.

5   Use a small screwdriver and remove the two-hose vacuum harness from the vacuum switch on the side of the control assembly.

6   At the back of the control assembly, use a screwdriver or needle-nose pliers and release the temperature and function cable snap-in flag from the white control bracket.

7   At the bottom of the control assembly, remove the temperature control cable (black cable with a blue snap-in flag) from the control by rotating the cable until the T-pin releases the cable.

8   Pull enough cable through the instrument opening until the function cable (white cable with black snap-in flag) can be held vertical to the control assembly, then remove the cable end from the function lever.

9   Remove the control assembly from the instrument panel.

1995 and later models

Refer to illustrations 10.11 and 10.12

10   Remove the radio (see Chapter 12) and the instrument panel surround (see Chapter 11).

11   Remove the four screws holding the control panel to the dashboard **(see illustration)**.

12   Pull the panel out enough to access the vacuum and electrical connectors **(see illustration)**. Tag and disconnect the connectors. **Note:** *On models with electronic Automatic Temperature Control, the vacuum connection is retained to the control head by two nuts.*

Installation

1994 and earlier models

13   Pull the control cables through the instrument panel opening by approximately eight inches.

14   Hold the control assembly up to the instrument panel with the control assembly face pointed toward the floor.

15   Carefully bend and attach the function cable (white) to the white plastic lever. Rotate the control assembly back to its normal position for installation, then snap the black cable flag into the control assembly bracket.

16   On the opposite side of the control assembly, attach the black temperature control cable with the blue snap-in flag to the blue plastic lever on the control assembly. Be sure that the end of the cable is seated securely with the T-top pin on the control assembly. Rotate the cable to its operating position and snap the blue cable flag into the control assembly bracket.

17   Install the electrical wire harness connectors.

18   Connect the dual terminal on the vacuum hose to the vacuum switch on the control assembly.

1995 and later models

19   Move the control panel close enough to the dashboard to make the connections.

20   Connect the electrical and vacuum connectors, noting the tags or identification you marked when removing the panel.

All models

19   Position the control assembly into the instrument panel and install the four mounting screws.

17   Complete the installation by installing the various trim panels and instrument panel items previously removed. Check for proper operation and on cable-control models, adjust the cables if necessary (see Section 11).

---

11 Heater control cables - check and adjustment

Check

1   Move the control lever all the way from left to right.

2   If the control lever stops before the end and bounces back, the cables are out of adjustment.

Adjustment

Refer to illustrations 11.4a and 11.4b

3   Squeeze the tabs on either side of the glove box door to disengage it, then let the door hang down to provide access to the control cables.

4   Working through the glove box opening, remove the cable jacket from its metal attaching clip on top of the heater assembly

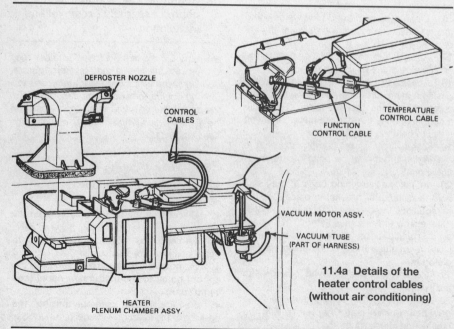

11.4a Details of the heater control cables (without air conditioning)

DEFROSTER NOZZLE

CONTROL CABLES

TEMPERATURE CONTROL CABLE

FUNCTION CONTROL CABLE

VACUUM MOTOR ASSY.

VACUUM TUBE (PART OF HARNESS)

HEATER PLENUM CHAMBER ASSY.

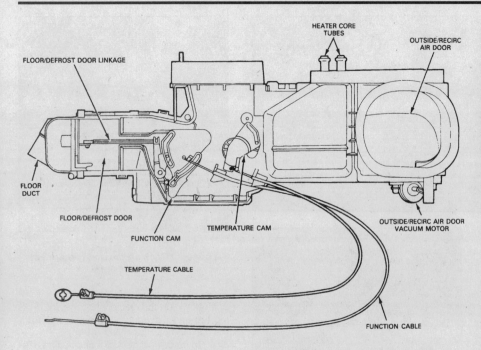

11.4b Details of the heater control cables (with air conditioning)

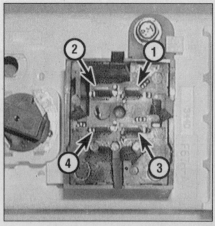

12.9 Blower motor switch terminal identification

(see illustrations). The cable ends should remain attached to the door cams and/or crank arms at this time.

5   To adjust the temperature control cable, set the temperature lever to the Cool position and hold it there.

6   Push gently on the black cable jacket to seat the blend door (push until you feel resistance).

7   Reinstall the cable to the clip by pushing the jacket into the clip from the top until it snaps in.

8   To adjust the function control cable, set the function control lever to Defrost and hold it there.

9   Pull on the white cam jacket until the cam travel stops.

10   Reinstall the cable to the clip by pushing the jacket into the clip from the top until it snaps in.

11   Run the system on High and actuate the levers, checking for proper operation. Readjust if necessary.

12   Install the glove box door.

12   Heater and air-conditioning blower motor and circuit - check

Refer to illustration 12.9

Note: The blower motor is switched on the ground-side of the circuit.

1   Check the fuse and all connections in the circuit for looseness and corrosion.

2   Make sure the battery is fully charged.

3   With the transmission in Park, the parking brake securely set, turn the ignition switch to the On position. It isn't necessary to start the vehicle.

4   Switch the heater controls to FLOOR and the blower speed to HI. Listen at the ducts to hear if the blower is operating. If it is, then switch the blower speed to LO and listen again. Try all the speeds.

5   The blower motor resistor assembly is located on the heater housing in the engine compartment, next to the blower motor (see illustration 13.8). There are three resistor elements mounted on the resistor board to provide low and medium blower speeds (HI bypasses the resistor). The blower operates continuously, anytime the ignition switch is On and the mode switch is in any position other than Off. A thermal limiter resistor is integrated into the circuits to prevent heat damage to the components. If the thermal limiter circuit has been opened as a result of excessive heat, it should be replaced only with the identical replacement part. **Note:** Do not replace your blower resistor with a resistor that does not incorporate the thermal limiter.

6   With the resistor removed from the vehicle, visually check the limiter for damage, indicated by the material melting out between the contacts of the limiter. Check the resistor block for continuity between terminals. There should be continuity between all terminals (with varying resistance at each set). If any of the resistor elements do not pass the tests, replace the blower resistor.

7   Locate the electrical connector at the blower motor. Backprobe the pink/white wire terminal; there should be at least 10 volts with the mode switch in any position other than Off and the ignition switch On. If not, there is a problem in the circuit from the fuse panel to the heater/air conditioning control

panel, or from the control panel to the blower.

8   If there is voltage at the feed wire, but the blower does not operate, backprobe the black wire and connect it to a known good chassis ground with a jumper wire. If the blower now operates there is a problem in the ground circuit (the blower resistors and switch are in the ground side of the circuit). If it still doesn't operate, replace the blower motor.

9   If the blower operates, but not at all speeds and you have already checked the blower resistor, refer to Section 10 and remove the heater/air conditioning control panel. Disconnect the electrical connector from the back of the blower speed switch and test the terminals for continuity (see illustration). In the Low position, there should be no continuity; in Medium/Low position, there should be continuity between terminals 2 and 3; in Medium/High position, there should be continuity between 2, 3 and 4; and in HI position, there should be continuity between terminals 1, 2 and 4. If the continuity is not as described, replace the blower speed switch.

10   Locate the blower motor relay, in the relay box behind the right headlight (see Chapter 12). There are four pins on the connector. Check for resistance between the pink/white wire terminal in the connector, and the pink/white wire at the blower motor. If resistance is correct (less than 5 ohms) and the blower doesn't operate, replace the blower relay.

Rear air circulation unit

Refer to illustration 12.13

13   Some 1996 and later models are equipped with an optional console air circulation system that delivers conditioned air to the rear seat passengers. A separate control panel operates the system. The panel is behind a cover at the back of the console; the panel also controls the rear seat sound system. The driver controls the main function, either heating or cooling, while the rear passengers choose the volume and direction of

3

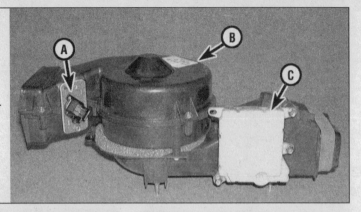

12.13 Console blower housing components

A   Blower resistor
B   Blower motor (in housing)
C   Console blower controller

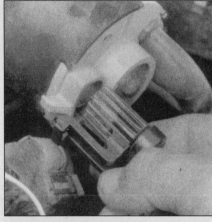

13.3 Push down on the electrical connector locking tab and disconnect the connector

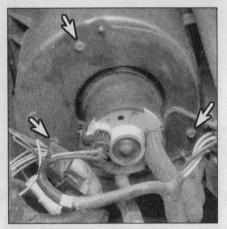

13.6 Remove the three screws securing the heater blower motor

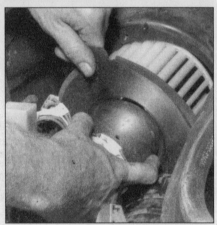

13.7 Carefully pull the blower assembly from the heater assembly

airflow to the rear seat area. A blower motor for the rear seats is located within the air ducts in the console **(see illustration)**.

14   Refer to Chapter 11 for removal of the console trim to expose the blower motor and its resistor, which can be tested as outlined above for the front blower and resistor. Once the console trim is removed and the electrical connectors disconnected, the console can be pulled toward the rear of the vehicle, disengaging it from tabs on the floor. Pull the console and blower housing out of the vehicle and remove the nuts holding the blower housing to the console.

13   Heater and air-conditioning blower motor - replacement

Note: *The blower motor is located in the right side of the engine compartment.*
1   Disconnect the negative cable from the battery.

1994 and earlier models

Refer to illustrations 13.3, 13.6, 13.7 and 13.8
2   Remove the air cleaner (see Chapter 4).
3   Working in the engine compartment,

push down on the electrical connector locking tab and disconnect the connector from the blower motor **(see illustration)**.
4   On models without air conditioning, disconnect the small rubber cooling tube from the blower motor.
5   On models with air conditioning:
  a) *Working in the passenger compartment, remove the single nut from the bottom of the plenum. The nut is located just to the right of the heater core access cover.*
  b) *Working in the engine compartment, disconnect the electrical connector from the blower motor resistor.*
  c) *Disconnect the vacuum hose from the check valve.*
  d) *Disconnect the vacuum line from the intake manifold and remove it from the routing channel.*
6   Remove the three screws securing the blower motor to the heater blower assembly **(see illustration)**.
7   Being very careful not to damage the gasket, pull the blower motor out of the heater assembly **(see illustration)**.
8   If necessary, remove the push nut on the motor shaft and remove the blower wheel from the motor **(see illustration)**.
9   Installation is the reverse of the removal Steps with the addition of the following:
  a) *Replace the gasket between the blower motor and heater assembly if its condition is in doubt. This gasket must be good to prevent moisture from entering the assembly.*
  b) *Make sure the electrical connector is fully seated and that it "clicks" into place.*

1995 and later models

Refer to illustrations 13.11 and 13.12
10   Remove the coolant recovery/windshield washer reservoir (see Section 6) for access to the blower motor.
11   Disconnect the electrical connector at

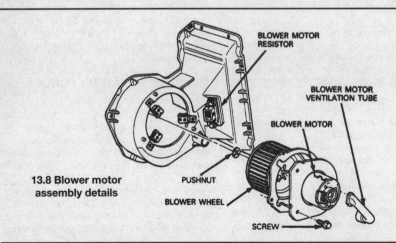

BLOWER MOTOR RESISTOR
BLOWER MOTOR VENTILATION TUBE
BLOWER MOTOR
PUSHNUT
BLOWER WHEEL
SCREW

13.8 Blower motor assembly details

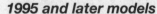

13.11 Disconnect the electrical connector (A) and the vent hose (B) from the blower motor

13.12 Remove the four mounting screws (arrows)

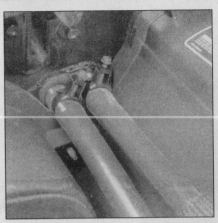

14.3 Disconnect the heater hoses from the fittings at the firewall

the blower motor and the vent hose **(see illustration)**. Disconnect the cruise control servo and set it aside.
12   Remove the four screws holding the blower motor **(see illustration)**.

14   Heater core - replacement

Warning: *On models equipped with airbags, always disable the airbag system before*

working in the vicinity of the impact sensors, steering column or instrument panel to avoid the possibility of accidental deployment of the airbag, which could cause personal injury (see Chapter 12).

1994 and earlier models

Refer to illustrations 14.3, 14.7 and 14.8
1   Disconnect the cable from the negative battery terminal.
2   Drain the cooling system (see Chapter 1).

3   Working within the engine compartment, loosen the clamps on the heater hoses at the engine compartment side of the firewall **(see illustration)**.
4   Twist the hoses and carefully separate them from the heater core tubes.
5   Plug or cap the heater core tubes to prevent coolant from spilling into the passenger compartment when the heater core is removed.
6   Place a plastic sheet on the floor of the vehicle to prevent stains if the coolant spills.
7   Working in the passenger compartment, remove the four screws securing the heater core access cover to the plenum assembly **(see illustration)**.
8   Carefully pull the heater core to the rear and down and remove it from the plenum assembly **(see illustration)**.
9   Installation is the reverse of the removal procedure. Fill the cooling system (see Chapter 1).
10   Run the engine, check for leaks and test the heater for proper operation.

1995 and later models

Refer to illustrations 14.14, 14.16, 14.17, 14.18 and 14.19
Warning: *The air conditioning system is under high pressure. Do not loosen any hose*

3

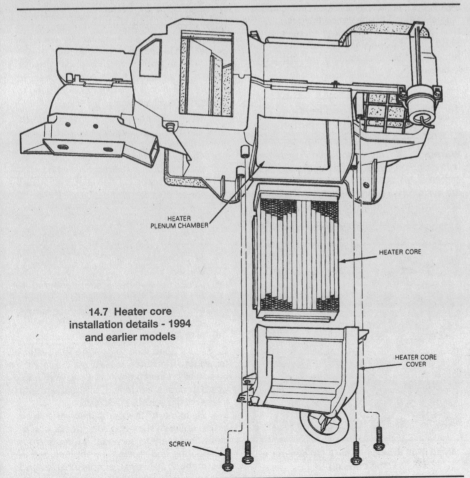

HEATER PLENUM CHAMBER

HEATER CORE

HEATER CORE COVER

14.7 Heater core installation details - 1994 and earlier models

SCREW

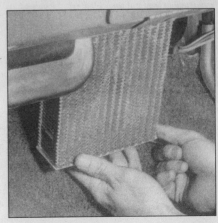

14.8 Pull the heater core to the rear and down to remove it from the plenum

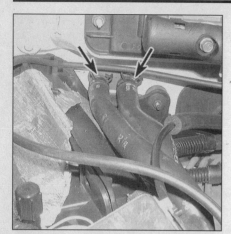

14.14 On 1995 and later models, disconnect the heater hoses (arrows) at the firewall

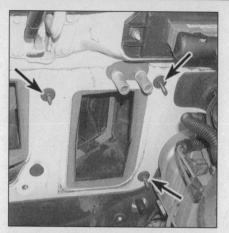

14.16 After removing the evaporator case, remove the three nuts (arrows) from the studs

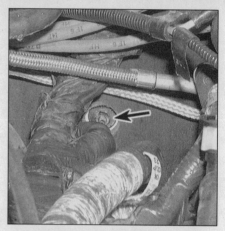

14.17 Remove this nut (arrow) in the center of the firewall

fittings or remove any components until after the system has been discharged. Air conditioning refrigerant should be properly discharged into an EPA-approved recovery/recycling unit at a dealer service department or an automotive air conditioning repair facility. Always wear eye protection when disconnecting air conditioning system fittings.

11   Take the vehicle to a dealer or air-conditioning repair facility and have the refrigerant recovered.

12   Disconnect the negative battery cable.

13   Refer to Section 18 and remove the evaporator housing assembly, with evaporator, accumulator/drier and blower motor attached, from the firewall.

14   Disconnect the two heater hoses from the firewall, clamp the hoses shut, and plug the heater core tubes **(see illustration)**.

15   Refer to Chapter 11 and remove the instrument panel.

16   Remove the three nuts on the engine side of the firewall that are on the same studs used to hold the evaporator housing to the firewall **(see illustration)**. These nuts are not accessible until the evaporator housing is removed.

17   Remove the nut behind the center of the

engine on the firewall **(see illustration)**.

17   Refer to Chapter 6 and remove the PCM (computer) and PCM heat sink.

18   Pull the air distribution housing away from the firewall. Remove the screws holding the heater core cover to the housing **(see illustration)**.

19   Pull the heater core from the housing, being careful not to tear the foam insulation **(see illustration)**.

20   Installation is the reverse of the removal procedure. Make sure the foam insulation is in place around the heater core pipes.

21   Take the vehicle back to a dealer or air-conditioning repair facility and have the refrigerant system evacuated, recharged and leak tested.

15   Air conditioning and heating system - check and maintenance

Warning: *The air conditioning system is under high pressure. Do not loosen any hose fittings or remove any components until after the system has been discharged. Air conditioning refrigerant should be properly dis-*

charged into an EPA-approved recovery/recycling unit at a dealer service department or an automotive air conditioning repair facility. Always wear eye protection when disconnecting air conditioning system fittings.

Note: *1993 and earlier air conditioning systems use R-12 refrigerant. For 1994 and later models, the air conditioning system was changed to use the new, "environmentally friendly" R-134a refrigerant. Each system uses similar components and locations but components are NOT interchangeable. All discharging of refrigerant, for part replacement or maintenance, should be done by an approved air conditioning facility with the proper refrigerant recovery equipment.*

1   The following maintenance checks should be performed on a regular basis to ensure that the air conditioner continues to operate at peak efficiency.

a) Check the compressor drivebelt. If it's worn or deteriorated, replace it (see Chapter 1).

b) Check the system hoses. Look for cracks, bubbles, hard spots and deterioration. Inspect the hoses and all fittings for oil bubbles and seepage. If there's any evidence of wear, damage or leaks, replace the hose(s).

c) Inspect the condenser fins for leaves, bugs and other debris. Use a "fin comb" or compressed air to clean the condenser.

d) Check the wire harness for correct routing, broken wires, damaged insulation, etc. Make sure the harness connections are clean and tight.

e) Maintain the correct refrigerant charge.

2   It's a good idea to operate the system for about 10 minutes at least once a month, particularly during the winter. Long term non-use can cause hardening, and subsequent failure, of the seals.

3   Because of the complexity of the air conditioning system and the special equipment necessary to service it, in-depth troubleshooting and repairs are not included in this manual. However, simple checks and

14.18 Heater core housing cover screws (arrows) - 1995 and later models

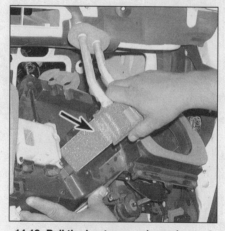

14.19 Pull the heater core (arrow) out of the housing

15.7 Place an accurate thermometer in the center dash vent, turn the air conditioning on and check the output temperature

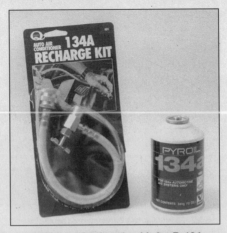

15.10 A basic charging kit for R-134a systems is available at most auto parts stores - it must say R-134a (not R-12) to be compatible with late-model air-conditioning systems

15.13 Add R-134a only to the low-side port (arrow) - the procedure is easier if you wrap the can with a warm, wet towel to prevent icing

component replacement procedures are provided in this Chapter. For more complete information on the air conditioning system, refer to the *Haynes Automotive Heating and Air Conditioning Manual*.

3   Because of the complexity of the air conditioning system and the special equipment required to effectively work on it, accurate troubleshooting of the system should be left to a professional technician. One probable cause for poor cooling that can be determined by the home mechanic is low refrigerant charge. Should the system lose its cooling ability, the following procedure will help you pinpoint the cause.

Check

Refer to illustration 15.7

4   Warm the engine up to normal operating temperature.

5   Place the air conditioning temperature selector at the coldest setting and put the blower at the highest setting. Open the doors (to make sure the air conditioning system doesn't cycle off as soon as it cools the passenger compartment).

6   After the system reaches operating temperature, feel the two pipes connected to the evaporator at the firewall.

7   The pipe (thinner tubing) leading from the condenser outlet to the evaporator should be cold, and the evaporator outlet line (the thicker tubing that leads back to the compressor) should be slightly colder (3 to 10 degrees F). If the evaporator outlet is considerably warmer than the inlet, the system needs a charge. Insert a thermometer in the center air distribution duct **(see illustration)** while operating the air conditioning system - the temperature of the output air should be 35 to 40 degrees F below the ambient air temperature (down to approximately 40 degrees F). If the ambient (outside) air temperature is very high, say 110 degrees F, the duct air temperature may be as high as 60 degrees F, but generally the air conditioning is 35 to 40 degrees F cooler than the ambient air.

8   If the air isn't as cold as it used to be, the system probably needs a charge. Further inspection or testing of the system is beyond the scope of the home mechanic and should be left to a professional.

Adding refrigerant

1993 and earlier models

9   Because of environmental regulations, R-12 refrigerant is not available for home-mechanic use. Have the system discharged, evacuated, charged and leak tested by a qualified shop. Use only refrigerant oil compatible with your system. Refrigerant oils for use with refrigerant R-12 are not compatible with other oils. **Note:** *If an earlier model has been retrofitted for R-134a refrigerant, it should be clearly marked on the receiver-drier or a decal on the engine fan shroud.*

1994 and later models

Refer to illustrations 15.10 and 15.13

10   Buy an automotive charging kit at an auto parts store. A charging kit includes a 12-ounce can of R-134a refrigerant, a tap valve and a short section of hose that can be attached between the tap valve and the system low side service valve **(see illustration)**. Because one can of refrigerant may not be sufficient to bring the system charge up to the proper level, it's a good idea to buy a couple of additional cans.

11   Connect the charging kit by following the manufacturer's instructions.

12   Back off the valve handle on the charging kit and screw the kit onto the refrigerant can, making sure first that the O-ring or rubber seal inside the threaded portion of the kit is in place. **Warning:** *Wear protective eye wear when dealing with pressurized refrigerant cans.*

13   Remove the dust cap from the low-side charging port and attach the quick-connect fitting on the kit hose **(see illustration)** . **Warning:** *DO NOT hook the charging kit hose to the system high side! The fittings on the*

charging kit are designed to fit **only** on the low side of the system.

14   Warm the engine to normal operating temperature and turn on the air conditioning. Keep the charging kit hose away from the fan and other moving parts.

15   Turn the valve handle on the kit until the stem pierces the can, then back the handle out to release the refrigerant. You should be able to hear the rush of gas. Add refrigerant to the low side of the system until both the outlet and the evaporator inlet pipe feel about the same temperature . Allow stabilization time between each addition. **Warning:** *Never add more than two cans of refrigerant to the system.* The can may tend to frost up, slowing the procedure. Wet a shop towel with hot water and wrap it around the bottom of the can to keep it from frosting.

16   Put your thermometer back in the center register and check that the output air is getting colder.

17   When the can is empty, turn the valve handle to the closed position and release the connection from the low-side port. Replace the dust cap.

18   Remove the charging kit from the can and store the kit for future use with the piercing valve in the UP position, to prevent inadvertently piercing the can on the next use.

Heating systems

Refer to illustration 15.23

19   If the air coming out of the heater vents isn't hot, the problem could stem from any of the following causes:

a)   *The thermostat is stuck open, preventing the engine coolant from warming up enough to carry heat to the heater core. Replace the thermostat (see Section 3).*

b)   *A heater hose is blocked, preventing the flow of coolant through the heater core. Feel both heater hoses at the firewall. They should be hot. If one of them is cold, there is an obstruction in one of the hoses or in the heater core, or the heater control valve is shut. Detach the*

3

15.23 Make sure the evaporator drain tube (arrow) on the firewall, below the bottom of the evaporator housing, is clear

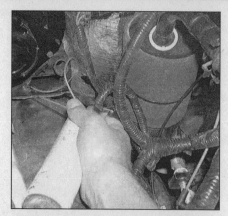

15.27 Remove the blower motor resistor and spray the disinfectant over the evaporator core

16.3 Unhook the lock-tab and pull the connector off (left arrow); then pry the retainer (right arrow) out of the compressor with a trim panel removal tool or equivalent

hoses and back flush the heater core with a water hose. If the heater core is clear but circulation is impeded, remove the two hoses and flush them out with a water hose.

c) If flushing fails to remove the blockage from the heater core, the core must be replaced. (see Section 14).

20   If the blower motor speed does not correspond to the setting selected on the blower switch, the problem could be a bad fuse, circuit, control panel or blower resistor (see Section 1).

21   If there isn't any air coming out of the vents:

a) Turn the ignition ON and activate the fan control. Place your ear at the heating/air conditioning register (vent) and listen. Most motors are audible. Can you hear the motor running?

b) If you can't (and have already verified that the blower switch and the blower motor resistor are good), the blower motor itself is probably bad (see Section 13).

22   If the carpet under the heater core is damp, or if antifreeze vapor or steam is coming through the vents, the heater core is leaking. Remove it (see Section 14) and install a new unit (most radiator shops will not repair a leaking heater core).

23   Inspect the drain hose from the heater/evaporator assembly at the bottom of the evaporator/blower motor housing on the right side of the firewall, make sure it is not clogged **(see illustration)**. If there is a humid mist coming from the system ducts, this hose may be plugged with leaves or road debris.

Eliminating air-conditioning odors

Refer to illustration 15.27

24   Unpleasant odors that often develop in air-conditioning systems are caused by the growth of a fungus, usually on the surface of the evaporator core. The warm, humid environment there is a perfect breeding ground for mildew to develop.

25   The evaporator core on most vehicles is

difficult to access, and factory dealerships have a lengthy, expensive process for eliminating the fungus by opening up the evaporator case and using a powerful disinfectant and rinse on the core until the fungus is gone. You can service your own system at home, but it takes something much stronger than basic household germ-killers or deodorizers.

26   Aerosol disinfectants for automotive air-conditioning systems are available in most auto parts stores, but remember when shopping for them that the most effective treatments are also the most expensive. The basic procedure for using these sprays is to start by running the system in the RECIRC mode for ten minutes with the blower on its highest speed. Use the highest heat mode to dry out the system and keep the compressor from engaging by disconnecting the wiring connector at the compressor (see Section 16).

27   The disinfectant can usually comes with a long spray hose. Remove the blower motor (see Section 12), point the nozzle inside the hole and towards the evaporator core, and spray according to the manufacturer's recommendations **(see illustration)**. Try to cover the whole surface of the evaporator core, by aiming the spray up, down and sideways. Follow the manufacturer's recommendations for the length of spray and waiting time between applications.

28   Once the evaporator has been cleaned, the best way to prevent the mildew from coming back again is to make sure your evaporator housing drain tube is clear **(see illustration 15.23)**.

16   Air conditioning compressor - removal and installation

Refer to illustrations 16.3, 16.6 and 16.8

Warning: *The air conditioning system is under high pressure. Do not loosen any hose fittings or remove any components until after the system has been discharged. Air conditioning refrigerant should be properly discharged into an EPA-approved recovery/recycling unit at a dealer service depart-*

ment or an automotive air conditioning repair facility. Always wear eye protection when disconnecting air conditioning system fittings. **Note:** *The accumulator/drier (see Section 18) and the orifice tube (see Section 20) should be replaced whenever the compressor is replaced.*

1   Have the air conditioning system discharged and recovered (see **Warning** above). **Note:** *If you know the compressor is to be replaced, it's a good idea to have the professional who discharges your system perform a flushing procedure also, to remove any contaminants, dirt or metal that may corrupt the new compressor.*

2   Disconnect the negative cable from the battery.

3   Disconnect the electrical connector from the compressor and detach the harness retainer from the back of the compressor **(see illustration)**.

4   Remove the drivebelt (see Chapter 1).

5   On 2001 4.0L SOHC V6 engine models, remove the bolts and detach the power steering fluid reservoir from the bracket. Remove the bolts and detach the reservoir bracket.

6   Remove the bolt that secures the refrig-

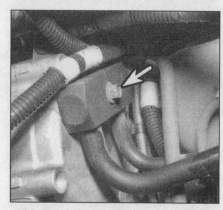

16.6 The refrigerant lines are secured to the rear of the compressor by one bolt (arrow)

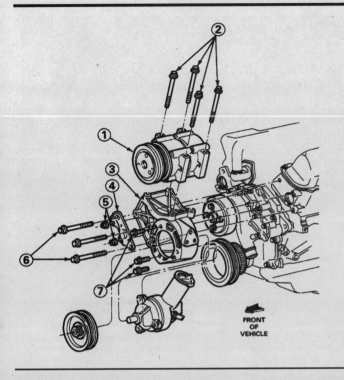

16.8 Air conditioning compressor and related components (4.0L pushrod V6 shown)

1. Compressor
2. Mounting bolts
3. Air conditioning compressor and power steering pump bracket
4. Power steering pump support brace
5. Support brace-to-timing chain cover nuts
6. Power steering pump bolts
7. Power steering pump bolts

17 Air conditioning condenser - removal and installation

Refer to illustration 17.6

Warning: *The air conditioning system is under high pressure. Do not loosen any hose fittings or remove any components until after the system has been discharged. Air conditioning refrigerant should be properly discharged into an EPA-approved recovery/recycling unit at a dealer service department or an automotive air conditioning repair facility. Always wear eye protection when disconnecting air conditioning system fittings.*

Note: *The accumulator/drier (see Section 18) and the orifice tube (see Section 20) should be replaced whenever the condenser is replaced.*

1   Have the air conditioning system discharged and recovered (see **Warning** above).
2   Remove the battery (see Chapter 5).
3   Drain the cooling system (see Chapter 1).
4   Remove the radiator (see Section 4). On 1998 and later models with a 5.0L V8 engine, release the plastic clips at the bottom of the radiator holding the condenser.
5   Disconnect the refrigerant lines from the condenser. On 1996 and earlier models, this requires a spring-lock coupling tool similar to that used for fuel injection lines (see Chapter 4). You remove the metal clips from the connection, then use the tool to release the coupling. On later models the condenser lines have bolted flanges.
6   Working from below, remove the mounting nuts from the condenser studs (see illustration). On 1996 and 1997 models, the condenser bolts (no studs) to the radiator support, and on 1998 and later models, the condenser has plastic squeeze-tabs that fit into plastic receptors on the body.
7   Lift the condenser out of the vehicle and plug the lines to keep dirt and moisture out.

erant lines to the rear of the compressor (see illustration). Plug the open fittings to prevent entry of dirt and moisture.
7   On some models, the power steering lines are bolted to a bracket on the side of the compressor. Unbolt the bracket.
8   Unbolt the compressor from the mounting brackets and lift it out of the vehicle (see illustration).
9   If a new compressor is being installed, follow the directions with the compressor regarding the draining of excess oil prior to installation. Some manufacturers ship new or remanufactured compressors with oil and some ship them without. Drain the old compressor oil into a graduated container. If the amount drained was less than 3 ounces,

add 3 ounces of new oil to the new compressor. If the amount drained was 3 to 5 ounces, add that amount plus 1 extra ounce, and if the drained amount was over 5 ounces, add only that amount of new oil into the new compressor.
10   The clutch may have to be transferred from the original to the new compressor.
11   Installation is the reverse of removal with the following additions:

a) *Replace all O-rings with new ones specifically made for air conditioning system used (R-12 or R-134a) and lubricate them with the proper refrigerant oil.*
b) *Have the system evacuated, recharged and leak tested by the shop that discharged it.*

3

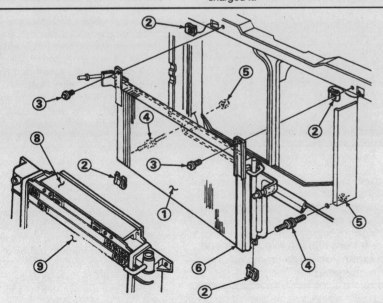

17.6 Typical air conditioning condenser and related components

1. Condenser
2. U-nut
3. Bolt
4. Stud and washer
5. Nut and washer
6. Seal (some automatic transmission models use two seals)
7. Condenser bottom seal (automatic transmission models only)
8. Condenser top seal
9. Radiator

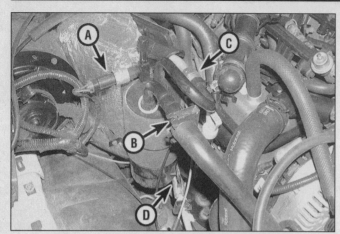

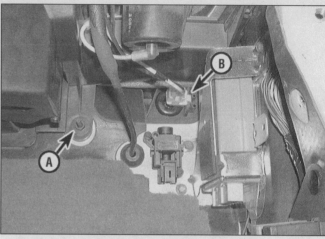

18.4 Disconnect the electrical connector at the switch (A), use a spring-lock coupling tool on the line (B) to the refrigerant manifold, use backup wrenches to disconnect the line to the evaporator (C), and remove the bracket bolt (D)

19.5 Remove this nut (A) and disconnect the vacuum connector (B) inside the vehicle

8   If the original condenser will be reinstalled, store it with the line fittings on top to prevent oil from draining out.
9   If a new condenser is being installed, pour 1 ounce of refrigerant oil into it prior to installation.
10   Reinstall the components in the reverse order of removal. Be sure the rubber pads are in place under the condenser. Be sure the seals are in place around the condenser.
11   Have the system evacuated, recharged and leak tested by the shop that discharged it.

18   Air conditioning accumulator/drier - removal and installation

Warning: *The air conditioning system is under high pressure. Do not loosen any hose fittings or remove any components until after the system has been discharged. Air conditioning refrigerant should be properly discharged into an EPA-approved recovery/recycling unit at a dealer service department or an automotive air conditioning repair facility. Always wear eye protection when disconnecting air conditioning system fittings.*

Removal
Refer to illustration 18.4
1   Have the air conditioning system discharged and recovered (see **Warning** above).
2   Disconnect the negative cable from the battery.
3   Unplug the electrical connector from the pressure switch near the top of the accumulator. Unscrew the pressure switch.
4   Disconnect the refrigerant line from the accumulator **(see illustration)**. This requires a spring lock coupling tool of the type used for fuel injection system lines (see Chapter 4).
5   Use a backup wrench to hold the fitting that connects the accumulator to the evaporator core, then disconnect the fitting **(see**

illustration 18.4). Plug the open fittings to prevent entry of dirt and moisture.
6   Remove the mounting bracket screw and the screw that holds the evaporator tube to the accumulator bracket, then lift the accumulator out.
7   If a new accumulator is being installed, remove the Schrader valve and pour the oil out into a measuring cup, noting the amount. Add fresh refrigerant oil to the new accumulator equal to the amount removed from the old unit, plus 2 ounces.

Installation
8   Loosely position the bracket on the new accumulator.
9   Connect the accumulator to the evaporator core, using a new O-ring lubricated with clean refrigerant oil. At the same time, align the bracket with the slot between the evaporator case flanges.
10   Using a backup wrench, tighten the fitting securely.
11   Install the screw that holds the bracket.
12   Tighten the accumulator bracket and install the clip that holds the evaporator inlet tube to the bracket.
13   Place a new O-ring, lubricated with clean refrigerant oil, on the pressure switch nipple on the accumulator.
14   Install the pressure switch. If it has a metal base, tighten it to the torque listed in this Chapter's Specifications. If it has a plastic base, tighten it by hand only. Connect the pressure switch electrical connector.
15   Reconnect the negative cable to the battery.
16   Take the vehicle to the shop that discharged the air conditioning system. Have the system recharged and tested for leaks.

19   Air conditioning evaporator and expansion valve - removal and installation

Warning: *The air conditioning system is*

under high pressure. Do not loosen any hose fittings or remove any components until after the system has been discharged. Air conditioning refrigerant should be properly discharged into an EPA-approved recovery/recycling unit at a dealer service department or an automotive air conditioning repair facility. Always wear eye protection when disconnecting air conditioning system fittings.

Removal
Refer to illustrations 19.5, 19.6 and 19.7
1   The evaporator core is located in a housing on the engine compartment side of the firewall. **Note:** *Before replacing an evaporator core, determine for certain that the core is leaking by having a leak test performed with special equipment at dealer service department or automotive air conditioning repair facility.*
2   Read all of the Steps before beginning this procedure. If you are going to replace the evaporator, have the refrigerant discharged and recovered at a dealer or air-conditioning service shop. **Note:** *Whenever the evaporator core is replaced with a new one, the accumulator/drier will also have to be replaced (see Section 18).*
3   Disconnect the refrigerant lines at the firewall **(see illustration 18.4)**.
4   Refer to Sections 6 and 13 and remove the coolant recovery tank and disconnect the wiring to the blower motor.
5   Inside the vehicle, remove the one nut and disconnect the vacuum connector from the evaporator housing **(see illustration)**.
6   On the engine-compartment side of the firewall, remove the three nuts from the housing studs **(see illustration, and illustration 14.16)**. Pull the blower motor/evaporator housing assembly out of the vehicle.
7   Out of the vehicle, remove the 11 screws holding the two halves of the case together **(see illustration)**. On 5.0L V8 models, remove the heat shield first.
8   Remove the evaporator core from the case.

19.6 Remove these three nuts (arrows, plus a lower nut visible in illustration 14.16) and pull the blower motor/evaporator housing out of the vehicle

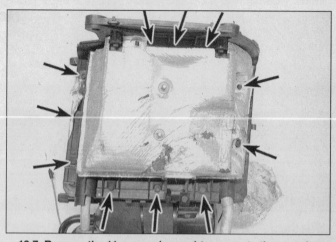

19.7 Remove the 11 screws (arrows) to separate the case for evaporator core removal

Installation

9 Installation is the reverse of the removal process. Add 3 ounces of new refrigerant oil to the (new) accumulator/drier inlet tube when a new evaporator core is installed. Also, before the lines are reconnected, it 's a good idea to replace the evaporator core orifice (see Section 20).

10 Have the system evacuated, recharged and leak tested by the dealer service department or an air conditioning repair facility.

20 Air conditioning expansion (orifice) tube - removal and installation

Warning: *The air conditioning system is under high pressure. Do not loosen any hose fittings or remove any components until after the system has been discharged. Air conditioning refrigerant should be properly discharged into an EPA-approved recovery/recycling unit at a dealer service depart-*

ment or an automotive air conditioning repair facility. Always wear eye protection when disconnecting air conditioning system fittings.
Note: *Whenever the expansion tube is replaced, the accumulator-drier should also be replaced (see Section 15).*

Removal

Refer to illustrations 20.2 and 20.3
1 Have the air conditioning system discharged and the refrigerant recovered (see **Warning** above). Disconnect the cable from the negative terminal of the battery.
2 Disconnect the condenser-to-evaporator refrigerant line at the firewall **(see illustration)**.
3 The expansion tube is a tube with a fixed-diameter orifice and a mesh filter at each end **(see illustration)**. When you separate the pipe at the fitting you will see one end of the expansion tube inside the pipe leading to the evaporator. Use a special removal tool (available at auto parts stores) to remove the expansion tube. **Caution:** *Pull the core straight out, do not twist it.*

4 The expansion tube acts to meter the refrigerant, changing it from high-pressure liquid to low-pressure gas. It is possible to reuse the expansion tube if:

a) *The screens aren't plugged with grit or foreign material*
b) *Neither screen is torn*
c) *The plastic housing over the screens is intact*
d) *The brass orifice inside the plastic housing is unrestricted*

Installation

5 Installation is the reverse of removal. Be sure to insert the expansion tube (using the special tool) with the shorter end in first, toward the evaporator, and lubricate the refrigerant line and the expansion tube with clean refrigerant oil to aid assembly. **Caution:** *Always use a new O-ring when installing the expansion tube.*
6 Retighten the fitting and refrigerant line, then have the system evacuated, recharged and leak-tested by the shop that discharged it.

3

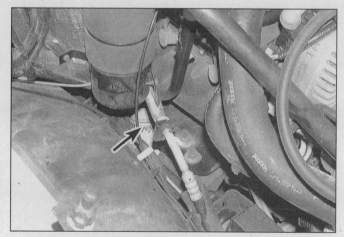

20.2 Disconnect the condenser-to-evaporator line (arrow) at the firewall

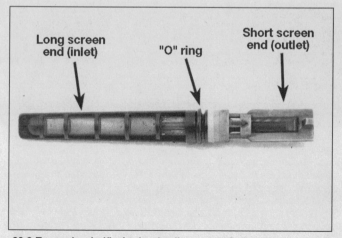

20.3 Expansion (orifice) tube details - a new O-ring must be used when installing it

Long screen end (inlet)   "O" ring   Short screen end (outlet)

Notes

Chapter 4
Fuel and exhaust systems

Contents

4

Specifications

Fuel pressure

Fuel system pressure (at idle)	
Vacuum hose attached	30 to 45 psi
Vacuum hose detached	40 to 50 psi
Fuel system hold pressure (after 5 minutes)	30 to 40 psi
Fuel pump pressure (maximum)	65 psi

Injector resistance
13.5 to 19 ohms

Torque specifications
Ft-lbs (unless otherwise indicated)

Upper intake manifold mounting bolts	
4.0L pushrod engine	15 to 18
4.0L SOHC engine	53 to 62 in-lbs
5.0L engine	12 to 18
Throttle body mounting nuts	
1996 and earlier (4.0L pushrod engine)	70 to 106 in-lbs
1996 (5.0L engine)	15 to 22
1997 and later (4.0L pushrod engine)	72 to 108 in-lbs
1997 and later (4.0L SOHC engine)	53 to 62
1997 and later (5.0L engine)	12 to 18
Fuel pressure regulator mounting bolts	
4.0L pushrod engine	70 to 97 in-lbs
4.0L SOHC engine	76 to 103 in-lbs
5.0L engine	27 to 44 in-lbs
Fuel rail mounting bolts/studs	
1996 and earlier (4.0L pushrod engine)	89 to 124 in-lbs
1997 and later (4.0L pushrod engine)	105 to 141 in-lbs
1997 and later (4.0L SOHC engine)	14 to 19
1996 and later (5.0L engine)	71 to 106 in-lbs
Accelerator cable bracket bolts	15 to 22
Accelerator cable bracket screws (4.0l SOHC engine)	19 to 25 in-lbs

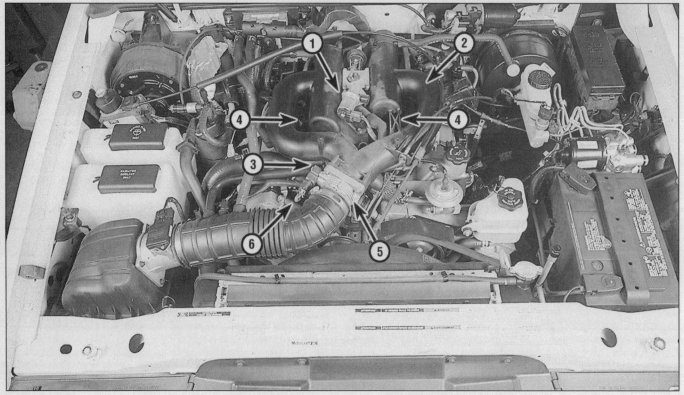

1.1a Fuel injection components - 4.0L SOHC engine

1	IAC valve	4	Fuel rails and fuel injectors (not visible)
2	Air intake plenum	5	Throttle body
3	Fuel pressure regulator (not visible)	6	TPS sensor

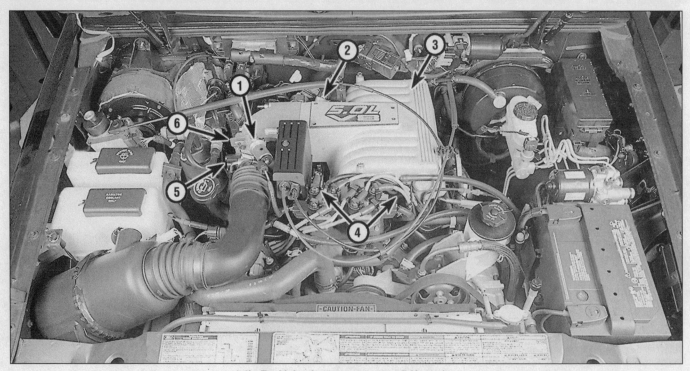

1.1b Fuel injection components - 5.0L engine

1	IAC valve	4	Fuel rails and fuel injectors (not visible)
2	Fuel pressure regulator	5	TPS sensor
3	Air intake plenum	6	Throttle body

1   General information

Refer to illustrations 1.1a and 1.1b

The fuel system consists of a fuel tank, an electric fuel pump (located in the fuel tank), a fuel pump relay, the fuel rail and fuel injectors, an air cleaner assembly and a throttle body unit. All models are equipped with a Sequential Electronic Fuel Injection (SEFI) system **(see illustrations)**.

Sequential Electronic Fuel Injection (SEFI) system

Sequential Electronic Fuel Injection uses timed impulses to inject the fuel directly into the intake port of each cylinder according to its firing order. The injectors are controlled by the Powertrain Control Module (PCM). The PCM monitors various engine parameters and delivers the exact amount of fuel required into the intake ports. The throttle body serves only to control the amount of air passing into the system. Because each cylinder is equipped with its own injector, much

better control of the fuel/air mixture ratio is possible.

Fuel pump and lines

Fuel is circulated from the fuel tank to the fuel injection system, and back to the fuel tank, through a pair of metal lines running along the underside of the vehicle. An electric fuel pump and fuel level sending unit is located inside the fuel tank. A vapor return system routes all vapors back to the fuel tank through a separate return line.

The fuel pump relay is equipped with a primary and secondary voltage circuit. The primary circuit is controlled by the PCM and the secondary circuit is linked directly to battery voltage from the ignition switch. With the ignition switch ON (engine not running), the PCM will ground the relay for one second. During cranking, the PCM grounds the fuel pump relay as long as the camshaft position sensor (CMP) sends its position signal (see Chapter 6). If there are no reference pulses, the fuel pump will shut off after two or three seconds.

The inertia switch will disable the fuel pump circuit in the event of collision. The inertia switch is a cylindrical magnet with a steel ball that will release (breakaway) and trip a shutdown lever when the vehicle inertia reaches a certain peak value.

Exhaust system

The exhaust system includes a pair of exhaust manifolds, a diverter (Y) pipe(s), an oxygen sensor, dual catalytic converters, a muffler and a tail pipe. Later model vehicles are fitted with two upstream (before catalytic converter) and two downstream (after catalytic converter) oxygen sensors,

The catalytic converters are an emission control device added to the exhaust system to reduce pollutants. A single-bed converter is used in combination with a three-way (reduction) catalyst. Refer to Chapter 6 for more information regarding the catalytic converters or oxygen sensors.

2   Fuel pressure relief procedure

Warning: *Gasoline is extremely flammable, so take extra precautions when you work on any part of the fuel system. Don't smoke or allow open flames or bare light bulbs near the work area, and don't work in a garage where a natural gas-type appliance (such as a water heater or a clothes dryer) with a pilot light is present. Since gasoline is carcinogenic, wear latex gloves when there's a possibility of being exposed to fuel, and, if you spill any fuel on your skin, rinse it off immediately with soap and water. Mop up any spills immediately and do not store fuel-soaked rags where they could ignite. The fuel system is under constant pressure, so, if any fuel lines are to be disconnected, the fuel pressure in the system must be relieved first. When you perform any kind of work on the fuel system, wear safety glasses and have a Class B type fire extinguisher on hand.*

Note: *After the fuel pressure has been relieved, it's a good idea to lay a shop towel over any fuel connection to be disassembled, to absorb the residual fuel that may leak out when servicing the fuel system.*

1   There are two methods for relieving the fuel system pressure; the easiest and most accessible is using a special fuel pressure gauge with a bleed-off valve. This special tool can be purchased at an automotive parts or tool company. In the event the tool is not available, locate the inertia switch and disable the fuel pump.

Fuel pressure gauge bleeding method

Refer to illustrations 2.2a, 2.2b, 2.2c and 2.3

2   Locate the fuel pressure test port on the fuel rail and connect the fuel pressure gauge to the Schrader valve **(see illustrations)**.

3   Direct the bleed-off hose into a metal cup or suitable container for gasoline storage

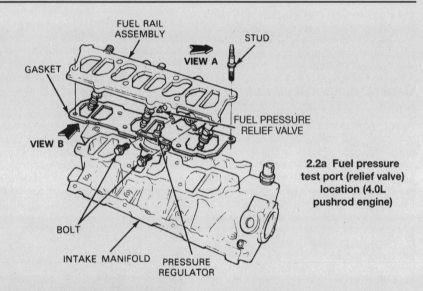

2.2a Fuel pressure test port (relief valve) location (4.0L pushrod engine)

2.2b Fuel pressure test port location (4.0L SOHC engine)

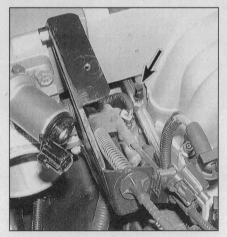

2.2c Fuel pressure test port location (5.0L engine)

4

2.3 Connect a fuel pressure gauge (equipped with a drain hose) to the test port connector and bleed the fuel into a suitable container

2.7 The inertia shut off switch is mounted to the floor pan on the passenger side of the vehicle below the heater unit - disconnect the electrical connector (A) to disable the fuel pump - To energize the fuel pump, simply plug the connector back into the switch and push the reset button (B) if necessary

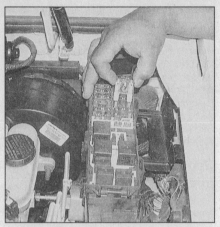

3.2 Remove the fuel pump fuse from the engine compartment power distribution box - make sure it is not blown

(see illustration).

4   Turn the valve and allow the excess fuel to bleed into the container.

5   Close the valve, remove the fuel pressure gauge and cap the test port. Disconnect the cable from the negative terminal of the battery before performing any work on the fuel system.

Inertia switch method

Refer to illustration 2.7

6   The fuel pump switch - more commonly referred to as the "inertia switch" - which shuts off fuel to the engine in the event of a collision, affords a simple and convenient means by which fuel pressure can be relieved before servicing fuel injection components.

7   Disconnect the electrical connector from the inertia shut off switch (see **illustration**).

8   Start the engine and allow it to run until it stops. This should take only a few seconds.

9   The fuel system pressure is now relieved. When you're finished working on the fuel system, simply plug the electrical connector back into the switch. If the inertia switch was "popped" (activated) during this procedure, push the reset button on the top of the switch (see **illustration 2.7**).

---

3   Fuel pump/fuel pressure - check

Warning: *Gasoline is extremely flammable, so take extra precautions when you work on any part of the fuel system. See the* **Warning** *in Section 2.*

Note 1: *To perform the fuel pressure test, you will need to obtain a fuel pressure gauge and adapter set (fuel line fittings).*

Note 2: *The fuel pump will operate as long as the engine is cranking or running and the PCM is receiving ignition reference pulses from the electronic ignition system. If there are no reference pulses, the fuel pump will shut off after two or three seconds.*

Note 3: *After the fuel pressure has been relieved, it's a good idea to lay a shop towel over any fuel connection to be disassembled, to absorb the residual fuel that may leak out when servicing the fuel system.*

General electrical circuit check

Refer to illustrations 3.2 and 3.3

1   Should the fuel system fail to deliver the proper amount of fuel, or any fuel at all, inspect it as follows. Remove the fuel filler cap. Have an assistant turn the ignition key to the On position (engine not running) while you listen at the fuel filler opening. You should hear a whirring sound that lasts for a couple of seconds.

2   If you don't hear anything, check the fuel pump fuse (see Chapter 12). If the fuse is blown, replace it and see if it blows again (see **illustration**). If it does, trace the fuel pump circuit for a short. Refer to the wiring diagrams at the end of Chapter 12 for additional wiring schematics.

3   If the fuse is OK, remove the fuel pump relay and check for battery voltage to the fuel pump relay connector with the ignition key in the OFF position. Then turn the ignition key to the ON position and check for battery voltage from the PCM power relay (see **illustration**). **Note 1:** *The fuel pump relay is located in the engine compartment power distribution box on all models.* **Note 2:** *The inertia switch is an electrical device wired into the fuel pump circuit that will shut down power to the fuel pump in an accident. Be sure to check that the inertia switch has not "popped" and is in working order if the fuel pump is not receiving the proper voltage (see Section 2).* **Note 3:** *The fuel pump relay is equipped with a primary and secondary voltage circuit. The primary circuit is controlled by the PCM and the*

secondary circuit provides battery voltage to the fuel pump as the relay is energized. With the ignition switch ON (engine not running), the PCM will ground the relay for several seconds. During cranking, the PCM grounds the fuel pump relay as long as the camshaft position sensor (CMP) sends its position signal (see Chapter 6). If there are no reference pulses, the fuel pump will shut off after two or three seconds. Refer to the wiring diagrams at the end of Chapter 12 for additional information on the wiring color designations for the fuel pump relay.

4   If there is no voltage present at the relay connector, check the fuse(s) and the wiring circuit for the fuel pump relay and/or PCM power relay (see Chapter 12). If voltage is present at the relay connector, check the relay (see Chapter 12). If the relay is OK, check for battery voltage at one terminal of

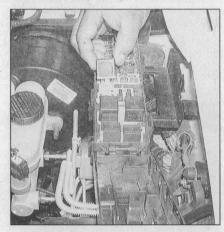

3.3 Remove the fuel pump relay and check for battery voltage at one terminal with the ignition key in the OFF position (this terminal is HOT at all times) - then, have an assistant cycle the ignition key ON/OFF while you check for voltage at one of the remaining four terminals; there should be voltage for about two to three seconds at one of the remaining four terminals

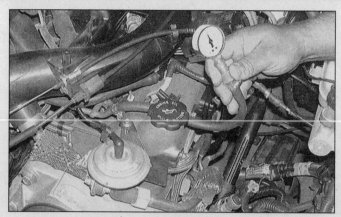

3.7 If you don't have the correct adapter, it is possible to remove the Schrader valve from the fitting and install a standard fuel pressure gauge, using a hose clamp

4.5 A hairpin clip type push-connect fitting - the clip is being pointed to with a screwdriver

the fuel pump harness connector located near the fuel tank. If there is no voltage reaching the fuel pump connector check for an open circuit between the fuel pump and the relay. If there is battery power at the fuel pump connector and the fuel pump still does not work, check for continuity to ground at two terminals of the fuel pump harness connector. If the ground is OK, replace the fuel pump.

Operating pressure check

Refer to illustration 3.7

5   Relieve the fuel system pressure (see Section 2).

6   Detach the cable from the negative terminal of the battery.

7   Remove the cap from the fuel pressure test port and attach a fuel pressure gauge **(see illustration)**. If you don't have the correct adapter for the test port, remove the Schrader valve and connect the gauge hose to the fitting, using a hose clamp.

8   Attach the cable to the negative terminal of the battery.

9   Start the engine.

10   Check the fuel pressure at idle. Compare your readings with the values listed in this Chapter's Specifications. Disconnect the vacuum hose from the fuel pressure regulator and watch the fuel pressure gauge - the fuel pressure should increase considerably as soon as the hose is disconnected. If it doesn't, check for a vacuum signal to the fuel pressure regulator (see Step 15).

11   If the fuel pressure is low, pinch the fuel return line shut and watch the gauge. If the pressure doesn't rise, the fuel pump is defective or there is a restriction in the fuel feed line. If the pressure rises sharply, replace the pressure regulator. **Note:** *If the vehicle is equipped with a nylon fuel return line (or fuel lines made up of steel or other rigid material), it will be necessary to install a special fuel testing harness between the fuel rail and the return line. This can be made up from compatible fuel line connectors (available at the dealer parts department and some auto parts stores), fuel hose and hose clamps.*

12   If the fuel pressure is too high, turn the engine off. Disconnect the fuel return line and blow through it to check for a blockage. If there is no blockage, replace the fuel pressure regulator.

13   Hook up a hand-held vacuum pump to the port on the fuel pressure regulator.

14   Read the fuel pressure gauge with vacuum applied to the fuel pressure regulator and also with no vacuum applied. The fuel pressure should decrease as vacuum increases (and increase as vacuum decreases).

15   Connect a vacuum gauge to the pressure regulator vacuum hose. Start the engine and check for vacuum. If there isn't vacuum present, check for a clogged hose or vacuum port. If the amount of vacuum is adequate, replace the fuel pressure regulator.

16   Turn the ignition switch to OFF, wait five minutes and recheck the pressure on the gauge. Compare the reading with the hold pressure listed in this Chapter's Specifications. If the hold pressure is less than specified:

a) The fuel lines may be leaking.

b) The fuel pressure regulator may be allowing the fuel pressure to bleed through to the return line

c) A fuel injector (or injectors) may be leaking.

d) The fuel pump may be defective.

---

4   Fuel lines and fittings - general information

Warning: *Gasoline is extremely flammable, so take extra precautions when you work on any part of the fuel system. See the* **Warning** *in Section 2.*

Note: *The majority of fuel line fittings on these vehicles require special disconnect tools to allow removal of the fuel lines. These special disconnect tools are available at most auto part stores.*

Push-connect fittings - disassembly and reassembly

1   Ford uses two different push-connect fitting designs. Fittings used with 3/8 and

5/16-inch diameter lines have a "hairpin" type clip; fittings used with 1/4-inch diameter lines have a "duck bill" type clip. The procedure used for releasing each type of fitting is different. The clips should be replaced whenever a connector is disassembled.

2   Disconnect all push-connect fittings from fuel system components such as the fuel filter, the fuel charging assembly, the fuel tank, etc. before removing the assembly.

3/8 and 5/16-inch fittings (hairpin clip)

Refer to illustration 4.5

3   Inspect the internal portion of the fitting for accumulations of dirt. If more than a light coating of dust is present, clean the fitting before disassembly.

4   Some adhesion between the seals in the fitting and the line will occur over a period of time. Twist the fitting on the line, then push and pull the fitting until it moves freely.

5   Remove the hairpin clip from the fitting by bending the shipping tab down until it clears the body. Then, using nothing but your hands, spread each leg about 1/8-inch to disengage the body and push the legs through the fitting. Remember, don't use any tools to perform this part of the procedure. Finally, pull lightly on the triangular end of the clip and work it clear of the line and fitting **(see illustration)**.

6   Grasp the fitting and hose and pull it straight off the line.

7   Do not reuse the original clip in the fitting. A new clip must be used.

8   Before reinstalling the fitting on the line, wipe the line end with a clean cloth. Inspect the inside of the fitting to ensure that it's free of dirt and/or obstructions.

9   To reinstall the fitting on the line, align them and push the fitting into place. When the fitting is engaged, a definite click will be heard. Pull on the fitting to ensure that it's completely engaged. To install the new clip, insert it into any two adjacent openings in the fitting with the triangular portion of the clip pointing away from the fitting opening. Using your index finger, push the clip in until the legs are locked on the outside of the fitting.

4

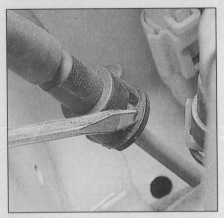

4.10 A push-connect fitting with a duck bill clip

4.13 Remove the safety clamp

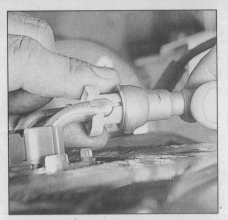

4.14 Disassembling a duck bill clip fitting using the special tool

1/4-inch fittings (duck bill clip)

Refer to illustrations 4.10, 4.13 and 4.14

10   The duck bill clip type fitting consists of a body, spacers, O-rings and the retaining clip **(see illustration)**. The clip holds the fitting securely in place on the line. One of the two following methods must be used to disconnect this type of fitting.

11   Before attempting to disconnect the fitting, check the visible internal portion of the fitting for accumulations of dirt. If more than a light coating of dust is evident, clean the fitting before disassembly.

12   Some adhesion between the seals in the fitting and line will occur over a period of time. Twist the fitting on the line, then push and pull the fitting until it moves freely.

13   Remove the safety clamp from the fuel line.**(see illustration)**.

14   The preferred method used to disconnect the fitting requires a special tool available at most auto part stores. To disengage the line from the fitting, align the slot in the push-connect disassembly tool with either tab on the clip (90-degrees from the slots on the side of the fitting) and insert the tool **(see illustration)**. This disengages the duck bill from the line. **Note:** *Some fuel lines have a secondary bead which aligns with the outer surface of the clip. The bead can make tool insertion difficult. If necessary, use the alternative disassembly method described in Step 16. Holding the tool and the line with one hand, pull the fitting off.* **Note:** *Only moderate effort is necessary if the clip is properly disengaged. The use of anything other than your hands should not be required.*

15   After disassembly, inspect and clean the line sealing surface. Also inspect the inside of the fitting and the line for any internal parts that may have been dislodged from the fitting. Any loose internal parts should be immediately reinstalled (use the line to insert the parts).

16   The alternative disassembly procedure requires a pair of small adjustable pliers. The pliers must have a jaw width of 3/16-inch or less.

17   Align the jaws of the pliers with the openings in the side of the fitting and com-

press the portion of the retaining clip that engages the body. This disengages the retaining clip from the body (often one side of the clip will disengage before the other - both sides must be disengaged).

18   Pull the fitting off the line. **Note:** *Only moderate effort is required if the retaining clip has been properly disengaged. Do not use any tools for this procedure.*

19   Once the fitting is removed from the line end, check the fitting and line for any internal parts that may have been dislodged from the fitting. Any loose internal parts should be immediately reinstalled (use the line to insert the parts).

20   The retaining clip will remain on the line. Disengage the clip from the line bead to remove it. Do not reuse the retaining clip - install a new one!

21   Before reinstalling the fitting, wipe the line end with a clean cloth. Check the inside of the fitting to make sure that it's free of dirt and/or obstructions.

22   To reinstall the fitting, align it with the line and push it into place. When the fitting is engaged, a definite click will be heard. Pull on the fitting to ensure that it's fully engaged.

23   Install the new replacement clip by inserting one of the serrated edges on the duck bill portion into one of the openings.

4.26a If the spring lock couplings are equipped with safety clips, pry them off with a small screwdriver

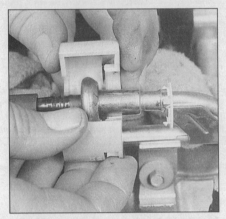

4.26b Open the spring-loaded halves of the spring lock coupling tool and place it in position around the coupling, then close it

Push on the other side until the clip snaps into place.

Spring lock couplings - disassembly and reassembly

Refer to illustrations 4.26a, 4.26b and 4.26c

24   The fuel supply and return lines used on SEFI engines utilize spring lock couplings at the engine fuel rail end instead of plastic push-connect fittings. The male end of the spring lock coupling, which is girded by two O-rings, is inserted into a female flared end engine fitting. The coupling is secured by a garter spring which prevents disengagement by gripping the flared end of the female fitting. A cup-tether assembly provides additional security.

25   To disconnect the 1/2-inch spring lock coupling supply fitting or the 3/8-inch return fitting, you will need to obtain the appropriate spring lock coupling tool from your local auto parts store. The tools come in two different sizes 3/8 and 1/2-inch, which correspond with the diameter of the fuel supply and return lines.

26   Study the accompanying illustrations carefully before detaching either spring lock coupling fitting **(see illustrations)**.

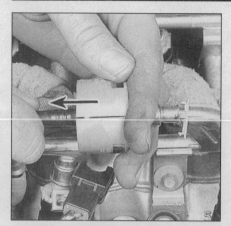

4.26c To disconnect the coupling, push the tool into the cage opening to expand the garter spring and release the female fitting, then pull the male and female fittings apart

5 Fuel tank - removal and installation

Refer to illustrations 5.5, 5.6, 5.8, 5.10a and 5.10b

Warning: *Gasoline is extremely flammable, so take extra precautions when you work on any part of the fuel system. See the* **Warning** *in Section 2.*

Note 1: *Don't begin this procedure until the gauge indicates that the tank is empty or nearly empty. If the tank must be removed when it's full (for example, if the fuel pump malfunctions), siphon any remaining fuel from the tank prior to removal.*

Note 2: *This procedure requires a special fuel line disconnect tool which is available at most auto part stores. Some models may be equipped with a fuel line disconnect tool which is installed on the vehicle about ten inches behind the fuel filter.*

1   Unless the vehicle has been driven far enough to completely empty the tank, it's a good idea to siphon the residual fuel out before removing the tank from the vehicle. **Warning:** *DO NOT start the siphoning action by mouth ! Use a siphoning kit (available at most auto parts stores).*

2   Relieve the fuel pressure (see Section 2).

3   Disconnect the cable from the negative terminal of the battery.

4   Raise the vehicle and support it securely on jackstands. **Caution:** *If the vehicle is equipped with Automatic Ride Control (ARC), make sure the air suspension switch is turned to the OFF position before the vehicle is raised to prevent damage to the system components (see Chapter 10).*

5   On 4WD models, remove the shield, skid plate and the front retaining strap **(see illustration)**.

6   Loosen the hose clamps and disconnect the fuel filler and vent hose at the fuel tank **(see illustration)**.

7   Remove the fuel tank ground strap.

8   On later models, disconnect the fuel

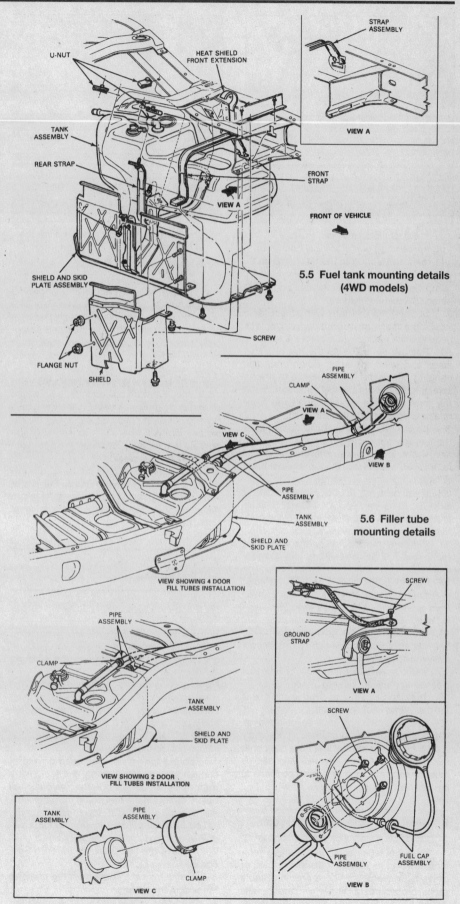

5.5 Fuel tank mounting details (4WD models)

5.6 Filler tube mounting details

4

5.8 On later models, the fuel pump/sending unit electrical connector (arrow) is attached to the left frame rail

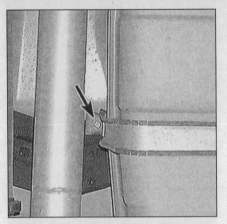

5.10a Remove the bolt (arrow) from the rear fuel tank strap and pivot it out of the way

5.10b Remove the retaining bolts (arrows) at the front of the tank, then slide the tank rearward enough to clear the retaining bracket

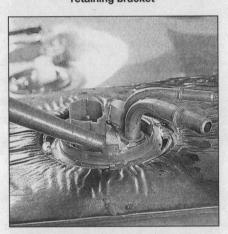

7.6a Use a brass punch to turn the locking ring (early models)

pump electrical connector located on the left frame rail **(see illustration)**.

9   Support the fuel tank with a floor jack. Position a piece of wood between the jack head and the fuel tank to protect the tank.

10   Remove the bolt from the rear retaining strap and pivot the strap down until it is hanging out of the way. Remove the bolts securing the front of the fuel tank, then move the tank rearward until it clears the front retaining bracket **(see illustrations)**.

11   Lower the tank enough to disconnect the fuel line connectors from the fuel pump and vapor valve. **Note:** *The fuel feed and return lines and the vapor return line are three different diameters, so reattachment is simplified. If you have any doubts, however, clearly label the three lines and the fittings. Be sure to plug the hoses to prevent leakage and contamination of the fuel system.*

12   On early models, disconnect the fuel pump and fuel gauge sending unit connectors.

13   Remove the tank from the vehicle. **Warning:** *Store the tank in a safe place, away from sparks and open flames (read the Warning at the beginning of this Section).*

14   Installation is the reverse of removal. The manufacturer recommends that new tank retaining strap bolts be used.

15   If you're replacing the tank, or having it cleaned or repaired, refer to Section 6.

16   Refer to Section 7 to remove and install the fuel pump or sending unit.

17   Installation is the reverse of removal. Clean engine oil can be used as an assembly aid when pushing the fuel filler hose back onto the fuel tank. **Warning:** *Fuel tank heat shields must be reinstalled in their original positions to prevent heat damage to the fuel tank.*

6   Fuel tank - cleaning and repair

1   The fuel tanks equipped in these vehicles are made of non corrosive steel or polyethylene plastic. There are no reliable repair procedures available to correct leaks or damage on polyethylene plastic fuel tanks.

Fuel tank replacement is the only approved service. Any repairs on steel fuel tanks or filler necks should be carried out by a professional who has experience in this critical and potentially dangerous work. **Warning:** *Even after cleaning and flushing of the fuel tank, explosive fumes can remain and ignite.*

2   If the fuel tank is removed from the vehicle, it should not be placed in an area where sparks or open flames could ignite the fumes coming out of the tank. Be especially careful inside garages where a natural gas-type appliance is located, because the pilot light could cause an explosion.

3   Whenever the fuel tank is steam-cleaned or otherwise serviced, the vapor valve assembly should be replaced. All grommets and seals must be replaced to prevent possible leakage.

7   Fuel pump - removal and installation

Refer to illustrations 7.6a, 7.6b, 7.6c, 7.7, 7.9, 7.10 and 7.11

Warning: *Gasoline is extremely flammable, so take extra precautions when you work on any part of the fuel system. See the* **Warning** *in Section 2.*

Note: *This procedure requires a special fuel line disconnect tool which is available at most auto part stores. Some models may be equipped with a fuel line disconnect tool which is installed on the vehicle about ten inches behind the fuel filter.*

1   Unless the vehicle has been driven far enough to completely empty the tank, it's a good idea to siphon out the residual fuel before removing the fuel pump from the vehicle. **Warning:** *DO NOT start the siphoning action by mouth! Use a siphoning kit (available at most auto parts stores).*

2   Relieve the fuel pressure (refer to Section 2).

3   Disconnect the cable from the negative terminal of the battery.

4   Raise the vehicle and support it securely

on jackstands. **Caution:** *If the vehicle is equipped with Automatic Ride Control (ARC), make sure the air suspension switch is turned to the OFF position before the vehicle is raised to prevent damage to the system components (see Chapter 10).*

5   Remove the fuel tank from the vehicle (see Section 5). Detach the fuel feed line and return lines from the fuel pump module (see Section 4).

6   On early models, using a brass punch or wood dowel only, tap the fuel pump module lock ring counterclockwise until it's loose **(see illustration)**. On later models, remove the fuel pump mounting bolts **(see illustrations)**.

7   Lift the fuel pump/sending unit assembly out of the tank **(see illustration)**. **Caution:** *The fuel level float and sending unit are delicate. Don't bump them into the lock ring during removal or the accuracy of the sending unit may be affected.*

8   Inspect the condition of the O-ring around the opening of the tank. If it is dried, cracked or deteriorated, replace it.

9   Remove the fuel pump mounting clamp bolt **(see illustration)**.

10   To separate the fuel pump from the

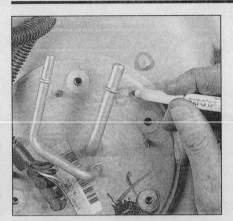

7.6b On later models, use paint or a marker to highlight the alignment marks on the fuel pump assembly

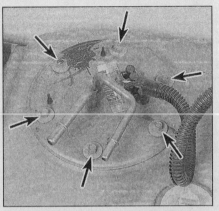

7.6c Remove the mounting bolts (arrows) from the fuel pump assembly

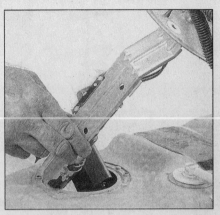

7.7 Carefully angle the fuel pump out of the fuel tank without damaging the fuel strainer

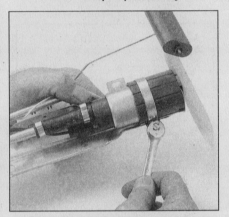

7.9 Loosen the fuel pump mounting clamp, then slide it down past the end of the mounting bracket

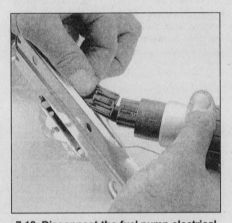

7.10 Disconnect the fuel pump electrical connector and the fuel line lower hose clamp from the fuel pump

a new O-ring, it may be necessary to push down on the lock ring until the locking cams slide under the retaining tangs. On later models, align the marks on the fuel pump with the marks on the fuel tank, then install the mounting bolts and tighten them to the torque listed in this Chapter's Specifications.

16   The remainder of the installation is the reverse of removal.

8   Fuel level sending unit - check and replacement

Check

Refer to illustrations 8.3 and 8.5

1   Raise the vehicle and support it securely on jackstands. **Warning:** *Some models covered by this manual are equipped with self leveling suspension systems. Always disconnect electrical power to the suspension system before lifting or towing the vehicle (see Chapter 10). Failure to perform this procedure may result in unexpected shifting or movement of the vehicle which could cause personal injury.*

2   Disconnect the fuel pump/sending unit electrical connector **(see illustration 5.8)**.

3   Position the ohmmeter probes into the electrical connector and check the resistance **(see illustration)**. Use the 200-ohm scale on the ohmmeter.

assembly, remove the fuel hose lower clamp and disconnect the electrical connector from the fuel pump **(see illustration)**.

11   Remove the strainer from the lower end of the fuel pump **(see illustration)**. If it's dirty, remove it, clean it with solvent and blow it out with compressed air. If it's too dirty to be cleaned, replace it. **Note:** *Most new fuel pump assemblies come equipped with a new strainer.*

12   Installation of the fuel pump to the sending unit assembly is the reverse of removal.

13   Clean the fuel pump mounting flange and the tank mounting surface and seal ring groove. Apply a thin coat of heavy grease to the new seal ring to hold it in place during assembly.

14   Position the O-ring around the opening in the fuel tank and guide the fuel pump/sending unit assembly into the tank.

15   On early models, turn the lock ring clockwise until the locking cams are fully engaged by the retaining tangs. **Note:** *If you've installed*

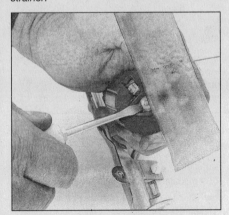

7.11 Remove the C-clip from the base of the fuel pump then separate the strainer from the fuel pump

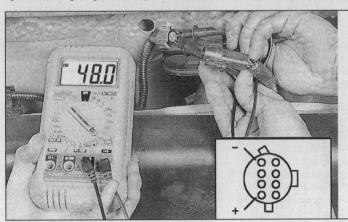

8.3 Using an ohmmeter, probe the terminals of the fuel sending unit connector to check the resistance

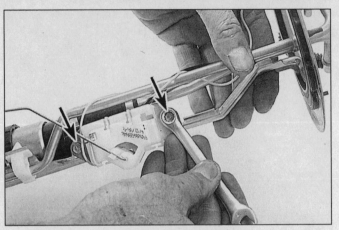

8.5 A more accurate check of the fuel level sending unit can be performed with the assembly on the bench. Connect the ohmmeter probes to the connector and check the resistance of the sending unit with the float positioned on "empty" and "full". Check for a smooth change in resistance as the float is moved between the positions

8.8 Remove the fuel level sending unit bolts (arrows) from the fuel pump assembly frame, then detach the electrical connectors

4   With the fuel tank completely full, the resistance should be about 160.0 ohms. With the fuel tank nearly empty, the resistance of the sending unit should be about 15.0 ohms.
5   If the readings are incorrect, replace the sending unit. **Note:** *A more accurate check of the sending unit can be made by removing it from the fuel tank and checking its resistance while manually operating the float arm* **(see illustration)**.

Replacement

Refer to illustration 8.8

6   Remove the fuel tank (see Section 5).
7   Remove the fuel pump module from the fuel tank (see Section 7).
8   Remove the sending unit mounting bolt(s) **(see illustration)** and electrical connectors from the fuel pump module.
9   Installation is the reverse of removal.

9   Air cleaner housing - removal and installation

1   Detach the cable from the negative terminal of the battery.

1994 and earlier models

Refer to illustration 9.2

2   Loosen the hose clamps on the air cleaner outlet tube assembly **(see illustration)**. Remove the tube from the throttle body and the air cleaner housing. Cover the throttle body inlet to prevent the entry of foreign matter.
3   Disconnect the vacuum lines at the bimetal sensor in the air cleaner cover.
4   Disconnect the electrical connector at the Mass Air Flow (MAF) sensor.
5   Remove the screws securing the air cleaner assembly upper half. Label and disconnect the two small hoses from the upper case half and remove the upper half and the air cleaner filter.

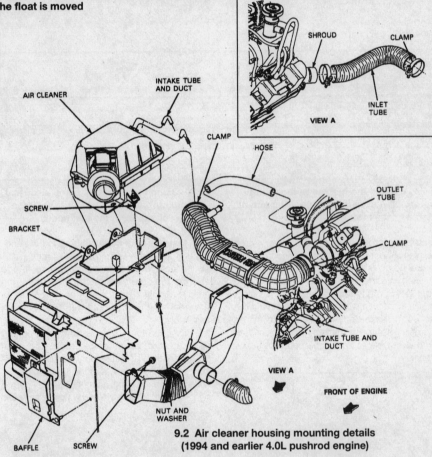

9.2 Air cleaner housing mounting details (1994 and earlier 4.0L pushrod engine)

6   To remove the remainder of the assembly, perform the following:

a) *Squeeze the protrusions on the air cleaner hot air tube and disconnect the tube from the air cleaner housing.*
b) *Remove the bolt securing the air cleaner intake tube assembly to the air cleaner intake deflector next to the radiator panel. Remove the intake tube from the lower half of the air cleaner housing.*
c) *Remove the screw securing the lower half of the air cleaner housing to the mounting bracket. Remove the lower half of the air cleaner housing.*

7   Installation is the reverse of the removal. Tighten all hose clamps and fittings securely.

1995 and later models

Refer to illustrations 9.8a and 9.8b

8   Disconnect the IAT sensor and the MAF

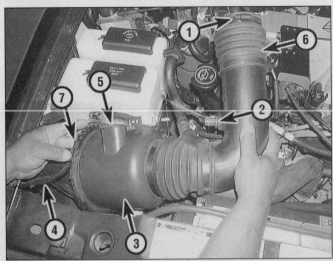

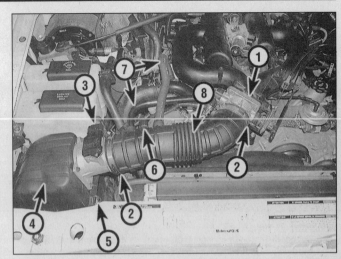

9.8a Air cleaner housing mounting details (1995 and later 4.0L pushrod and 5.0L engines)

1	Retaining clamp	5	Mass Airflow (MAF)
2	Intake Air Temperature		sensor
	(IAT) sensor	6	Air cleaner outlet tube
3	Air cleaner housing	7	Air cleaner housing spring
4	Intake air duct		clip

9.8b Air cleaner housing mounting details (1997 and later 4.0L SOHC engine)

1	Throttle body	6	Intake Air Temperature
2	Retaining clamps		(IAT) sensor
3	Mass Airflow (MAF) sensor	7	Crankcase ventilation
4	Air cleaner housing		tubes
5	Air cleaner housing spring	8	Air cleaner outlet tube
	clip		

10.1a On 4.0L SOHC engines, remove the screws (arrows) and the accelerator cable splash shield

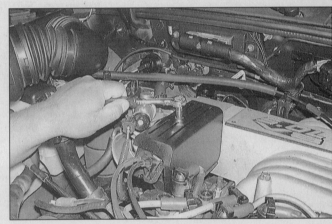

10.1b On 5.0L engines, remove the bolt and the accelerator cable splash shield

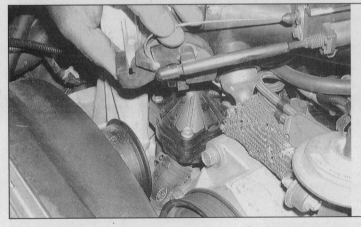

10.2a On 4.0L SOHC engines, rotate the accelerator cable until the slot in the throttle lever aligns with the cable and slide it out of the bracket

11 To remove the lower half (intake air duct) of the air cleaner housing, carefully lift upward on the lower half to release the rubber grommets securing the housing to the fenderwell.

12 Installation is the reverse of removal.

10 Accelerator cable - removal, installation and adjustment

Removal

Refer to illustrations 10.1a, 10.1b, 10.2a, 10.2b, 10.3 and 10.4

1 Remove the protective shield from the throttle body **(see illustrations)**.

2 Detach the accelerator cable from the throttle lever **(see illustrations)**.

sensor wiring **(see illustrations)**.

9 Remove the crankcase ventilation hose and the retaining clamps from the air cleaner outlet tube. Detach the air outlet tube from

the throttle body and the MAF sensor.

10 Disconnect the air cleaner housing spring clips, then lift the upper half of the air cleaner assembly from the vehicle.

4

10.2b On 4.0L pushrod and 5.0L engines it will be necessary to unclip the accelerator cable (A) from the throttle lever, then separate the cables from the cable bracket (B)

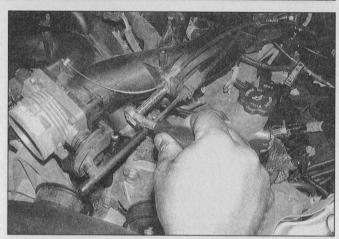

10.3 On 4.0L SOHC engines, remove the cable retaining bolt and separate the accelerator cable from the cable bracket

10.4 Working under the dash, pull the cable end from the accelerator pedal recess and lift it through the slot (arrow)

3   Separate the accelerator cable from the cable bracket **(see illustration)**.
4   Working under the dash, pull the cable end out from the accelerator pedal arm **(see illustration)**.
5   Disconnect any cable clips or brackets securing the accelerator cable.
6   Remove the cable through the firewall from the engine compartment side.

Installation

7   Installation is the reverse of removal. Be sure the cable is routed correctly and the grommet seats completely in the firewall.
8   If necessary, at the engine compartment side of the firewall, apply sealant around the accelerator cable to prevent water from entering the passenger compartment.

Adjustment

9   Measure the freeplay by firmly gripping the cable and pressing it down from the cable housing. There should be a slight amount of cable freeplay.
10   If there is no cable freeplay or the cable is binding and not allowing the throttle lever to completely close, replace the cable.

---

11   Fuel injection system - general information

Sequential Electronic Fuel Injection (SEFI)

The Sequential Electronic Fuel Injection (SEFI) system is a multi-point fuel injection system. On the SEFI system, fuel is metered into each intake port in sequence with the engine firing order in accordance with engine demand through one injector per cylinder mounted on a tuned intake manifold. The intake manifold incorporates an air intake plenum to aid in air flow and distribution. Each engine uses a slightly different plenum design and fuel rail arrangement. All air intake plenums bolt to the intake manifold which sits directly in the middle of the engine block.

4.0L SOHC engines are equipped with a Variable Induction (VIS) System. The (VIS) system consists of an intake manifold tuning valve mounted to the top of the air intake plenum and a vacuum control solenoid. When the engine reaches approximately 3,000 rpm, the PCM energizes the vacuum control solenoid which then controls the amount of manifold vacuum applied to the intake manifold tuning valve and the length of the time the valve is opened. As the IMT valve opens it allows both sides of the incoming intake air charge to blend together which increases high-end performance.

The Sequential Electronic Fuel Injection system incorporates an on-board Electronic Engine Control (EEC-IV 1995 and earlier and EEC-V 1996 and later) computer that accepts inputs from various engine sensors to compute the required fuel flow rate necessary to maintain a prescribed air/fuel ratio throughout the entire engine operational range. The computer then outputs a command to the fuel injectors to meter the approximate quantity of fuel. The system automatically senses and compensates for changes in altitude, load and speed. **Note:** *The computer termi-*

nology has changed from Electronic Control Module (ECM) to the Powertrain Control Module (PCM) due to standardization of the Self Diagnosis system within the automotive industry.

The fuel delivery systems include an electric in-tank fuel pump which forces pressurized fuel through a series of metal and plastic lines and an inline fuel filter/reservoir to the fuel charging manifold assembly. The SEFI system uses a single high-pressure pump mounted inside the tank.

The fuel rail assembly incorporates an electrically actuated fuel injector directly above each intake port. When energized, the injectors spray a metered quantity of fuel into the intake air stream.

A constant fuel pressure is supplied to the injectors. A pressure regulator is positioned downstream from the fuel injectors. Excess fuel passes through the regulator and returns to the fuel tank through a fuel return line.

On the SEFI system, each injector is energized once every other crankshaft revolution in sequence with engine firing order. The period of time that the injectors are energized (known as "on time" or "pulse width") is controlled by the PCM. Air entering the engine is sensed by speed, pressure and temperature sensors. The outputs of these sensors are processed by the PCM. The computer determines the needed injector pulse width and outputs a command to the injector to meter the exact quantity of fuel.

---

12   Fuel injection system - check

Warning: *Gasoline is extremely flammable, so take extra precautions when you work on any part of the fuel system. See the **Warning** in Section 2.*
Note: *The following procedure is based on the assumption that the fuel pump is working and the fuel pressure is adequate (see Section 3).*

12.7 Use a stethoscope or screwdriver to determine if the injectors are working properly - they should make a steady clicking sound that rises and falls as engine speed changes

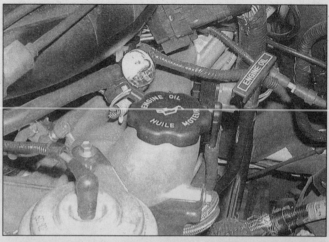

12.8 Install the fuel injector test light or "noid light" into the fuel injector electrical connector and confirm that it blinks when the engine is cranked or running

General checks

1   Check all electrical connectors that are related to the system. Loose electrical connectors and poor grounds can cause many problems that resemble more serious malfunctions.
2   Check to see that the battery is fully charged, as the control unit and sensors depend on an accurate supply voltage in order to properly meter the fuel.
3   Check the air filter element - a dirty or partially blocked filter will severely impede performance and economy (see Chapter 1).
4   If a blown fuse is found, replace it and see if it blows again. If it does, search for a grounded wire in the harness to the fuel pump (see Chapter 12).

System checks

Refer to illustrations 12.7, 12.8 and 12.9

5   Check the condition of the vacuum hoses connected to the intake manifold.
6   Remove the air intake duct from the throttle body and check for dirt, carbon or other residue build-up in the throttle body, particularly around the throttle plate. **Caution:** *The throttle body on these models is coated with a sludge-resistant material designed to protect the bore and throttle plate. Do not attempt to clean the interior of the throttle body with carburetor or other spray cleaners. This throttle body is designed to resist sludge accumulation and cleaning with a solvent may impair the performance of the engine.*
7   With the engine running, place an automotive stethoscope against each injector, one at a time, and listen for a clicking sound, indicating operation **(see illustration)**. If you don't have a stethoscope, you can place the tip of a long screwdriver against the injector and listen through the handle.
8   If an injector isn't functioning (not clicking), purchase a special injector test light (sometimes called a "noid" light) and install it

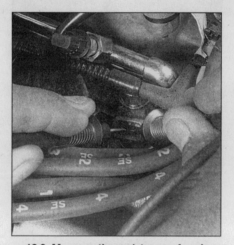

12.9 Measure the resistance of each injector - the readings should be within the range listed in this Chapter's Specifications

into the injector electrical connector **(see illustration)**. Start the engine and check to see if the noid light flashes. If it does, the injector is receiving proper voltage. If it doesn't flash, further diagnosis should be performed by a dealer service department or other properly equipped repair facility.
9   With the engine OFF and the fuel injector electrical connectors disconnected, measure the resistance of each injector **(see illustration)**. Check the Specifications listed in this Chapter for the correct injector resistance.
10   The remainder of the system checks can be found in Chapter 6.

---

13 Throttle body and upper intake manifold - removal and installation

---

Warning: *Wait until the engine is completely cool before beginning this procedure.*

13.3 Remove the electrical connector (arrow) from the Throttle Position Sensor (TPS)

Caution: *The throttle body is coated with a sludge-resistant material designed to protect the bore and throttle plate. Do not attempt to clean the interior of the throttle body. The throttle body is designed to resist sludge accumulation and cleaning may impair the performance of the engine.*

Throttle body

Refer to illustrations 13.3, 13.5 and 13.7

1   Detach the cable from the negative terminal of the battery.
2   Remove the accelerator cable splash shield, the accelerator cable (see Section 10) and cruise control cable (if equipped) from the throttle body.
3   Detach the Throttle Position Sensor (TPS) electrical connector **(see illustration)**.
4   Remove the air cleaner outlet tube (see Section 9).
5   On 5.0L engines, drain the cooling system until the level is below the EGR spacer, then remove the Idle Air Control (IAC) valve

4

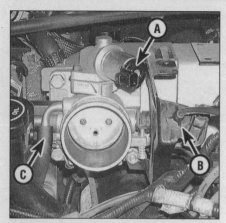

13.5 On 5.0L engines, disconnect the IAC valve connector (A), the accelerator cable bracket (B) and the crankcase ventilation hose (C)

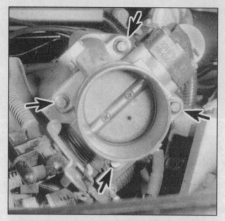

13.7 Remove the throttle body mounting nuts (4.0L pushrod engine shown, all other models similar)

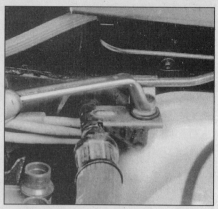

13.18 Unbolt the air conditioning hose bracket from the upper intake manifold and move it aside - DO NOT disconnect any refrigerant lines

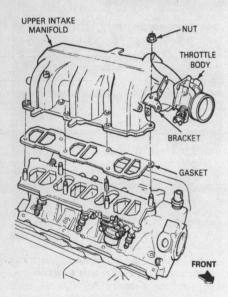

13.19a Upper intake manifold mounting details (4.0L pushrod engine)

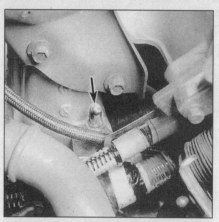

13.19b Remove the cap from the front stud (arrow) before unscrewing the nut

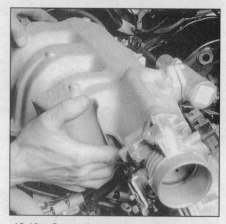

13.19c Grasp the upper manifold at front and rear ends and lift it off the engine

electrical connector, the crankcase ventilation hose and the accelerator cable bracket from the throttle body **(see illustration)**. Position the accelerator cables and bracket out of the way.

6   On 4.0L pushrod engines, disconnect the canister purge hose from underneath the throttle body.

7   Remove the throttle body mounting bolts and the throttle body from the upper intake manifold **(see illustration)**.

8   Remove and discard the throttle body gasket.

9   Installation is the reverse of removal. Be sure to clean the gasket mating surfaces. If scraping is necessary, be careful not to damage the gasket surfaces or allow material to drop into the manifold. Install a new gasket and place the throttle body into position, then tighten the throttle body mounting bolts to the torque listed in this Chapter's Specifications.

Upper intake manifold
Removal

10   Disconnect the cable from the negative terminal of the battery.

11   Detach the air cleaner outlet tube from the air cleaner and the throttle body **(see Section 9)**.

12   Remove the accelerator cable splash shield from the accelerator linkage. Remove the accelerator cable bracket and disconnect the accelerator cable from the throttle body (see Section 10).

13   Disconnect all remaining hoses and electrical connectors from the throttle body (see Steps 1 through 6).

4.0L pushrod engine
Refer to illustrations 13.18, 13.19a, 13.19b and 13.19c

14   Disconnect the electrical connector from the Idle Air Control (IAC) valve.

15   Label and disconnect the vacuum lines from the upper intake manifold.

16   Detach the PCV valve from the right valve cover (see Chapter 6).

17   Detach the spark plug wires from the loom at the back of the upper intake manifold.

18   Remove the bolt that secures the air conditioning refrigerant line at the upper rear part of the manifold **(see illustration)**. DO NOT disconnect any refrigerant lines! On later model 4.0L pushrod engines, it will be necessary to remove the EGR tube from the left exhaust manifold and the EGR valve. Also disconnect the vacuum hoses and electrical connection from the EGR backpressure transducer and the EVR solenoid.

19   Remove six nuts that secure the upper intake manifold **(see illustrations)**. Lift the upper manifold and throttle body together off the fuel rail **(see illustration)**.

4.0L SOHC engine (models through 2000)
Refer to illustrations 13.20 and 13.22

20   Disconnect the electrical connectors from the Idle Air Control (IAC) valve, the engine vacuum regulator and the variable induction solenoid valve **(see illustration)**.

21   Clearly label, then detach the vacuum hoses from the intake manifold tuning valve, the PCV valve and at the rear of the upper intake manifold.

22   Remove the upper intake manifold mounting bolts in the reverse order of the tightening sequence **(see illustration)**. Separate the upper intake manifold from the lower

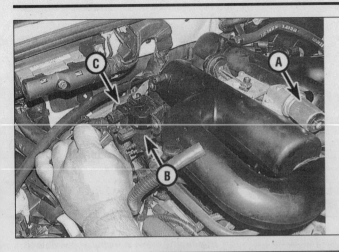

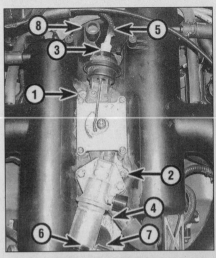

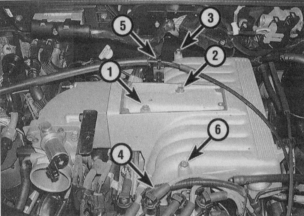

13.20 Disconnect the electrical connectors from the Idle Air Control (IAC) valve (A), the engine vacuum regulator (B) and the variable induction solenoid valve (C)

13.22 Upper intake manifold tightening sequence (4.0L SOHC engine)

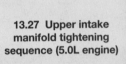

13.27 Upper intake manifold tightening sequence (5.0L engine)

intake manifold and remove it from the vehicle.

5.0L engine
Refer to illustration 13.27

23   Make sure to drain the cooling system until the level is below the EGR spacer as described in Step 5, then remove the coolant hoses from the EGR spacer (if equipped).

24   Remove the vacuum hose and electrical connection from the EGR valve (if equipped). Clearly label, then detach the remaining vacuum hoses from the upper intake manifold. Unclip the electrical wiring attached to intake manifold bracket.

25   Remove the ignition coil and mounting bracket (see Chapter 5).

26   Remove the 5.0L emblem and mounting screws from the top of the upper intake manifold.

27   Remove the upper intake manifold mounting bolts in the reverse order of the tightening sequence **(see illustration)**. Separate the upper intake manifold from the lower intake manifold and remove it from the vehicle.

Installation

28   Installation is the reverse of removal. Be sure to clean and inspect the mounting faces of the lower intake manifold (see Chapter 2A) and the upper intake manifold before positioning the new gasket(s) onto the lower intake mounting face. The use of alignment

studs may be helpful. Install the upper intake manifold and throttle body assembly onto the lower intake manifold. Ensure the gasket remains in place (if alignment studs are not used). Install the upper intake manifold retaining bolts/nuts and tighten them to the torque listed in this Chapter's Specifications, using the correct sequence **(see illustrations 13.22 and 13.27)**.

One-piece intake manifold (2001 models)

Removal

29   Disconnect the cable from the negative terminal of the battery.

30   Remove the bolts and detach the shield above the air filter outlet pipe.

31   Disconnect the mass air flow sensor wiring connector and move the wiring aside.

32   Loosen the clamp screws and remove the air filter outlet pipe.

33   Disconnect the idle air control valve and the throttle position sensor electrical connectors.

34   Disconnect the vacuum hose from the exhaust gas recirculation (EGR) valve.

35   Detach the accelerator and cruise control (if equipped) cables from the throttle body. Detach the cables from the bracket and move the cables aside.

36   Loosen the fitting and disconnect the exhaust manifold-to-EGR valve tube.

37   Disconnect the vacuum regulator electrical connector and vacuum hoses from the EGR valve.

38   Disconnect the vacuum hose from the brake booster.

39   Detach the spark plug wires and remove them from the valve cover clips.

40   Remove the wiring harness bracket bolt and move the harness aside.

41   Detach the accelerator cable clip from the bracket.

42   Disconnect the ignition coil pack electrical connector.

43   Remove the bolts that secure the ignition coil pack mounting bracket. Move the assembly aside.

44   Clearly label and disconnect the vacuum hoses from the intake manifold.

45   Remove the power train control module (PCM) harness retaining bolt. Remove the bolt and carefully disconnect the electrical connector from the PCM.

46   Remove the nut and disconnect the ground wire.

47   Disconnect the electrical connector clip and move the connector aside.

48   Remove the six bolts that secure the intake manifold. Lift up and remove the intake manifold from the cylinder heads. If it's stuck, tap it lightly with a soft-faced hammer to break the gasket seal.

Installation

49   Clean away all traces of old gasket material. Remove oil and dirt with a cloth and solvent, such as lacquer thinner.

50   Install the intake manifold onto the cylinder heads, making sure the bolt holes are aligned, install the bolts and tighten them finger-tight.

51   Tighten the bolts to the torque listed in this Chapter's Specifications, starting with the center bolts and working towards the ends.

52   The remainder of installation is the reverse of the removal Steps.

4

14.5a   On 4.0L pushrod engines, remove the fuel return line (A) from the top of the pressure regulator, then remove the mounting bolts (B)

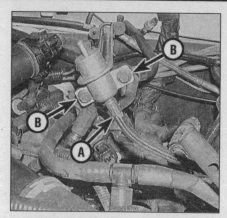

14.5b   On 4.0L SOHC engines, the passenger side fuel rail has to be separated from the lower intake manifold to allow access to the fuel return line (A) and the regulator mounting bolts (B)

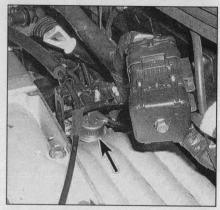

14.6   On 5.0L engines, the fuel pressure regulator (arrow) is easily accessible at the rear of the manifold

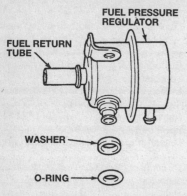

14.7   Fuel pressure regulator installation details (4.0L pushrod engine)

14   Fuel pressure regulator - removal and installation

Refer to illustrations 14.5a, 14.5b, 14.6 and 14.7

1   Relieve fuel system pressure (see Section 2).

2   Disconnect the cable from the negative terminal of the battery.

3   On 4.0L SOHC engines, remove the upper intake manifold (see Section 13) then detach the passenger side fuel rail and fuel injector assembly from the lower intake manifold (see Section 15).

4   On 4.0L pushrod and 4.0L SOHC engines, disconnect the vacuum line from the pressure regulator.

5   On 4.0L pushrod and 4.0L SOHC engines, remove the fuel return line from the pressure regulator and the mounting bolts which secure the regulator to the fuel rail (**see illustrations**).

6   On 5.0L engines, simply remove the vacuum line from the pressure regulator and the regulator mounting bolts (**see illustration**).

7   Remove the regulator from the fuel rail and discard the O-ring and washer if equipped (**see illustration**).

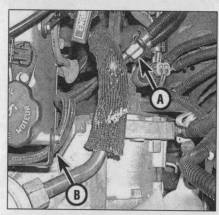

15.4   On 4.0L SOHC engines, it will be necessary to disconnect the fuel feed and return line connections (A) and the fuel line retaining brackets (B) leading to the fuel rail assembly

8   Install new O-rings and washer (if equipped) on the pressure regulator and lubricate them with a light coat of oil.

9   Installation is the reverse of removal. Tighten the pressure regulator mounting bolts to the torque listed in this Chapter's Specifications. Check the fuel return line for kinks or worn spots and replace as needed.

15   Fuel rail and injectors - removal and installation

Refer to illustrations 15.4, 15.5a, 15.5b, 15.5c, 15.6 and 15.8

Removal

1   Relieve the fuel pressure (see Section 2).

2   Disconnect the cable from the negative terminal of the battery.

3   Remove the upper intake manifold from the lower intake manifold (see Section 13).

4   Disconnect the fuel feed and return lines from the fuel rail assembly (**see illustration**).

15.5a   On 4.0L pushrod engines, use a Torx socket and ratchet to remove the fuel rail mounting studs

Note: *Refer to Section 4 for additional information on disconnecting fuel lines.*

5   Disconnect the fuel injector connectors and remove the fuel rail retaining bolts/studs (**see illustrations**).

6   Using a rocking, side-to-side motion, carefully lift the fuel rail and the fuel injectors as an assembly from the lower intake manifold (**see illustration**).

7   To remove a fuel injector from the fuel rail, simply rock the injector body from side-to-side and detach it from the fuel rail.

8   Inspect the injector O-rings (two per injector) for signs of deterioration (**see illustration**). Replace as required. **Note:** *As long as you have the fuel rail off, it's a good idea to replace all of the O-rings.*

9   Inspect the injector plastic "hat" (covering the injector pintle) and washer for signs of deterioration. Replace as required. If the hat is missing, look for it in the intake manifold.

Installation

10   Ensure that the injector caps are clean and free of contamination.

11   Lubricate the new O-rings with light grade oil and install two on each injector. **Caution:** *Do not use silicone grease. It will clog the injectors.*

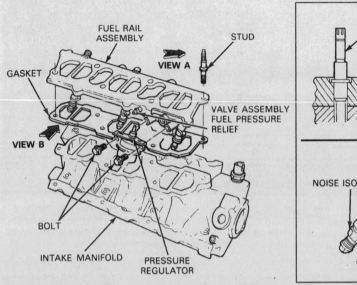

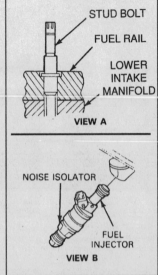

15.5c Fuel rail mounting bolt locations (arrows) (two on each side) - 5.0L engine shown, 4.0L SOHC engine similar

15.5b Fuel rail and fuel injector installation details (4.0L pushrod engine)

12   Using a light twisting motion, install the injector(s) into the lower intake manifold.

13   Place the fuel rail over each of the injectors and seat the injectors into the fuel rail. Ensure that the injectors are well seated in the fuel rail assembly. **Note:** *It may be easier to seat the injectors in the fuel rail and then seat the entire assembly in the lower intake manifold.*

14   On 4.0L SOHC and 5.0L engines, secure the fuel rail assembly with the four retaining bolts. On 4.0L pushrod engines, install the lower gasket and the fuel rail onto the lower intake manifold, then install the fuel rail retaining studs. Tighten the fuel rail mounting bolts/studs to the torque listed in this Chapter's Specifications.

15   The remainder of installation is the reverse of removal.

16 Variable Induction (VIS) System (1997 and later 4.0L SOHC engine)

General information

1   The VIS system controls the air intake charge by opening or closing the Intake Manifold Tuning Valve located directly on top of the upper intake manifold. By closing the IMT valve, the dual plenum design sends the air intake charge through a single corridor into the intake system of the engine. Above 3,000 rpm, the IMT valve is opened, allowing the intake pulses to blend together at the intake manifold, thereby creating a more efficient air/fuel intake charge for the additional rpm and engine load.

2   The VIS system is difficult to check and requires a special SCAN tool to access the PCM for information and operating conditions. Have the system diagnosed by a dealer

service department or other qualified repair shop.

Replacement

Refer to illustration 16.4

3   Remove the accelerator cable splash shield.

4   Detach the vacuum hose and the retaining screws from the IMT valve **(see illustration)**.

5   Remove the IMT valve from the upper intake manifold.

6   Installation is the reverse of removal.

17 Exhaust system servicing - general information

Refer to illustrations 17.1a, 17.1b, 17.1c and 17.1d

Warning: *Inspection and repair of exhaust system components should be done only after enough time has elapsed after driving*

15.6 Using a rocking, side-to-side motion, carefully lift the fuel rail and the fuel injectors from the lower intake manifold - on 4.0L SOHC engines, it will be necessary to perform this task to the left and right side fuel rails

the vehicle to allow the system components to cool completely. Also, when working under the vehicle, make sure it is securely supported on jackstands.

1   The exhaust system consists of the exhaust manifold(s), the catalytic converter(s), the muffler, the tailpipe and all connecting

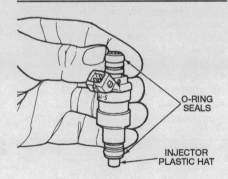

15.8 Each injector uses two O-rings - the injector pintle is protected by a plastic "hat"

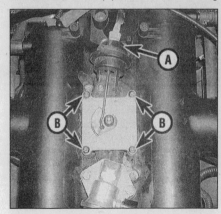

16.4 Remove the IMT valve vacuum hose (A) and retaining screws (B)

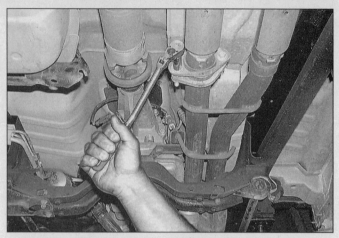

17.1a   Be sure to apply penetrating lubricant to the exhaust flange nuts (arrows) before attempting to remove them

17.1b   Remove the exhaust flange retaining bolts from the center section of the exhaust system

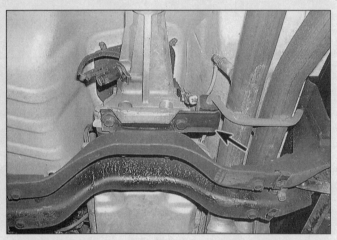

17.1c   First remove the transmission crossmember (see Chapter 7) then remove the exhaust system bracket assembly (arrow)

17.1d   Check the condition of the rubber mounts (arrows) that support the weight of the exhaust system

pipes, brackets, hangers and clamps **(see illustrations)**. The exhaust system is attached to the body with mounting brackets and rubber hangers. If any of the parts are improperly installed, excessive noise and vibration will be transmitted to the body.

2   Conduct regular inspections of the exhaust system to keep it safe and quiet. Look for any damaged or bent parts, open seams, holes, loose connections, excessive corrosion or other defects which could allow exhaust fumes to enter the vehicle. Deteriorated exhaust system components should not be repaired; they should be replaced with new parts.

3   If the exhaust system components are extremely corroded or rusted together, welding equipment will probably be required to remove them. The convenient way to accomplish this is to have a muffler repair shop remove the corroded sections with a cutting torch. If, however, you want to save money by doing it yourself (and you don't have a welding outfit with a cutting torch), simply cut off the old components with a hacksaw. If you have compressed air, special pneumatic cutting chisels can also be used. If you do decide to tackle the job at home, be sure to wear safety goggles to protect your eyes from metal chips and work gloves to protect your hands.

4   Here are some simple guidelines to follow when repairing the exhaust system:

a)   *Work from the back to the front when removing exhaust system components.*

b)   *Apply penetrating oil to the exhaust system component fasteners to make them easier to remove.*

c)   *Use new gaskets, hangers and clamps when installing exhaust systems components.*

d)   *Apply anti-seize compound to the threads of all exhaust system fasteners during reassembly.*

e)   *Be sure to allow sufficient clearance between newly installed parts and all points on the underbody to avoid overheating the floor pan and possibly damaging the interior carpet and insulation. Pay particularly close attention to the catalytic converter and heat shield.*

Chapter 5
Engine electrical systems

Contents

Specifications

General

Battery voltage	
Engine off	12 volts
Engine running	14 to 15 volts
Firing order	
V6 engines	1-4-2-5-3-6
5.0L V8 engine	1-3-7-2-6-5-4-8

Ignition system

Ignition coil resistance	
Primary resistance	0.3 to 1.0 ohms
Secondary resistance	6.5 to 11.5 K-ohms
Spark plug wire resistance	5,000 ohms per foot
Ignition timing (base setting)	10-degrees BTDC with SPOUT disconnected

Charging system

Alternator brush length	
New	1/2 inch
Minimum	1/4 inch

**2.1 Removing the cable from a battery post with a wrench -
sometimes special battery pliers are required for this procedure if
corrosion has caused deterioration of the nut hex (always remove
the ground cable first and hook it up last!)**

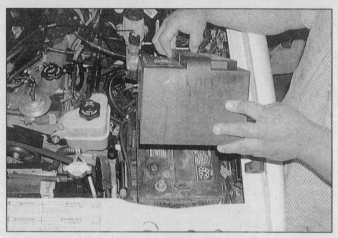

**2.2 On 1997 and later models remove the heat shield from the top
of the battery**

1   General information

The engine electrical systems include all ignition, charging and starting components. Because of their engine-related functions, these components are considered separately from chassis electrical devices like the lights, instruments, etc.

Be very careful when working on the engine electrical components. They are easily damaged if checked, connected or handled improperly. The alternator is driven by an engine drivebelt which could cause serious injury if your hands, hair or clothes become entangled in it with the engine running. Both the starter and alternator are connected directly to the battery and could arc or even cause a fire if mishandled, overloaded or shorted out.

Never leave the ignition switch on for long periods of time with the engine off. Don't disconnect the battery cables while the engine is running. Correct polarity must be maintained when connecting battery cables from another source, such as another vehi-cle, during jump starting. Always disconnect the negative cable first and hook it up last or the battery may be shorted by the tool being used to loosen the cable clamps.

Additional safety related information on the engine electrical systems can be found in Safety first near the front of this manual. It should be referred to before beginning any operation included in this Chapter.

2   Battery - removal and installation

Refer to illustrations 2.1, 2.2, 2.3a, 2.3b and 2.5

1   Disconnect both cables from the battery terminals **(see illustration)**. **Warning:** *Always disconnect the negative cable first and hook it up last or the battery may be shorted by the tool being used to loosen the cable clamps.* **Note:** *When the battery is disconnected and reconnected, the vehicle may experience abnormal driving symptoms while the computer (PCM) relearns its adaptive strategy. The vehicle may need to be driven 10 miles or more to regain smooth operation.*

2   If equipped, remove the heat shield from the top of the battery **(see illustration)**.

3   Remove the bolt and wedge from the battery tray **(see illustrations)**.

4   Lift out the battery. Be careful - it's heavy. Special lifting straps that attach to the battery posts are available at auto parts stores - lifting and moving the battery is

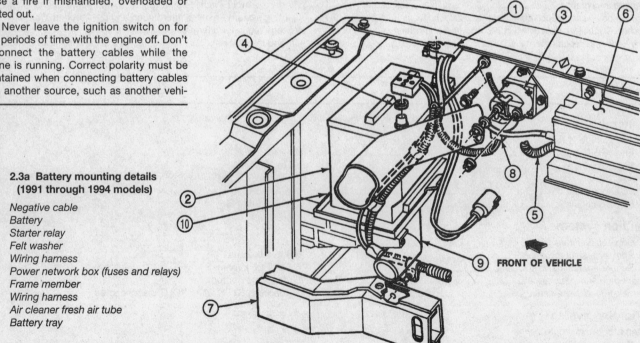

**2.3a Battery mounting details
(1991 through 1994 models)**

1   *Negative cable*
2   *Battery*
3   *Starter relay*
4   *Felt washer*
5   *Wiring harness*
6   *Power network box (fuses and relays)*
7   *Frame member*
8   *Wiring harness*
9   *Air cleaner fresh air tube*
10   *Battery tray*

FRONT OF VEHICLE

2.3b Remove the bolt (arrow) and the battery hold down clamp which secures the base of the battery to the battery tray

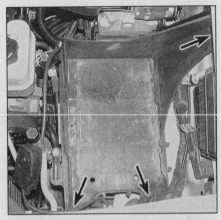

2.5 On 1995 and later models, the battery tray is mounted on the driver's side of the vehicle and is fastened by three retaining bolts (arrows) - On 1994 and earlier models, the battery tray is mounted on the passenger side of the vehicle and is fastened by four retaining bolts (two upper and two lower retaining bolts)

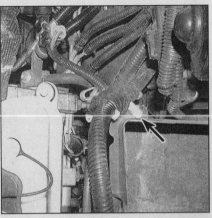

4.4a Detach any battery cable ties or retaining clips (arrow)

4.4b The positive battery cable is fastened to the starter relay (arrows) . . .

much easier if you use one.

5   If necessary for access to other components, remove the bolts that secure the battery tray (see illustration).

6   Installation is the reverse of removal.

3   Battery - emergency jump starting

Refer to the *Booster battery (jump) starting* procedure at the front of this manual.

4   Battery cables - check and replacement

Refer to illustrations 4.4a, 4.4b, 4.4c, 4.4d and 4.4e

1   Periodically inspect the entire length of each battery cable for damage, cracked or burned insulation and corrosion. Poor battery cable connections can cause starting problems and decreased engine performance.

2   Check the cable-to-terminal connections at the ends of the cables for cracks,

loose wire strands and corrosion. The presence of white, fluffy deposits under the insulation at the cable terminal connection is a sign that the cable is corroded and should be replaced. Check the terminals for distortion, missing mounting bolts and corrosion.

3   When replacing the cables, always disconnect the negative cable first and hook it up last or the battery may be shorted by the tool used to loosen the cable clamps. Even if only the positive cable is being replaced, be sure to disconnect the negative cable from the battery first.

4   Disconnect and remove the cable (see illustrations). Make sure the replacement cable is the same length and diameter.

5   Clean the threads of the relay or ground connection with a wire brush to remove rust and corrosion. Apply a light coat of petroleum jelly to the threads to prevent future corrosion.

6   Attach the cable to the relay or ground connection and tighten the mounting nut/bolt

securely.

7   Before connecting the new cable to the battery, make sure that it reaches the battery post without having to be stretched. Clean the battery posts thoroughly and apply a light coat of petroleum jelly to prevent corrosion (see Chapter 1).

8   Connect the positive cable first, followed by the negative cable.

4.4c . . . and the starter (arrow)

4.4d The negative battery cable is fastened to the inner fenderwell (arrow) . . .

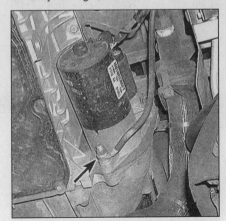

4.4e . . . the starter motor lower retaining bolt (arrow), the engine block or the frame rail depending on the model year of the vehicle

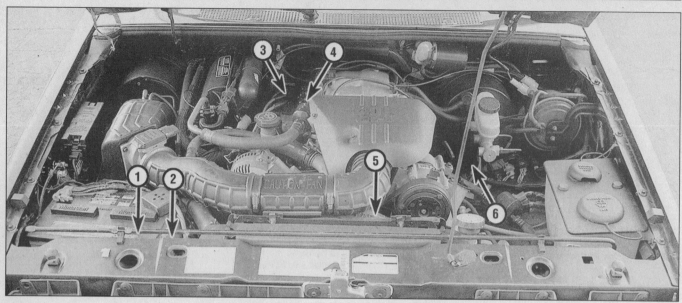

5.1a Ignition system components - 4.0L pushrod V6 engine

1   *Ignition Control Module (fastened to the back-side of the radiator support)*
2   *Shorting bar (in ICM wiring harness)*
3   *Ignition coil pack*

4   *Spark plug wires*
5   *Crankshaft position (CKP) sensor (located on front of engine block)*
6   *Spark plugs (left bank)*

5   Ignition system - general information

Refer to illustrations 5.1a, 5.1b and 5.1c

1   All models covered by this manual are equipped with an Electronic Distributorless Ignition System (EDIS). The EDIS system is a complete electronically controlled ignition system that does not incorporate a distributor. The EDIS system consists of a crankshaft position sensor (CKP), an externally mounted ignition control module (1991 through 1995 models), ignition coil pack, an EEC-IV or EEC-V Powertrain Control Module (PCM), the spark plug wires and the spark plugs **(see illustrations)**. The EDIS system features a waste-spark method of spark distribution. Each cylinder is paired with its companion cylinder in the firing order (1-5, 2-6, 3-4 [V6 engine]) or (1-6, 3-5, 4-7, 2-8 [V8 engine]).

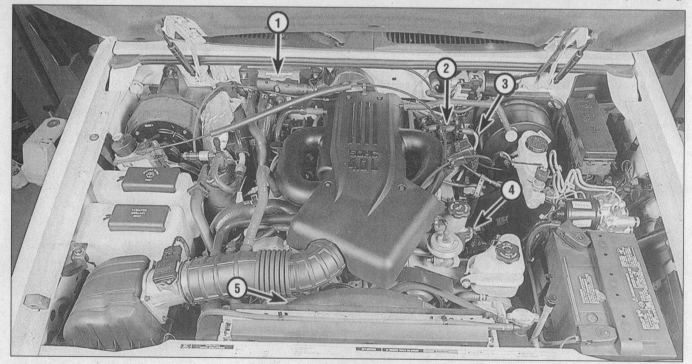

5.1b Ignition system components - 4.0L SOHC V6 engine

1   *Powertrain Control Module (PCM)*
2   *Ignition coil pack*
3   *Spark plug wires*

4   *Spark plugs (left bank)*
5   *Crankshaft position (CKP) sensor (located on front of engine block)*

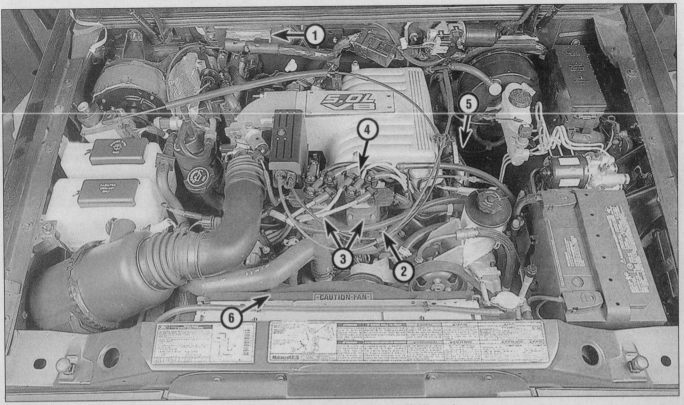

5.1c Ignition system components - 5.0L V8 engine

1 Powertrain Control Module (PCM) (under cowl)
2 Shorting bar
3 Ignition coil pack
4 Spark plug wires
5 Spark plugs (left bank)
6 Crankshaft position (CKP) sensor (located on front of engine block)

The cylinder under compression fires simultaneously with its opposing cylinder, which is on the exhaust stroke. Since the cylinder on the exhaust stroke requires very little of the available voltage to fire its plug, the majority of the voltage is used to fire the plug under compression. V6 engines are equipped with one ignition coil pack which contains three separate coils. V8 engines are equipped with two coil packs which contain four separate coils (two coils in each coil pack). When either type of coil is triggered it supplies ignition voltage to two (companion) cylinders at the same time.

2 This ignition system does not have any moving parts (no distributor) and all engine timing and spark distribution is handled electronically. This system has fewer parts that require replacement and provides more accurate spark timing than conventional distributor type ignition systems. During engine operation, the ignition control module receives signals from the crankshaft position sensor and the PCM to determine the proper spark advance and firing sequence of the ignition coils. **Note:** *1996 and later systems are not equipped with an ignition control module. The PCM also functions as the controller of the ignition coil(s) primary circuit which was basically the job of the ignition control module on earlier models.*

3 The CKP sensor is mounted on the engine front cover and triggered by a trigger wheel on the front of the crankshaft. The trigger wheel has 35 evenly spaced teeth and one gap where a 36th tooth would be. This gap allows the CKP sensor to signal the PCM when the crankshaft is 90 degrees before TDC for cylinders 1 and 5 on V6 engines or cylinders 1 and 6 on V8 engines. The PCM then computes actual TDC or any number of degrees before or after TDC. The CKP sensor is a magnetic pickup, or variable reluctance, sensor that produces a direct-current sine wave voltage signal. The sensor is similar to a distributor pickup in a distributor-type electronic ignition system. Besides providing information on crankshaft position, the CKP sensor delivers the engine speed signal to the PCM. Refer to Chapter 6 for more information on the PCM and the CKP sensor.

6 Ignition system - check

Refer to illustrations 6.3, 6.6 and 6.9
Warning: *Because of the very high secondary (spark plug) voltage generated by the ignition system, extreme care must be taken when this check is done.*

1 A "calibrated ignition tester" is required for this check, they are available at most auto parts stores. Several different types are available, be sure to use a type calibrated for the

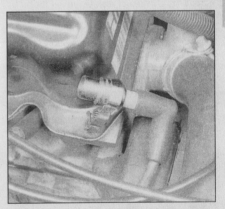

6.3 To use a calibrated ignition tester, simply disconnect a spark plug wire, clip the tester to a convenient ground (like a valve cover bolt) and operate the starter - if there is enough power to fire the plug, sparks will be visible between the electrode tip and the tester body

high-voltage ignition system on these models.

2 If the engine cranks but won't start, disconnect the spark plug wire from any spark plug and attach it to the calibrated ignition tester.

3 Clip the tester to a bolt or metal bracket on the engine **(see illustration)**, crank the engine and watch the end of the tester to see if consistent, bright, well-defined sparks occur.

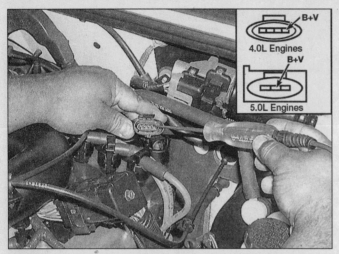

6.6 Disconnect the electrical connector from the ignition coil and check for battery voltage to the coil with the ignition key On

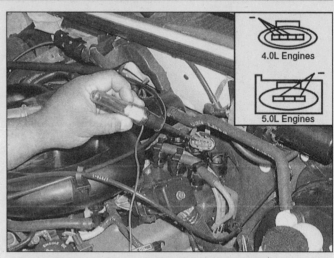

6.9 Connect an LED test light to the positive battery terminal and the coil negative (-) terminals on the ignition coil harness connectors and watch for a blinking light when the engine is cranked

4   If sparks occur, sufficient voltage is reaching the plug to fire it (repeat the check at the remaining plug wires to verify that each ignition coil pack is functioning. However, the plugs themselves may be fouled, so remove and check them as described in Chapter 1 or install new ones.

5   If no sparks or intermittent sparks occur, check for a bad spark plug wire by swapping wires or by using an ohmmeter to measure the resistance between the spark plug wire terminal ends. There should be approximately 5000 ohms of resistance per foot.

6   If the problem isn't caused by the spark plug wire, check for battery voltage to the ignition coil with the ignition key ON (engine not running). Attach a 12 volt test light to the battery negative (-) terminal or other good ground. Disconnect the coil electrical connector and check for power at the positive (+) terminal **(see illustration)**. Battery voltage should be available. If there is no battery volt-

age, check the fuse that protects the ignition circuit (see Chapter 12 for additional information on the fuses and the wiring diagrams).

7   Check the primary and secondary resistances of the ignition coils (see Section 7).

8   Check the ignition coil electrical connectors for dirt, corrosion and damage.

9   If battery voltage is available to the ignition coils, attach an LED test light to the battery positive (+) terminal and each negative (-) terminal to the coil (on the vehicle harness side) **(see illustration)**, then crank the engine (be sure to check each negative terminal, one at a time). Confirm that the test light flashes. This test checks for the trigger signal (ground) from the computer or Ignition Control Module (if equipped). If a trigger signal is present at the coil, the computer and/or the Ignition Control Module are functioning properly. **Caution:** *Use only an LED test light to avoid damaging the PCM.*

10   If the test light does not flash, check the

crankshaft position sensor (see Chapter 6) and the Ignition Control Module (if equipped) (see Section 8). If the crankshaft sensor and the Ignition Control Module (if equipped) are good, have the PCM checked by a dealer service department or other qualified automotive repair shop.

7   Ignition coil - check and replacement

Check

Refer to illustrations 7.1 and 7.2

1   With the ignition off, disconnect the electrical connector(s) from the coil. Connect an ohmmeter across the coil positive (+) terminal and each negative (-) terminal **(see illustration)**. The resistance should be as listed in this Chapter's Specifications. If not, replace the coil.

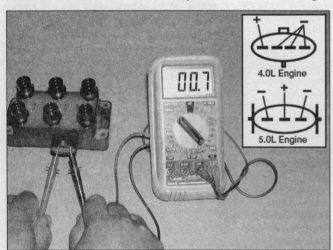

7.1 To check the primary resistance of the ignition coil, connect the probes to the positive (+) terminal and each negative (-) terminal of the coil. The resistance should be the same for each check

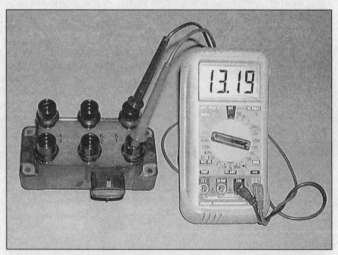

7.2 Check the coil secondary resistance by probing the paired companion cylinders (1-5, 2-6, 3-4 on V6 engines or 1-6, 5-3, 4-7, 2-8 on V8 engines)

7.4 Squeeze the locking tabs, then remove the plug wires from the coil pack using a twisting and pulling motion

7.5 Disconnect the electrical connector from the coil pack

7.6a Coil pack mounting screws (arrows) - V6 engines (4.0L SOHC shown)

2   Connect an ohmmeter between the secondary terminals (see illustration) (the one that the spark plug wires connect to) of each coil pack. The resistance should be as listed in this Chapter's Specifications. If continuity does not exist, replace the coil. Note: The ignition coil pack(s) contains three separate coils on V6 models and four separate coils on V8 models. Each coil is paired with to its companion cylinder. Be sure to check resistance across the appropriate secondary terminals for each coil (1-5, 4-3, 2-6 [V6 engines] or 1-6, 3-5, 4-7, 2-8 [V8 engine]).

Replacement

Note: Although it is not necessary to remove the coil mounting bracket when replacing the coil pack, some procedures in this manual require removing the ignition coil and the coil mounting bracket as an assembly for access to other components such as valve cover or intake manifold removal.

Ignition coil pack

Refer to illustrations 7.4, 7.5, 7.6a and 7.6b
3   Disconnect the cable from the negative

terminal of the battery.
4   Check to see if all the spark plug wires are numbered. If not, clearly label each plug wire with a piece of tape marked to match the numbers listed on the coil terminals. Then disconnect all the spark plug wires by squeezing the locking tabs and twisting while pulling upward (see illustration). Do not pull on the wires.
5   Disconnect the ignition coil pack electrical connector (see illustration).
6   Remove the screws that secure the coil pack to the coil mounting bracket (see illustrations).
7   Installation is the reverse of the removal procedure with the following additions:

a) Before installing the spark plug wire connector into the ignition coil, coat the entire interior of the rubber boot with silicone dielectric compound.
b) Insert each spark plug wire into the proper terminal of the ignition coil. Push the wire into the terminal and make sure the boots are fully seated and both locking tabs are engaged properly.

Ignition coil and mounting bracket

Refer to illustrations 7.9, 7.10 and 7.12
8   Perform Steps 3 through 5 listed above in the ignition coil pack replacement procedure.
9   Detach the radio interference capacitor (if equipped) from the coil mounting bracket (see illustration).
10   On 4.0L pushrod engines, detach the plastic clip that secures the fuel injection wiring harness to the coil mounting bracket. Then remove the bolt holding the air conditioning hose to the intake manifold plenum. Move the air conditioning tube enough to provide working access to the upper coil mounting bracket bolts. Note: It may be possible to remove the upper coil mounting bracket bolts with a socket and universal joint adapter instead of removing the air conditioning hose bolt. Remove the bolts securing the ignition coil mounting bracket to the engine (see illustration).
11   On 4.0L SOHC engines, the ignition coil mounting bracket is welded to the valve cover, and therefore unnecessary to remove for access to other components.

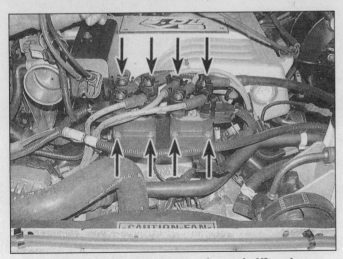

7.6b Coil pack mounting screws (arrows) - V8 engine

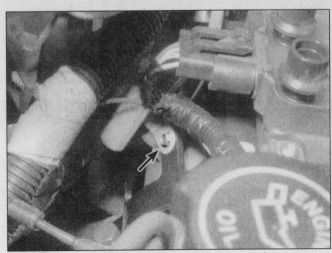

7.9 If equipped, remove the radio interference capacitor from the coil mounting bracket

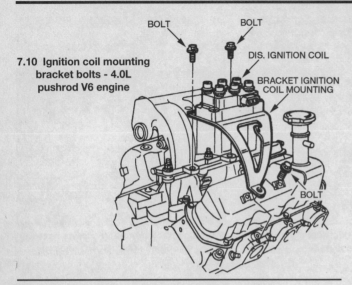

7.10 Ignition coil mounting bracket bolts - 4.0L pushrod V6 engine

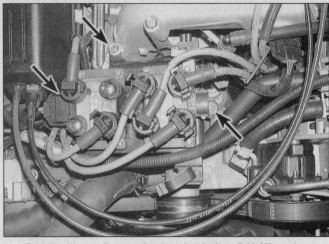

7.12 Ignition coil mounting bracket bolts - 5.0L V8 engine

12   On 5.0L engines, disconnect the hoses and wiring from the EGR vacuum solenoid, then remove the bolts securing the ignition coil mounting bracket to the engine **(see illustration)**.

13   Installation is the reverse of the removal.

8   Ignition Control Module (ICM) (1995 and earlier models) - check and replacement

Caution: *The Ignition Control Module is a delicate and relatively expensive electronic component. Failure to follow the step-by-step procedures could result in damage to the*

module and/or other electronic devices, including the PCM. Additionally, all devices under computer control are protected by a Federally mandated warranty. Check with your dealer concerning this warranty before attempting to diagnose and replace the module yourself.

Note: *This test assumes that the coil trigger signal has been checked as described in Section 6 and is not present at the ignition coil harness side connector. Special electronic test equipment is required to thoroughly test the ICM and related circuits. This test verifies the integrity of the most important circuits and assumes that, if the circuits are good the module is most likely defective.*

Check

Refer to illustrations 8.2 and 8.9

1   The Ignition Control Module is mounted to the radiator support behind the battery.

2   Check for power to the ICM. Using a voltmeter, probe terminal number 8 (red wire [+]) and check for battery voltage **(see illustration)**. With the ignition ON (engine not running), there should be battery voltage. If battery voltage is not present, refer to the wiring diagrams at the end of Chapter 12 and check the PCM power relay and related wiring for an open or shorted circuit.

3   Turn the ignition OFF. Using an ohmmeter, check for continuity between terminal 9 of the ICM connector and battery ground. Continuity should exist. If no continuity exists, refer to the wiring diagrams at the end of Chapter 12 and check the ground circuit for an open.

4   Check the resistance of the crankshaft position (CKP) sensor on terminals number 5 and number 6. It should be between 2000 and 3000 ohms. If it isn't, check the crankshaft position sensor (see Chapter 6).

5   Check the circuit from the ignition module to the SPOUT connector on terminal number 3 (pink wire [+]) for a complete circuit. Continuity should exist.

6   If the test results are correct, replace the ignition module with a new one.

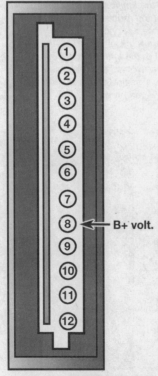

8.2 Ignition Control Module (ICM) connector terminal guide

1   *Profile Ignition Pick-up (PIP) signal*
2   *Ignition Diagnostic Module (IDM) signal*
3   *Spark Output (SPOUT) input*
4   *Ignition ground*
5   *Crankshaft Position (CKP) sensor - negative*
6   *Crankshaft Position (CKP) sensor - positive*
7   *Shield ground*
8   *Battery positive (from PCM power relay)*
9   *Battery ground*
10   *Ignition coil No. 1*
11   *Ignition coil No. 3*
12   *Ignition coil No. 2*

8 → B+ volt.

36051-5-10.28 HAYNES

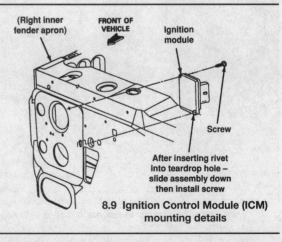

(Right inner fender apron)   FRONT OF VEHICLE   Ignition module

Screw

After inserting rivet into teardrop hole – slide assembly down then install screw

8.9 Ignition Control Module (ICM) mounting details

9.3 Shorting bar (arrow) location on a 5.0L V8 engine

Replacement

7   Remove the battery (see Section 2).
8   Disconnect the electrical connector from the ignition module.
9   Remove the screw securing the Ignition Control Module to the radiator support **(see illustration)**. Slide the assembly up to remove the rivet from the positioning hole in the panel.
10   Installation is the reverse of the removal.

9   Ignition timing - check

Refer to illustration 9.3
Note 1: *This procedure checks the base timing on 4.0L pushrod V6 and 5.0L V8 models only, the 4.0L SOHC engine is not equipped with a SPOUT connector. The SPOUT controls the ignition signal from the computer. Removing the shorting bar from the SPOUT connector will place the timing in the base timing mode, eliminating computer control. Do not remove the shorting bar from the SPOUT connector except for checking ignition base timing.*
Note 2: *Timing cannot be adjusted, therefore the purpose of this check is to verify that the computer is controlling the ignition timing and*

11.3 To measure charging voltage, attach the voltmeter leads to the battery terminals, start the engine and record the voltage reading

that the base setting is correct. In most cases, the ignition system can be checked (see Section 6) but if the base setting remains incorrect, the PCM (computer) is defective. Take the vehicle to a dealer service department or other qualified repair shop to verify and repair the ignition system problem(s).
Note 3: *On 1991 through 1995 4.0L pushrod V6 models, the SPOUT connector is located near the Ignition Control Module. On 1996 and later 4.0L pushrod V6 models, the SPOUT connector is located near the PCM. On 5.0L V8 models, the SPOUT connector is located near the ignition coils.*
1   Apply the parking brake and block the wheels.
2   Start the engine and warm it up. Once it has reached operating temperature, turn it off.
3   Remove the shorting bar from the SPOUT connector **(see Illustration)**.
4   Connect an inductive timing light and a tachometer in accordance with the manufacturer's instructions. **Caution:** *Make sure that the timing light and tachometer wires don't hang anywhere near the cooling fan or they may become entangled in the fan blades when the fan begins to rotate.*
5   Locate the timing marks on the crankshaft pulley (see Chapter 2).
6   Start the engine again. Place the transmission in DRIVE (parking brake applied). Turn off all accessories (heater, air conditioner, etc.).
7   Point the timing light at the pulley timing marks and note whether the specified timing mark is aligned with the timing pointer on the front of the timing chain cover. Refer to the Specifications listed in this Chapter.
8   If the proper mark isn't aligned with the stationary pointer, have the PCM checked by a dealer service department or other qualified repair shop.
9   Turn off the engine.
10   Insert the shorting bar into the SPOUT connector.
11   Remove the timing light and tachometer from the engine compartment.

10   Charging system - general information and precautions

The charging system includes the alternator, a voltage regulator (mounted on the backside of the alternator), a charge indicator or warning light, the battery, a large fuse (called a mega fuse) or fusible link (on some models) and the wiring between all the components. The charging system supplies electrical power for the ignition system, the lights, the radio, etc. The alternator is driven by a drivebelt at the front of the engine.
The purpose of the voltage regulator is to limit the alternator's voltage to a preset value. This prevents power surges, circuit overloads, etc., during peak voltage output. On integral voltage regulator systems, a solid state regulator is housed inside a plastic module mounted on the alternator itself.

These models are equipped a Motorcraft a 130 amp output rated alternator. The voltage regulator can be removed from the backside of the alternator but the alternator must be removed from the engine first (see Section 12).
The charging system is protected by a series of large fuses (MEGA fuses) or fusible links which are located at the engine compartment fuse box. In the event of charging system problems, check the fuses and fusible links for damage or broken contacts.
The charging system doesn't ordinarily require periodic maintenance. However, the drivebelt, battery and wires and connections should be inspected at the intervals outlined in Chapter 1.
Be very careful when making electrical circuit connections to a vehicle equipped with an alternator and note the following:

a) *When reconnecting wires to the alternator from the battery, be sure to note the polarity.*
b) *Before using arc welding equipment to repair any part of the vehicle, disconnect the wires from the alternator and the battery terminals.*
c) *Never start the engine with a battery charger connected.*
d) *Always disconnect both battery cables before using a battery charger (negative cable first, positive cable last).*

11   Charging system - check

Refer to illustrations 11.3, 11.7a and 11.7b
1   If a malfunction occurs in the charging circuit, do not immediately assume that the alternator is causing the problem. First, check the following items:

a) *The battery cables where they connect to the battery. Make sure the connections are clean and tight.*
b) *The battery electrolyte specific gravity. If it is low, charge the battery.*
c) *Check the external alternator wiring and connections.*
d) *Check the drivebelt condition and tension (see Chapter 1).*
e) *Check the alternator mounting bolts for tightness.*
f) *Run the engine and check the alternator for abnormal noise.*
g) *Check the fusible links (if equipped) exiting the engine compartment fuse box (see Chapter 12). If they're burned, determine the cause and repair the circuit*
h) *Refer to wiring diagrams in Chapter 12 and check all the fuses in series with the charging system. The location of the fuses may vary from year to year but the designations are the same.*

2   Using a voltmeter, check the battery voltage with the engine off. It should be approximately 12 volts.
3   Start the engine and check the battery voltage again. It should now be approximately 14 to 15 volts **(see illustration)**.

5

11.7a Alternator terminal identification - 4.0L pushrod V6 engine

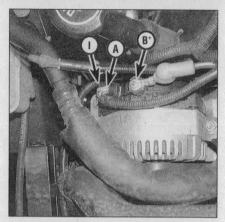

11.7b Alternator terminal identification - 4.0L SOHC V6 and 5.0L V8 engines

12.4a Depending on the model year of the vehicle, 4.0L pushrod V6 engines use three vertically mounted retaining bolts (arrows) at the top of the alternator or two horizontally mounted retaining bolts at the front of the alternator

4   Turn ON the headlights. The voltage should drop and then come back up if the charging system is working properly.

5   If the voltage reading is more than the specified charging voltage, replace the voltage regulator (see Section 13).

6   If the voltage reading is less than the specified charging voltage, check the alternator as follows:

7   Using a voltmeter and working on the backside of the alternator, backprobe the "B+" terminal. There should be 12 volts present with the ignition key OFF **(see illustrations)**.

8   With the ignition key ON (engine not running), backprobe each terminal. There should be 12 volts at the "A" terminal, one volt at the "I" terminal and 12 volts at the "B+" terminal.

9   Start the engine, then raise the engine to 2000 RPM, and backprobe each terminal again. There should be 14.0 to 14.7 volts at the "A" terminal and "B+" terminal and 13.0 to 14.0 volts at the "I" terminal.

10   If the voltages are not as specified, check the wiring harness. If the wiring harness is not defective, replace the alternator.

11   If you suspect that there is a voltage drain on the battery while the vehicle is sitting in the driveway, remove the battery positive terminal, install a test light between the bat-

tery terminal and the cable (series connection) and observe the test light. If the light is brightly illuminated, there is a circuit that continues to be activated. **Note:** *The test light will glow dimly because of the parasitic drain of the ECM, radio, clock, etc.*

12   Carefully remove the fuses one-by-one that govern accessories such as radio, blower motor, trunk lights, etc. until the test light goes out. Trace the short circuit in the particular fused circuit and repair the problem. Recheck the electrical system as described.

13   If all the fuses are pulled out and the test light remains lit, remove the output cable at the rear of the alternator then unplug all the connectors from the backside of the alternator. If the test light goes out, then there is an internal drain in the alternator or voltage regulator. Replace the alternator.

12   Alternator - removal and installation

Refer to illustrations 12.4a and 12.4b

1   Detach the cable from the negative terminal of the battery.

2   Unplug the electrical connectors from

the alternator.

3   Remove the drivebelt (see Chapter 1).

4   Remove the bolts and separate the alternator from the engine **(see illustrations)**.

5   Installation is the reverse of removal.

6   After the alternator is installed, install the drivebelt and reconnect the cable to the negative terminal of the battery.

13   Voltage regulator/alternator brushes - removal and installation

Removal

1   Remove the alternator (see Section 12).

2   Set the alternator on a clean workbench.

4.0L pushrod V6 models

Refer to illustrations 13.3a, 13.3b and 13.3c

3   Remove the four voltage regulator mounting screws and detach the voltage regulator from the alternator **(see illustrations)**. Detach the rubber plugs and remove the

12.4b Alternator mounting bolts (arrows) - 1996 and later 4.0L SOHC V6 and 5.0L V8 engines

13.3a To detach the voltage regulator/brush holder assembly, remove the four screws

13.3b Lift the assembly from the alternator

brush lead retaining screws and nuts to separate the brush leads from the holder **(see illustration)**.

4   Note the relationship of the brushes to the brush holder assembly, then move both brushes. Don't lose the springs.

5   If you're installing a new voltage regulator, insert the old brushes into the brush holder of the new regulator. If you're installing new brushes, insert them into the brush holder of the old regulator. Make sure the springs are properly compressed and the brushes are properly inserted into the recesses in the brush holder, then install the brush lead retaining screws and nuts.

4.0L SOHC V6 and 5.0L V8 models

Refer to illustrations 13.6 and 13.7

6   Remove the rear cover mounting screws and detach the rear cover from the alternator **(see illustration)**.

7   Remove three voltage regulator mounting screws and detach the voltage regulator from the alternator **(see illustration)**. **Note:** *The alternator brushes on these models are soldered to the voltage ... the voltage regulato... replaced as a unit.*

Installation

Refer to illustration 13...

8   Insert a short sec...
ened paper clip, throu...
age regulator (see illu...
brushes in the retracte...
lator installation.

9   Carefully install the...
the brushes don't hang u...

10   Install the voltage r...
tighten them securely.

11   Remove the wire or p...

12   Install the alternator ...

---

14   Starting system - ...
information and pr...

The function of the start... system is to crank the engine fast enough to start it. The system is composed of the starter motor, starter relay, battery, switch and connecting wires.

Turning the ignition key to the Start position actuates the starter relay through the starter control circuit. The starter relay then connects the battery to the starter. The battery supplies the electrical energy to the starter motor, which does the actual work of cranking the engine.

Vehicles equipped with an automatic transmission have a Neutral start switch in the starter control circuit, which prevents operation of the starter unless the shift lever is in Neutral or Park. The circuit on vehicles with a manual transmission prevents operation of the starter motor unless the clutch pedal is depressed.

Never operate the starter motor for more than 15 seconds at a time without pausing to

13.3c To remove the brushes from the voltage regulator/brush holder assembly, detach the rubber plugs from the two brush lead screws and remove both screws (arrows)

...l for at least two minutes. ...king can cause overheating, ...sly damage the starter.

...tor and circuit - in-
...ck

...nosing starter problems, ...ry is fully charged.

...motor doesn't turn at all ...operated, make sure the ...al or Park.

...battery is charged and ...he battery and starter ...e secure.

...r spins but the engine ...n the drive assembly in ...ipping and the starter ...d (see Section 16).

...tch is actuated, the ...operate at all but the ...tes (clicks), then the ...lies with either the battery, the

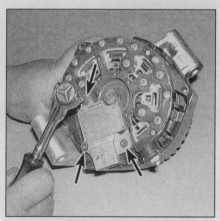

13.7 Detach the mounting screws (arrows) and remove the voltage regulator/ brush assembly from the alternator

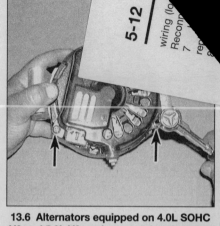

13.6 Alternators equipped on 4.0L SOHC V6 and 5.0L V8 engines, require removing the rear cover mounting screws (arrows) and rear cover to access the voltage regulator/brush assembly

starter solenoid contacts or the starter motor connections.

5   If the starter relay doesn't click when the ignition switch is actuated, a discharged battery is the most likely cause. If the battery is fully charged, the starter relay circuit may be open or the relay itself could be defective. Check the starter relay circuit or replace the relay (see Section 18).

6   To check the starter relay circuit, remove the push-on electrical connector from the relay. Make sure that the connection is clean and secure and the relay bracket is grounded. If the connections are good, check the operation of the relay with a jumper wire. To do this, place the transmission in Park (automatic) or Neutral (manual) and apply the parking brake. Connect a jumper wire between the battery positive terminal and the relay main terminal (BAT side) on the starter relay. If the starter motor now operates, the starter relay is okay. The problem is in the ignition switch, Neutral start switch, starter/clutch switch, or in the starting circuit

5

13.8 Before installing the voltage regulator/brush holder assembly, insert a paper clip as shown to hold the brushes in place during installation - after installation, simply pull the paper clip out

...ok for open or loose connections). ...ect the push-on electrical connector. ...f the starter motor still doesn't operate, ...place the starter solenoid (see Section 17).

If the starter motor cranks the engine at an abnormally slow speed, first make sure the battery is fully charged and all terminal connections are clean and tight. Also check the connections at the starter solenoid and battery ground. Eyelet terminals should not be easily rotated by hand. Also check for a short to ground. If the engine is partially seized, or has the wrong viscosity oil in it, it will crank slowly.

16 Starter motor - removal and installation

Refer to illustration 16.4

1 Detach the cable from the negative terminal of the battery.

2 Raise the vehicle and support it securely on jackstands. **Caution:** *If the vehicle is equipped with Automatic Ride Control (ARC), make sure the air suspension switch is turned to the OFF position before the vehicle is raised to prevent damage to the system components (see Section 20).*

3 Remove the starter solenoid cover (if equipped). Disconnect the large cable from the terminal on the starter motor and the solenoid terminal connections.

4 Remove the starter motor mounting bolt and nut **(see illustration)** and detach the starter from the engine.

5 Installation is the reverse of removal.

17 Starter solenoid - replacement

Refer to illustrations 17.3a and 17.3b

1 Remove the starter from the engine (see Section 16).

2 Remove the electrical connector from the solenoid M terminal.

3 Remove the solenoid mounting bolts and separate the solenoid from the starter body **(see illustrations)**.

4 Installation is the reverse of removal.

18 Starter relay - removal and installation

Refer to illustrations 18.2 and 18.6

1 Disconnect the negative terminal of the battery.

2 The starter relay is mounted in the engine compartment between the battery and the power distribution box **(see illustration)**.

3 Label the wires and the terminals on the relay to prevent mix-up during installation.

4 Disconnect the ignition feed wire (push-on connector) from the starter relay.

5 Remove the positive battery cable and the starter feed cable retaining nuts, then detach the wires from the starter relay.

6 Remove the relay mounting bolts and detach the relay from the inner fenderwell **(see illustration)**.

7 Installation is the reverse of removal.

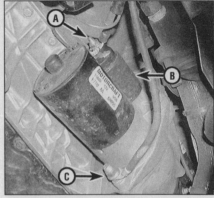
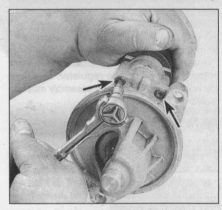

16.4 Remove the electrical connectors (A), the upper mounting bolt (B) and the lower mounting nut (C), then detach the starter from the transmission bellhousing

17.3a Remove the solenoid mounting bolts (arrows)

17.3b Separate the solenoid from the starter

18.2 The starter relay is mounted in the engine compartment next to the battery

18.6 Remove the starter relay mounting bolts (arrows) to detach it from the inner fenderwell

Chapter 6
Emissions and engine control systems

Contents

6

1   General information

Refer to illustrations 1.1a, 1.1b, 1.1c and 1.7

To prevent pollution of the atmosphere from incompletely burned and evaporating gases, and to maintain good driveability and fuel economy, a number of emission control systems are incorporated **(see illustrations)**. They include the:

Electronic Engine Control system
Evaporative Emission Control System (EVAP)
Positive Crankcase Ventilation (PCV) system
Exhaust Gas Recirculation (EGR) system
Catalytic converter

All of these systems are linked, directly or indirectly, to the emission control system.

The Sections in this Chapter include general descriptions, checking procedures within the scope of the home mechanic and component replacement procedures (when possible) for each of the systems listed above.

Before assuming that an emissions control system is malfunctioning, check the fuel and ignition systems carefully. The diagnosis of some emission control devices requires specialized tools, equipment and training. If checking and servicing become too difficult or if a procedure is beyond your ability, consult a dealer service department. Remember, the most frequent cause of emissions problems is simply a loose or broken vacuum hose or wire, so always check the hose and wiring connections first.

This doesn't mean, however, that emission control systems are particularly difficult to maintain and repair. You can quickly and easily perform many checks and do most of the regular maintenance at home with common tune-up and hand tools. **Note:** *Because of a Federally mandated warranty which covers the emission control system components, check with your dealer about warranty coverage before working on any emissions-related systems. Once the warranty has expired, you may wish to perform some of the component checks and/or replacement procedures in this Chapter to save money.*

Control System Component Locator

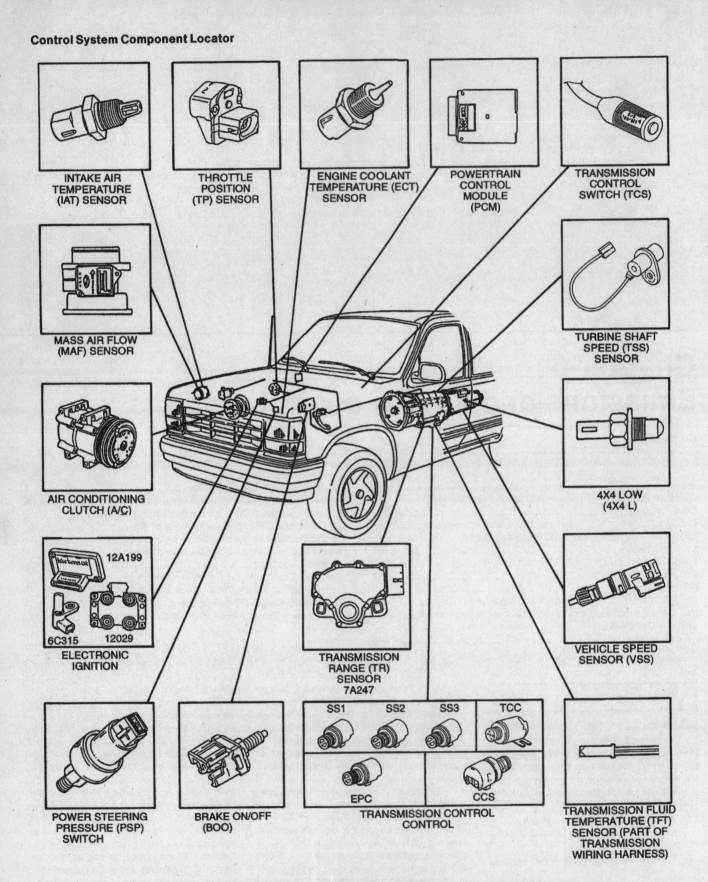

1.1a Typical emission and engine control system components - 4.0L pushrod V6 engine

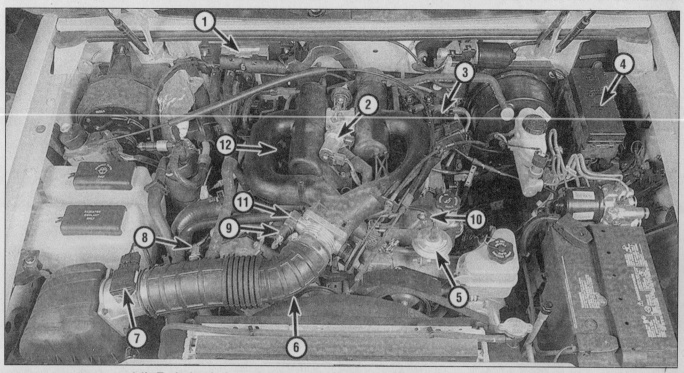

1.1b Typical emission and engine control system components - 4.0L SOHC V6 engine

1	Powertrain Control Module (PCM) (under cowl)	6	Crankshaft Position (CKP) sensor (mounted on front of engine next to harmonic balancer)	9 Throttle Position (TPS) sensor
2	IAC valve			10 Camshaft Position (CMP) sensor
3	Ignition coil pack	7	MAF sensor	11 Engine Coolant Temperature (ECT) (under throttle body)
4	Power distribution box	8	IAT sensor	12 Knock sensor (KS)
5	EGR valve			

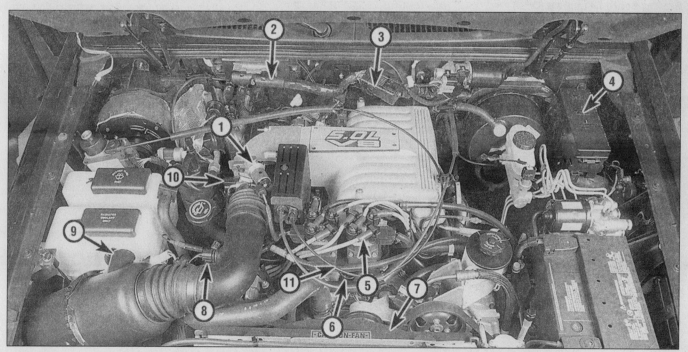

1.1c Typical emission and engine control system components - 5.0L V8 engine

1	IAC valve	5	Ignition coil pack	9 MAF sensor
2	Powertrain Control Module (PCM) (under cowl)	6	Camshaft Position (CMP) sensor	10 Throttle Position (TPS) sensor
3	PCV valve (at rear of intake manifold)	7	Crankshaft Position (CKP) sensor (next to harmonic balancer)	11 Engine Coolant Temperature (ECT) (on thermostat housing)
4	Power distribution box	8	IAT sensor	

6

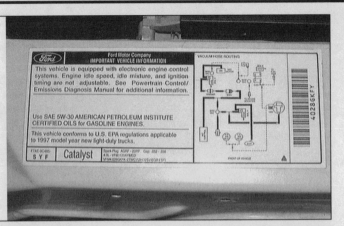

1.7 The Vehicle Emission Control Information (VECI) label is located in the engine compartment and contains information on the emission devices on your vehicle, vacuum line routing, etc.

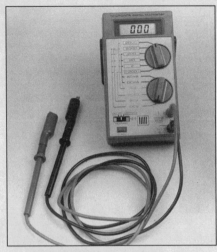

2.1 Digital multimeters can be used for testing all types of circuits; because of their high impedance, they are much more accurate than analog meters for measuring voltage in low-voltage computer circuits

Pay close attention to any special precautions outlined in this Chapter. It should be noted that the illustrations of the various systems may not exactly match the system installed on the vehicle you're working on because of changes made by the manufacturer during production or from year-to-year.

A Vehicle Emissions Control Information (VECI) label is located in the engine compartment **(see illustration)**. This label contains important emissions specifications and adjustment information, as well as a vacuum hose schematic with emissions components identified. When servicing the engine or emissions systems, the VECI label in your particular vehicle should always be checked for up-to-date information.

2 On Board Diagnosis (OBD) system and trouble codes

Note: *1995 and earlier vehicles are equipped with OBD-I self diagnosis system, while 1996 and later vehicles are equipped with OBD-II self diagnosis system. Both systems require the use of a Scan tool to access trouble codes. However, many of the information sensor checks and replacement procedures*

do apply to both systems. Because OBD-I and OBD-II systems require a special SCAN tool to access the trouble codes, have the vehicle diagnosed by a dealer service department or other qualified automotive repair facility if the proper SCAN tool is not available. The codes indicated in the text are designed and mandated by the EPA for all 1995 and earlier OBD-I and 1996 and later OBD-II vehicles produced by automobile manufacturers. These generic trouble codes do not include the manufacturer's specific trouble codes. Consult a dealer service department for additional information. Refer to the troubleshooting tips in the beginning of this manual to gain some insight to the most likely causes to a problem.

Diagnostic tool information

Refer to illustrations 2.1, 2.2 and 2.4

1   A digital multimeter is necessary for checking fuel injection and emission related components **(see illustration)**. A digital volt-ohmmeter is preferred over the older style analog multimeter for several reasons. The analog multimeter cannot display the volts-ohms or amps measurement in hundredths and thousandths increments. When working with electronic circuits which are often very

low voltage, this accurate reading is most important. Another good reason for the digital multimeter is the high impedance circuit. The digital multimeter is equipped with a high resistance internal circuitry (10 million ohms). Because a voltmeter is hooked up in parallel with the circuit when testing, it is vital that none of the voltage being measured should be allowed to travel the parallel path set up by the meter itself. This dilemma does not show itself when measuring larger amounts of voltage (9 to 12 volt circuits) but if you are measuring a low voltage circuit such as the oxygen sensor signal voltage, a fraction of a volt may be a significant amount when diagnosing a problem. Obtaining the diagnostic trouble codes is one exception where using an analog voltmeter is necessary.

2   Hand-held scanners are the most pow-

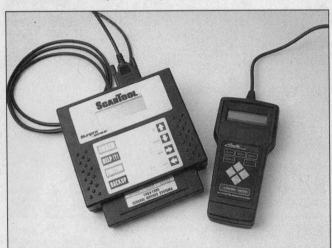

2.2 Scanners like the Actron Scantool and the AutoXray XP240 are powerful diagnostic aids - programmed with comprehensive diagnostic information, they can tell you just about anything you want to know about your engine management system

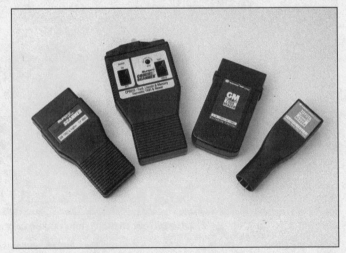

2.4 Trouble code tools simplify the task of extracting the trouble codes on OBD-I systems

erful and versatile tools for analyzing engine management systems used on later model vehicles **(see illustration)**. Early model scanners handle codes and some diagnostics for many OBD-I systems. Each brand scan tool must be examined carefully to match the year, make and model of the vehicle you are working on. Often interchangeable cartridges are available to access the particular manufacturer (Ford, GM, Chrysler, etc.). Some manufacturers will specify by continent (Asia, Europe, USA, etc.).

3   With the arrival of the Federally mandated emission control system (OBD-II), a specially designed scanner has been developed. Several tool manufacturers have released OBD-II scan tools for the home mechanic. Ask the parts salesman at a local auto parts store for additional information concerning dates and costs.

4   Another type of code reader is available at parts stores **(see illustration)**. These tools simplify the procedure for extracting codes from the engine management computer on OBD-I systems by simply "plugging in" to the diagnostic connector on the vehicle wiring harness. **Note:** *Some diagnostic connectors are located under the dash, kick panel or glovebox while others are located in the engine compartment.*

OBD system general description

5   All 1995 and earlier vehicles are equipped with the Electronic Engine Control (EEC) IV (OBD-I) system. All 1996 and later engines and powertrain combinations described in this manual are equipped with the EEC V (OBD-II) system. Both systems consist of an onboard computer, known as the Powertrain Control Module (PCM), and information sensors, which monitor various functions of the engine and send data to the PCM. Based on the data and the information programmed into the computer's memory, the PCM generates output signals to control various engine functions via control relays, solenoids and other output actuators.

6   The PCM is the "brain" of the Electronic Engine Control (EEC) system. It receives data from a number of sensors and other electronic components (switches, relays, etc.). Based on the information it receives, the PCM generates output signals to control various relays, solenoids and other actuators. The PCM is specifically calibrated to optimize the emissions, fuel economy and driveability of the vehicle.

7   Because of a Federally mandated warranty which covers the EEC-IV and EEC-V system components and because any owner-induced damage to the PCM, the sensors and/or the control devices may void the warranty, it isn't a good idea to attempt diagnosis or replacement of the PCM at home while the vehicle is under warranty. Take the vehicle to a dealer service department if the PCM or a system component malfunctions.

Information sensors

8   **Heated Oxygen sensors (HO2S)** - The HO2S generates a voltage signal that varies with the difference between the oxygen content of the exhaust and the oxygen in the surrounding air.

9   **Crankshaft position (CKP) sensor** -. The CKP sensor provides information on crankshaft position and the engine speed signal to the PCM.

10   **Camshaft position (CMP) sensor** - The CMP sensor produces a signal which the PCM uses to identify number 1 cylinder and to time the sequential fuel injection.

11   **Engine coolant temperature (ECT) sensor** - The ECT monitors engine coolant temperature and sends the PCM a voltage signal that affects PCM control of the fuel mixture, ignition timing, and EGR operation.

12   **Intake Air Temperature (IAT) sensor** - The IAT provides the PCM with intake air temperature information. The PCM uses this information to control fuel flow, ignition timing, and EGR system operation.

13   **Throttle Position Sensor (TPS)** - The TPS senses throttle movement and position, then transmits a voltage signal to the PCM. This signal enables the PCM to determine when the throttle is closed, in a cruise position, or wide open.

14   **Mass airflow (MAF) sensor** - The MAF sensor measures the molecular mass of the intake airflow entering the engine. The MAF sensor, along with the IAT sensor, provide mass airflow and air temperature information for the most precise fuel metering.

15   **Vehicle speed sensor (VSS)** - The vehicle speed sensor provides information to the PCM to indicate vehicle speed.

16   **Knock sensor (KS)** – The knock sensor is a piezoresistive microphone that detects the sound of engine detonation, or "pinging". The PCM uses the input signal from the knock sensor to recognize detonation and retard spark advance to avoid engine damage. The knock sensor is used only on the 4.0L SOHC V6 engine.

17   **EGR backpressure sensor** - The EGR backpressure sensor is used to monitor the rate and flow of exhaust gas recirculation into the intake system.

18   **Fuel tank pressure sensor** - The fuel tank pressure sensor used on 1996 and later OBD-II vehicles is part of the evaporative emission control system and is used to monitor vapor pressure in the fuel tank. The PCM uses this information to turn on and off the purge valves and solenoids of the evaporative emission system.

19   **Power steering pressure (PSP) switch** - The PSP sensor is used to increase transmission hydraulic line pressure during low-speed vehicle maneuvers.

20   **Brake On-Off (BOO) switch** - This switch is used when the driver applies the brakes, to disengage the cruise control and change fuel metering and spark advance during deceleration.

21   **Transmission sensors** - In addition to the vehicle speed sensor, the PCM receives input signals on 1995 and later vehicles from the following sensors inside the transmission or connected to it: (a) the turbine shaft speed sensor, (b) the transmission fluid temperature sensor, and (c) the transmission range sensor.

22   **A/C clutch control switch** When battery voltage is applied to the air conditioning compressor solenoid, a signal is sent to the PCM, which interprets the signal as an added load created by the compressor and increases engine idle speed accordingly to compensate.

Output actuators

23   **PCM power relay** - The main PCM power relay is activated by the ignition switch and supplies battery power to the PCM and the Electronic Engine Control (EEC) system when the switch is in the Start or Run position. Refer to Chapter 12 or your owner's manual for more information on relay location.

24   **Fuel pump relay** - The fuel pump relay is activated by the PCM with the ignition switch in the Start or Run position. When the ignition switch is turned on, the relay is activated to supply initial line pressure to the fuel system. For more information on fuel pump check and replacement, refer to Chapter 4.

25   **Fuel injectors** - The PCM opens the fuel injectors individually in firing order sequence. The PCM also controls the time the injector is open, called the "pulse width." The pulse width of the injector (measured in milliseconds) determines the amount of fuel delivered. For more information on the fuel delivery system and the fuel injectors, including injector replacement, refer to Chapter 4.

26   **Ignition Control Module (ICM)** - The ICM triggers the ignition coils and determines proper spark advance based on inputs from the PCM. An externally mounted ICM is used on all 1995 and earlier vehicles. All other models use an ignition module which is incorporated into the PCM. Refer to Chapter 5 for more information on the Ignition Control Module.

27   **Idle air control (IAC) valve** - The IAC valve controls the amount of air to bypass the throttle plate when the throttle valve is closed or at idle position. The IAC valve opening and the resulting airflow is controlled by the PCM. Refer to Chapter 4 for more information on the IAC valve.

28   **EGR vacuum solenoid** - The EGR vacuum solenoid is controlled by the PCM to regulate the opening of the vacuum-operated EGR valve.

29   **Canister purge valve** - The evaporative emission canister purge valve is a solenoid valve, operated by the PCM to purge the fuel vapor canister and route fuel vapor to the intake manifold for combustion.

30   **Canister vent solenoid** - The evaporative emission canister vent solenoid is operated by the PCM during the OBD-II evaporative emission monitor (see Section 3) and during an emission test of the evaporative system.

6

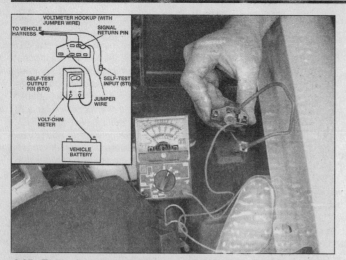 

2.35a To read any stored trouble codes, connect a voltmeter to the Diagnostic Test connector as shown, then connect a jumper wire between the self-test input and pin number 2 on the larger connector - turn the ignition key ON (engine not running) and watch the voltmeter needle or CHECK ENGINE light on later models

2.35b The EEC-IV (OBD-I) diagnostic connector is located behind the fuse box on the passenger side inner fenderwell

Obtaining OBD-I system codes

Refer to illustrations 2.35a and 2.35b

31   The diagnostic codes for the EEC-IV systems are arranged in such a way that a series of tests must be completed in order to extract ALL the codes from the system. If one portion of the test is performed without the others, there may be a chance the trouble code that will pinpoint a problem in your particular vehicle will remain stored in the PCM without detection. The tests start first with a Key On, Engine Off (KOEO) test followed by a computed timing test then finally a Engine Running (ER) test. Here is a brief overview of the code extracting procedures of the EEC-IV system followed by the actual test:

Quick Test - Key On Engine Off (KOEO)

32   The following tests are all included with the key on, engine off:

Self test codes - These codes are accessed on the test connector by using a jumper wire and an analog voltmeter or the factory diagnostic tool called the Star tester. These codes are also called *Hard Codes.*

Separator pulse codes - After the initial Hard Codes, the system will flash a code 111 and then will flash a series of Soft (or Continuous Memory) Codes.

Continuous Memory Codes - These codes indicate a fault that may or may not be present at the time of testing. These codes usually indicate an intermittent failure. Continuous Memory codes are stored in the system and they will flash after the normal Hard Codes. These codes are three digit codes (1994 and 1995). These codes can indicate chronic or intermittent problems. Also called *Soft Codes.*

Fast codes - These codes are transmitted 100 times faster than normal codes and can only be read by a Star Tester from Ford or an equivalent SCAN tool.

Engine running codes (KOER) or (ER)

33   **Running tests** - These tests make it possible for the PCM to pick-up a diagnostic trouble code that cannot be set while the engine is in KOEO. These problems usually occur during driving conditions. Some codes are detected by cold or warm running conditions, some are detected at low rpm or high rpm and some are detected at closed throttle or WOT.

I.D. Pulse codes - These codes indicate the type of engine (4, 6 or 8 cylinder) or the correct module and Self Test mode access.

Computed engine timing test - This engine running test determines base timing for the engine and starts the process of allowing the engine to store running codes.

Wiggle test - This engine running test checks the wiring system to the sensors and output actuators as the engine performs.

Cylinder balance test - This engine running test determines injector balance as well as cylinder compression balance. **Note:** *This test should be performed by a dealer service department.*

Beginning the test

34   Position the parking brake ON, Shift lever in PARK, block the drive wheels and turn off all electrical loads (air conditioning, radio, heater fan blower etc.). Make sure the engine is warmed to normal operating temperature (if possible).

35   Perform the KOEO tests:

a) *Turn the ignition key off for at least 10 seconds*

b) *Locate the Diagnostic Test connector inside the engine compartment. Install the voltmeter leads onto the battery and pin number 4 (STO) of the test connector (see illustrations). Install a jumper wire from the test terminal to pin number 2 of the Diagnostic Test terminal (STI).*

c) *Turn the ignition key ON (engine not running) and observe the needle sweeps on the voltmeter. For example on code 123, the voltmeter will sweep once and then pause. There will be a two second pause between digits and then there will be two distinct sweeps of the needle to indicate the second digit of the code number. There will be another pause and then three distinct sweeps of the meter. It is possible to also observe the CHECK ENGINE light flash the codes. Additional codes will be separated by a four second pause and then the indicated sweeps on the voltmeter. Be aware that the code sequence may continue into the continuous memory codes (read further).* **Note:** *These models will flash the CHECK ENGINE light on the dash in place of the voltmeter.*

36   Interpreting the continuous memory codes:

a) *After the KOEO codes are reported, there will be a short pause and any stored Continuous Memory codes will appear in order. Remember that the "Separator" code is 111. The computer will not enter the Continuous Memory mode without flashing the separator pulse code. The Continuous Memory codes are read the same as the initial codes or "Hard Codes". Record these codes onto a piece of paper and continue the test.*

37   Perform the Engine Running (ER) tests.

a) *Remove the jumper wires from the Diagnostic Test connector to start the test.*

b) *Run engine until it reaches normal operating temperature.*

c) *Turn the engine OFF for at least 10 seconds.*

d) *Install the jumper wire onto Diagnostic Test connector (see illustrations 2.35a and 2.35b) and start the engine.*

e) *Observe that the voltmeter or CHECK ENGINE light will flash the engine identification code. This code indicates 1/2 the number of cylinders of the engine. For example, 4 flashes represent an 8 cylinder engine, or 3 flashes represent a six cylinder engine.*

f) *Within 1 to 2 seconds of the I.D. code, turn the steering wheel at least 1/2 turn and release. This will store any power steering pressure switch trouble codes.*

g) *Depress the brake pedal and release. Note: Perform the steering wheel and brake pedal procedure in succession immediately (1 to 2 seconds) after the I.D. codes are flashed.*

h) *Observe all the codes and record them on a piece of paper. Be sure to count the sweeps or flashes very carefully as you jot them down.*

38 On some models the PCM will request a Dynamic Response check. This test quickly checks the operation of the TPS, MAF or MAP sensors in action. This will be indicated by a code 1 or a single sweep of the voltmeter needle (one flash on CHECK ENGINE light). This test will require the operator to simply full throttle ("goose") the accelerator pedal for one second. DO NOT throttle the accelerator pedal unless it is requested.

39 The next part of this test makes sure the system can advance the timing. This is called the Computed Timing test. After the last ER code has been displayed, the PCM will advance the ignition timing a fixed amount and hold it there for approximately 2 minutes.

Use a timing light to check the amount of advance. The computed timing should equal the base timing plus 20 BTDC. The total advance should equal 27 to 33 degrees advance. If the timing is out of specification, have the system checked at a dealer service department. **Note:** *Remember to remove the SPOUT from the connector as described in the ignition timing procedure in Chapter 5. This will remove the computer from the loop and give base timing.*

40 Finally perform the Wiggle Test. This test can be used to recreate a possible intermittent fault in the harness wiring system.

a) *Use a jumper wire to ground the STI lead on the Diagnostic Test connector* (see illustration 2.35a).

b) *Turn the ignition key ON (engine not running).*

c) *Now deactivate the self test mode (remove the jumper wire) and then immediately reactivate the self-test mode. Now the system has entered Continuous Monitor Test Mode.*

d) *Carefully wiggle, tap or remove any suspect wiring to a sensor or output actuator. If a problem exists, a trouble code will be stored that indicates a problem with the circuit that governs the particular component. Record the codes that are indicated.*

e) *Next, enter Engine Running Continuous Monitor Test Mode to check for wiring problems only when the engine is running. Start first by deactivating the Diagnostic Test connector and turning the*

ignition key OFF. Now start the engine and allow it to idle.

f) *Use a jumper wire to ground the STI lead on the Diagnostic Test connector* (see illustration 2.35a). *Wait ten seconds and then deactivate the test mode and reactivate it again (install jumper wire). This will enter Engine Running Continuous Monitor Test Mode.*

g) *Carefully wiggle any suspect wiring to a sensor or output actuator. If a problem exists, a trouble code will be stored that indicates a problem with the circuit that governs the particular component. Record the codes that are indicated.*

41 If necessary, perform the Cylinder Balance Test. This test must be performed by a dealer service department.

Clearing OBD-I system codes

42 To clear the codes from the PCM memory, start the KOEO self test diagnostic procedure **(see illustration 2.35a)** and install the jumper wire into the Diagnostic Test connector. When the codes start to display themselves on the voltmeter or CHECK ENGINE light, remove the jumper wire from the Diagnostic Test connector. This will erase any stored codes within the system. **Caution:** *Do not disconnect the battery from the vehicle to clear the codes. This will erase stored operating parameters from the KAM (Keep Alive Memory) and cause the engine to run rough for a period of time while the computer relearns the information.*

OBD-I Trouble Codes

Code	Test Condition*	Probable Cause
11	O,C,R	Pass (separator code)
12	R	RPM not within Self-test upper limit
13	R	RPM not within Self-test lower limit
14	C	Profile Ignition Pick-up circuit fault
15	O	Read Only Memory test failed
15	C	Keep Alive Memory test failed
16	R	RPM too low to perform Oxygen Sensor/fuel test
18	C	Loss of TACH input to PCM; SPOUT circuit grounded
18	R	SPOUT circuit open
19	O	Failure in EEC reference voltage
21	O,R	Coolant Temperature Sensor out of range
22	O,C	Manifold Absolute/Baro Pressure Sensor out of range
23	O,R	Throttle Position Sensor out of range
24	O,R	Intake Air Temperature sensor out of range
26	O,R	Mass Air Flow Sensor out of range
29	C	No input from Vehicle Speed Sensor
31	O,C,R	EGR Valve Position Sensor out of range (low)

6

OBD-I Trouble Codes (continued)

Code	Test Condition*	Probable Cause
32	O,C,R	EGR valve not seated; closed voltage low
33	C,R	EGR valve not opening; Insufficient flow detected
34	O,C,R	EGR Valve Pressure Transducer/Position Sensor sonic voltage above closed limit
35	O,C,R	EGR Valve Pressure Transducer/Position Sensor voltage out of range (high)
41	R	Heated Oxygen Sensor circuit indicates system lean, right side
41	C	No Heated Oxygen Sensor switch detected, right side
42	R	Heated Oxygen Sensor circuit indicates system rich, right side
44	R	Secondary Air system inoperative, right side
45	R	Secondary Air upstream during Self-test
45	C	Distributorless Ignition system (DIS) coil pack circuit failure
46	R	Secondary Air not by-passed during Self-test
51	O,C	Coolant Temperature sensor circuit open
53	O,C	Throttle Position sensor out of range (high)
54	O,C	Intake Air Temperature sensor circuit open
56	O,C	Mass Air Flow sensor out of range (high)
61	O,C	Coolant Temperature sensor circuit grounded
63	O,C	Throttle Position sensor circuit out of range (low)
64	O,C	Intake Air Temperature sensor circuit grounded
66	C	Mass Air Flow sensor circuit out-of-range (low)
67	O	Neutral Drive Switch circuit open
72	R	Insufficient Mass Air Flow change during Dynamic Response Test
73	R	Insufficient Throttle Position output during Dynamic Response Test
74	R	Brake On/Off switch failure
75	R	Brake On/Off circuit failure
77	R	Wide Open Throttle not sensed during Self-test
79	O	Air conditioning on during self-test
81	O	Secondary Air Injection Diverter (AIRD) circuit failure
82	O	Secondary Air Injection Bypass (AIRB) circuit failure
84	O	EGR Vacuum Regulator circuit failure
85	O	Canister Purge circuit failure
87	O,C	Primary Fuel Pump circuit failure
91	R	Heated oxygen sensor indicates system lean, left side
91	C	No heated oxygen sensor switching indicated, left side
92	R	Heated oxygen sensor indicates system rich, left side
94	R	Torque Converter Clutch (TCC) solenoid circuit failure
95	O,C	Fuel Pump circuit open, PCM to motor
96	O,C	Fuel Pump circuit open, Battery to PCM
98	R	Hard Fault present
111	O,C,R	Pass
112	O,R	Intake Air Temperature sensor circuit indicates circuit grounded/above 245 degrees F

Code	Test Condition*	Probable Cause
113	O,R	Intake Air Temperature sensor circuit indicates open circuit/below -40 degrees F
114	O,R	Intake Air Temperature sensor out of self-test range
116	O,R	Coolant Temperature sensor out of self-test range
117	O,C	Coolant Temperature circuit below minimum voltage/indicates above 245 degrees F
118	O,C	Coolant Temperature sensor circuit above maximum voltage/ indicates below -40 degrees F
121	O,C,R	Throttle Position sensor out of self-test range
122	O,C	Throttle Position sensor below minimum voltage
123	O,C	Throttle Position sensor above maximum voltage
124	C	Throttle Position Sensor voltage higher than expected
125	C	Throttle Position Sensor voltage lower than expected
126	O,C,R	MAP/BARO sensor higher than expected
128	C	MAP sensor vacuum hose damaged or disconnected
129	R	Insufficient Manifold Absolute Pressure/Mass Air Flow change during Dynamic Response Check
136	R	Heated oxygen sensor indicates lean condition, left side
137	R	Heated oxygen sensor indicates rich condition, left side
139	C	No heated oxygen sensor switching detected, left side
144	C	No heated oxygen sensor switching detected, right side
157	R,C	Mass Air Flow Sensor below minimum voltage
158	O,C,R	Mass Air Flow Sensor above maximum voltage
159	O,R	Mass Air Flow Sensor out of self-test range
167	R	Insufficient Throttle Position Sensor change during Dynamic Response Check
171	C	Heated oxygen sensor unable to switch, right side
172	R,C	Heated oxygen sensor indicates lean condition, right side
173	R,C	Heated oxygen sensor indicates rich condition, right side
174	C	Heated oxygen sensor switching slow, right side
175	C	Heated oxygen sensor unable to switch, left side
176	C	Heated oxygen sensor indicates lean condition, left side
177	C	Heated oxygen sensor indicates rich condition, left side
178	C	Heated oxygen sensor switching slow, left side
179	C	Adaptive Fuel lean limit reached at part throttle, system rich, right side
181	C	Adaptive Fuel rich limit reached at part throttle, right side
182	C	Adaptive Fuel lean limit reached at idle, right side
183	C	Adaptive Fuel rich limit reached at idle, right side
184	C	Mass Air Flow higher than expected
185	C	Mass Air Flow lower than expected
186	C	Injector Pulse-width higher than expected
187	C	Injector Pulse-width lower than expected
188	C	Adaptive Fuel lean limit reached, left side
189	C	Adaptive Fuel rich limit reached, left side
211	C	Profile Ignition Pick-up circuit fault
212	C	Ignition module circuit failure/SPOUT circuit grounded

6

OBD-I Trouble Codes (continued)

Code	Test Condition*	Probable Cause
213	R	SPOUT circuit open
214	C	Cylinder identification (CID) circuit failure
215	C	PCM detected coil 1 primary circuit failure
216	C	PCM detected coil 2 primary circuit failure
217	C	PCM detected coil 3 primary circuit failure
218	C	Loss of ignition diagnostic monitor (IDM) signal - left side (dual plug EI)
222	C	Loss of ignition diagnostic monitor (IDM) signal - right side (dual plug)
223	C	Loss of dual plug Inhibit (DPI) control (dual plug)
224	C	PCM detected coil 1,2,3 or 4 primary circuit failure (dual plug EI)
225	C	Knock sensor not detected during dynamic response test KOER
226	O	Ignition Diagnostic Module (IDM) signal not received (EI)
232	C	PCM detected coil 1,2,3 or 4 primary circuit failure (EI)
311	R	Secondary Air System inoperative, right side
312	R	Secondary Air not by-passed
313	R	Secondary Air inoperative, left side
327	O,C,R	EGR Valve Pressure Transducer/Position Sensor circuit below minimum voltage
328	O,C,R	EGR valve position sensor voltage below closed limit
332	C,R	EGR valve opening not detected
334	O,C,R	EGR valve position sensor voltage above closed limit
335	O	EGR Sensor voltage out-of-range
336	R	EGR circuit higher than expected
337	O,C,R	EGR Valve Pressure Transducer/Position Sensor circuit above maximum voltage
341	O	Octane adjust service pin open
411	R	Unable to control RPM during Low RPM Self-test
412	R	Unable to control RPM during High PRM Self-test
452	C	No input from Vehicle Speed Sensor
511	O	Read Only Memory test failed - replace PCM
512	C	Keep Alive Memory test failed
513	O	Internal voltage failure in PCM
519	O	Power steering pressure switch (PSP) circuit open (1993 and 1994 only)
521	R	Power steering pressure switch (PSP) circuit did not change states (1993 and 1994 only)
522	O	Manual Lever Position (MLP) sensor circuit open/vehicle in gear
528	O	Clutch pedal position (CPP) circuit failure
536	C,R	Brake ON/OFF (BOO) circuit failure/not activated during the KOER
538	R	Insufficient change in RPM/operator error in Dynamic Response Check
538	R	Invalid cylinder balance test due to throttle movement during test (1995 and later models only)
538	R	Invalid cylinder balance test due to CID circuit failure (1995 and later models)
539	O	Air conditioning on during Self-test
542	O,C	Fuel Pump circuit open; PCM to motor
543	O,C	Fuel Pump circuit open; Battery to PCM

Code	Test Condition*	Probable Cause
551	O	Idle Air Control (IAC) circuit failure KOEO
552	O	Air Management 1 circuit failure
552	O	Secondary Air Injection Bypass (AIRB) circuit failure
553	O	Secondary Air Injection Diverter (AIRB) circuit failure
556	O,C	Primary Fuel Pump circuit failure
558	O	EGR Vacuum Regulator circuit failure
565	O	Canister Purge circuit failure
566	O	3-4 shift solenoid circuit failure KOEO (A4LD transmission)
569	O	Auxiliary Canister Purge (AUX-CANP) circuit failure
617	C	1-2 shift error
618	C	2-3 shift error
619	C	3-4 shift error
621	O,C	Shift Solenoid 1 (SS 1) circuit failure KOEO
622	O	Shift Solenoid 2 (SS2) circuit failure KOEO
624	O,C	Electronic Pressure Control (EPC) circuit failure
625	O,C	Electronic Pressure Control (EPC) driver open in PCM
626	O	Coast Clutch Solenoid (CCS) circuit failure KOEO
628	C	Excessive converter clutch slippage
629	O,C	Torque Converter Clutch (TCC) solenoid circuit failure
631	O	Transmission Control Indicator Lamp (TCIL) circuit failure KOEO
632	R	Transmission Control Switch (TCS) circuit did not change states during KOER
633	O	4X4L switch closed during KOEO
634	O,C,R	Manual Lever Position (MLP) sensor voltage higher or lower than expected
636	O,R	Transmission Fluid Temp (TFT) higher or lower than expected
637	O,C	Transmission Fluid Temp (TFT) sensor circuit above maximum voltage/ -40°F (-40°C) indicated / circuit open
638	O,C	Transmission Fluid Temp (TFT) sensor circuit below minimum voltage/ 290°F (143°C) indicated / circuit shorted
639	C,R	Insufficient input from Transmission Speed Sensor (TSS)
641	O	Shift Solenoid 3 (SS3) circuit failure KOEO
643	O	Coast Clutch Solenoid (CCS) circuit failure KOEO
652	O	Torque Converter Clutch (TCC) solenoid circuit failure
654	O	Transmission Range (TR) sensor not indicating PARK during KOEO
655	O	Transmission Range (TR) sensor not indicating NEUTRAL during KOEO
656	C	Torque Converter Clutch continuous slip error
657	C	Transmission fluid temperature (TFT) overheating
667	C	Transmission range (TR) circuit voltage above minimum voltage
668	C	Transmission range (TR) circuit voltage above maximum voltage
691	C	4X4 LOW switch open or short circuit present (1995 and later models)
692	C	Transmission state does not match calculated ratio (1995 and later models)
998	O	Hard fault present

* O = Key On, Engine Off; C = Continuous Memory; R = Engine Running

6

Obtaining OBD-II system codes

Refer to illustration 2.44

43   On OBD-II systems, the PCM will illuminate the Malfunction Indicator Light on the dash if it recognizes a component fault for two consecutive drive cycles. It will continue to set the light until the PCM does not detect any malfunction for three or more consecutive drive cycles. Because the OBD-II system requires a SCAN tool to reset the light, if the tool is not available for diagnostics, have the system checked by a dealer service department or other qualified repair facility.

44   The diagnostic codes for the EEC-V (OBD-II) systems can be extracted from the PCM using a special SCAN tool that is programmed to interface with this new system by plugging into the DLC **(see illustration)**. If the tool is not available, have the vehicle checked at a dealer service department.

Clearing OBD-II system codes

45   To clear the codes from the PCM memory, install the OBD-II SCAN tool, scroll the menu for the function that describes "CLEARING CODES' and follow the prescribed method for that particular SCAN tool. If necessary, have the codes cleared by a dealer service department or other qualified repair facility. **Caution:** *Do not disconnect the battery from the vehicle to clear the codes. This will erase stored operating parameters from the memory and cause the engine to run rough for a period of time while the computer relearns the information.*

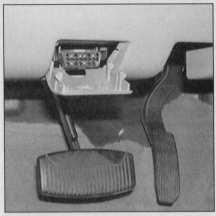

2.44 Typical Data Link Connector (DLC) on an EEC-V (OBD-II) equipped vehicle

OBD-II Trouble Codes

Code	Probable cause
P0102	Mass Airflow (MAF) sensor circuit low input
P0103	Mass Airflow (MAF) sensor circuit high input
P0112	Intake Air Temperature (IAT) sensor circuit low input
P0113	Intake Air Temperature (IAT) sensor circuit high input
P0117	Electronic Coolant Temperature (ECT) sensor circuit low input
P0118	Electronic Coolant Temperature (ECT) sensor circuit high input
P0121	In range Throttle Position Sensor (TPS) fault
P0122	Throttle Position Sensor (TPS) circuit low input
P0123	Throttle Position Sensor (TPS) circuit high input
P0131	Upstream heated O2 sensor circuit low voltage (Bank 1)
P0133	Upstream heated O2 sensor circuit slow response (Bank 1)
P0135	Upstream heated O2 sensor heater circuit fault (Bank 1)
P0136	Downstream heated O2 sensor fault (Bank 1)
P0141	Downstream heated O2 sensor heater circuit fault (Bank 1)
P0151	Upstream heated O2 sensor circuit low voltage (Bank 2)
P0153	Upstream heated O2 sensor circuit slow response (Bank 2)
P0155	Upstream heated O2 sensor heater circuit fault (Bank 2)
P0156	Downstream heated O2 sensor fault (Bank 2)
P0161	Downstream heated O2 sensor heater circuit fault (Bank 2)
P0171	System Adaptive fuel too lean (Bank 1)
P0172	System Adaptive fuel too rich (Bank 1)
P0174	System Adaptive fuel too lean (Bank 2)
P0172	System Adaptive fuel too rich (Bank 2)
P0191	Injector Pressure sensor system performance
P0192	Injector Pressure sensor circuit low input
PO193	Injector Pressure sensor circuit high input
PO301	Cylinder no. 1 misfire detected
PO302	Cylinder no. 2 misfire detected

Code	Probable cause
PO303	Cylinder no. 3 misfire detected
PO304	Cylinder no. 4 misfire detected
PO305	Cylinder no. 5 misfire detected
PO306	Cylinder no. 6 misfire detected
PO307	Cylinder no. 7 misfire detected
PO308	Cylinder no. 8 misfire detected
PO325	Knock sensor circuit fault
PO326	Knock sensor circuit performance
PO351	Ignition coil no. 1 primary circuit fault
PO352	Ignition coil no. 2 primary circuit fault
PO353	Ignition coil no. 3 primary circuit fault
PO354	Ignition coil no. 4 primary circuit fault
PO355	Ignition coil no. 5 primary circuit fault
PO356	Ignition coil no. 6 primary circuit fault
PO357	Ignition coil no. 7 primary circuit fault
PO358	Ignition coil no. 8 primary circuit fault
PO400	EGR flow fault
PO401	EGR insufficient flow detected
PO402	EGR excessive flow detected
PO420	Catalyst system efficiency below threshold (Bank 1)
PO421	Catalyst system efficiency below threshold (Bank 1)
PO430	Catalyst system efficiency below threshold (Bank 2)
PO431	Catalyst system efficiency below threshold (Bank 2)
PO442	EVAP small leak detected
PO443	EVAP VMV circuit fault
PO452	EVAP fuel tank pressure sensor low input
PO453	EVAP fuel tank pressure sensor high input
PO500	VSS fault
PO505	IAC valve system fault
PO603	PCM Keep Alive Memory test error
PO605	PCM Read Only Memory test error

3 Powertrain Control Module (PCM) - replacement

Refer to illustrations 3.4a and 3.4b
Warning: *Some models covered by this manual are equipped with airbags. Always disconnect the negative battery cable, then the positive battery cable and wait 2 minutes before working in the vicinity of the impact sensors, steering column or instrument panel to avoid the possibility of accidental deployment of the airbag, which could cause personal injury (see Chapter 12).*

Note: *Because of a Federally mandated warranty which covers the emission control system components, check with your dealer about warranty coverage before working on any emissions-related systems. Once the warranty has expired, you may wish to perform some of the component checks and/or replacement procedures in this Chapter to save money.*

1   Disconnect the cable from the negative terminal of the battery, then the positive cable and wait at least two minutes before proceeding.

2   On 1994 and earlier models, the PCM is located behind the right side kick panel in the passenger compartment. On 1995 and later models, the PCM is located on the right rear corner of the engine compartment in the cowl.

3   To remove the PCM on 1994 and earlier models, remove the right kick panel and door sill plate (if necessary). Then detach the bolts from the mounting bracket and lower the PCM to access the electrical connectors. Disconnect the electrical connectors from the PCM and remove it from the vehicle.

4   On 1995 and later models, remove the bolt that retains the electrical connector to

3.4a Remove the PCM bracket mounting bolts (arrows) (1995 and later models)

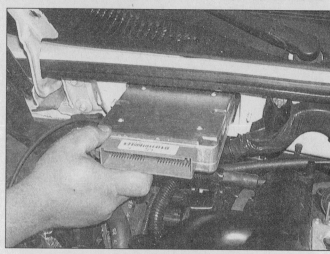

3.4b The PCM is removed through the engine compartment (1995 and later models)

the PCM. Detach the PCM bracket retaining bolts and cover, then carefully slide the PCM out of the cowl **(see illustrations)**.
5   Installation is the reverse of removal. **Note:** *Avoid any static electricity damage to the computer by using gloves and a special anti-static pad to store the PCM on once it is removed.*

4   Throttle Position Sensor (TPS) - check and replacement

Check

Refer to illustration 4.3
Note: *If the following tests indicate that a sensor is good, and not the cause of a driveability problem or DTC, check the wiring harness and connectors between the sensor and the PCM for an open or short circuit. If no problems are found, have the vehicle checked by a dealer service department or*

other qualified repair shop.
1   The Throttle Position Sensor (TPS) is a variable-resistance potentiometer, mounted on the side of the throttle body and connected to the throttle shaft **(see illustrations 1.1a, 1.1b and 1.1c)**. It senses throttle movement and position, then transmits a voltage signal to the PCM. This signal enables the PCM to determine when the throttle is closed, in a cruise position, or wide open. A defective TPS can cause surging, stalling, rough idle and other driveability problems because the PCM thinks the throttle is moving when it is not. The Electronic Engine Control (EEC) system can detect several different TPS problems and set trouble codes to indicate the specific fault (see Section 2).
2   First disconnect the sensor connector and check the TPS reference voltage. Insert the voltmeter positive (+) probe into the reference voltage terminal of the connector and the negative (-) probe into the ground terminal. With the ignition On but the engine not run-

ning, the meter should read 5.0 ± 0.1 volts.
3   Reconnect the sensor connector and backprobe the signal terminal of the sensor connector with the positive (+) lead of your voltmeter. Backprobe the ground terminal with the meter negative (-) lead **(see illustration)**.
4   Turn the ignition On but do not start the engine. The meter should read less than 1.5 volt with the throttle closed.
5   Open the throttle (or have a helper depress the accelerator) until the throttle is wide open. The voltmeter reading should increase smoothly and steadily to approximately 5.0 volts.
6   If TPS-related driveability problems continue but these general tests don't indicate a TPS fault, have the sensor tested by a dealer service department or other qualified repair shop. A TPS often develops a voltage signal dropout of such short duration that it can't be seen on a voltmeter. The PCM can see such a signal fault and a driveability problem will result.

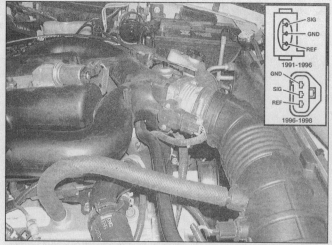

4.3 With the TPS connected backprobe the TPS using straight pins or other suitable probes on SIG (+) and GND (-). Check the signal voltage, it should be 0.5 to 1.5 volt at closed throttle. Rotate the accelerator completely to wide open throttle and confirm the voltage increases steadily to 4.0 to 5.0 volts

4.7 Detach the screws and disengage the TPS from the throttle shaft (4.0L SOHC V6 shown, 4.0L pushrod V6 and 5.0L V8 engines similar)

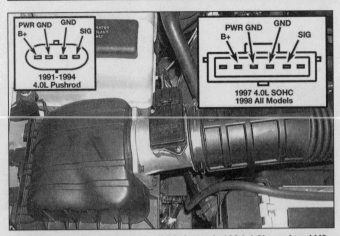

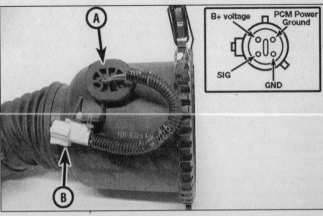

5.3a MAF sensor location - 1991 through 1994 4.0L pushrod V6
engine; On 1997 4.0L SOHC V6 engines and all 1998 engines the
sensor is located in the air intake duct - Check for battery voltage
to the B+ terminal on the MAF sensor (key ON engine not running)

5.3b The MAF sensor (A) on 1995 through 1997 4.0L pushrod V6
and 5.0L V8 engines is located in the air cleaner housing - unplug
the connector (B) from the bottom of the housing and check for
battery voltage to the B+ terminal on the MAF sensor
(key ON engine not running)

Replacement

Refer to illustration 4.7

7   The TPS sensor is not adjustable. On 2001 models, remove the bolts and detach the snow shield above the TPS. Unplug the electrical connector, remove the two retaining screws **(see illustration)** and remove the TPS from the throttle body.

8   Install the new sensor, making sure it engages the throttle shaft correctly. Reconnect the electrical connector.

5   Mass Airflow (MAF) sensor - check and replacement

Check

Refer to illustrations 5.3a and 5.3b

Note: *If the following tests indicate that a sensor is good, and not the cause of a driveability problem or DTC, check the wiring harness and connectors between the sensor and the PCM for an open or short circuit. If no problems are found, have the vehicle checked by a dealer service department or other qualified repair shop.*

1   The mass airflow (MAF) sensor is installed in the air intake duct **(see illustrations 1.1a, 1.1b and 1.1c)**. This sensor uses a hot-wire sensing element to measure the molecular mass (or weight) of air entering the engine. The air passing over the hot wire causes it to cool, and the sensor converts this temperature change into an analog voltage signal to the PCM. The PCM in turn calculates the required fuel injector pulse width to obtain the necessary air/fuel ratio. A defective MAF sensor can cause surging, stalling, rough idle and other driveability problems. The Electronic Engine Control (EEC) system can detect several different MAF sensor problems and set trouble codes to indicate the specific fault (see Section 2).

2   To check for power to the MAF sensor, disconnect the MAF sensor electrical connector.

3   Connect the positive (+) lead of your voltmeter to the B+ terminal of the harness connector; connect the meter negative (-) lead to the sensor connector ground terminal **(see illustrations)**.

4   Turn the ignition On but do not start the engine. The meter should read more than 10 volts or close to battery voltage.

5   Reconnect the electrical connector and use straight pins or other suitable probes to backprobe the MAF signal (+) and ground (-) terminals with the voltmeter. Start the engine and check the voltage, it should be 0.5 to 0.7 volts at idle.

6   Increase the engine rpm. The signal MAF voltage should increase to about 1.5 to 3.0 volts. It is impossible to simulate driving conditions in the driveway, but it is necessary to watch the voltmeter for an increase in signal voltage as the engine speed is raised. The engine is not under load, but signal voltage should vary slightly.

7   If you suspect a defective MAF sensor, stop the engine and disconnect the MAF harness connector. Using an ohmmeter, probe the MAF signal (+) and ground (-) terminals. If the hot-wire element inside the sensor has been damaged, the ohmmeter will show an open circuit (infinite resistance).

8   If the voltage readings are correct, refer to the wiring diagrams and check the wiring harness for open circuits or a damaged harness. **Note:** *If MAF-related driveability problems continue but these general tests don't indicate a MAF fault, have the sensor tested by a dealer service department or other qualified repair shop. A MAF sensor can develop voltage signal problems that can't be seen on a voltmeter. The PCM can see such signal faults and a driveability problem will result.*

Replacement

Refer to illustrations 5.9a, 5.9b, 5.10a and 5.10b

Note: *The plastic MAF sensor body and the metal air duct on which it is mounted are an assembly that must be replaced as a unit. Do not try to separate the sensor body from the metal duct.*

9   Disconnect the electrical connector from the MAF sensor and remove the air intake duct. On 1995 through 1997 4.0L pushrod V6 and 5.0L V8 engines, remove the air cleaner housing, then detach the MAF sensor and the air cleaner housing **(see illustrations)**.

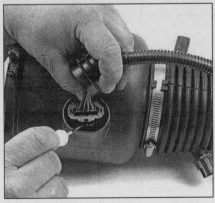

5.9a On vehicles equipped with cylindrical type air cleaners, unplug the connector from the sensor . . .

5.9b . . . then separate the MAF sensor from the air cleaner housing

10   Remove the four nuts that secure the sensor to the air cleaner housing and remove the clamp that secures the sensor to the intake air duct **(see illustrations)**.
11   Install and connect the new sensor.

6   Intake Air Temperature (IAT) sensor - check and replacement

Check

Refer to illustration 6.2
Note: *If the following tests indicate that a sensor is good, and not the cause of a drive-ability problem or DTC, check the wiring harness and connectors between the sensor and the PCM for an open or short circuit. If no problems are found, have the vehicle checked by a dealer service department or other qualified repair shop.*
1   The IAT sensor is a thermistor that changes resistance as temperature changes. The sensor is installed in the intake air duct to sense air temperature **(see illustrations 1.1a, 1.1b and 1.1c)**. As temperature increases, sensor resistance decreases and vice versa. The PCM uses this information to compute the intake temperature and fine tune fuel metering. A problem in the IAT sensor circuit will set a trouble code (see Section 2). The fault may be in the circuit wiring or connections or in the sensor itself.
2   With the engine cool, disconnect the IAT sensor and use an ohmmeter to measure resistance across the two terminals of the sensor. For example, at 68-degrees F the resistance should be approximately 37,300 ohms **(see illustration)**.
3   Next, start the engine and warm it up until it reaches operating temperature. Turn the engine off, disconnect the sensor and measure the resistance again. It should be lower. If the sensor resistance doesn't change as described, replace it.
4   If the resistance values of the sensor are correct, check for reference voltage from the

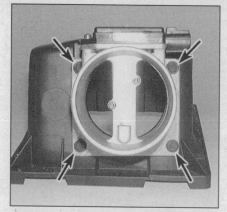

5.10a On vehicles equipped with box type air cleaners, disconnect the air inlet duct, remove the nuts (arrows) and separate the MAF sensor from the air cleaner housing (air cleaner housing removed for clarity)

PCM to the sensor connector. The open-circuit voltage at the sensor connector should be approximately 5 volts.

Replacement

Refer to illustration 6.5
5   Disconnect the electrical connector, then carefully remove the IAT sensor from the air intake duct **(see illustration)**. Be careful not to damage any of the plastic parts.
6   Install and connect the new sensor.

7   Engine Coolant Temperature (ECT) sensor - check and replacement

Check

Refer to illustration 7.2
Note 1: *If the following tests indicate that a sensor is good, and not the cause of a drive-ability problem or DTC, check the wiring harness and connectors between the sensor and*

5.10b MAF sensor retaining bolts (arrows) on a cylindrical type air cleaner housing

the PCM for an open or short circuit. If no problems are found, have the vehicle checked by a dealer service department or other qualified repair shop.
Note 2: *Before condemning an ECT sensor, check the coolant level in the system.*
1   Like the IAT sensor, the ECT sensor is a thermistor, which is a variable resistor that changes its resistance as temperature changes. The sensor is installed in the engine cooling system by the thermostat housing **(see illustrations 1.1a, 1.1b and 1.1c)** to sense coolant temperature. As coolant temperature increases, sensor resistance decreases and vice versa. The PCM uses this information to compute the engine operating temperature. A problem in the ECT sensor circuit will set a trouble code (see Section 2). The fault may be in the circuit wiring or connections or in the sensor itself.
2   Some engines have two almost identical coolant temperature sensor units **(see illustration)** mounted next to each other. One is the sender for the instrument panel temperature gauge, the other is the ECT sensor for the Electronic Engine Control (EEC) system. The temperature sender connector for the instrument panel gauge has a tan or brown

TEMPERATURE (DEGREES-F)	RESISTANCE (K-OHMS)
248	1.18
230	1.56
212	2.07
194	2.80
176	3.84
158	5.37
140	7.70
122	10.97
104	16.15
86	24.27
68	37.30
50	58.75

6.2 Intake Air Temperature (IAT) and Engine Coolant Temperature (ECT) sensors approximate temperature vs. resistance values

6.5 Disconnect the electrical connector, then carefully pull the IAT sensor from the air intake duct

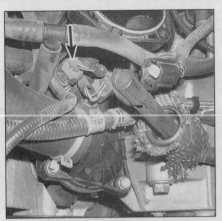

7.2 The ECT sensor (arrow) has a two-wire connector, and the plastic shell is usually gray (4.0L SOHC V6 engine shown)

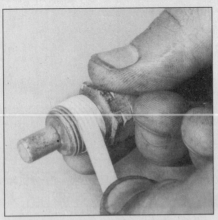

7.6 Wrap the threads of the ECT sensor with Teflon tape before installing it

7.7 Disconnect the electrical connector (A), then unscrew the ECT sensor (B) from the engine (5.0L V8 engine shown)

plastic body, the ECT sensor connector has a gray plastic body.

3   Disconnect the ECT sensor and use an ohmmeter to measure resistance across the two terminals of the sensor. At 65 degrees F, resistance should be approximately 40,500 ohms (see illustration 6.2).

4   Next, start the engine and warm it up until it reaches operating temperature. The resistance should be lower. For example, at 180 to 220-degrees F resistance should be 3,800 to 1,840 ohms.

5   If the resistance values of the sensor are correct, check for reference voltage from the PCM to the sensor connector. The open-circuit voltage at the sensor connector should be approximately 5 volts.

Replacement

Refer to illustrations 7.6 and 7.7
Warning: *Wait until the engine is completely cool before performing this procedure.*

6   Before installing the new sensor, wrap the threads with Teflon sealing tape to prevent leakage and thread corrosion (see illustration).

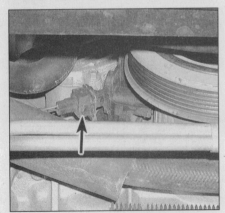

8.3 Disconnect the crankshaft position sensor connector (arrow) - approximately 1.5 volts should be present on one of the sensor wires with the key On and the engine Off (4.0L SOHC V6 shown, 4.0L pushrod V6 and 5.0L V8 similar)

7   Disconnect the electrical connector, then unscrew the ECT sensor from the engine (see illustration). Install the new sensor as quickly as possible to minimize coolant loss. Tighten the sensor securely and reconnect the electrical connector.

8   Check the coolant level as described in Chapter 1, adding some, if necessary. Start the engine and allow it to reach normal operating temperature, then check for coolant leaks. Check the coolant level in the expansion tank after the engine has warmed up and then cooled down again.

8   Crankshaft position (CKP) sensor - check and replacement

Check

Refer to illustration 8.3
Note: *If the following tests indicate that a sensor is good, and not the cause of a driveability problem or DTC, check the wiring harness and connectors between the sensor and the PCM for an open or short circuit. If no problems are found, have the vehicle checked by a dealer service department or other qualified repair shop.*

1   The crankshaft position CKP) sensor is mounted on the engine front cover, next to a toothed trigger wheel (see illustrations 1.1a, 1.1b and 1.1c). The trigger wheel has 35 evenly spaced teeth and one gap where a 36th tooth would be. The gap lets the CKP sensor signal the PCM when the crankshaft is 60-degrees before TDC for cylinders 1 and 5. The PCM then computes actual TDC or any number of degrees before or after TDC and uses this information to control ignition spark advance. The CKP sensor also provides the engine speed signal to the PCM and is part of the misfire monitor circuit integrated into the PCM.

2   The Electronic Engine Control (EEC) system can detect different CKP sensor problems and set a trouble code to indicate the specific fault (see Section 2).

3   Disconnect the CKP sensor connector (see illustration) and turn the ignition On but do not start the engine. Use a voltmeter to check for voltage between the sensor connector and ground as shown on the wiring diagrams. Approximately 1.5 volts should be present on one of the sensor wires with the key On and the engine Off.

4   Disable the fuel system as described in the fuel pressure relief section of Chapter 4 (this will enable the engine to be cranked over without it starting). Connect a voltmeter to the CKP sensor, set the meter on the AC scale, and check for voltage pulses as you crank the engine. **Caution:** *Keep voltmeter leads away from the drivebelt and rotating engine parts while cranking the engine.*

5   If no pulsing voltage signal is produced, replace the crankshaft sensor.

Replacement

Refer to illustration 8.7

6   Be sure the ignition is off and disconnect the sensor electrical connector (see illustration 8.3).

7   Remove the retaining bolts and remove the sensor from the engine front cover (see illustration).

8   Installation is the reverse of removal.

8.7 Crankshaft position sensor retaining bolts (arrows) (5.0L V8 shown)

6

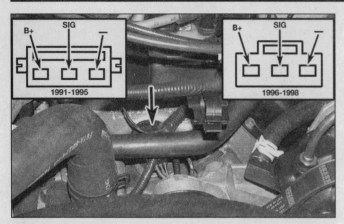

9.2a The camshaft position sensor (arrow) is located at the front of the engine block on 5.0L V8 engines and at the rear of the block on 4.0L pushrod V6 engines

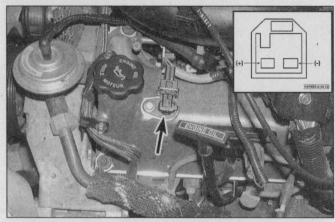

9.2b The camshaft position sensor (arrow) on 4.0L SOHC V6 engines is located on the left valve cover

9 Camshaft position (CMP) sensor - check and replacement

Check

Refer to illustrations 9.2a and 9.2b

Note: *If the following tests indicate that a sensor is good, and not the cause of a drive-ability problem or DTC, check the wiring harness and connectors between the sensor and the PCM for an open or short circuit. If no problems are found, have the vehicle checked by a dealer service department or other qualified repair shop.*

1   The camshaft position (CMP) sensor sends a voltage signal pulse to the PCM that indicates when the number 1 piston is approaching TDC on the compression stroke. The PCM uses this signal to synchronize and sequence the fuel injectors.

2   The CMP sensor on the 4.0L and 5.0L V8 engines is a Hall-effect switch, mounted on the top of the engine where the distributor was installed on earlier versions of this engine **(see illustration)**. The CMP sensor on the 4.0L SOHC V6 engine is a magnetic pickup, or variable-reluctance, sensor, mounted on the left side valve cover, and triggered by a special tooth on the exhaust camshaft **(see illustration)**.

4.0L pushrod V6 and 5.0L V8 engines

3   Disconnect the CMP sensor connector and turn the ignition On but do not start the engine. Use a voltmeter to check for voltage between the sensor connector and ground **(see illustration 9.2a)**, the reading should be more than 10 volts. If it isn't, trace and repair the circuit between the CMP sensor and the PCM.

4   Disable the fuel system as described in the fuel pressure relief section of Chapter 4 (this will enable the engine to be cranked over without it starting). Set the voltmeter on the DC scale, reconnect the CMP sensor connector, then backprobe the sensor connector and check for voltage pulses as you crank

the engine. The meter should show a pulsing DC-voltage reading of approximately 5 volts, once for each sensor revolution.

5   If the CMP sensor doesn't produce a pulsating voltage signal but battery voltage is present in Step 3, refer to the wiring diagrams and check the sensor ground connection. If the ground is OK, then replace the sensor.

4.0L SOHC V6 engine

Refer to illustration 9.9

6   Disconnect the CMP sensor connector and turn the ignition On but do not start the engine. Use a voltmeter to check for voltage between the sensor connector and ground as shown on the wiring diagrams. Approximately 1.5 volts should be present on one of the sensor wires with the key on and the engine off **(see illustration 9.2b)**.

7   Connect a voltmeter to the CMP sensor, set the meter on the AC scale, and check for voltage pulses as you crank the engine. **Caution:** *Keep the voltmeter leads away from the drivebelt and rotating engine parts while cranking the engine.*

8   If no pulsing voltage signal is produced, replace the CMP sensor.

9   As an alternative, you can remove the

CMP sensor from the engine and test it as follows:

a) *Remove the CMP sensor from the engine and place it on a workbench.*

b) *Connect the voltmeter to both terminals of the CMP sensor and set the meter on the AC-voltage scale. Pass a steel object across the tip of the sensor and look for voltage pulses on the meter* **(see illustration)**.

c) *If no pulsing voltage signal is produced, replace the camshaft sensor.*

Replacement

4.0L pushrod V6 and 5.0L V8 engines

1995 and earlier models

Refer to illustrations 9.10 and 9.13

Note: *The CMP sensor is mounted on a drive unit. If you are replacing only the CMP sensor, remove the sensor screws, detach it from the drive assembly, and install the new sensor. You do not have to remove the drive assembly from the engine. Many engine repair procedures, however, require removal of the drive assembly, in which case it will be necessary to perform the following procedure.*

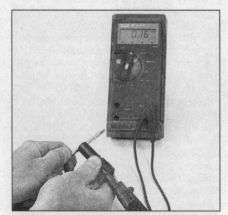

9.9 Working on the bench, pass a metal object close to the tip of the CMP sensor and see if an AC voltage is produced

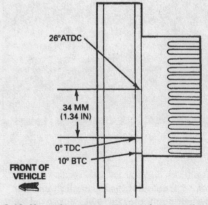

9.10 If one is not present, make a mark on the crankshaft damper exactly 34mm (1.34 inches) After Top Dead Center, which is 26-degrees ATDC

be approximately battery voltage). Now turn the sensor clockwise until the voltmeter reads 0-volts. Finally, rotate the sensor slowly counterclockwise and stop exactly when the voltmeter reading changes from 0-volts to a positive voltage reading.

19   Detach the voltmeter and tighten the hold-down bolt securely.

1996 and later models

Refer to illustrations 9.23, 9.27a and 9.27b

Note: *The CMP sensor is mounted on a drive unit. If you are replacing only the CMP sensor, remove the sensor screws, detach it from the drive assembly, and install the new sensor. You do not have to remove the drive assembly from the engine. Many engine repair procedures, however, require removal of the drive assembly. In these cases, you must time the drive assembly when you reinstall it. This procedure requires a Ford special tool to align the sensor properly. Read the entire procedure and obtain the necessary tool before beginning.*

20   Refer to Chapter 2A and position the number 1 piston at TDC.

21   Disconnect the cable from the negative terminal of the battery.

22   Mark the relative position of the CMP sensor electrical connector so the assembly can be oriented properly during installation. (This is necessary only if the drive assembly will be removed.) Disconnect the electrical connector from the CMP sensor. Remove the screws and remove the sensor from the drive assembly.

23   To remove the drive assembly, remove the hold-down bolt and clamp and lift the drive out of the engine. **(see illustration).**

24   Place the alignment tool onto the drive assembly and align the vane of the synchronizer with the radial slot in the tool.

25   Turn the tool on the drive assembly until the boss on the tool engages the notch on the drive housing.

26   Lubricate the gear, the thrust washer, and the lower bearing of the drive with clean engine oil.

27   Make sure that the oil pump intermediate shaft is aligned in the oil pump correctly, then insert the assembly into the engine so that the drive gear engages the camshaft

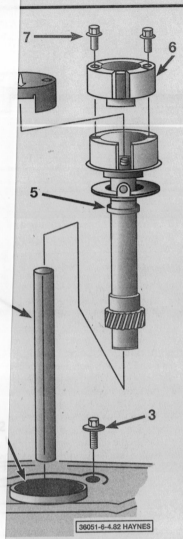

Exploded view of the CMP sensor and drive assembly with the alignment tool

Alignment tool
Engine block
Hold-down clamp and bolt
Oil pump intermediate shaft
Drive assembly
CMP sensor
Bolt

16   Rotate the engine two revolutions and return the engine to TDC compression for cylinder number 1. This will take up any slack in the timing chain. Now continue to turn the engine until it is positioned at 26-degrees ATDC.

17   Reconnect the cable to the negative terminal of the battery, then turn the ignition key to the On position. Connect the positive (+) probe of a high-impedance (10-meg ohms) digital voltmeter to the backside of the center terminal of the sensor (the dark blue wire with the orange stripe). **Note:** *If the probes of your voltmeter are too large, you can use a pin or a straightened-out paper clip to backprobe the connector. Connect the negative probe to a good ground.*

18   Rotate the sensor counterclockwise until voltage reads on the voltmeter (it should

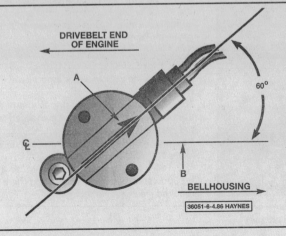

9.27a On 4.0L pushrod V6 engines, the arrow on the tool (A) must point 60-degrees counterclockwise from the engine centerline (B)

DRIVEBELT END OF ENGINE

60°

BELLHOUSING

6

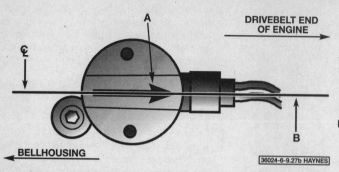

9.27b On 5.0L V8 engines, the arrow on the tool (A) must point towards the front of the engine, parallel to the engine centerline (B)

36024-6-9.27b HAYNES

9.33 Camshaft position sensor retaining bolt (arrow) (4.0L SOHC V6 engine)

gear and the oil pump intermediate shaft. On 4.0L engines, rotate the sensor assembly so that the arrow on the tool points 60-degrees counterclockwise from the engine centerline **(see illustration)**. On 5.0L V8 engines, rotate the sensor assembly so that the arrow on the tool points towards the front of the vehicle, parallel to the engine centerline **(see illustration)**. This should be the orientation point for the connector that you marked before removing the sensor drive assembly.

28   Check the position of the electrical connector on the sensor to make sure it is aligned with the mark you made during removal. If it isn't oriented correctly, do not rotate the drive assembly to reposition it. Doing so will result in the fuel system being out of time with the engine and possible engine damage. If the connector is not oriented properly, repeat the installation procedure.

29   Install the hold-down clamp and bolt and tighten it securely. Remove the positioning tool.

30   Install the CMP sensor and tighten the screws securely.

31   Connect the sensor electrical connector and reconnect the battery ground cable.

4.0L SOHC V6 engine

Refer to illustration 9.33

32   Disconnect the electrical connector from the sensor.

33   Remove the retaining bolt and remove then CMP sensor from the valve cover **(see illustration)**.

34   Installation is the reverse of removal.

10.4 Power steering pressure switch location (4.0L SOHC V6 engine)

10   Power Steering Pressure (PSP) switch - check and replacement

Check

Refer to illustration 10.4

1   The power steering pressure (PSP) switch is a normally closed switch, mounted in the high pressure line on some models. When steering system pressure reaches a high-pressure setpoint, the PSP switch opens and sends a signal to the PCM that the PCM uses to maintain engine idle speed during parking maneuvers. The Electronic Engine Control (EEC) system can detect switch problems and set trouble codes to indicate specific faults (see Section 2).

2   Check the operation of the PSP switch if the engine stalls during parking or if the engine idles continuously at high rpm.

3   Refer to the wiring diagrams at the end of this manual to identify the functions of connector terminals.

4   Disconnect the PSP switch connector and connect an ohmmeter to the terminals on the switch body **(see illustration)**.

5   Start the engine and let it idle.

6   Turn the steering wheel to point the front wheels straight ahead and read the ohmmeter. It should indicate continuity of close to zero ohms.

7   Turn the steering wheel to either side and watch the ohmmeter. The PSP switch should open as the wheel nears the steering stop on either side, and the meter should indicate an open circuit (infinite resistance).

8   If the switch fails either test, replace it. If the switch is OK, troubleshoot the engine idle control operation if high idle speed or stalling problems continue.

Replacement

9   Raise the vehicle and support it securely on jackstands.

10   Disconnect the electrical connector from the switch and unscrew the switch from the pressure line.

11   Install and connect the new switch and lower the vehicle to the ground.

12   Refer to Chapter 10 and bleed air from the power steering system. Add fluid as required (see Chapter 1).

11   Oxygen sensor (O2S) - check and replacement

Check

Refer to illustrations 11.1a and 11.1b

Note: *If the following tests indicate that a sensor is good, and not the cause of a driveability problem or DTC, check the wiring harness and connectors between the sensor and the PCM for an open or short circuit. If no problems are found, have the vehicle checked by a dealer service department or other qualified repair shop.*

1   The oxygen in the exhaust reacts with the O2S to produce a voltage output that varies from 0.1 volt (high oxygen, lean mixture) to 0.9 volt (low oxygen, rich mixture). The upstream O2S in the exhaust system provides a feedback signal to the PCM that indicates the amount of leftover oxygen in the exhaust **(see illustrations)**. The PCM monitors this variable voltage continuously to determine the required fuel injector pulse width and to control the engine air/fuel ratio. A mixture ratio of 14.7 parts air to 1 part fuel is the ideal ratio for minimum exhaust emissions, as well as the best combination of fuel economy and engine performance. Based on

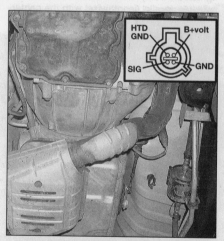

11.1a Location of the upstream oxygen sensors (4.0L SOHC V6 engine shown, other engines similar)

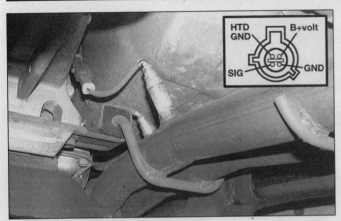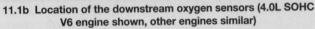

11.1b Location of the downstream oxygen sensors (4.0L SOHC V6 engine shown, other engines similar)

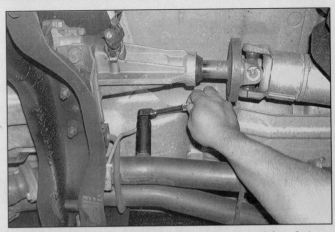

11.12 Removing an oxygen sensor using a special socket

O2S signals, the PCM tries to maintain this air/fuel ratio of 14.7:1 at all times.

2   The downstream O2S in the exhaust system has no effect on PCM control of the air/fuel ratio. This sensor is identical to the upstream sensor and operates in the same way. The PCM uses their signals, however, for the catalyst monitor system. A downstream O2S will produce a slower fluctuating voltage signal that reflects the lower oxygen content in the postcatalyst exhaust. **Note:** *Some earlier models are only equipped with one (upstream) O2S while later model vehicles are either equipped with one upstream and one downstream O2S or two upstream and two downstream O2S.*

3   An O2S produces no voltage when it is below its normal operating temperature of about 600-degrees F. During this warm-up period, the PCM operates in an open-loop fuel control mode. It does not use the O2S signal as a feedback indication of residual oxygen in the exhaust. Instead, the PCM controls fuel metering based on the inputs of other sensors and its own programs.

4   Proper operation of an O2S depends on four conditions:

a) *Electrical* - The low voltages generated by the sensor require good, clean connections which should be checked whenever a sensor problem is suspected or indicated.

b) *Outside air supply* - The sensor needs air circulation to the internal portion of the sensor. Whenever the sensor is installed, make sure the air passages are not restricted.

c) *Proper operating temperature* - The PCM will not react to the sensor signal until the sensor reaches approximately 600-degrees F. This factor must be considered when evaluating the performance of the sensor.

d) *Unleaded fuel* - Unleaded fuel is essential for proper operation of the sensor.

5   The Electronic Engine Control (EEC) system can detect several different HO2S problems and set a trouble code to indicate the specific fault (see Section 2). When an HO2S fault occurs that sets a DTC, the PCM

will disregard the O2S signal voltage and revert to open-loop fuel control as described previously. **Note:** *Refer to the wiring diagrams in Chapter 12 to identify circuit functions by wire color coding for the following tests.* **Caution:** *The O2S is very sensitive to excessive circuit loads and circuit damage of any kind. Carefully backprobe the wires in the connector shell with straight pins or similar devices (see Chapter 12 for additional information on how to backprobe a connector). Do not puncture the O2S wires or try to backprobe the sensor itself. Use only a digital voltmeter to test an O2S (see Chapter 12).*

6   Turn the ignition On but do not start the engine. Connect your voltmeter negative (-) lead to a good ground and the positive (+) lead to the SIG wire at the O2S connector **(see illustrations 11.1a and 11.1b).** The meter should read approximately 400 to 450 millivolts (0.40 to 0.45 volt). If it doesn't, trace and repair the circuit from the sensor to the PCM.

7   Start the engine and let it warm up to normal operating temperature; again check the O2S signal voltage.

a) *Voltage from an upstream sensor should range from 100 to 900 millivolts (0.1 to 0.9 volt) and switch actively between high and low readings.*

b) *Voltage from a downstream sensor should also read between 100 to 900 millivolts (0.1 to 0.9 volt) but it should not switch actively. The downstream O2S voltage may stay toward the center of its range (about 400 millivolts) or stay for relatively longer periods of time at the upper or lower limits of the range.*

8   Also check the battery voltage supply to the O2S heating circuits. Move the voltmeter negative (-) lead to the HTR GND terminal and the positive (+) lead to the B+Volt terminal of the sensor connector **(see illustrations 11.1a and 11.1b).** With the ignition On, the meter should read more than 10 volts. Battery voltage is supplied to the sensors through a relay for only about three seconds when the engine is not running. Have an assistant turn the ignition On while you read the voltmeter. Refer to the wiring diagrams in

Chapter 12 for more information on the circuits and relays.

Replacement

Refer to illustration 11.12

9   The exhaust pipe contracts when cool, and the O2S may be hard to loosen when the engine is cold. To make sensor removal easier, start and run the engine for a minute or two; then shut it off. Be careful not to burn yourself during the following procedure. Also observe these guidelines when replacing an O2S.

a) *The sensor has a permanently attached pigtail and electrical connector which should not be removed from the sensor. Damage or removal of the pigtail or electrical connector can harm operation of the sensor.*

b) *Keep grease, dirt and other contaminants away from the electrical connector and the louvered end of the sensor.*

c) *Do not use cleaning solvents of any kind on the oxygen sensor.*

d) *Do not drop or roughly handle the sensor.*

10   Raise the vehicle and place it securely on jackstands.

11   Disconnect the electrical connector from the sensor.

12   Using a suitable wrench or specialized O2 sensor socket, unscrew the sensor from the exhaust manifold **(see illustration).**

13   Anti-seize compound must be used on the threads of the sensor to aid future removal. The threads of most new sensors will be coated with this compound. If not, be sure to apply anti-seize compound before installing the sensor.

14   Install the sensor and tighten it securely.

15   Lower the vehicle and reconnect the electrical connector for the sensor.

12 Knock sensor - general information

1   A knock sensor is used on the 4.0L SOHC V6 engine to detect engine detona-

6

13.1 Location of the Vehicle Speed Sensor (VSS) (arrow)

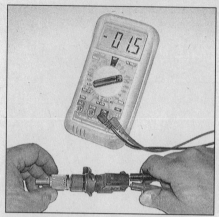

13.5 Remove the VSS and check for a pulsing AC voltage signal as the VSS gear is turned

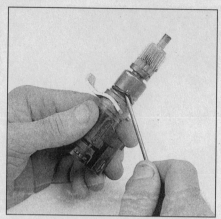

13.10 Inspect the sensor O-ring – install a new one if damaged or if you're installing a new sensor

tion, or pinging. The sensor produces a fluctuating output voltage which increases with the severity of the knock. The signal goes to the PCM, which will retard ignition timing to stop the detonation. The knock sensor is located on the right cylinder head by the intake manifold **(see illustration 1.1b)**.

2   Knock sensor operation and any associated problems can be monitored best with an OBD-II scan tool. If your engine suffers from detonation or pinging, have it tested with the necessary equipment by a dealer service department or other properly equipped repair facility.

---

13   Vehicle Speed Sensor (VSS) - check and replacement

Check

Refer to illustrations 13.1 and 13.5

Note: *If the following tests indicate that a sensor is good, and not the cause of a driveability problem or DTC, check the wiring harness and connectors between the sensor and the PCM for an open or short circuit. If no problems are found, have the vehicle checked by a dealer service department or other qualified repair shop.*

1   The vehicle speed sensor (VSS) is pickup coil (variable-reluctance) sensor mounted on the transmission **(see illustration)**. It produces an AC voltage sine wave, the frequency of which is proportional to vehicle speed. The PCM uses the sensor input signal for several different engine and transmission control functions. The VSS signal also drives the speedometer on the instrument panel. A defective VSS can cause various driveability and transmission problems. The Electronic Engine Control (EEC) system can detect sensor problems and set trouble codes to indicate specific faults (see Section 2).

2   Refer to the wiring diagrams in Chapter 12 to identify the functions of connector terminals.

3   Disconnect the VSS connector and turn the ignition On but do not start the engine.

Use a voltmeter to check for voltage between the sensor connector and ground as shown on the wiring diagrams. Approximately 1.5 volts should be present on one of the sensor wires with the key on and the engine off.

4   Remove the VSS from the vehicle as described below.

5   Connect a voltmeter to the VSS, set the meter on the AC scale, and check for voltage pulses as you spin the sensor drive gear **(see illustration)**.

6   If no pulsing voltage signal is produced, replace the sensor.

Replacement

Refer to illustration 13.10

7   Raise the vehicle and support it securely on jackstands.

8   Disconnect the electrical connector from the VSS.

9   Remove the hold-down bolt and clamp and remove the VSS from the transaxle **(see illustration 13.1)**.

10   Inspect the O-ring on the sensor **(see illustration)** and replace it if damaged. If you are installing a new sensor, use a new O-ring.

11   Installation is the reverse of removal.

---

14   Transmission Range (TR) sensor - general information

---

1   The Transmission Range (TR) sensor is mounted on the transaxle and senses the position of the gear selector as chosen by the driver. The TR sensor contains a series of resistors and switch contacts. Depending on the gear selector position, the sensor contacts route current through different combinations of resistors and produce voltage signals of different levels. The sensor sends these signals to the PCM, which uses them for a number of engine and transmission control operations.

2   The TR sensor input affects operation of the EGR system, idle speed control, and transaxle torque converter lockup. The sensor also takes the place of the neutral safety

switch used on older automatic transmissions. If the gear selector is not in Park or Neutral, the TR sensor will not let the starter motor operate. The Electronic Engine Control (EEC) system can detect sensor problems and set trouble codes to indicate specific faults (see Section 2).

3   Refer to Chapter 7B for the check and replacement procedures.

---

15   Brake On-Off (BOO) switch - check

---

1   The brake on-off switch (also called the brake pedal position switch or brake light switch) tells the PCM when the brakes are being applied. The switch closes when brakes are applied and opens when the brakes are released. The switch is mounted on the brake pedal.

2   The brake light circuit is controlled by this switch, and burned-out bulbs or other circuit problems will cause the engine to idle roughly. Therefore, check the BOO switch operation when troubleshooting any rough-idle problems.

3   Refer to the wiring diagrams at the end of this manual to identify the functions of connector terminals.

4   Disconnect the switch connector and connect your voltmeter positive (+) lead to the connector terminal that provides battery positive (B+) voltage to the switch. Connect the negative (-) meter lead to a good ground.

5   Turn the ignition On and read the meter. It should indicate more than 10 volts or close to battery voltage.

6   Connect an ohmmeter to the switch terminals and manually open and close the switch. The meter should alternate from continuity to an open-circuit reading as the switch is opened and closed.

7   Also check continuity from the switch to the brake light bulbs. Replace any burned-out bulbs or damaged wire looms.

8   Replacement of the switch is covered in Chapter 9.

16.2 Disconnect the electrical connector from the IAC valve and check for voltage with the ignition key On (engine not running)

16 Idle Air Control (IAC) valve – check and replacement

Caution: *The throttle body on these models is coated with a sludge resistant material designed to protect the bore and throttle plate. Do not attempt to clean the interior of the throttle body. The throttle body is designed to resist sludge accumulation and cleaning may impair the performance of the engine.*

Note: *The minimum idle speed is pre-set at the factory and is not adjustable under normal circumstances. If the idle fluctuates, stalls, idles high or speeds out of control, follow these quick checks to determine if the IAC valve is damaged. Because idle problems involve possible air leaks, fuel injector problems, malfunctioning TPS, PCM problems, etc. have the IAC valve and system diagnosed by a dealer service department or other qualified repair shop.*

Check

Refer to illustration 16.2

1   The Idle Air Control (IAC) valve controls the amount of air that bypasses the throttle valve, which controls the engine idle speed. This output actuator is mounted on the throttle body and is controlled by voltage pulses sent from the PCM (computer). The IAC valve within the body moves in or out, allowing more or less intake air into the system. To increase idle speed, the PCM extends the IAC valve from the seat and allows more air to bypass the throttle bore. To decrease idle speed, the PCM retracts the IAC valve towards the seat, reducing the air flow.

2   To check the system, first check for the voltage signal from the PCM. Turn the ignition key On (engine not running) and with a voltmeter, probe the wires of the IAC valve electrical connector (harness side). It should be approximately 10.5 to 12.5 volts **(see illustration)**. This indicates that the IAC valve is receiving the proper signal from the PCM.

3   If the IAC valve is receiving proper volt-

16.7 IAC valve mounting bolts (4.0L pushrod V6 engine shown, all other engines similar)

age, check the condition of the valve itself. Measure the resistance across the terminals on the IAC valve. There should be 6.0 to 13.0 ohms. If the resistance is incorrect, replace the IAC valve.

4   Check the IAC valve for an internal short circuit. Measure resistance from either terminal to the IAC body. There should be 10,000 ohms or greater. If less, the internal circuitry is grounding against the case; replace the IAC valve.

5   Next, remove the valve (proceed to Step 7) and check the pintle for excessive carbon deposits. If necessary, clean it with a soft rag. Also clean the valve housing to remove any deposits.

Replacement

Refer to illustration 16.7

6   Unplug the electrical connector from the IAC valve **(see illustrations 1.1a, 1.1b and 1.1c)**.

7   Remove the two valve attaching screws and withdraw the valve assembly from the throttle body **(see illustration)**.

8   Check the condition of the O-ring. If it's hardened or deteriorated, replace it.

9   Clean the sealing surface and the bore of the throttle body assembly with a shop rag or soft cloth to ensure a good seal. **Caution:** *The IAC valve itself is an electrical component and must not be soaked in any liquid cleaner, as damage may result.*

10   Position the new O-ring on the IAC valve. Lubricate the O-ring with a light film of engine oil.

11   Install the IAC valve and tighten the screws securely.

12   Plug in the electrical connector at the IAC valve assembly.

17 Positive Crankcase Ventilation (PCV) system

Refer to illustration 17.1

1   The Positive Crankcase Ventilation (PCV) system reduces hydrocarbon emissions by scavenging crankcase vapors. It does this by circulating fresh air from the air cleaner through the crankcase, where it mixes with blow-by gases and is then rerouted through a PCV valve to the intake manifold **(see illustration)**.

2   The PCV system consists of a replaceable PCV valve, a crankcase ventilation filter and the connecting hoses.

3   To maintain idle quality, the PCV valve restricts the flow when the intake manifold vacuum is high. If abnormal operating conditions arise, the system is designed to allow excessive amounts of blow-by gases to flow back through the crankcase vent tube into the air cleaner to be consumed by normal combustion.

6

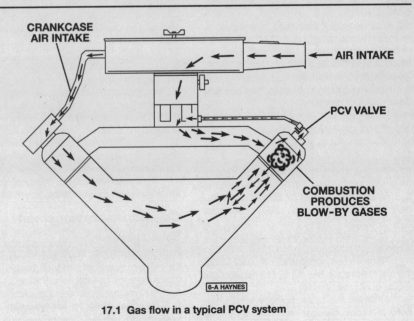

17.1 Gas flow in a typical PCV system

4   Checking and replacement of the PCV valve and filter is covered in Chapter 1.

18   Inlet air temperature control system (1994 and earlier models)

General description

1   The inlet air temperature control system provides heated intake air during warmup, then maintains the inlet air temperature within a 70-degree F to 105-degree F operating range by mixing warm and cool air. This allows leaner fuel/air mixture settings for the EFI system, which reduces emissions and improves driveability.
2   Two fresh air inlets - one warm and one cold - are used. The balance between the two is controlled by intake manifold vacuum, cold weather modulator and a bimetal sensor. A vacuum motor, which operates a heat duct valve in the air cleaner, is controlled by the vacuum switch.
3   When the underhood temperature is cold, warm air radiating off the exhaust manifold is routed by a shroud which fits over the manifold up through a hot air inlet tube and into the air cleaner (**see illustration 8.2 in** Chapter 4). This provides warm air for the EFI, resulting in better driveability and faster warmup. As the underhood temperature rises, a heat duct valve is gradually closed by a vacuum motor and the air cleaner draws air through a cold air duct instead. The result is a consistent intake air temperature.
4   A temperature vacuum switch mounted in the air cleaner cover monitors the temperature of the inlet air heated by the exhaust manifold. A bimetal disc in the temperature vacuum switch orients itself in one of two positions, depending on the temperature. One position allows vacuum through a hose to the motor; the other position blocks vacuum.
5   The vacuum motor itself is regulated by a Cold Weather Modulator (CWM), mounted in the side of the air cleaner housing assembly, between the temperature vacuum switch and the motor, which provides the motor with a range of graduated positions between fully open and fully closed.

System checks

Note: *Make sure the engine is cold before beginning this test.*
6   Always check the vacuum source and the integrity of all vacuum hoses between the source and the vacuum motor before beginning the following test. Do not proceed until they're okay.
7   Apply the parking brake and block the wheels.
8   Remove components as necessary from the air intake duct in order to see the vacuum motor door (see Chapter 4).
9   Observe the vacuum motor door position - it should be open. If it isn't, it may be binding or sticking. Make sure it's not rusted

18.18 To remove the bimetal sensor, detach the hoses and pry off the retaining clip with a small screwdriver

in an open or closed position by attempting to move it by hand. If it's rusted, it can usually be freed by cleaning and oiling the hinge. If it fails to work properly after servicing, replace it.
10   If the vacuum motor door is okay but the motor still fails to operate correctly, check carefully for a leak in the hose leading to it. Check the vacuum source to and from the bimetal sensor and the cold weather modulator as well. If no leak is found, replace the vacuum motor (see Step 23).
11   Start the engine. If the duct door has moved or moves to the "heat on" (closed to fresh air) position, go to Step 15.
12   If the door stays in the "heat off" (closed to warm air) position, place a finger over the bimetal sensor bleed. The duct door must move rapidly to the "heat on" position. If the door doesn't move to the "heat on" position, stop the engine and replace the vacuum motor (see Step 23). Repeat this Step with the new vacuum motor.
13   With the engine off, allow the bimetal sensor and the Cold Weather Modulator to cool completely.
14   Restart the engine. The duct door should move to the "heat on" position. If the door doesn't move or moves only partially, replace the bimetal sensor (see Step 18).
15   Start and run the engine briefly (less than 15 seconds). The duct door should move to the "heat on" position.
16   Shut off the engine and watch the duct door. It should stay in the "heat on" position for at least two minutes.
17   If it doesn't stay in the "heat on" position for at least two minutes, replace the CWM.

Component replacement

Bimetal sensor

Refer to illustrations 18.18 and 18.21
18   Clearly label and detach both vacuum hoses from the bimetal sensor (one is coming from the vacuum source at the manifold and the other is going to the vacuum motor underneath the air cleaner housing) **(see illustration)**.

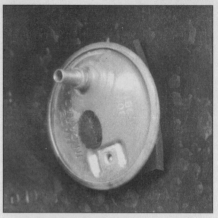

18.21 The bimetal sensor is mounted inside the air cleaner housing

19   Remove the air cleaner housing cover (see Chapter 4).
20   Pry the sensor retaining clip off with a screwdriver **(see illustration 18.18)**.
21   Remove the bimetal sensor **(see illustration)**.
22   Installation is the reverse of removal procedure.

Vacuum motor

23   Remove the air intake duct (see Chapter 4). Place the assembly on a workbench.
24   Detach the vacuum hose from the motor.
25   Drill out the vacuum motor retaining strap rivet.
26   Remove the motor.
27   Installation is the reverse of removal. Use a sheet metal screw of the appropriate size to replace the rivet.

19   Exhaust Gas Recirculation (EGR) system

General description

Refer to illustrations 19.2a and 19.2b
1   The EGR system is used to lower oxides of nitrogen (NOx) emission levels caused by high combustion temperatures. The EGR recirculates a small amount of exhaust gases into the intake manifold. The additional mixture lowers the temperature of combustion thereby reducing the formation of NOx compounds.
2   The EGR flow rate is determined by monitoring the pressure across a fixed metering orifice as exhaust gasses pass through it. The more simplified system called the Pressure Feedback (PFE) system **(see illustration)**, monitors only the downstream (after) pressure after the exhaust gasses have passed through the metering orifice. This backpressure coefficient is relayed to the PCM and the correct amount of EGR (duty cycle) is applied to the EGR vacuum regulator control (EVR). The more complex system called the Differential Pressure Feedback (DPFE) system **(see illustration)** monitors upstream (before) and downstream (after)

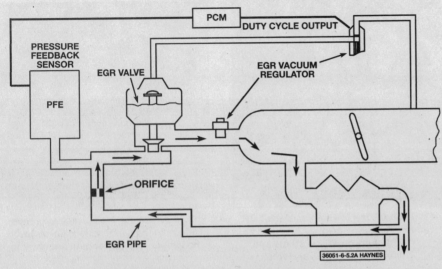

19.2a Typical Pressure Feedback EGR (PFE) system

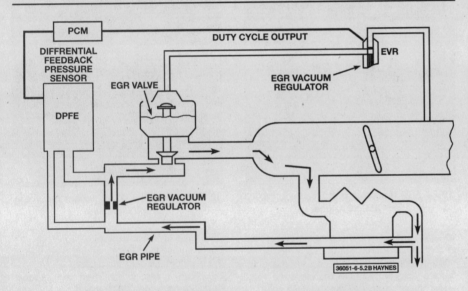

19.2b Typical Differential Pressure Feedback EGR (DPFE) system

exhaust backpressure. By calculating the difference between the two pressures, the PCM determines exactly the EGR flow rate at all driving conditions. The DPFE is more accurate in that the computer does not have to guess at the upstream pressure coefficient to determine EGR flow rate as the engine drives through various road conditions such as hard acceleration, downshifting, engine misfire, poor fuel combustion, etc. All these conditions will cause the exhaust backpressure to vary and requires more strict and responsive EGR control to limit Nox emission levels.

3   Different engine options are equipped with different EGR systems:

a) *All 1995 and earlier models and 1996 5.0L V8 models are equipped with the Pressure Feedback Exhaust Gas Recirculation (PFE) system. This system relies upon the PCM for EGR control. The*

control module (PCM) calculates the desired flow of exhaust gases into the combustion chamber and subsequently controls the EGR valve position with the EGR vacuum regulator (EVR) using an analog voltage signal or duty cycle (On/Off time). A duty cycle of 50-percent would hold the EGR valve half way open.

b) *1996 and later models (except 1996 5.0L V8 engine) are equipped with the Differential Pressure Feedback EGR (DPFE) system. This operates in the same way as the PFE except it monitors the pressure drop across the orifice allowing for a more precise measurement of the exhaust gas pressures.*

4   The Electronic Engine Control (EEC) system can detect a variety of different EGR system problems and set codes to indicate the specific trouble area (see Section 2).

Component checks

5   Too much EGR flow tends to weaken combustion, causing the engine to run rough or stop. When EGR flow is excessive, the engine can stop after a cold start or at idle after deceleration, the vehicle can surge at cruising speeds or the idle may be rough. If the EGR valve remains constantly open, the engine may not idle at all.

6   Too little or no EGR flow allows combustion temperatures to get too high during acceleration and load conditions. This can cause spark knock (detonation), engine overheating or emission test failure.

7   The following checks will help you pinpoint problems in the EGR system. Where the procedure says to lift up on the EGR valve diaphragm, it's a good idea to wear a heat-resistant glove to prevent burns.

EGR valve

8   The EGR valve is controlled by a EGR vacuum regulator (EVR) which allows vacuum to pass when energized. The PCM energizes the EVR to turn on the EGR. The PCM controls the EGR when three conditions are present: engine coolant is above 113-degrees F, the TPS is at part throttle and the MAF sensor is in its mid-range.

9   Make sure the vacuum hoses are in good condition and hooked up correctly.

10   To perform a leakage test, hook up a vacuum pump to the EGR valve. Apply a vacuum of 5 to 6 in-Hg to the valve. The vacuum pump should hold vacuum.

11   If access is possible, position your finger tip under the vacuum diaphragm and apply vacuum to the EGR valve. You should feel movement of the EGR diaphragm. **Warning:** *The EGR valve becomes very hot during engine operation - it's a good idea to wear a glove when performing this check.*

12   Remove the EGR valve (see Step 20) and clean the inlet and outlet ports with a wire brush or scraper. Do not sandblast the valve or clean it with gasoline or solvents. These liquids will destroy the EGR valve diaphragm.

13   If the specified conditions are not met, replace the EGR valve.

EGR control system

Refer to illustrations 19.16, 19.17 and 19.18

14   If an EGR system trouble code is displayed, there are several possibilities for EGR failure. Engine coolant temperature (ECT) sensor, TPS, MAF sensor, TCC system and the engine rpm govern the parameters the EGR system use for distinguishing the correct ON time.

15   These systems use an Electronic Vacuum Regulator (EVR) to control the amount of exhaust gas through the EGR valve. The regulator valve is normally open (engine at operating temperature) and the vacuum source is a ported signal. The PCM uses a controlled "pulse width" or electronic signal to turn the EGR ON and OFF (the "duty cycle"). The duty cycle should be zero percent (no EGR) when

6

19.16 Working on the harness side of the Electronic Vacuum Regulator electrical connector, check for battery voltage

19.17 Check the resistance of the EGR vacuum regulator (arrow) - it should be 20 to 70 ohms on PFE systems and 26 to 40 ohms DPFE systems

19.18 Check for the correct reference voltage to the DPFE sensor (arrow) with the ignition key ON (engine not running)

19.23 Disconnect the vacuum hose (A) and the EGR pipe retaining nut (B) from the EGR valve

19.24 EGR mounting bolts (arrows) - 4.0L SOHC V6 engine

19.37 Remove the DPFE sensor mounting bolts (arrows)

in Park or Neutral, when the TPS input is below the specified value or when Wide Open Throttle (WOT) is indicated.

16   To check the EGR vacuum regulator, disconnect the electrical connector to the EGR vacuum regulator, turn the ignition key ON (engine not running) and check for battery voltage to the solenoid **(see illustration)**. Battery voltage should be present.

17   Next, use an ohmmeter and check the resistance of the EGR vacuum regulator. It should be between 20 and 70 ohms on a pressure feedback system and between 26 to 40 ohms on a differential pressure feed-back system **(see illustration)**.

18   Check for reference voltage to the DPFE sensor. With the ignition key on (engine not running), check for voltage on the harness side of the electrical connector **(see illustration)** on terminal VREF. It should be between 4.0 and 6.0 volts.

19   Check the operation of the Differential Pressure Feedback (DPFE) sensor. **Note:** *The DPFE sensor on the Differential Pressure Feedback EGR systems have two exhaust lines hooked into the EGR tube.* Check for signal voltage to the PCM. Backprobe the SIG and the GND terminals and check for a

voltage signal while the engine is running first at cold temperatures and then at warm operating temperatures. With the engine cold there should be NO EGR therefore the voltage should be approximately 0.75 to 1.25 volts (black plastic housing type) or 0.35 to 0.8 volts (aluminum housing type). As the engine starts to warm and EGR is signaled by the computer, voltage values should increase to approximately 4.0 to 6.0 volts. If the test results are incorrect, recheck the vacuum source to DPFE, if the vacuum source to sensor is ok, replace the DPFE sensor.

Component replacement

EGR valve

Refer to illustrations 19.23 and 19.24

20   When buying a new EGR valve, make sure that you have the right EGR valve. Use the stamped code located on the top of the EGR valve.

21   Detach the cable from the negative terminal of the battery.

22   Detach the vacuum line from the EGR valve.

23   Remove the EGR pipe from the exhaust manifold **(see illustration)**.

24   Remove the mounting bolts securing the

EGR valve **(see illustration)**.

25   Remove the EGR valve and gasket from the manifold. Discard the gasket.

26   With a wire wheel, buff the exhaust deposits from the EGR valve mounting surface on the manifold and, if you plan to use the same valve, the mounting surface of the valve itself. Look for exhaust deposits in the valve outlet. Remove deposit build-up with a screwdriver. **Caution:** *Never wash the valve in solvents or degreaser - both agents will permanently damage the diaphragm. Sand blasting is also not recommended because it will affect the operation of the valve.*

27   If the EGR passage contains an excessive build-up of deposits, clean it out with a wire wheel. Make sure that all loose particles are completely removed to prevent them from clogging the EGR valve or from being ingested into the engine.

28   Installation is the reverse of removal.

EGR vacuum regulator

29   Detach the cable from the negative terminal of the battery.

30   Unplug the electrical connector from the solenoid.

31   Clearly label and detach both vacuum hoses.

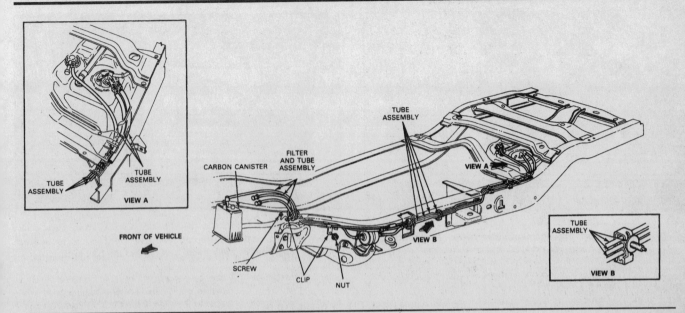

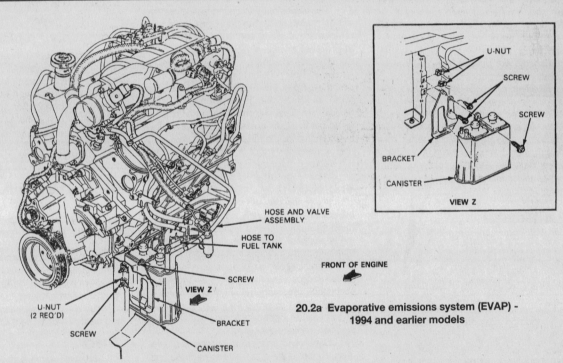

20.2a Evaporative emissions system (EVAP) -
1994 and earlier models

6

32   Remove the solenoid mounting screw and remove the solenoid.
33   Installation is the reverse of removal.

DPFE sensor

Refer to illustration 19.37
34   Detach the cable from the negative terminal of the battery.
35   Unplug the electrical connector from the sensor.
36   Clearly label and detach both vacuum hoses.
37   Remove the sensor mounting nuts **(see illustration)** and remove the assembly.
38   Installation is the reverse of removal.

20   Evaporative Emissions Control System (EVAP)

General description

Refer to illustrations 20.2a, 20.2b and 20.2c
1   This system is designed to trap and store fuel vapors that evaporate from the fuel tank, throttle body and intake manifold during non-operation or idling, store them in the charcoal canister and then route them into the combustion chamber to be burned during engine operation.
2   On early models, the Evaporative Emission Control System (EVAP) consisted of a charcoal-filled canister, a canister purge valve and the lines connecting the canister to the vapor control valve and the fuel tank. With the introduction of OBD-II (EEC-V) systems on later models, the EVAP system became electronically controlled and consisted of a charcoal-filled canister and the lines connecting the canister to the fuel tank, fuel vapor management valve (VMV), a fuel tank pressure sensor, fuel filler cap, fuel vapor valve, ported vacuum and intake manifold vacuum **(see illustrations)**.
3   On 1994 and earlier models, the fuel vapor trapped in the gas tank is vented through a valve in the top of the tank. From the valve, the vapor is routed through a single

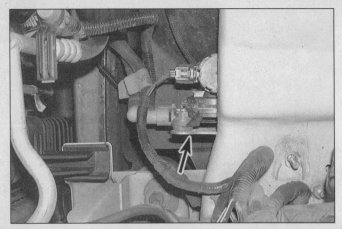

20.2b The Fuel Vapor Management Valve (arrow) is mounted on the left frame rail on 1995 and later models

20.2c The charcoal canister, on 1995 and later models, is located underneath the vehicle just behind the fuel tank

line to a carbon canister located in the engine compartment near the radiator, where it's stored until the next time the engine is started. On 1995 and later systems, fuel vapors are transferred from the fuel tank, throttle body and intake manifold to a canister where they are stored when the engine is not operating. When the engine is running, the fuel vapors are purged from the canister by a vapor management valve (VMV) which is PCM controlled, and consumed in the normal combustion process. The fuel tank pressure sensor relays the inside fuel tank pressure to the PCM which in turn regulates the EVAP system purge control system.

4   1995 and later systems can detect a variety of different EVAP system problems and set codes to indicate the specific trouble area (see Section 2).

System checks

5   On 1994 and earlier models, There are no moving parts and nothing to wear in the canister. Check for loose, missing, cracked or broken fittings and inspect the canister for cracks and other damage. If the canister is damaged, replace it. Check for fuel smells around the vehicle. Make sure the gas cap is in good condition and properly installed.

6   On 1995 and later models, poor idle, stalling and poor driveability can be caused by an inoperative vapor management valve, a damaged canister, split or cracked hoses or hoses connected to the wrong tubes.

7   Evidence of fuel loss or fuel odor can be caused by fuel leaking from fuel lines or the throttle body, a cracked or damaged canister, an inoperative vapor management valve (VMV), disconnected, misrouted, kinked, deteriorated or damaged vapor or control hoses or an improperly seated air cleaner or air cleaner gasket.

8   Inspect each hose attached to the canister for kinks, leaks and breaks along its entire length. Repair or replace as necessary.

9   Inspect the canister. If it is cracked or damaged, replace it.

10   Look for fuel leaking from the bottom of the canister. If fuel is leaking, replace the

canister and check the hoses and hose routing.

Excessive pressure in fuel tank

11   The easiest way to check for excess fuel vapor pressure in the fuel tank is simply remove the gas cap and listen for the sound of pressure release similar to a flat tire or air compressor discharge. If the weather is extremely hot, take into account for the extra pressure from the heated molecules. The most accurate test is performed using a SCAN tool. This will run a series of checks using the fuel tank pressure sensor and other output actuators to detect excess pressure. If the proper SCAN tool is not available, have the vehicle diagnosed by a dealer service department or other qualified automotive repair facility.

12   If excess pressure is detected, check the canister fuel vapor hose and inlet port for blockage or collapsed hoses. Also inspect the hoses near the VMV and between the fuel tank and body for kinks and damage.

13   Also, check the evaporative emission valve. Remove the fuel tank (see Chapter 4) and check the evaporative emission valve to make sure the passage through the orifice is open to atmospheric pressure. If it is plugged, replace the valve.

14   Check the charcoal canister for tightness. Remove the close-off line near the canister and install a hand-held pressure pump. Pump the canister to approximately 2.5 psi and confirm that pressure vents through the close-off line. Now install the pump directly at the canister and apply 2.5 psi. The pressure should hold steady.

Fuel vapor odor in engine compartment

15   Check the hoses around the VMV for damage or incorrectly routed lines. Correct if necessary.

16   Check the VMV for battery voltage. With the ignition key ON (engine not running), check for battery voltage on the harness side of the VMV harness connector (see illustration 20.2b). If there is no battery voltage available, have the PCM diagnosed by a dealer

service department or other qualified automotive repair facility. If reference voltage is available to the VMV, have the VMV checked using a SCAN tool at a dealer service department or other qualified repair facility.

Component replacement

17   Clearly label, then detach, all vacuum lines from the canister.

18   Loosen the canister mounting clamp bolt and remove the canister (see illustrations 20.2a and 20.2c).

19   Installation is the reverse of removal.

21 Catalytic converter

General description

1   The catalytic converter is an emission control device added to the exhaust system to reduce pollutants from the exhaust gas stream. A single-bed converter design is used in combination with a three-way (reduction) catalyst. The catalytic coating on the three-way catalyst contains platinum and rhodium, which lowers the levels of oxides of nitrogen (NOx) as well as hydrocarbons (HC) and carbon monoxide (CO).

Check

2   The test equipment for a catalytic converter is expensive and highly sophisticated. If you suspect that the converter on your vehicle is malfunctioning, take it to a dealer or authorized emissions inspection facility for diagnosis and repair.

3   Whenever the vehicle is raised for servicing of underbody components, check the converter for leaks, corrosion and other damage. If damage is discovered, the converter should be replaced.

Replacement

4   Because the converter is part of the exhaust system, converter replacement requires removal of the exhaust pipe assembly (see Chapter 4). Take the vehicle, or the exhaust system, to a dealer service department or a muffler shop.

Chapter 7 Part A
Manual transmission

Contents

Specifications

General

Transmission type	M5OD 5-speed synchromesh
Lubricant type	See Chapter 1

Torque specifications

	Ft-lbs
1991 through 1996 models	
Transmission-to-clutch housing bolt or nut	18 to 38
Mount-to-crossmember nut	65 to 85
Mount-to-transmission bolt	60 to 80
Crossmember-to-frame bracket nut and bolt	65 to 85
1997 and later models	
Transmission-to-clutch housing bolt or nut	36 to 50
Transmission mount bolts/nuts	64 to 81
Crossmember-to-frame bolts/nuts	39 to 53

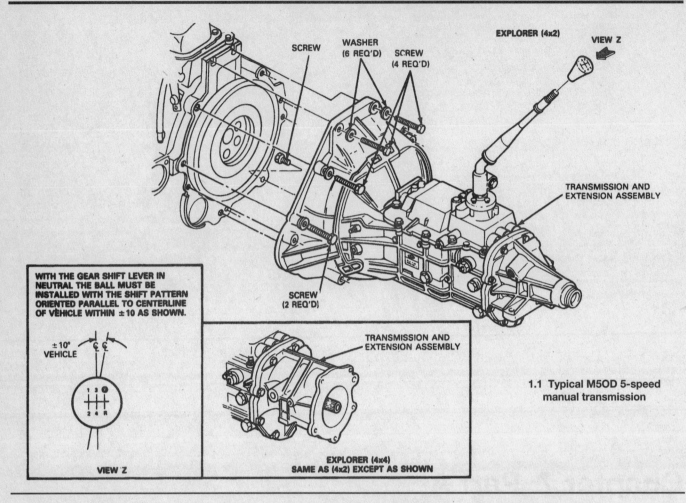

WITH THE GEAR SHIFT LEVER IN NEUTRAL THE BALL MUST BE INSTALLED WITH THE SHIFT PATTERN ORIENTED PARALLEL TO CENTERLINE OF VEHICLE WITHIN ±10 AS SHOWN.

±10°
VEHICLE

VIEW Z

SCREW (6 REQ'D)
WASHER (6 REQ'D)
SCREW (4 REQ'D)

EXPLORER (4x2)   VIEW Z

TRANSMISSION AND EXTENSION ASSEMBLY

SCREW (2 REQ'D)

TRANSMISSION AND EXTENSION ASSEMBLY

EXPLORER (4x4)
SAME AS (4x2) EXCEPT AS SHOWN

1.1 Typical M5OD 5-speed manual transmission

1   General information

Refer to illustration 1.1

All vehicles covered in this manual come equipped with a 5-speed manual transmission **(see illustration)** or an automatic transmission. All information on the manual transmission is included in this Part of Chapter 7. Information on the automatic transmission can be found in Part B of this Chapter.

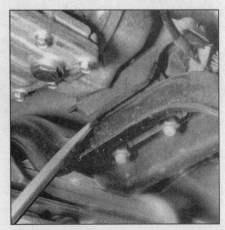

3.2 Pry up on the transmission mount and check for excessive looseness

Due to the complexity, unavailability of replacement parts and the special tools necessary, internal transmission repair procedures are not recommended for the home mechanic.

Depending on the expense involved in having a faulty transmission overhauled, it may be an advantage to consider replacing the unit with either a new or rebuilt one. Your local dealer or transmission shop should be able to supply you with information concerning cost, availability and exchange policy. Regardless of how you decide to remedy a transmission problem, you can still save a lot of money by removing and installing the unit yourself.

2   Shift lever - removal and installation

1996 and earlier models

1   Place the transmission in Neutral.
2   Carefully pull up the carpeting and remove the shifter boot retaining bolts.
3   Remove the bolt that secures the shift lever to the transmission **(see illustration 1.1).**
4   Pull the shift lever straight up and off the transmission.

5   Installation is the reverse of the removal Steps.

1997 through 2000 models

6   Pull the outer rubber boot up on the shifter handle.
7   Remove the bolt holding the shift lever to the transmission, and pull the lever off.
8   Remove the four screws holding the inner shifter boot to the floor.
9   Installation is the reverse of the removal Steps.

2001 models

10   Remove the two screws holding the floor console and rubber boot. Raise the floor console and rubber boot and secure it in this position.
11   Remove the nut on the right side of the eccentric stud holding the shift lever to the transmission. Install the nut onto the left side of the eccentric stud. Tighten the nut and withdraw the eccentric stud from gearshift lever.
12   Remove the gearshift lever, boot and floor console as an assembly.
13   Remove the four screws holding the inner shifter boot to the floor.
14   Installation is the reverse of the removal Steps.

3.4 Remove the two bolts attaching the transmission mount to the frame crossmember and the two bolts securing the mount to the transmission extension housing (arrows) - early model shown

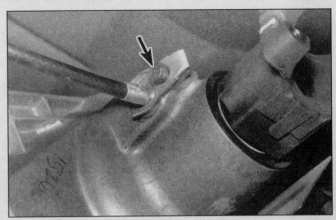

4.7 Remove the bolt and retainer securing the speedometer and disconnect the speedometer cable from the transmission or transfer case (1995 and earlier model)

3 Transmission mount - check and replacement

Refer to illustrations 3.2 and 3.4

1 Insert a large screwdriver or pry bar into the space between the transmission extension housing and the frame crossmember and pry up.

2 The transmission should not move significantly away from the mount **(see illustration)**. If it does, the mount should be replaced.

3 Place a jack under the transmission with a wood block on top of it to protect the transmission case. Apply a slight amount of jack pressure to support the weight of the transmission.

4 Remove the two nuts securing the mount to the frame crossmember and the two bolts securing the mount to the transmission extension housing **(see illustration)**. **Note:** *On 1997 and later models, there is an exhaust mount bracket under the right-side transmission mount bolt. After removing that bolt, you may have to pry the bracket aside slightly to remove the mount.*

5 Raise the transmission with the jack until the studs on the bottom of the transmission mount clear the crossmember. Remove the mount and install a new one.

6 Installation is the reverse of the removal Steps.

4 Manual transmission - removal and installation

Removal

Refer to illustrations 4.7 and 4.15

1 Disconnect the negative cable from the battery.

2 From inside the vehicle, remove the shift lever (see Section 2).

3 Drain the transmission lubricant (see Chapter 1).

4 Raise the vehicle and support it securely on jackstands. **Caution:** *If the vehicle is equipped with Automatic Ride Control (ARC), make sure the air suspension switch is turned*

to the OFF position before the vehicle is raised to prevent damage to the system components (see Chapter 10).

5 Remove the driveshaft(s) (see Chapter 8). On 2WD models, install a plug in the transmission rear extension housing to prevent lubricant leakage.

6 Disconnect the hydraulic fluid line from the clutch release cylinder (see Chapter 8) and plug the line to prevent fluid spillage.

7 On 1995 and earlier models, disconnect the speedometer cable from the transfer case or transmission **(see illustration)**. On 1996 and later models, disconnect the electrical connector from the Vehicle Speed Sensor.

8 Disconnect the electrical connector from the backup light switch. Detach the wiring harness from the transmission and position it aside.

9 Remove the starter motor (see Chapter 5).

10 Place a jack under the engine and protect the oil pan with a wood block. Apply a slight amount of jack pressure to support the rear of the engine.

11 On 4WD models, remove the transfer case (see Chapter 7 Part C).

12 Remove the exhaust crossover pipe (see Chapter 4).

13 Place a transmission jack under the transmission. Apply a slight amount of jack pressure to support the transmission, using a strap or chain to secure the transmission to the jack.

14 Remove the bolts and nuts securing the transmission mount to the frame crossmember and transmission **(see illustration 3.4)**.

15 Remove the nuts/bolts securing the crossmember to the frame side rails and remove the crossmember and the mount **(see illustration)**. Refer to Chapter 7B for crossmember mounting details on later models.

16 Slowly lower the jack until you can remove the bolts securing the transmission to the engine. Several long extensions may have to be used to reach the topmost transmission-to-engine bolts.

17 Carefully pull the transmission toward the rear and work it free from the locating dowels. Pull the transmission straight back until the input shaft clears the clutch assembly.

18 Lower the transmission jack and remove the transmission.

Installation

19 Installation is the reverse of the removal Steps with the following additions:

a) *Secure the transmission to a transmission jack and position it under the vehicle.*

7A

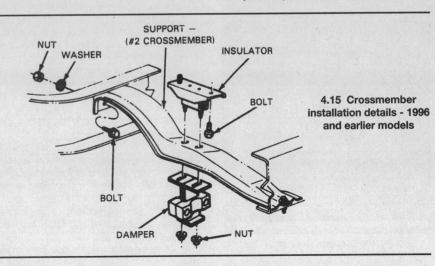

4.15 Crossmember installation details - 1996 and earlier models

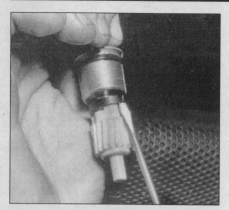

5.3 Pry the retaining clip from the pinion gear and slide the gear off of the speedometer drive adapter

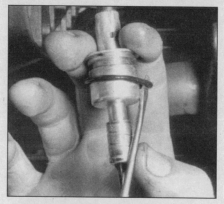

5.4 To replace the O-ring, pry off the old O-ring with a small screwdriver

6.3 Remove the extension housing seal with a seal removal tool

b) *Start the transmission input shaft into the clutch disc. Align the shaft splines with the clutch disc splines and move the transmission forward.*

c) *Tighten the bolts and nuts to the torque listed in this Chapter's Specifications.*

5   Speedometer/VSS pinion gear and seal - removal and installation

Note: *1996 and later models do not use a conventional speedometer cable, they use an electronic speedometer and a Vehicle Speed Sensor (VSS).*

Removal

Refer to illustrations 5.3 and 5.4

1   On 1995 and earlier models, disconnect the speedometer cable from the transmission **(see illustration 4.7)**. On 1996 and later models, disconnect the electrical connector from the VSS and remove the mounting bolt.

2   Pull the speedometer pinion gear adapter or VSS straight out of the transfer case or extension housing.

3   Use a small screwdriver and remove the retaining clip from the pinion gear, then slide the gear off the adapter **(see illustration)**.

4   If necessary, use a small screwdriver

6.7 Install the extension housing oil seal with a large socket or section of pipe with the proper diameter

and remove the O-ring from the retaining groove **(see illustration)**. Discard the O-ring.

Installation

5   Lubricate the new O-ring with transmission lubricant and install it in the retaining groove. Make sure it's seated correctly.

6   Install the pinion gear and the retaining clip. Make sure the clip is properly seated in the groove.

7   Install the speedometer pinion gear adapter or VSS in the transfer case or rear extension housing and secure it with the bolt and bracket.

6   Extension housing oil seal (2WD models) - replacement

Refer to illustrations 6.3 and 6.7

1   Raise the vehicle and support it securely on jackstands. **Caution:** *If the vehicle is equipped with Automatic Ride Control (ARC), make sure the air suspension switch is turned to the OFF position before the vehicle is raised to prevent damage to the system components (see Chapter 10).*

2   Place a drain pan under the extension housing, some transmission lubricant may drain out when the driveshaft slip joint is withdrawn from the extension housing. Remove the driveshaft (see Chapter 8).

3   Use a seal removal tool (available at most auto parts stores) to remove the oil seal from the end of the extension housing **(see illustration)**. If the special tool is not available, a thin blade screwdriver or chisel may be used to drive the oil seal out, but be extremely careful not to damage the extension housing.

4   Inspect the sealing surface on the driveshaft slip joint for scoring or burrs that may damage the new oil seal. If the yoke is damaged in any way, replace it.

5   Inspect the extension housing bore for burrs. If found, remove them with emery cloth or medium grit wet-and-dry sandpaper. Use a clean cloth dipped in solvent and remove any sanding residue from the bore.

6   Apply silicone sealant to the outside diameter of the oil seal and apply multi-purpose grease to the seal lip.

7   Using a seal driver, large socket or sec-

tion of pipe, install a new oil seal and make sure it is completely seated and square to the housing **(see illustration)**.

8   Install the driveshaft (see Chapter 8).

9   Lower the vehicle and check the transmission lubricant level. Top up if necessary (see Chapter 1).

7   Transmission overhaul - general information

Overhauling a manual transmission is a difficult job for the do-it-yourselfer. It involves the disassembly and reassembly of many small parts. Numerous clearances must be precisely measured and, if necessary, changed with select-fit spacers and snap-rings. As a result, if transmission problems arise, it can be removed and installed by a competent do-it-yourselfer, but overhaul should be left to a transmission repair shop. Rebuilt transmissions may be available - check with your dealer parts department and auto parts stores. At any rate, the time and money involved in an overhaul is almost sure to exceed the cost of a rebuilt unit.

Nevertheless, it's not impossible for an inexperienced mechanic to rebuild a transmission if the special tools are available and the job is done in a deliberate step-by-step manner so nothing is overlooked.

The tools necessary for an overhaul include internal and external snap-ring pliers, bearing puller, slide hammer, set of pin punches, dial indicator and possibly a hydraulic press. In addition, a large, sturdy workbench and a large vise or transmission stand will be required.

During disassembly of the transmission, make careful notes of how each piece comes off, where it fits in relation to other pieces and what holds it in place.

Before taking the transmission apart for repair, it will help if you have some idea what area of the transmission is malfunctioning. Certain problems can be closely tied to specific areas in the transmission, which can make component examination and replacement easier. Refer to the *Troubleshooting* section at the front of this manual for information regarding possible sources of trouble.

Chapter 7 Part B
Automatic transmission

Contents

Specifications

General

Transmission type	
1991 through 1994 models	A4LD
1995 and 1996 V6 models	4R55E
1996 and later V8 models	4R70W
1997 and later V6 models	5R55E
Fluid type and capacity	See Chapter 1

Torque specifications

	Ft-lbs (unless otherwise indicated)
1995 and earlier models	
Transmission-to-engine bolts	28 to 38
Torque converter-to-driveplate nuts	20 to 34
Neutral start switch	84 to 120 in-lbs
1996 and later models	
Transmission-to-engine bolts	
V6 models	30 to 41
V8 models	
1996 models	37 to 50
1997 and later models	30 to 41
Torque converter-to-driveplate nuts	
V6 models	22 to 30
V8 models	20 to 34
Transmission range sensor mounting bolts	
V6 models	71 to 97 in-lbs
V8 models	62 to 89 in-lbs
Extension housing bolts	
V6 models	27 to 39
V8 models	18 to 22
VSS sensor mounting bolt	
V6 models	63 to 89 in-lbs
V8 models	98 to 115 in-lbs

1 General information

All vehicles covered in this manual come equipped with a five-speed manual transmission or an automatic transmission. All information on the automatic transmissions is included in this Part of Chapter 7. Information on the manual transmission can be found in Part A of this Chapter.

Due to the complexity of the automatic transmissions covered in this manual and the need for specialized equipment to perform most service operations, this Chapter contains only general diagnosis, routine maintenance, adjustment and removal and installation procedures.

Models up through 1994 used an A4LD 4-speed automatic transmission. The 1995 and 1996 V6 models used the 4-speed, electronically-controlled 4R55E transmission, while the V8 models (1996 and later) used a 4R70W 4-speed automatic transmission. Models after 1996 equipped with V6 engines use a 5R55E five-speed electronic transmission.

If the transmission requires major repair work, it should be taken to a dealer service department or an automotive or transmission repair shop for troubleshooting with an electronic scan tool. You can, however, remove and install the transmission yourself and save that labor expense, even if the repair work is done by a transmission shop.

2   Diagnosis - general

Note: *Automatic transmission malfunctions may be caused by five general conditions: poor engine performance, improper adjustments, hydraulic malfunctions, mechanical malfunctions or malfunctions in the computer or its signal network. Diagnosis of these problems should always begin with a check of the easily repaired items: fluid level and condition (see Chapter 1) and shift linkage adjustment. Next, perform a road test to determine if the problem has been corrected or if more diagnosis is necessary. If the problem persists after the preliminary tests and corrections are completed, additional diagnosis should be done by a dealer service department or transmission repair shop using a scan tool to diagnose the electronics of the transmission. Refer to the* Troubleshooting *section at the front of this manual for information on symptoms of transmission problems.*

Preliminary checks

1   Drive the vehicle to warm the transmission to normal operating temperature.
2   Check the fluid level as described in Chapter 1:

   a) *If the fluid level is unusually low, add enough fluid to bring the level within the designated area of the dipstick, then check for external leaks (see below).*
   b) *If the fluid level is abnormally high, drain off the excess, then check the drained fluid for contamination by coolant. The presence of engine coolant in the automatic transmission fluid indicates that a failure has occurred in the internal radiator walls that separate the coolant from the transmission fluid (see Chapter 3).*
   c) *If the fluid is foaming, drain it and refill the transmission, then check for coolant in the fluid or a high fluid level.*

3   Check the engine idle speed. **Note:** *If the engine is malfunctioning, do not proceed with the preliminary checks until it has been repaired and runs normally.*
4   Inspect the shift linkage (see Section 3). Make sure that it's properly adjusted and that the linkage operates smoothly.

Fluid leak diagnosis

5   Most fluid leaks are easy to locate visually. Repair usually consists of replacing a seal or gasket. If a leak is difficult to find, the following procedure may help.
6   Identify the fluid. Make sure it's transmission fluid and not engine oil or brake fluid (automatic transmission fluid is a deep red color when new and may turn light brown after the vehicle has been driven some distance).
7   Try to pinpoint the source of the leak. Drive the vehicle several miles, then park it over a large sheet of cardboard. After a minute or two, you should be able to locate the leak by determining the source of the fluid

dripping onto the cardboard.
8   Make a careful visual inspection of the suspected component and the area immediately around it. Pay particular attention to gasket mating surfaces. A mirror is often helpful for finding leaks in areas that are hard to see.
9   If the leak still cannot be found, clean the suspected area thoroughly with a degreaser or solvent, then dry it.
10   Drive the vehicle for several miles at normal operating temperature and varying speeds. After driving the vehicle, visually inspect the suspected component again.
11   Once the leak has been located, the cause must be determined before it can be properly repaired. If a gasket is replaced but the sealing flange is bent, the new gasket will not stop the leak. The bent flange must be straightened.
12   Before attempting to repair a leak, check to make sure that the following conditions are corrected or they may cause another leak. **Note:** *Some of the following conditions cannot be fixed without highly specialized tools and expertise. Such problems must be referred to a transmission repair shop or a dealer service department equipped with an electronic scan tool.*

Gasket leaks

13   Check the pan periodically. Make sure the bolts are tight, no bolts are missing, the gasket is in good condition and the pan is flat (dents in the pan may indicate damage to the valve body inside).
14   If the pan gasket is leaking, the fluid level or the fluid pressure may be too high, the vent may be plugged, the pan bolts may be too tight, the pan sealing flange may be warped, the sealing surface of the transmission housing may be damaged, the gasket may be damaged or the transmission casting may be cracked or porous. If sealant instead of gasket material has been used to form a seal between the pan and the transmission housing, it may be the wrong sealant.

Seal leaks

15   If a transmission seal is leaking, the fluid level or pressure may be too high, the vent may be plugged, the seal bore may be damaged, the seal itself may be damaged or improperly installed, the surface of the shaft protruding through the seal may be damaged or a loose bearing may be causing excessive shaft movement.
16   Make sure the dipstick tube seal is in good condition and the tube is properly seated. Periodically check the area around the speedometer gear or sensor for leakage. If transmission fluid is evident, check the O-ring for damage.

Case leaks

17   If the case itself appears to be leaking, the casting is porous and will have to be repaired or replaced.
18   Make sure the oil cooler hose fittings are tight and in good condition.

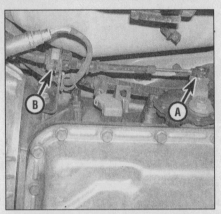

3.3 Pry the cable end from the ball-stud on the lever (A), then use a screwdriver to pull up the locking tab (B) at the cable bracket

Fluid comes out vent pipe or fill tube

19   If this condition occurs, the transmission is overfilled, there is coolant in the fluid, the case is porous, the dipstick is incorrect, the vent is plugged or the drain back holes are plugged.

3   Shift cable - adjustment

Refer to illustration 3.3

1   Raise the vehicle and support it securely on jackstands. **Caution:** *If the vehicle is equipped with Automatic Ride Control (ARC), make sure the air suspension switch is turned to the OFF position before the vehicle is raised to prevent damage to the system components (see Chapter 10).*
2   Have an assistant position the column shift selector lever in the Drive/Overdrive position and hold in place during adjustment. If working by yourself, hang a three-pound weight on the gear selector lever.
3   Working under the vehicle, pull down on the lock tab on the shift cable and remove the fitting from the manual shift lever ball stud with a screwdriver **(see illustration)**.
4   Position the transmission manual shift lever in the Drive/Overdrive position by moving the bellcrank lever all the way to the rear (counterclockwise), then forward three detents (clockwise).
5   Connect the cable end fitting to the manual lever ball stud.
6   Push the lock tab all the way down to lock the cable in the correctly adjusted position.
7   If used, remove the weight from the gear selector lever.
8   After adjustment, check for Park engagement. The control lever must move to the right when engaged in Park. Check the control lever in all detent positions with the engine running to ensure correct detent/transmission action and readjust if necessary.

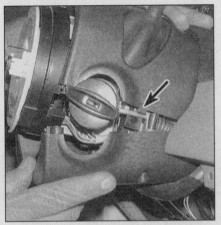

5.4 Remove the upper steering column shroud by pulling its tab out of the slot (arrow)

5.5 Use a punch (A) to drive out the shift lever pin (B) to remove the shift lever

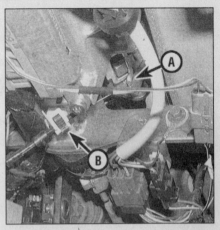

6.4 Disconnect the shift indicator cable (A) at the shift lever - (B) is the thumbwheel adjuster

9   Remove the jackstands and lower the vehicle to the ground.

10   Test drive the vehicle slowly at first to make sure the adjustment is correct.

4   Shift cable - removal and installation

1   Use a screwdriver and pry the end fitting from the steering column lever ball stud. **Note:** *You may have to work by feel, since the shift cable end and bracket are very hard to see under the dash.*

2   Remove the nuts securing the shift cable bracket to the steering column bracket and remove the bracket.

3   Raise the vehicle and support it securely on jackstands. **Caution:** *If the vehicle is equipped with Automatic Ride Control (ARC), make sure the air suspension switch is turned to the OFF position before the vehicle is raised to prevent damage to the system components (see Chapter 10).*

4   Working under the vehicle, pull down on the lock tab on the shift cable and remove the fitting from the manual lever ball stud with a screwdriver **(see illustration 3.3).**

5   Press in on the retaining tabs and disengage the cable from the transmission bracket. Slide the cable down from the transmission bracket and remove the cable.

6   Installation is the reverse of the removal procedure with the following additions:

a)   *Adjust the shift linkage (see Section 3).*

b)   *Remove the jackstands and lower the vehicle to the ground.*

5   Shift lever assembly - removal and installation

Refer to illustrations 5.4 and 5.5

Warning: *On models equipped with airbags, disconnect the negative battery cable, then the positive battery cable and wait two minutes before working in the vicinity of the*

impact sensors, steering column or instrument panel to avoid the possibility of accidental deployment of the airbag, which could cause personal injury (see Chapter 12).

1   Disconnect the negative cable (and positive cable, if equipped with airbags) from the battery.

2   Remove the driver's side knee bolster and upper and lower steering column covers (see Chapter 11).

3   Disconnect the electrical connector from the transmission control switch, if equipped (see Section 9).

4   Remove the boot by prying up one end and pulling the bottom out of a slot in the column **(see illustration).**

5   Drive the retaining pin from the shift lever base with a small punch and hammer **(see illustration).** Remove the shift lever and the clip. **Note:** *If the shift lever is to be removed from the vehicle, not just from the column, refer to Section 6 and disconnect the shift indicator cable.*

6   Installation is the reverse of removal. **Caution:** *Replace the shift lever pin with a new one any time the shift lever is removed.*

6   Shift indicator cable - replacement and adjustment

Warning: *On models equipped with airbags, disconnect the negative battery cable, then the positive battery cable and wait two minutes before working in the vicinity of the impact sensors, steering column or instrument panel to avoid the possibility of accidental deployment of the airbag, which could cause personal injury (see Chapter 12).*

Replacement

Refer to illustrations 6.4 and 6.5

1   Disconnect the negative battery cable (and positive cable, if equipped with airbags).

2   Refer to Chapter 11 and remove the driver's side lower instrument panel cover and the upper and lower steering column covers.

3   Remove the ignition lock cylinder (see Chapter 12).

4   The shift indicator cable attaches to the shift control lever on the top side of the steering column **(see illustration).**

5   Follow the cable routing from there to the instrument panel, removing any retaining clips along the way, and disconnect the upper end from the shift indicator in the instrument panel **(see illustration).**

6   Replacement is the reverse of the removal process.

Adjustment

7   Position the shift lever in the Overdrive position, which is two clicks counterclockwise from the most clockwise position. Observe the indicator, which has a "flag" around the indicated position (with standard instrument panels) or a pointer (electronic instrument panel).

8   Apply several pounds of pressure downward (clockwise) on the shifter. If the flag or pointer isn't correctly centered on the Overdrive symbol, adjust the thumbwheel on the shift indicator cable until it is centered **(see illustration 6.4).**

7B

6.5 Upper end of shift indicator cable (arrow)

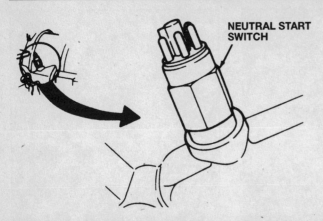

7.3 Neutral start switch - the switch is very fragile and the special tool is recommended for removal and installation

NEUTRAL START SWITCH

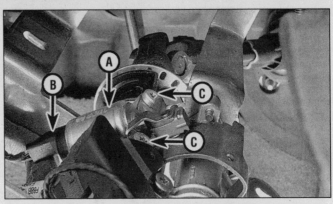

8.2 The interlock solenoid (A) on column-shift models is located near the bottom of the steering column (shown with column dropped from dash) - (B) is the electrical connector and (C) are the mounting screws

9   After adjustment, try the shift lever in all positions to make sure the indicator matches all the shift lever detents.

7   Neutral-start switch (1994 and earlier models) - removal, installation and adjustment

Refer to illustration 7.3

1   Disconnect the negative cable from the battery.
2   Disconnect the electrical connector from the switch on the transmission housing.
3   Carefully remove the switch **(see illustration)**. **Caution:** *It is easy to crush or puncture the walls of the switch. A special socket, designed to remove the switch without damaging it, may be available at your local auto parts store.*
4   Install the switch and tighten to the torque listed in this Chapter's Specifications. If available, use the same special tool used for removal to avoid damaging the switch.
5   Install the electrical connector.
6   Connect the negative cable to the battery.
7   Check that the engine will start only when the shift selector is in the Neutral or Park positions.

8   Brake/transmission shift interlock system (1995 and later models) - description, check and solenoid replacement

Warning: On models equipped with airbags, disconnect the negative battery cable, then the positive battery cable and wait two minutes before working in the vicinity of the impact sensors, steering column or instrument panel to avoid the possibility of accidental deployment of the airbag, which could cause personal injury (see Chapter 12).

Description
1   The brake transmission interlock system

prevents the shift lever from being moved out of Park unless the brake pedal is depressed simultaneously. When the car is started, a solenoid is energized, locking the shift lever in Park; when the brake pedal is depressed, the solenoid is deenergized, unlocking the shift lever so that it can be moved out of Park. **Note:** *Before making the following checks, refer to Chapter 9 and verify that the brake light switch is functioning properly, because it is part of the interlock circuit. Also check the 15-amp fuse in position 13 in the fuse panel below the instrument panel (see Chapter 12). This fuse serves the interlock solenoid.*

Check
Refer to illustration 8.2
2   Remove the driver's side lower instrument panel cover and upper and lower steering column covers (see Chapter 11). Using a flashlight, locate the brake/transmission shift interlock solenoid **(see illustration)**.
3   Verify that the brake/transmission shift interlock solenoid operates as follows:

a) *When the ignition key is in the Lock position, the solenoid plunger should be in and you should not be able to move the shift lever, even with the brake pedal applied.*
b) *With the ignition key is turned to the Off position, the solenoid plunger should be out and you should be able to move the shift lever to any gear position without applying the brake pedal.*
c) *When the ignition key is turned to the Run position, the solenoid should go back into the solenoid and you should not be able to move the shift lever; except with the brake pedal applied, the solenoid plunger should be released and you should be able to move the shift lever out of Park into any gear.*
d) *Place the key in the Lock position, then turn the key to the Acc (accessory) position. The solenoid plunger should go in a little further than it does when the key is turned to Lock, and you should not be able to move the shift lever (even with the brake applied).*

4   If the solenoid doesn't operate as described above, unplug the electrical connector from the solenoid, apply battery voltage to the solenoid and verify that it "clicks" on, then open the circuit and verify that the solenoid clicks off. If the solenoid doesn't operate as described, replace it.
5   Check for battery voltage (10 or more volts) at the connector at the terminal for the wire from the fuse panel (gray/yellow).
6   Step on the brake pedal and check for voltage at the wire from the brake light switch (light green). There should be battery voltage when the pedal is depressed, but less than one volt with the pedal released.
6   Using an ohmmeter, check the resistance at the black wire on the solenoid connector. If the resistance is greater than 5 ohms, there is a problem in the ground circuit.

Solenoid replacement
7   Disable the airbag system (see Chapter 12). Remove the driver's side lower instrument panel cover and upper and lower steering column covers (see Chapter 11).
8   Lower the steering column and support it (see Chapter 10).
9   Unplug the electrical connector from the solenoid, and remove the solenoid mounting screws **(see illustration 8.2)**. Remove the solenoid.
10 Installation is the reverse of the removal procedure.

9   Transmission control switch (1996 and later models) - check and replacement

Refer to illustrations 9.3a and 9.3b
1   The Transmission Control Switch located on the shift lever allows the driver to turn the Overdrive capability On or Off. In normal driving the Overdrive is always turned On.
2   Check the 10-amp fuse (designated "V") in the engine compartment fuse panel.

9.3a Pry the cover (arrow) off the end of the shift lever top expose the transmission control switch

9.3b The transmission control switch simply pulls out of the shift lever - check the pins for continuity when the button is depressed

3   Pry off the cover on the end of the shift lever and pull out the transmission control switch **(see illustrations)**.

4   Using a voltmeter, check for battery voltage at one of the pin sockets in the end of the shift lever, with the ignition key On (engine not running).

5   Using an ohmmeter, check for continuity across the pins when the switch is depressed. If continuity is not indicated, replace the transmission control switch.

6   Press the new switch into the lever, making sure the pins are aligned with the sockets. Press the cover back on.

---

10   Transmission range sensor (1995 and later models) - check and replacement

Check

Refer to illustration 10.4

1   With the ignition key Off, remove the electrical connector from the transmission range sensor. **Caution:** *Disconnect the connector by hand, do not use a tool to pry it off.*

2   Use a small mirror to examine the connector socket on the sensor for loose, corroded or bent pins.

3   Major testing of the transmission range sensor and harness requires a scan tool, normally used by a dealer service department. However, you can check for basic continuity between some of the terminals on the sensor with an ohmmeter or continuity tester.

4   Check for continuity between the indicated terminals at the sensor, with the connector disconnected **(see illustration)**.

   a) *There should be continuity between terminals 1 and 4 in Park or Neutral.*
   b) *There should be continuity between terminals 5 and 8 in Neutral.*
   c) *There should be continuity between terminals 2 and 3 in Reverse.*

5   If one or more of these tests fail, replace the transmission range sensor.

6   Further testing should be performed with a scan tool.

Replacement

Refer to illustration 10.13

7   Disconnect the negative cable from the battery.

8   Shift the transmission into Neutral.

9   Raise and support the front of the vehicle securely on jackstands. **Caution:** *If the vehicle is equipped with Automatic Ride Control (ARC), make sure the air suspension switch is turned to the OFF position before the vehicle is raised to prevent damage to the system components (see Chapter 10).*

10   Refer to Section 4 and remove the shift cable from the shift lever and the cable bracket.

11   Make a paint or scribe mark on the transmission shift lever, relative to its position on the splined shaft from the transmission when in Neutral, then remove the nut and take off the shift lever.

12   Disconnect the electrical connector from the transmission range sensor.

13   Remove the bolts and detach the sensor **(see illustration)**.

14   Before installing the sensor, make sure the transmission is in the Neutral position. Align the flats on the switch with the flats in the shaft and position the sensor onto the shaft.

15   Install the bolts finger-tight. Precise adjustment of the sensor requires a special tool. If the tool is not available, adjust the sensor as follows:

   a) *Connect the electrical sensor to the sensor, connect the negative battery cable and turn the ignition key On (engine not running).*
   b) *Shift the transmission into Reverse.*
   c) *Rotate the sensor until the back-up lights come on and tighten the mounting bolts.*

16   The remainder of installation is the reverse of removal. **Note:** *Use a new nut when installing the shift lever.*

17   Apply the parking brake and verify that the engine will start only in Neutral or Park.

7B

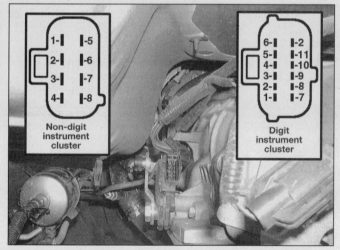

10.4 Transmission range sensor terminal identification

10.13 Mark and remove the lever (A), then remove the two mounting bolts (B) and take off the transmission range sensor

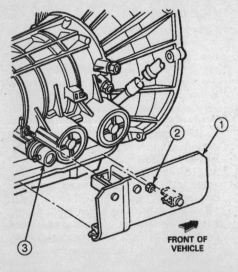

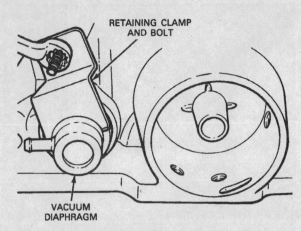

11.2 Details of the vacuum diaphragm and heat shield

1   Heat shield
2   Nut
3   Vacuum diaphragm

FRONT OF VEHICLE

11.4 Remove the retaining bolt and clamp, then pull the vacuum diaphragm out of the transmission

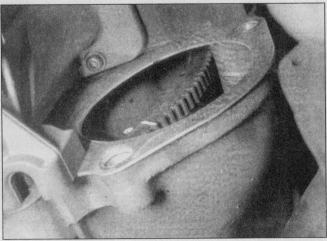

12.6 Mark one of the torque converter studs with white paint and make a corresponding mark on the driveplate so the torque converter and driveplate can be reassembled in the same relative positions

12.10 Disconnect the shift cable from the manual shift lever (A) and the kickdown cable from the transmission downshift lever (B) (early model shown)

11   Vacuum diaphragm (1994 and earlier models) - removal and installation

Refer to illustrations 11.2 and 11.4
1   Raise the vehicle and support it securely on jackstands.
2   Carefully pry the heat shield from the transmission pan, then slide it forward to gain access to the vacuum diaphragm **(see illustration)**.
3   Disconnect the hose from the vacuum diaphragm.
4   Remove the bolt and the retaining clamp and pull the diaphragm out of the transmission case **(see illustration)**.
5   Remove the control rod from the transmission case.
6   Installation is the reverse of the removal procedure, but be sure to install a new O-ring on the vacuum diaphragm and tighten the clamp bolt securely.

12   Automatic transmission - removal and installation

Removal

Refer to illustrations 12.6, 12.10, 12.11 and 12.16
1   Disconnect the negative cable from the battery.
2   Raise the vehicle and support it securely on jackstands. **Caution:** *If the vehicle is equipped with Automatic Ride Control (ARC), make sure the air suspension switch is turned to the OFF position before the vehicle is raised to prevent damage to the system components (see Chapter 10).*
3   Drain the transmission fluid (see Chapter 1), then reinstall the pan.
4   On 4WD models, remove the transfer case (see Chapter 7 Part C).
5   Unbolt the starter and tie it up out of the way (see Chapter 5). The starter mounting hole provides access to the torque converter

nuts. On 5.0L V8 models, remove the two bolts and the converter housing cover.
6   Mark the torque converter and one of the studs with white paint so they can be installed in the same position **(see illustration)**.
7   Remove the four torque converter-to-driveplate nuts. Turn the crankshaft for access to each nut. Turn the crankshaft in a clockwise direction only (as viewed from the front).
8   On 2WD models, remove the driveshaft (see Chapter 8). Plug the end of the extension housing to prevent the entry of dirt and to catch any residual transmission fluid.
9   On early 2WD models, remove the speedometer cable retaining bracket and disconnect the speedometer cable from the transmission (see Chapter 7A). On later models, disconnect the electrical connector from the VSS.
10   Disconnect the shift cable from the transmission shift lever and the kickdown cable from the transmission kickdown lever

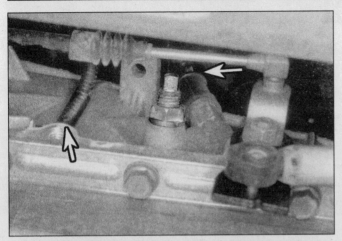

12.11 Disconnect the electrical connectors from the various components on the transmission (arrows)

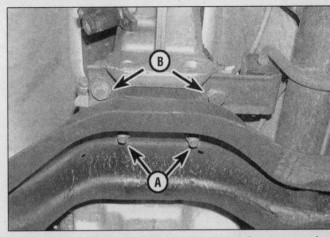

12.13a Remove the transmission mount nuts at the crossmember (A), and the bolts at the transmission (B)

on early models **(see illustration)**.

11   Disconnect the electrical connectors for the neutral safety switch, torque converter lockup solenoid, transmission range sensor, etc. and secure the transmission electrical harness aside **(see illustration)**.

12   Place a transmission jack beneath the transmission and apply jack pressure to slightly raise the transmission. Use safety chains to help steady the transmission on the jack.

13   Remove the transmission mount and crossmember **(see illustrations)**. Refer to Chapter 7B for crossmember mounting details on early models.

14   Support the engine with a jack. Use a block of wood under the oil pan to spread the load.

15   Slightly lower the engine and transmission.

16   Disconnect the transmission oil cooler lines from the transmission and plug them to prevent the entry of dirt **(see illustration)**. Use a flare nut wrench to avoid rounding off the nuts.

17   Remove the lower transmission-to-engine bolts.

18   Remove the transmission fluid dipstick tube from the transmission. **Note:** *On some models, the upper end of the dipstick tube is bolted to the back of one of the cylinder heads. Be sure to remove this bolt before separating the transmission from the engine.*

19   Make sure the transmission is securely mounted on the transmission jack. Using a long extension and universal-joint socket, remove the two upper transmission-to-engine bolts.

20   Carefully move the transmission to the rear to disengage it from the engine block dowel pins and make sure the torque converter is detached from the driveplate. Secure the torque converter to the transmission so it won't fall out during removal.

Installation

21   Installation is the reverse of the removal Steps with the following additions:

a)   *Install the converter to the transmission, making sure the converter hub is fully engaged in the pump gear.*

b)   *Rotate the converter to align the bolt drive lugs and the drain plug with the*

holes in the driveplate.

c)   *With the transmission secured to the jack, raise it into position. Be sure to keep it level so the torque converter does not slide forward and disengage from the pump gear.*

d)   *Turn the torque converter to line up the studs with the holes in the driveplate. The white paint mark on the torque converter and the stud made in Step 5 must line up.*

e)   *Move the transmission forward carefully until the dowel pins and the torque converter are engaged.*

f)   *When installing the driveplate-to-converter nuts, position the driveplate so the pilot hole is in the six o'clock position. First install one nut through the pilot hole and tighten it, then install the remaining nuts. Tighten the nuts to the torque listed in this Chapter's Specifications.*

g)   *Adjust the shift cable (see Section 3).*

h)   *Lower the vehicle.*

l)   *Fill the transmission with the specified fluid (see Chapter 1), run the engine and check for fluid leaks.*

7B

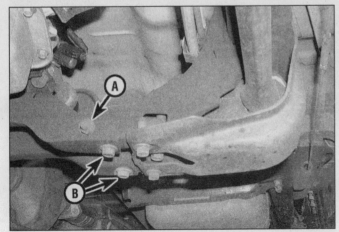

12.13b On 1997 and later models, remove the throughbolt on each side (A), then the two crossmember bolts (B) from each side and remove the crossmember

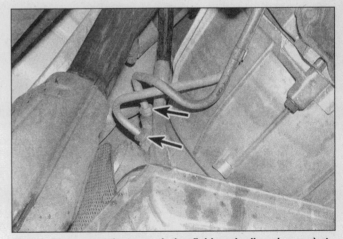

12.16 Disconnect the transmission fluid cooler lines (arrows) at the transmission (later model shown)

Notes

Chapter 7 Part C
Transfer case

Contents

Specifications

Lubricant type .... See Chapter 1

Torque specifications

Ft-lbs (unless otherwise indicated)

Cam plate **(see illustration 2.2)**	
Bolt A	70 to 90
Bolt B	31 to 42
Breather vent	72 to 168 in-lbs
Upper shift control and heat shield bolts	27 to 37
Transfer case-to-transmission bolts	25 to 35
Yoke nut	150 to 180
Skid plate-to-frame bolt	22 to 30
Speedometer cable bolt	20 to 25 in-lbs

1 General information

Refer to illustration 1.1a and 1.1b

Ford Explorers are equipped with either manually-shifted or electronically-shifted transfer cases.

The Borg Warner 13-54 is available in models that are controlled either mechanically or electronically. This model transfer case is equipped on manual and automatic transmission models. The electronic mode selector is mounted on the dash. The 13-54 transfer case is a part time system **(see illustrations)**. The unit transfers power from the transmission to the rear axle and, when activated, also to the front driveaxle. The unit is lubricated by a positive displacement oil pump that channels oil through drilled holes in the rear output shaft. The pump turns with the rear output shaft and allows towing of the vehicle at maximum legal road speeds for extended distances without disconnecting the front or rear driveshafts.

The Borg Warner 44-05 or C-Trac is controlled either by an electronic shift motor or by the GEM (Generic Electronic Module). This model transfer case is equipped onto manual and automatic transmission models. The electronic mode selector is mounted on the dash. The Borg Warner 44-05 or C-Trac uses an electromechanical clutch to shift the front driveshafts. Sensors determine the

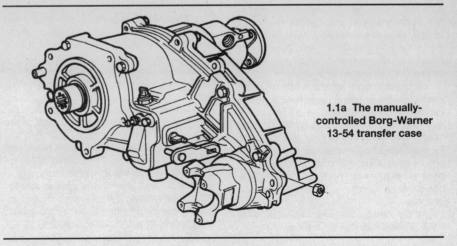

1.1a The manually-controlled Borg-Warner 13-54 transfer case

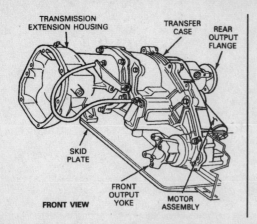

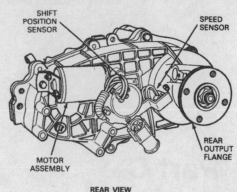

1.1b The electronically-controlled Borg-Warner 13-54 transfer case

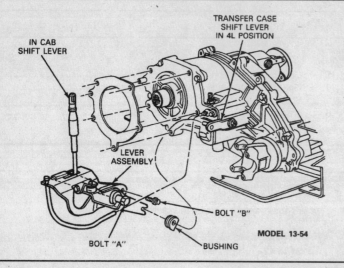

MODEL 13-54

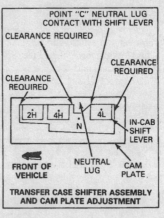

2.2 Shift linkage adjustment points (manual shift transfer case)

driveshaft speed and relay information to the electric clutch to activate the front driveshaft depending upon road conditions (wet, damp, snow, sleet etc.). This system uses a GEM (Generic Electronic Module) to control automatic shifting. The automatic selection only occurs in 4WD High. When the electric shift selector on the dash is selected for 4WD Low, a cam selects the proper gear.

In 1997 and later models with 5.0L V8 engines, an optional transfer case is available called AWD (All-Wheel-Drive). With this transfer case, the vehicle uses full-time four-wheel drive, with no driver controls. Also beginning with 1997 models, the C-Trac electronic transfer case described above is called Automatic Four-Wheel-Drive (A4WD), and the driver can select one of three different driving modes from a dash control panel.

Although each of the systems differ internally, the external components on the transfer cases are similar.

Due to the complexity of the transfer cases covered in this manual and the need for specialized equipment to perform most service operations, this chapter contains only general diagnosis, routine maintenance, adjustment and removal and installation procedures.

If the transfer case requires major repair work, it should be taken to a dealer service department or an automotive or transmission repair shop, particularly with the electronic models, which require a scan tool to properly diagnose. You can, however, remove and install the transfer case yourself and save the expense of that labor, even if the repair work is done by a transmission shop.

2 Linkage adjustment - manual-shift transfer case

Refer to illustration 2.2

1   The linkage should be adjusted if the transfer case won't engage properly or whenever the transfer case control assembly is removed from the vehicle.

2   Lift up the shift boot so the upper surface of the cam plate is visible **(see illustration)**.

3   Loosen bolts A and B one turn each.

4   Move the shift lever on the side of the transfer case to the 4L position (down).

5   Pivot the cam plate clockwise around bolt A until the bottom chamfered corner of the neutral lug just touches the forward right edge of the shift lever on the transfer case.

6   While holding the cam plate from turning, tighten bolt A, then bolt B to the torque listed in this Chapter's Specifications.

7   Move the shift lever inside the vehicle to all positions and check for positive engagement. In the 2H, 4H and 4L positions, there should be clearance between the cam plate and the shift lever on the side of the transfer case.

8   When the linkage is adjusted properly, install the shift lever boot.

3 Shift lever - removal and installation

Warning: On models equipped with airbags, disconnect the negative battery cable, then the positive battery cable and wait two minutes before working in the vicinity of the impact sensors, steering column or instrument panel to avoid the possibility of accidental deployment of the airbag, which could cause personal injury (see Chapter 12).

Removal

Refer to illustration 3.4

1   Place the transfer case shift lever in the 4L position.

2   **Note:** *Don't perform Step 2 unless the shift ball, boot or lever is to be replaced.* Remove the plastic insert from the shift ball. Warm the ball with a heat gun or portable hair

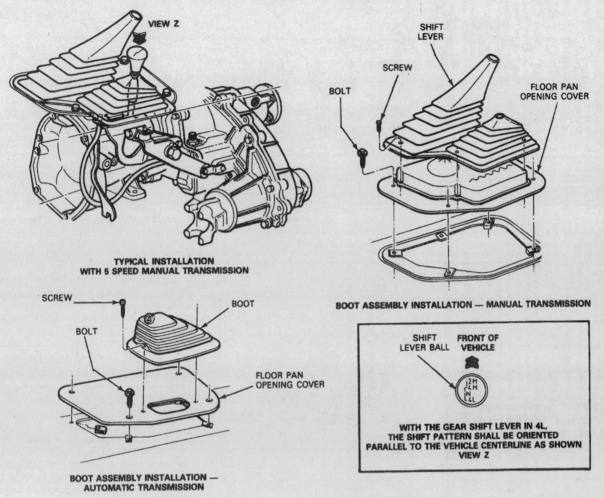

3.4 Manual control transfer case shift lever assembly details

dryer to 140 to 180-degrees F and drive the ball off the lever with a soft-faced hammer. Be careful not to damage the finish on the shift lever.

3   Remove the rubber boot and the floor pan cover.

4   Disconnect the vent hose from the control lever (see illustration).

5   Unscrew the shift lever from the control lever.

6   Remove the large and small housing bolts securing the shifter to the extension housing. Remove the control lever and bushing.

Installation

7   Adjust the linkage (see Section 2).

8   Install the vent assembly so the white mark on the housing is indexed into the notch in the shifter, if applicable. **Note:** *The upper end of the vent hose should be 3/4-inch above the top of the shifter and positioned just under the vehicle floor.*

9   Install the floor pan cover and the rubber boot.

10   If the shift ball was removed, warm it with a heat gun or portable hair dryer to 140

to 180-degrees F and carefully drive the ball onto the lever with a soft-faced hammer. Install the plastic insert into the shift ball. **Note:** *The shift ball must be driven onto the lever until all of the knurled portion of the shaft is covered.*

11   Check the transfer case for proper operation.

4   Electronic shift controls - removal and installation

Refer to illustration 4.4

Warning: On models equipped with airbags, disconnect the negative battery cable, then the positive battery cable and wait two minutes before working in the vicinity of the impact sensors, steering column or instrument panel to avoid the possibility of accidental deployment of the airbag, which could cause personal injury (see Chapter 12).

1   Disconnect the negative cable (and the positive cable on models with airbags) from the battery.

2   Remove the instrument cluster trim panel (see Chapter 12).

3   Remove the screws securing the shift control assembly and partially pull the assembly out from the instrument panel.

4   Disconnect the electrical connector from the switch assembly and remove the assembly (see illustration).

7C

4.4 Lift the connector latch and pull the electrical connector from the electronic shift control switch

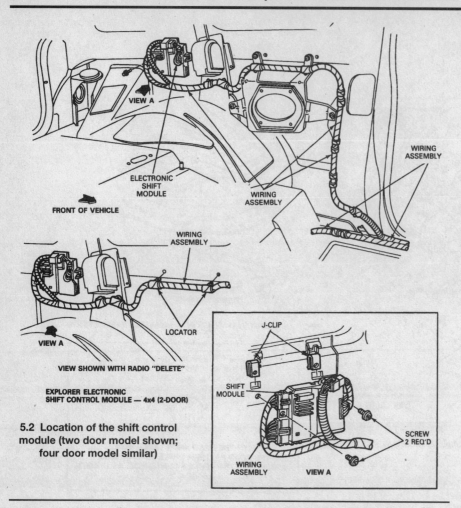

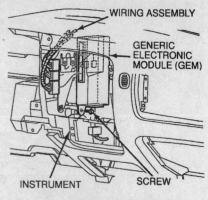

5.7 Location of the Generic Electronic Module (GEM) on 1995 and later models

5.2 Location of the shift control module (two door model shown; four door model similar)

5   Installation is the reverse of the removal Steps.

5   Electronic shift control module - removal and installation

Warning: *On models equipped with airbags, disconnect the negative battery cable, then the positive battery cable and wait two minutes before working in the vicinity of the impact sensors, steering column or instrument panel to avoid the possibility of accidental deployment of the airbag, which could cause personal injury (see Chapter 12).*

1   Disconnect the negative cable (and the positive cable on models with airbags) from the battery.

1991 through 1994 models

Refer to illustration 5.2

2   Working inside the vehicle, remove the trim panel above the left rear wheel well to gain access to the module. Remove the two bolts securing the control module and the bracket **(see illustration).**

3   Pull the control module part way out and disconnect the electrical connectors from the module. Remove the control module.

4   Installation is the reverse of removal.

1995 and later models

Refer to illustration 5.7

Note: *1995 and later models are equipped with a Generic Electronic Module (GEM). This module incorporates the functions of several different modules into one and offers diagnostic codes to locate failures (see a dealer service department for diagnosis).*

5   Remove the radio (see Chapter 12).

6   Remove the radio trim panel (see Chapter 12).

7   Slide the GEM off the retaining tabs and

toward the front of the vehicle **(see illustration).**

8   Disconnect the electrical connectors from the GEM and remove the unit from the dash.

9   Installation is the reverse of removal.

6   Output shaft oil seals, front and rear - removal and installation

Removal

Refer to illustrations 6.4 and 6.7

1   Raise the vehicle and support it securely on jackstands. **Caution:** *If the vehicle is equipped with Automatic Ride Control (ARC), make sure the air suspension switch is turned to the OFF position before the vehicle is raised to prevent damage to the system components (see Chapter 10).*

2   Drain the transfer case lubricant (see Chapter 1).

3   Refer to Chapter 8 and remove the front or rear driveshaft for access to the yoke and seal to be replaced.

4   On early models, remove the output shaft nut, washer, rubber seal and front yoke or rear flange **(see illustration). Note:** *Before removing the rear flange, make indexing marks on the flange, nut and shaft, to align*

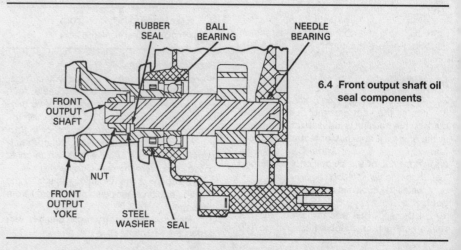

6.4 Front output shaft oil seal components

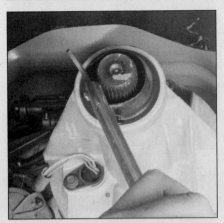

6.7 Pry out the seal with a seal removal tool (rear seal shown, front seal similar)

6.10 Use a large socket to drive the new seal in (rear seal shown, front similar)

7.3 Some models are equipped with a transfer case damper - this must be removed to remove the transfer case

7.5 Disconnect the shift motor connector (electronic shift models)

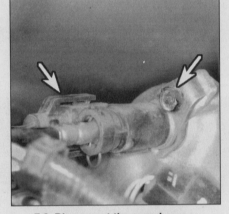

7.6 Disconnect the speed sensor connector (if equipped) and speedometer cable

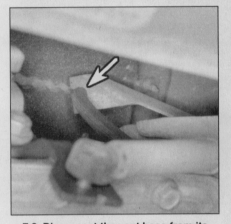

7.8 Disconnect the vent hose from its bracket (electronic shift models)

them on reassembly.

5 On later models, remove the snap-ring holding the front driveshaft yoke on the transfer case.

6 Remove the dust seal from the yoke opening of the transfer case.

7 Carefully pry out the oil seal with a seal removal tool **(see illustration)**.

Installation

Refer to illustration 6.10

8 Inspect the oil seal contact surface on the housing for scoring or burrs that may damage the new oil seal. If found, remove them with emery cloth or medium grit wet-and-dry sandpaper. Use a clean cloth dipped in solvent and remove any sanding residue from the bore.

9 Apply multipurpose grease to the oil seal lip. Position the seal into the output shaft housing bore and make sure the oil seal is not cocked in the bore.

10 Drive the oil seal into the output housing bore using a seal driver (available at most auto parts stores), or carefully drive the seal into the bore with a hammer and a large socket or section of pipe of the appropriate size **(see illustration)**.

11 Clean the transfer case output female

splines and apply multi-purpose grease to them. Insert the driveshaft splined shaft into the female splines.

12 Install the front yoke or rear flange onto the splines, then install the rubber seal, steel washer and nut (snap-ring on the front yoke on later models). Tighten the nut to the torque listed in this Chapter's Specifications.

13 Connect the front or rear driveshaft to the front or rear axle input yoke (see Chapter 8).

14 Refill the transfer case lubricant (see Chapter 1).

15 Remove the jackstands and lower the vehicle.

7 Transfer case - removal and installation

Removal

Refer to illustrations 7.3, 7.5, 7.6, 7.8 and 7.12

1 Raise the vehicle and support it securely on jackstands. **Caution:** *If the vehicle is equipped with Automatic Ride Control (ARC), make sure the air suspension switch is turned*

to the OFF position before the vehicle is raised to prevent damage to the system components (see Chapter 10).

2 On models so equipped, remove the bolts securing the skid plate to the frame and remove it.

3 On models so equipped, remove the damper from the transfer case **(see illustration)**.

4 Drain the lubricant from the transfer case (see Chapter 1).

5 On electronically-controlled models, squeeze the locking tabs (the part labeled PUSH) toward each other, then pull the electrical connector apart. This may take some effort, but it shouldn't be necessary to use tools to separate the connector halves. Disconnect the electrical connector from the transfer case motor and from the connector mounting bracket **(see illustration)**.

6 Disconnect the speed sensor electrical connector (if equipped) and speedometer cable from the transfer case **(see illustration)**.

7 Disconnect both driveshafts (see Chapter 8).

8 Disconnect the vent hose from the shift lever bracket (manual shift) or mounting bracket (electronic shift) **(see illustration)**.

7C

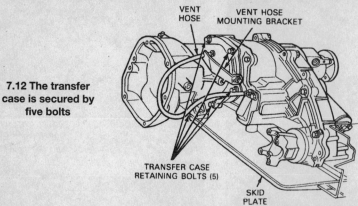

7.12 The transfer case is secured by five bolts

Warning: *The catalytic converter is next to the vent hose and is extremely hot after the engine has been run. Be careful when working around the converter and if possible allow the converter to cool down several hours prior to working around it.*

9   On manually-controlled models, remove the nut from the shift lever and remove the shift lever **(see illustration 3.4)**.

10   On manually-controlled models, remove the large bolt and the small bolt retaining the shifter to the extension housing. Pull on the control lever unit until the bushing slides off the transfer case shift lever pin.

11   Place a transmission jack beneath the transfer case and apply jack pressure to slightly raise the transfer case. Use safety chains to secure the transfer case to the jack.

12   Remove the five bolts holding the transfer case to the transmission extension housing **(see illustration)**.

13   Slide the transfer case rearward and off the transmission output shaft, then lower the transfer case from the vehicle. Remove and discard the gasket between the transfer case and the extension housing.

Installation

Refer to illustration 7.14

14   Installation is the reverse of the removal Steps with the following additions:

a)   *Install a new gasket between the transfer case and the extension housing.*

b)   *Slide the transfer case onto the transmission output shaft, making sure the splines align, until the transfer case seats over the dowel pin.*

c)   *Install the five bolts securing the transfer case and tighten to the torque listed in this Chapter's Specifications and in the torque sequence shown* **(see illustration)**. **Note:** *On manually controlled models, always tighten the large bolt securing the shifter to the extension housing before tightening the small bolt.*

d)   *Install the vent assembly so the white marking on the hose is positioned in the notch in the shift lever bracket (manual shift) or mounting bracket (electronic shift).* **Note:** *The upper end of the vent hose should be 3/4 inch above the top of the shifter on manually controlled models.*

e)   *Connect the rear driveshaft to the transfer case output flange. Tighten the bolts to the torque listed in the Chapter 8 Specifications.*

f)   *Refill the transfer case with the proper type of lubricant (see Chapter 1).*

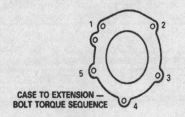

7.14 Tighten the transfer case mounting bolts in the sequence shown

Chapter 8
Clutch and drivetrain

Contents

Specifications

Clutch
Fluid type	See Chapter 1
Type	Single dry plate, diaphragm spring
Actuation	Hydraulic

Rear axle
Type	Integral carrier
Ring gear size	8.8-inch
Lubricant type	See Chapter 1
Driveshaft type	One-piece with single or double cardan type joints

Front axle
Type	Dana Model 35 IFS
Hubs (1994 and earlier)	Manual and automatic locking type
Lubricant type	See Chapter 1

Torque specifications
Ft-lbs (unless otherwise indicated)

Clutch
Pressure plate-to-flywheel bolts	17 to 23
Clutch master cylinder-to-firewall bolts	15 to 20
Clutch master cylinder reservoir bolts	16-23 in-lbs
Release cylinder-to-clutch housing bolts	13 to 17

Rear driveshaft
Driveshaft-to-transfer case bolts (4WD models only)	12 to 16
Driveshaft-to-rear axle bolts	
1991 through 1996	
With double-cardan U-joint	70 to 95
With single-cardan U-joint	8 to 15
1997 and later	
4WD models	70 to 95
2WD models	64 to 87

Front driveshaft
Driveshaft-to-transfer case flange bolts	
1991 through 1996	70 to 95
1997 and later	22
Driveshaft-to-differential yoke bolts	120 to 180 in-lbs

8

Torque specifications (continued)

Ft-lbs (unless otherwise indicated)

Rear axle

Cover bolts	See Chapter 1
Pinion shaft lock-bolt	15 to 30
Leaf spring U-bolt nuts	See Chapter 10
Rear shock absorber-to-axle bracket bolt or nut	See Chapter 10
Brake backing plate nuts	See Chapter 9

Front axle

1991 through 1994

Pivot bolt	120 to 150
Pivot bracket-to-frame nut	70 to 92
Spindle-to-knuckle nuts	40 to 50
Axle stud	190 to 230
Lower balljoint nut	95 to 110
Upper balljoint nut	85 to 100
Carrier to axle arm bolts	35 to 53
Carrier shear bolt	75 to 95
Front shock absorber-to-radius arm nut	See Chapter 10
Front spring seat nut	See Chapter 10
Front radius arm bracket bolts	See Chapter 10

1995 and later

Axle to frame bolts	45 to 59
Axle retainer throughbolts	45 to 59
A-arm bolts	See Chapter 10
Balljoint nuts	See Chapter 10
Wheel hub bolts	See Chapter 10

1   General information

The information in this Chapter deals with the components that transmit power to the wheels, except for the transmission and transfer case, which are dealt with in Chapter 7. For the purposes of this Chapter, these components are grouped into three categories; clutch, driveshaft and axles. Separate Sections within this Chapter offer general descriptions and checking procedures for components in each of the three groups.

Since nearly all the procedures covered in this Chapter involve working under the vehicle, make sure it's securely supported on sturdy jackstands or on a hoist where the vehicle can be easily raised and lowered. **Caution:** *If the vehicle is equipped with Automatic Ride Control (ARC), make sure the air suspension switch is turned to the OFF position before the vehicle is raised to prevent damage to the system components (see Chapter 10).*

2   Clutch - description and check

Refer to illustration 2.5

1   All models with a manual transmission use a single dry plate, diaphragm spring type clutch. The clutch disc has a splined hub which allows it to slide along the splines of the transmission input shaft. The clutch and pressure plate are held in contact by spring pressure exerted by the diaphragm in the pressure plate.

2   The clutch release system is operated by hydraulic pressure. The hydraulic release system consists of the clutch pedal, a master cylinder and fluid reservoir, the hydraulic line, an internal release cylinder and the clutch release (or throwout) bearing.

3   The internal release cylinder, mounted concentrically on the transmission input shaft, pushes directly on the release bearing. This eliminates the need for a release lever.

4   Terminology can be a problem when discussing the clutch components because common names are in some cases different from those used by the manufacturer. For example, the driven plate is also called the clutch plate or disc, the clutch release bearing is sometimes called a throwout bearing, the release cylinder is sometimes called the operating or slave cylinder.

5   Other than to replace components with obvious damage, some preliminary checks should be performed to diagnose clutch problems.

a) *The first check should be of the fluid level in the clutch master cylinder. If the fluid level is low, add fluid as necessary and inspect the hydraulic system for leaks. If the master cylinder reservoir has run dry, bleed the system as described in Section 10 and retest the clutch operation.*

b) *To check "clutch spin-down time," run the engine at normal idle speed with the transmission in Neutral (clutch pedal up - engaged). Disengage the clutch (pedal down), wait several seconds and shift the transmission into Reverse. No grinding noise should be heard. A grinding noise would most likely indicate a problem in the pressure plate or the clutch disc.*

c) *To check for complete clutch release, run the engine (with the parking brake applied to prevent movement) and hold the clutch pedal approximately 1/2-inch from the floor. Shift the transmission between 1st gear and Reverse several times. If the shift is hard or the transmission grinds, component failure is indicated. Check the release bearing travel* **(see illustration)**. *With the clutch pedal*

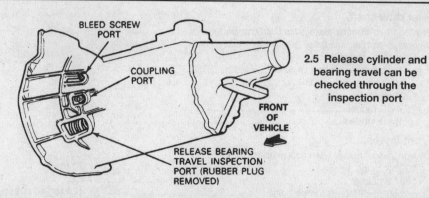

2.5 Release cylinder and bearing travel can be checked through the inspection port

BLEED SCREW PORT

COUPLING PORT

FRONT OF VEHICLE

RELEASE BEARING TRAVEL INSPECTION PORT (RUBBER PLUG REMOVED)

3.5a Loosen the pressure plate bolts (arrows) one turn at a time in rotation until they can be removed by hand

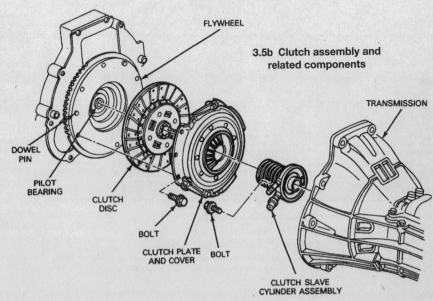

3.5b Clutch assembly and related components

FLYWHEEL

TRANSMISSION

DOWEL PIN

PILOT BEARING

CLUTCH DISC

BOLT

CLUTCH PLATE AND COVER

BOLT

CLUTCH SLAVE CYLINDER ASSEMBLY

depressed completely, the release cylinder should extend substantially. If it doesn't, check the fluid level in the clutch master cylinder (see Chapter 1). Bleed the system (see Section 10).

d) Visually inspect the pivot bushing at the top of the clutch pedal to make sure there is no binding or excessive play.

3 Clutch components - removal, inspection and installation

Warning: *Dust produced by clutch wear and deposited on clutch components may contain asbestos, which is hazardous to your health. DO NOT blow it out with compressed air and DO NOT inhale it. DO NOT use gasoline or petroleum-based solvents to remove the dust. Brake system cleaner should be used to flush the dust into a drain pan. After the clutch components are wiped clean with a rag, dispose of the contaminated rags and cleaner in a covered, marked container.*

Removal

Refer to illustrations 3.5a and 3.5b

1   Access to the clutch components is normally accomplished by removing the transmission (and transfer case on 4WD models), leaving the engine in the vehicle. If, of course, the engine is being removed for major overhaul, then check the clutch for wear and replace worn components as necessary. However, the relatively low cost of the clutch components compared to the time and trouble spent gaining access to them warrants their replacement anytime the engine or transmission is removed, unless they are new or in near perfect condition. The following procedures are based on the assumption the engine will stay in place.

2   Referring to Chapter 7 Part A and Part C, remove the transmission from the vehicle. Support the engine while the transmission is out. Preferably, an engine hoist should be used to support it from above. However, if a

jack is used underneath the engine, make sure a wood block is positioned between the jack and oil pan to spread the load. **Caution:** *The pick-up for the oil pump is very close to the bottom of the oil pan. If the pan is bent or distorted in any way, engine oil starvation could occur.*

3   To support the clutch disc during removal, install a clutch alignment tool through the clutch disc hub **(see illustration 3.13)**.

4   Carefully inspect the flywheel and pressure plate for indexing marks. The marks are usually an X, an O or a white letter. If they cannot be found, scribe marks yourself so the pressure plate and the flywheel will be in the same alignment during installation.

5   Turning each bolt only 1/4-turn at a time, loosen the pressure plate-to-flywheel bolts. Work in a criss-cross pattern until all spring pressure is relieved evenly. Then hold the pressure plate securely and completely remove the bolts, followed by the pressure plate and clutch disc **(see illustrations)**.

Inspection

Refer to illustrations 3.9, 3.11a and 3.11b

6   Ordinarily, when a problem occurs in the clutch, it can be attributed to wear of the clutch driven plate assembly (clutch disc). However, all components should be inspected at this time.

7   Inspect the flywheel for cracks, heat checking, grooves and other obvious defects. If the imperfections are slight, a machine shop can machine the surface flat and smooth, which is highly recommended regardless of the surface appearance. Refer to Chapter 2 for the flywheel removal and installation procedure.

8   Inspect the pilot bearing (see Section 6).

9   Inspect the lining on the clutch disc. There should be at least 1/16-inch of lining above the rivet heads. Check for loose rivets, distortion, cracks, broken springs and other obvious damage **(see illustration)**. As mentioned above, ordinarily the clutch disc is routinely replaced, so if in doubt about the condition, replace it with a new one.

8

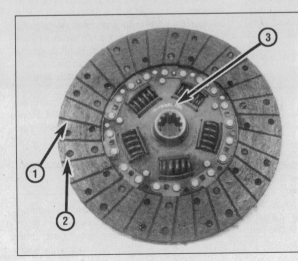

3.9 Typical clutch disc components

1   *Lining* - this will wear down in use

2   *Rivets* - these secure the lining and will damage the flywheel or pressure plate if allowed to contact the surfaces

3   *Marks* - "Flywheel side" or similar

3.11a Examine the pressure plate friction surface for score marks, cracks and evidence of overheating (bluing)

NORMAL FINGER WEAR

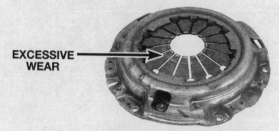

EXCESSIVE FINGER WEAR

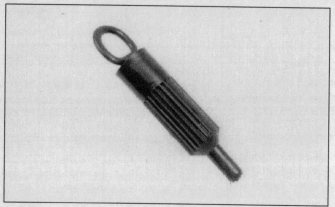

3.13 A clutch alignment tool (available at most auto parts stores) is required to center the clutch disc in the pressure plate

BROKEN OR BENT FINGERS

3.11b Replace the pressure plate if any of these conditions are noted

10   The release bearing should also be replaced along with the clutch disc (see Section 5).

11   Check the machined surfaces and the diaphragm spring fingers of the pressure plate **(see illustrations)**. If the surface is grooved or otherwise damaged, replace the pressure plate. Also check for obvious damage, distortion, cracking, etc. Light glazing can be removed with medium grit emery cloth. If a new pressure plate is required, new and factory-rebuilt units are available.

Installation

12   Before installation, clean the flywheel and pressure plate machined surfaces with lacquer thinner or acetone. It's important that no oil or grease is on these surfaces or the lining of the clutch disc. Handle the parts only with clean hands.

13   Position the clutch disc and pressure plate against the flywheel with the clutch held in place with an alignment tool **(see illustration)**. Make sure it's installed properly (most replacement clutch plates will be marked "flywheel side" or something similar - if not marked, install the clutch disc with the flat side of the hub toward the flywheel).

14   Tighten the pressure plate-to-flywheel bolts only finger tight, working around the pressure plate.

15   Center the clutch disc by ensuring the alignment tool extends through the splined hub and into the pilot bearing in the crankshaft. Wiggle the tool up, down or side-to-side as needed to bottom the tool in the pilot bearing. Tighten the pressure plate-to-flywheel bolts a little at a time, working in a criss-cross pattern to prevent distorting the cover. After all of the bolts are snug, tighten them to the torque listed in this Chapter's Specifications. Remove the alignment tool.

16   If removed, install the clutch release bearing as described in Section 5.

17   Install the transmission and all components removed previously. Tighten all fasteners to the torque values listed in this Chapter's Specifications.

4   Flywheel - inspection

1   Inspect the flywheel when the vehicle suffers from excessive transmission gear wear, transmission jumping out of gear, driveline vibration, clutch pedal vibration, clutch slippage, pilot bearing noise or release bearing noise.

2   Visually inspect the flywheel for signs of cracking, warpage, scoring or heat checking. If any of these conditions exist, the flywheel must be removed and resurfaced at an automotive machine shop (see Chapter 2).

5   Clutch release bearing - removal, inspection and installation

Warning: *Dust produced by clutch wear and deposited on clutch components may contain asbestos, which is hazardous to your health. DO NOT blow it out with compressed air and DO NOT inhale it. DO NOT use gasoline or petroleum-based solvents to remove the dust. Brake system cleaner should be used to flush the dust into a drain pan. After the clutch components are wiped clean with a rag, dispose of the contaminated rags and cleaner in a covered, marked container.*

Removal and inspection

Refer to illustration 5.2

1   Remove the transmission (see Chapter 7 Part A).

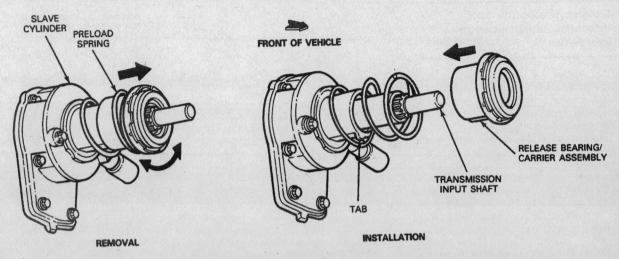

SLAVE CYLINDER
PRELOAD SPRING
FRONT OF VEHICLE
RELEASE BEARING/ CARRIER ASSEMBLY
TRANSMISSION INPUT SHAFT
TAB
REMOVAL
INSTALLATION

5.2 Release bearing removal and installation details

2   Within the clutch housing, twist the release bearing and carrier assembly until you feel resistance. Keep turning the assembly and the preload spring will push the release bearing off the release cylinder **(see illustration)**.
3   Hold the center of the bearing and rotate the outer portion while applying pressure. If the bearing doesn't turn smoothly or it's noisy, replace it with a new one. Wipe the bearing with a clean rag and inspect it for damage, wear and cracks. Don't immerse the bearing in solvent - it's sealed for life and to do so would ruin it.

Installation

4   Apply a light coat of high-temperature lithium based grease to the face of the release bearing, where it contacts the pressure plate diaphragm fingers.
5   Lubricate the inner bore of the bearing and bearing carrier with a synthetic long-life grease. Don't use petroleum-based grease.

6   Push the release bearing and carrier onto the slave cylinder until it bottoms out **(see illustration 5.2)**.
7   Install the transmission (see Chapter 7 Part A).

6   Pilot bearing - inspection and replacement

Refer to illustrations 6.1 and 6.5

1   The clutch pilot bearing is a needle roller type bearing which is pressed into the rear of the crankshaft **(see illustration)**. It is pre-lubricated and does not require additional lubrication. Its primary purpose is to support the front of the transmission input shaft. The pilot bearing should be inspected whenever the clutch components are removed from the engine. Due to its inaccessibility, if you are in doubt as to its condition, replace it with a new one. **Note:** *If the engine has been*

removed from the vehicle, disregard the following steps which do not apply.
2   Remove the transmission (refer to Chapter 7 Part A).
3   Remove the clutch components (see Section 3).
4   Inspect for any excessive wear, scoring, lack of grease, dryness or obvious damage. If any of these conditions are noted, the bearing should be replaced. A flashlight will be helpful to direct light into the recess.
5   Removal requires a special puller and slide hammer **(see illustration)**. Remove the bearing and clean the crankshaft recess.
6   To install the new bearing, lightly lubricate the outside surface with lithium-based grease, then drive it into the recess with a soft-face hammer. The seal must face out **(see illustration 6.1)**.
7   Install the clutch components, transmission and all other components removed previously, tightening all fasteners properly.

8

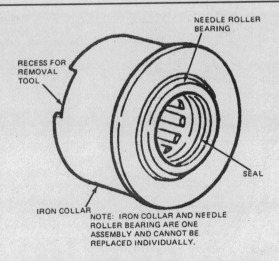

NEEDLE ROLLER BEARING
RECESS FOR REMOVAL TOOL
SEAL
IRON COLLAR
NOTE: IRON COLLAR AND NEEDLE ROLLER BEARING ARE ONE ASSEMBLY AND CANNOT BE REPLACED INDIVIDUALLY.

6.1 The pilot bearing seal must face out (towards the transmission) when the bearing is installed in the crankshaft

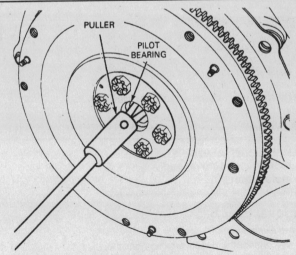

PULLER
PILOT BEARING

6.5 Removing the pilot bearing requires an internal puller connected to a slide hammer

7   Clutch hydraulic line quick-disconnect fittings - general information

Refer to illustration 7.2

1   The hydraulic lines are equipped with quick-disconnect fittings and a special tool (available at most auto parts stores) is required for separation.

2   Depress the white retainer within the quick-disconnect fitting with the special tool while pulling on the hydraulic line, then disconnect the male connector **(see illustration)**. To install, push the male connector into the female connector until it locks into place.

3   There should be no loss of hydraulic fluid during separation of the fittings so there is no need to bleed the system after a fitting has been disconnected unless shifting is hard or there's a lack of clutch reserve travel.

8   Clutch release cylinder - removal and installation

Refer to illustration 8.4
Note: *Prior to removing the release cylinder, disconnect the master cylinder pushrod from the clutch pedal. If not disconnected, permanent damage to the master cylinder will occur if the clutch pedal is depressed while the release cylinder is disconnected.*

1   Pry off the retainer bushing and disconnect the master cylinder pushrod from the clutch pedal.

2   Disconnect the hydraulic quick-disconnect fitting from the slave cylinder (see Section 7).

3   Remove the transmission (see Chapter 7 Part A).

4   Remove the bolts securing the release cylinder to the transmission and slide the release cylinder assembly off the transmission input shaft **(see illustration)**.

5   Installation is the reverse of the removal Steps with the following additions:

a) *Install the release cylinder onto the input shaft with the hydraulic line facing toward the left side of the transmission.*

b) *Install the release cylinder mounting bolts and tighten them to the torque listed in this Chapter's Specifications.*

c) *If necessary, bleed the clutch hydraulic system (see Section 10).*

9   Clutch master cylinder and reservoir - removal and installation

Removal

Refer to illustration 9.5

1   Disconnect the negative cable from the battery.

2   Working inside the vehicle, remove the spring clip or retainer bushing and disconnect the pushrod from the top of the clutch pedal.

3   Remove the clutch/starter interlock switch from the pushrod (see Section 12).

4   Disconnect the hydraulic line from the slave cylinder (see Section 7).

5   Within the engine compartment, remove the bolts securing the clutch master cylinder reservoir to the firewall **(see illustration)**. Be careful not to spill any of the fluid.

6   Working in the same area, remove the bolts securing the clutch master cylinder to the firewall. Be careful not to spill any of the fluid.

7   Remove the clutch master cylinder and reservoir from the engine compartment.

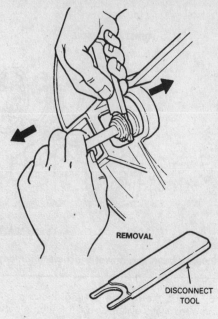

REMOVAL

DISCONNECT TOOL

7.2 A special tool is required to disconnect the hydraulic line fitting

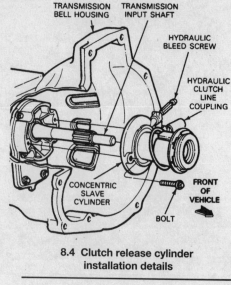

8.4 Clutch release cylinder installation details

Installation

8   Install the clutch master cylinder through the firewall and make sure the pushrod is positioned on the correct side of the clutch pedal so it can be installed onto the pedal's pivot pin.

9   Attach the master cylinder to the firewall and install the mounting bolts. Tighten the bolts to the torque listed in this Chapter's Specifications.

10   Attach the master cylinder reservoir to the firewall and install the mounting bolts, tightening them securely.

11   Connect the hydraulic line to the slave cylinder (see Section 7).

12   Working inside the vehicle, connect the pushrod to the clutch pedal and install the clip or retainer bushing.

13   Install the clutch/starter interlock switch (see Section 12).

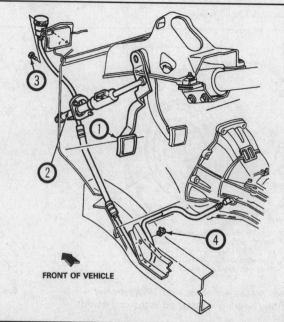

9.5 Typical clutch master cylinder installation details

1) *Clutch pedal*
2) *Master cylinder*
3) *Reservoir bolt*
4) *Clip*

FRONT OF VEHICLE

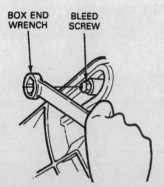

10.5 The bleeder valve is accessible through a port on the left side of the clutch housing

14   Connect the negative cable to the battery.
15   Fill the clutch master cylinder reservoir with brake fluid conforming to DOT 3 specifications and bleed the clutch system (see Section 10).

10 Clutch hydraulic system - bleeding

Refer to illustration 10.5
1   The hydraulic system should be bled to remove all air whenever air enters the system. This occurs if the fluid level has been allowed to fall so low that air has been drawn into the master cylinder. Under normal circumstances, air should not enter the system when the quick-disconnect hydraulic line fittings are disconnected. The procedure is very similar to bleeding a brake system, but depends mainly on gravity, rather than the pumping action of the pedal, for the bleeding effect.
2   Fill the master cylinder reservoir to the top with new brake fluid conforming to DOT 3 specifications. **Caution:** *Do not re-use any of the fluid coming from the system during the bleeding operation and don't use fluid which*

has been inside an open container for an extended period of time.
3   Raise the vehicle and place it securely on jackstands to gain access to the bleeder valve, which is located on the left side of the clutch housing. **Caution:** *If the vehicle is equipped with Automatic Ride Control (ARC), make sure the air suspension switch is turned to the OFF position before the vehicle is raised to prevent damage to the system components (see Chapter 10).*
4   Remove the dust cap which fits over the bleeder valve and push a length of clear plastic hose over the valve. Place the other end of the hose into a clear container.
5   Open the bleeder valve **(see illustration)**. Fluid will run from the master cylinder, down the hydraulic line, into the release cylinder and out through the clear plastic tube. Let the fluid run out until it is free of bubbles. **Note:** *Don't let the fluid level drop too low in the master cylinder, or air will be drawn into the hydraulic line and the whole process will have to be started over.*
6   Close the bleeder valve.
7   Have an assistant slowly depress the clutch pedal and hold it. Open the bleeder valve on the release cylinder, allowing fluid to flow through the clear plastic hose. Close the bleeder valve when the flow stops. Once closed, have the assistant release the pedal.
8   Slowly press and release the pedal five times, waiting for two seconds each time the pedal is released.
9   Fill the fluid reservoir to the step.
10   The clutch should now be completely bled. If it isn't (indicated by failure to disengage completely), repeat Steps 5 through 9.
11   Continue this process until all air is evacuated from the system, indicated by a solid stream of fluid being ejected from the bleeder valve each time with no air bubbles in the hose or container.
12   Install the dust cap and lower the vehicle. Check carefully for proper operation before placing the vehicle in normal service. Check the fluid level.

11 Clutch pedal - removal and installation

Refer to illustration 11.5
1   Disconnect the negative cable from the battery.
2   Disconnect the barbed end of the clutch/starter interlock switch rod from the clutch pedal. Remove the lockpin and remove the master cylinder pushrod from the clutch pedal. Remove the plastic bushing.
3   Remove the trim panel from the driver's sidewall.
4   Unbolt the parking brake and move it aside.
5   Pry the retainer from the clip on the pedal support bracket **(see illustration)**.
6   Slide the pedal out of the bracket.
7   Inspect the pedal bushings for excessive wear (sloppy fit on the pedal shafts). Replace the bushings if necessary. Always oil the bushings (new or old) with clean engine oil before reassembly.
8   Installation is the reverse of the removal Steps.

12 Clutch/starter interlock switch - removal and installation

Refer to illustration 12.2
1   Disconnect the negative cable from the battery.
2   Disconnect the electrical connector from the clutch/starter interlock switch, also called a clutch pedal position switch **(see illustration)**.
3   Rotate the switch so the plastic retainer is visible. Squeeze the retainer tabs together and slide the retainer rearward off the switch.
4   Detach the switch from the master cylinder pushrod and take it out.
5   Installation is the reverse of the removal Steps. Be sure the switch is securely seated on the master cylinder pushrod.

8

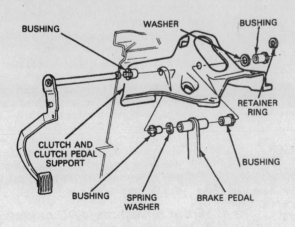

11.5 Typical clutch pedal installation details

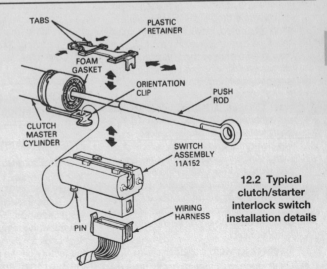

12.2 Typical clutch/starter interlock switch installation details

13   Driveshaft and universal joints - description and check

1   The driveshaft is a tube running between the transfer case or transmission and the rear axle or front axle. Universal joints are located at either end of the driveshaft(s) and permit power to be transmitted to the rear (and on 4WD models, the front) wheels at varying angles.

2   Some driveshafts feature a splined slip yoke at one end, which fits into the transmission extension housing. Others incorporate a slip yoke as part of the driveshaft. This arrangement allows the length of the driveshaft to change, which compensates for suspension movement.

3   On models with a slip yoke at the front of the driveshaft, an oil seal is used to prevent leakage of fluid and to keep dirt and contaminants from entering the transmission. If leakage is evident at the front of the driveshaft, replace the oil seal, referring to the procedures in Chapter 7.

4   The driveshaft assembly requires very little service. The universal joints on models not equipped with grease fittings are lubricated for life and must be replaced if problems develop. The driveshaft(s) must be removed from the vehicle for this procedure.

5   Since the driveshaft is a balanced unit, it's important that no undercoating, mud, etc. be allowed to stay on it. When the vehicle is raised for service it's a good idea to clean the driveshaft(s) and inspect it for any obvious damage. Also check that the small weights used to originally balance the driveshaft(s) are in place and securely attached. Whenever a driveshaft is removed it's important that it be reinstalled in the same relative position to preserve the balance.

6   Problems with the driveshaft(s) are usually indicated by a noise or vibration while driving the vehicle. A road test should verify if the problem is the driveshaft(s) or another vehicle component:

  a) *On an open road, free of traffic, drive the vehicle and note the engine speed (rpm) at which the problem is most evident.*
  b) *With this noted, drive the vehicle again, this time manually keeping the transmission in first, then second, then third gear ranges and running the engine up to the engine speed noted.*
  c) *If the noise or vibration decreased or was eliminated, visually inspect the driveshaft(s) for damage, material on the shaft which would effect balance, missing weights and damaged universal joints. Another possibility for this condition would be tires which are out-of-balance or a bent or damaged wheel(s).*
  d) *If the noise or vibration occurs at the same engine speed regardless of which gear the transmission is in, the driveshaft is not at fault.*

7   To check for worn universal joints:

  a) *On an open road, free of traffic, drive the*

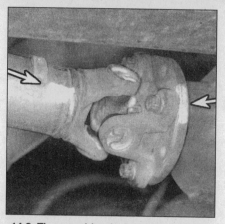

14.3 The rear driveshaft may have come from the factory with yellow paint marks that indicate the correct alignment of the driveshaft with the companion flange - if not, make new paint marks

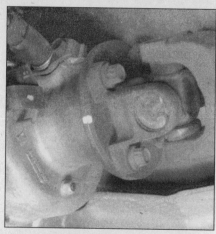

14.4 On 4WD models, make alignment marks on the driveshaft and transfer case flanges

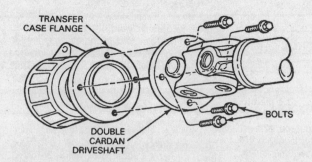

DOUBLE CARDAN 4x4 INSTALLATION

TRANSFER
CASE FLANGE

BOLTS

DOUBLE
CARDAN
DRIVESHAFT

14.9 Make alignment marks on the front driveshaft and transfer case flanges, then unbolt the driveshaft from the transfer case (double-cardan model shown)

vehicle slowly until the transmission is in High gear. Let off on the accelerator, allowing the vehicle to coast, then accelerate. A clunking or knocking noise will indicate worn universal joints.
  b) *Drive the vehicle at a speed of about 10 to 15 mph and then place the transmission in Neutral, allowing the vehicle to coast. Listen for abnormal driveline noises.*
  c) *Raise the vehicle and support it securely on jackstands. With the transmission in Neutral, manually turn the driveshaft(s), watching the universal joints for excessive play.*

14   Driveshafts - removal and installation

Rear driveshaft

Refer to illustrations 14.3 and 14.4

1   Disconnect the negative cable from the battery.

2   Raise the vehicle and support it securely on jackstands. Place the transmission in Neutral with the parking brake off.

3   Look for the yellow paint marks made at the factory on the axle companion flange and the driveshaft flange **(see illustration)**. If

these aren't visible, use a scribe, white paint or a hammer and punch to place marks on the driveshaft and the differential flange in line with each other. This is to make sure the driveshaft is reinstalled in the same position to preserve the balance.

4   On 4WD models, repeat Step 3 for the front end of the driveshaft **(see illustration)**.

5   Remove the bolts and lower the rear of the driveshaft. On 4WD models, remove the bolts and detach the front end from the transfer case. On 2WD models, slide the front end out of the transmission.

6   Installation is the reverse of the removal Steps with the following additions:

  a) *On 2WD models, slide the front of the driveshaft into the transmission.*
  b) *On 4WD models, position the front end of the driveshaft on the transfer case with the alignment marks aligned.*
  c) *Raise the rear of the driveshaft into position, checking to be sure the marks are in alignment. If not, make sure the transmission is still in Neutral, then turn the rear wheels to match the pinion flange and the driveshaft.*

Front driveshaft (4WD models)

Refer to illustrations 14.9 and 14.10

7   Using a scribe, white paint or a hammer and punch, place marks on the driveshaft

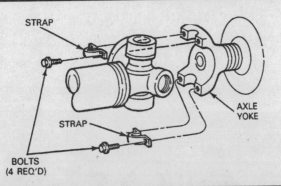

14.10 Make alignment marks on the front driveshaft and differential yoke - wrap the universal joint with tape after unbolting it so the bearings won't fall off

15.2 Remove the universal joint snap-rings with small pliers

and differential flanges in line with each other **(see illustration 14.3)**. This is to make sure the driveshaft is reinstalled in the same position to preserve the balance.

8   Make alignment marks on the driveshaft and transfer case flanges.

9   Unbolt the flange that secures the driveshaft double-cardan joint to the transfer case **(see illustration)**. **Note:** *On 1997 and later models, there is a constant-velocity (CV) joint at the transfer case, instead of a double-cardan.*

10   Remove the bolts and straps that secure the front end of the driveshaft to the differential yoke **(see illustration)**.

11   Wrap tape around the universal joint bearings at the axle end of the driveshaft so they won't fall off.

12   Installation is the reverse of the removal Steps with the following additions:

   a) *Raise the front of the driveshaft into position, checking to be sure the marks are in alignment. If not, make sure the transmission is still in Neutral, then turn the front wheels to align the marks.*

   b) *Remove the tape securing the bearing cups and install the straps and bolts. Tighten the bolts to the torque listed in this Chapter's Specifications.*

15   Single cardan type U-joint - overhaul

Refer to illustrations 15.2, 15.4a and 15.4b

Note: *A press or large vise will be required for this procedure. It may be a good idea to take the driveshaft to a driveline repair shop or machine shop where the universal joints can be replaced for you, normally at a reasonable charge.*

1   Remove the driveshaft (see Section 14).

2   Place the driveshaft in a sturdy vise and remove the snap-rings from the spider **(see illustration)**. **Caution:** *Never place the tube of the driveshaft in a vise. The tube can be distorted or a fracture could be started that will eventually ruin the shaft.*

3   Supporting the driveshaft, place it in position on either an arbor press or on a workbench equipped with a vise.

4   Place a piece of pipe or a large socket with the same inside diameter over one of the bearing cups. Position a socket which is of

slightly smaller diameter than the cup on the opposite bearing cup **(see illustration)** and use the vise or press to force the cup out (inside the pipe or large socket), stopping just before it comes completely out of the yoke. Use the vise or large pliers to work the cup the rest of the way out **(see illustration)**.

5   Transfer the sockets to the other side and press the opposite bearing cup out in the same manner.

6   After the bearings have been removed, lift the spider from the yoke and thoroughly clean all dirt and debris from the yokes on both ends of the driveshaft.

7   Pack the new universal joint bearings with grease. Ordinarily, specific instructions for lubrication will be included with the universal joint servicing kit and should be followed carefully.

8   Position the spider in the yoke and partially install one bearing cup in the yoke.

9   Start the spider into the bearing cup and then partially install the other cup. Align the spider and press the bearing cups into position, being careful not to damage the dust seals.

10   Install the snap-rings. If difficulty is encountered in seating the snap-rings, strike the driveshaft yoke sharply with a hammer. This will spring the yoke ears slightly and allow the snap-rings to seat in the groove. This should also be done if the joint feels tight after assembly.

11   Install the driveshaft (see Section 14).

16   Double-cardan type U-joint - overhaul

Note: *A press or large vise will be required for this procedure. It may be a good idea to take the driveshaft to a driveline repair shop or machine shop where the universal joints can be replaced for you, normally at a reasonable charge.* **Note:** *On 1997 and later 4WD models, the front driveshaft CV joint is not serviceable, if defective replace the driveshaft.*

1   Remove the driveshaft (see Section 14).

2   Use paint or small punch and hammer and mark the positions of the spiders, the center yoke and the centering socket as related to the stud yoke which is welded to the front of the driveshaft tube. The spiders must be assembled with the bosses in their original position to provide proper clearance.

3   Use a small screwdriver and pry out the snap-rings that attach the bearings in the front of the center yoke.

4   Supporting the driveshaft, place it in position on either an arbor press or on a workbench equipped with a vise.

5   Place a piece of pipe or a large socket with the same inside diameter over one of the

15.4a The universal joint bearing cups can be pressed out with a vise and sockets

15.4b The bearing cup can be extracted with locking pliers

8

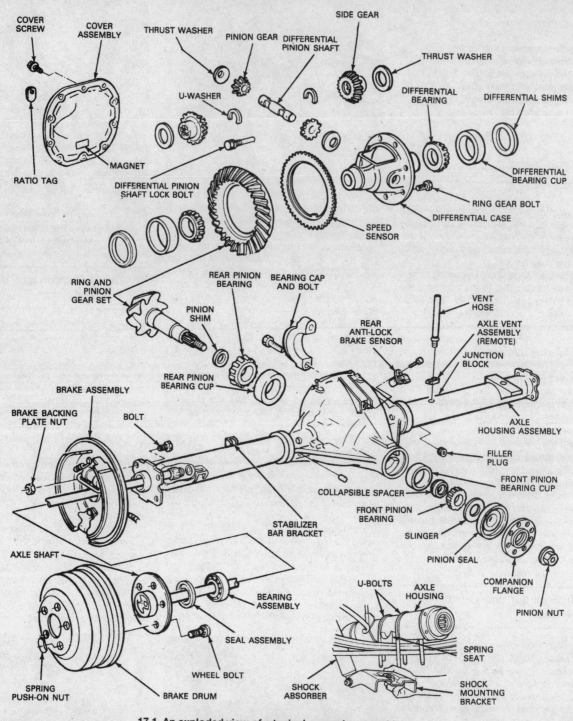

17.1 An exploded view of a typical rear axle assembly

bearing cups. Position a socket which is of slightly smaller diameter than the cup on the opposite bearing cup and use the vise or press to force the cup out (inside the pipe or large socket), stopping just before it comes completely out of the yoke **(see illustration 15.4a)**. Use the vise or large pliers to work the cup the rest of the way out **(see illustration 15.4b)**.

6   Transfer the sockets to the other side and press the opposite bearing cup out in the same manner.

7   Repeat Steps 6 and 7 and remove all bearings from the universal joints.

8   With the front bearing cups removed, remove the spider from the center yoke.

9   Pull the centering socket yoke off the center stud and remove the rubber seal from the centering ball stud.

10   Remove the snap-rings from the center yoke and from the driveshaft yoke.

11   Press or drive the bearing out as previously described.

12   With the bearing cups removed, remove the center yoke from the spider.

13   Remove the spider from the driveshaft yoke.

14   After the bearings have been removed, lift the spider from the yoke and thoroughly clean all dirt and debris from the yokes on

both ends of the driveshaft. If using a repair kit, install all of the parts supplied.

15   Pack the new universal joint bearings with grease. Ordinarily, specific instructions for lubrication will be included with the universal joint servicing kit and should be followed carefully.

16   Remove the clamps on the driveshaft dust boot seal and discard the clamps.

17   Note the orientation of the slip yoke to the driveshaft tube for installation during assembly. Mark the relationship of the slip yoke to the driveshaft tube.

18   Carefully pull the slip joint from the driveshaft, be careful not to damage the boot seal.

19   Inspect the spline area and thoroughly clean all foreign matter from the splines with solvent.

20   Lubricate the splines with multi-purpose grease.

21   Install the dust boot seal loosely on the driveshaft tube, then install the yoke into the driveshaft in its original position. Refer to marks made prior to removal.

22   Install new clamps on the dust boot seal and tighten securely.

23   To assemble the double cardan joints, position the spider in the driveshaft yoke. Make sure the spider bosses (or lubricating plugs) are in their original positions.

24   Start the spider into the bearing cup and then partially install the other cup. Align the spider and press the bearing cups into position, being careful not to damage the dust seals.

25   Install new snap-rings. If difficulty is encountered in seating the snap-rings, strike the driveshaft yoke sharply with a hammer. This will spring the yoke ears slightly and allow the snap-rings to seat in the groove.

26   Repeat Step 25 and 26 for the remaining bearings.

27   Pack the socket relief and the ball with multi-purpose grease, then position the center yoke over the spider ends and press in the bearing.

28   Install new snap-rings.

29   Install a new seal on the centering ball stud and position the centering socket yoke on the stud.

30   Place the front spider in the center yoke and make sure the spider bosses (or lubricating plugs) are in their original positions.

31   Start the spider into the bearing cup and then partially install the other cup. Align the spider and press the bearing cups into position, being careful not to damage the dust seals.

32   Install new snap-rings. If difficulty is encountered in seating the snap-rings, strike the driveshaft yoke sharply with a hammer. This will spring the yoke ears slightly and allow the snap-rings to seat in the groove. This should also be done if the joint feels tight after assembly.

33   If so equipped, lubricate the U-joints through the grease fittings with multi-purpose grease.

34   Install the driveshaft (see Section 14).

17   Rear axle - description and check

Refer to illustration 17.1

1   The rear axle assembly is a hypoid (the centerline of the pinion gear is below the centerline of the ring gear) semi-floating type **(see illustration)**. The differential carrier is a casting with a pressed-steel cover. The axle tubes are made of steel, pressed and welded into the carrier.

2   An optional Traction-Lok limited slip rear axle is also available. This differential allows for normal operation until one wheel loses traction. The unit utilizes multi-disc clutch packs and a speed-sensitive engagement mechanism which locks both axleshafts together, applying equal rotational power to both wheels.

3   In order to undertake certain operations, particularly replacement of the axleshafts, it's important to know the axle identification number. It's located on a small metal tag near one of the cover bolts.

4   Many times a problem is suspected in the rear axle area when, in fact, it lies elsewhere. For this reason, a thorough check should be performed before assuming a rear axle problem.

5   The following noises are those commonly associated with rear axle diagnosis procedures:

a) *Road noise is often mistaken for mechanical faults. Driving the vehicle on different surfaces will show whether the road surface is the cause of the noise. Road noise will remain the same if the vehicle is under power or coasting.*

b) *Tire noise is sometimes mistaken for mechanical problems. Tires which are worn or low on air pressure are particularly susceptible to emitting vibrations and noises. Tire noise will remain about the same during varying driving situations, where rear axle noise will change during coasting, acceleration, etc.*

c) *Engine and transmission noise can be deceiving because it will travel along the driveline. To isolate engine and trans-*

mission noises, make a note of the engine speed at which the noise is most pronounced. Stop the vehicle and place the transmission in Neutral and run the engine to the same speed. If the noise is the same, the rear axle is not at fault.

6   Overhaul and general repair of the rear axle is beyond the scope of the home mechanic due to the many special tools and critical measurements required. Thus, the procedures listed here will involve axleshaft removal and installation, axleshaft oil seal replacement, axleshaft bearing replacement and removal of the entire unit for repair or replacement.

18   Rear axle assembly - removal and installation

Removal

Refer to illustrations 18.6, 18.8, 18.9, 18.11 and 18.13

1   Loosen the lug nuts on the rear wheels.

2   Raise the rear of the vehicle, support it securely on jackstands and remove the rear wheels. Also remove the rear spare wheel/tire for more working room. **Caution:** *If the vehicle is equipped with Automatic Ride Control (ARC), make sure the air suspension switch is turned to the OFF position before the vehicle is raised to prevent damage to the system components (see Chapter 10).*

3   Remove the axleshafts (see Section 19), and disconnect the rear of the driveshaft from the axle (see Section 14).

4   Unbolt the brake line junction block from the axle housing and detach the brake lines from the clips on the axle housing. Disconnect the rear anti-lock electrical connector from the sensor on the axle housing (see Chapter 9). On models with Automatic Ride Control (ARC), disconnect the rear height sensor (see Chapter 10).

5   Remove the rear brake drums or discs/calipers (see Chapter 9).

6   Remove the nuts and bolts that secure the brake backing plate, or disc brake adapter and dust shield) to the axle **(see**

18.6 Remove the nuts that secure the brake backing plate to the axle housing (arrows)

8

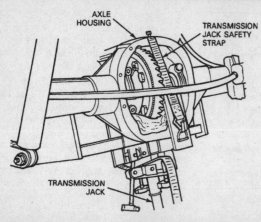

18.8 Place the axle housing on a transmission jack and secure it firmly - the axle housing is heavy and can cause injury if it falls

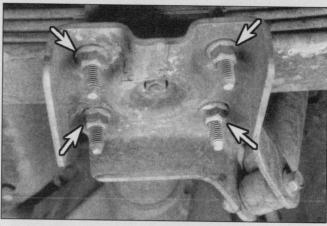

18.9 Remove the four nuts that secure the spring/shock mounting bracket to the axle

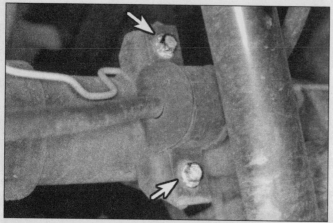

18.11 Unbolt the stabilizer bar brackets from the axle housing

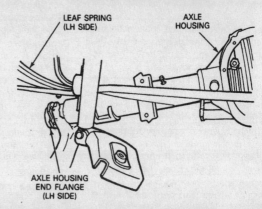

18.13 Tilt the axle housing as shown to clear the spring, then lower the jack and take the axle out from under the vehicle

illustration). Separate the backing plate from the axle and tie them up out of the way. **Caution:** *Be careful not to bend or kink the brake lines as you reposition the backing plate.*

7   Detach the vent hose from the axle housing. The vent hose is secured to the axle by a crimp-type clamp - you'll need to replace it with a screw-type clamp. Refer to Chapter 9 and release the front parking brake cable from the rear cables.

8   Place a transmission jack beneath the axle housing. Secure the axle housing to the jack with a safety strap or chain **(see illustration). Note:** *You'll need to tilt the axle housing on the jack to remove it. Secure the axle so it can't fall off the jack, but position the strap or chain so the axle can be tilted.*

9   Remove the U-bolt nuts that secure the spring/shock mounting bracket to the axle **(see illustration).** Detach the shock absorber from the bracket.

10   Lower the bracket away from the spring, then remove the U-bolts.

11   Unbolt the stabilizer bar bracket bolts from the axle housing **(see illustration).** Move the stabilizer bar out of the way. On models equipped with the 5.0L V8, remove the right-side anti-windup bar (see Chapter 10).

12   Raise the axle up off the spring. Move it

toward the passenger side of the vehicle until the driver's side end of the axleshaft clears the spring.

13   Tilt the driver's side of the axle housing down **(see illustration).** Lower the axle, move it toward the driver's side and remove it from beneath the vehicle.

Installation;

14   Installation is the reverse of the removal Steps. Be sure to tighten the fasteners to the torque values listed in this Chapter's Specifications. Tighten the wheel lug nuts to the torque listed in the Chapter 1 Specifications.

---

19   Axleshafts (rear), bearings and oil seals - removal and installation

Axleshaft removal

Refer to illustrations 19.6a, 19.6b, 19.6c, 19.7a, 19.7b and 19.8

1   Loosen the lug nuts on the rear wheels.

2   Raise the rear of the vehicle, support it securely on jackstands and remove the rear wheels. **Caution:** *If the vehicle is equipped with Automatic Ride Control (ARC), make sure the air suspension switch is turned to the OFF position before the vehicle is raised to*

prevent damage to the system components (see Chapter 10).

3   Remove the rear brake drums or calipers and discs (see Chapter 9).

4   Clean all dirt from the area surrounding the carrier cover.

5   Remove the cover and drain the differential lubricant (see Chapter 1).

6   Remove the differential shaft lock-bolt and differential pinion shaft **(see illustrations).** On Traction-Lok differentials, it isn't

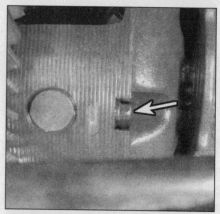

19.6a The pinion shaft lock-bolt fits in a recess in the differential housing

19.6b Remove the bolt with a thinwall socket . . .

19.6c . . . then pull the pinion shaft out - on Traction-Lok models, the spacer spring (arrow) need not be removed

19.7a Remove the C-lock from the end of each axleshaft - the differential lubricant is very slippery and makes the C-locks hard to hold with pliers, so be careful not to drop them

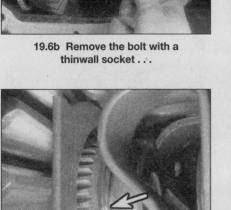

19.7b There's a rubber O-ring in the C-lock groove in the end of each axleshaft

19.8 Lift and pull the axleshaft out of the axle housing - be careful not to damage the oil seal

Using the slide hammer, withdraw the bearing and oil seal as a unit **(see illustration).**

10   Thoroughly clean the rear axle housing bore with a rag and solvent.

11   Lubricate the new bearing with rear axle lubricant and install the bearing into the housing bore. Use a bearing/seal driver tool, or large socket that matches the outer bearing race **(see illustration).** Tap the bearing in until it's into the full depth of its recess. **Caution:** *Do not tap on the inner bearing race while installing the bearing, as it will be damaged.*

12   Apply multi-purpose grease between the lips of the axleshaft seal.

13   Install a new axleshaft seal using the axle tube seal installation tool or a large socket that matches the seal diameter. **Caution:** *If the seal is installed incorrectly or gets cocked during installation it will be damaged, leading to component failure and an oil leak.*

necessary to remove the preload spring **(see illustration). Caution:** *If the axleshafts or pinion is rotated while the pinion shaft is out, the pinion gears could fall out and possibly be damaged. Reinsert the pinion shaft and lockbolt while the axleshafts are being removed.*

7   Push the axleshafts toward the center of the vehicle and remove the C-locks from the button end of the axleshafts **(see illustration).** Don't lose or damage the rubber O-ring which fits in the axle shaft groove under the C-lock **(see illustration).**

8   Carefully remove the axleshafts from the housing **(see illustration).** Don't damage the oil seal at the end of the housing.

Oil seal and bearing replacement

Refer to illustrations 19.9 and 19.11

9   Insert a rear axle bearing remover and slide hammer into the bore of the bearing and position it behind the bearing outer race.

Axleshaft installation

Refer to illustrations 19.14 and 19.15

14   Make sure the rubber O-ring is in the

8

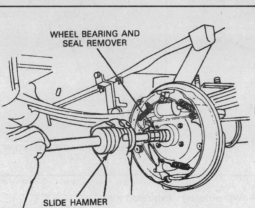

19.9 Remove the wheel bearing and seal with a puller and slide hammer

WHEEL BEARING AND SEAL REMOVER

SLIDE HAMMER

19.11 The wheel bearing and seal can be installed with a large socket or piece of pipe of the correct diameter if the special tool isn't available

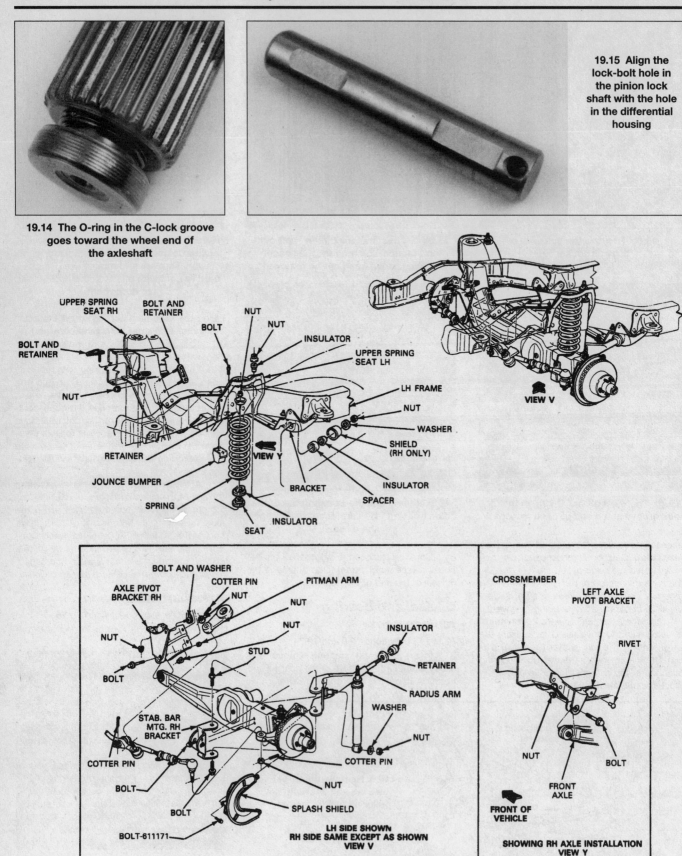

19.14 The O-ring in the C-lock groove goes toward the wheel end of the axleshaft

19.15 Align the lock-bolt hole in the pinion lock shaft with the hole in the differential housing

UPPER SPRING SEAT RH

BOLT AND RETAINER

BOLT

NUT

NUT

INSULATOR

UPPER SPRING SEAT LH

BOLT AND RETAINER

LH FRAME

NUT

NUT

WASHER

VIEW V

RETAINER

VIEW Y

SHIELD (RH ONLY)

JOUNCE BUMPER

BRACKET

INSULATOR

SPACER

SPRING

INSULATOR

SEAT

BOLT AND WASHER

COTTER PIN

PITMAN ARM

CROSSMEMBER

LEFT AXLE PIVOT BRACKET

AXLE PIVOT BRACKET RH

NUT

NUT

NUT

RIVET

NUT

INSULATOR

BOLT

STUD

RETAINER

RADIUS ARM

STAB. BAR MTG. RH BRACKET

WASHER

NUT

NUT

COTTER PIN

FRONT AXLE

COTTER PIN

NUT

BOLT

BOLT

BOLT

NUT

BOLT

SPLASH SHIELD

FRONT OF VEHICLE

BOLT-611171

LH SIDE SHOWN RH SIDE SAME EXCEPT AS SHOWN VIEW V

SHOWING RH AXLE INSTALLATION VIEW Y

20.7 The front axle assembly (1991 through 1994 4WD models) - exploded view

20.21 Use a large screwdriver or prybar (arrow) to carefully pry the CV joint out

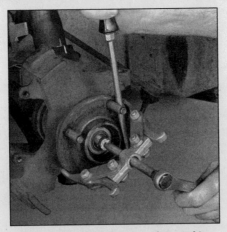

20.22 A two jaw puller can be used to press the driveaxle from the hub - do not hammer on the axle

20.23 After the driveaxle has been separated from the hub, pull out on the strut/knuckle assembly and remove the stub shaft from the hub

groove in the end of the axleshaft. Push the O-ring toward the outer end of the groove (toward the wheel) **(see illustration)**. Carefully insert the axle into the housing and install the C-lock on the button end of the axleshaft splines. Push the shaft out until the C-lock seats in the counterbore of the differential side gear.

15 Position the differential pinion shaft through the case and pinion gears, aligning the hole in the shaft with the lock-bolt hole **(see illustration)**. Apply thread-locking compound to the lock-bolt, install the bolt and tighten it to the torque listed in this Chapter's Specifications.

16 Clean the gasket mounting surface on the rear axle housing and cover of all old gasket material residue with lacquer thinner or gasket remover. Apply a continuous bead of silicone sealant. Run the bead to the inside of the bolt holes and install the cover and bolts.

17 Tighten the bolts in a criss-cross pattern to the torque listed in this Chapter's Specifications.

18 Refill the axle with the correct quantity and grade of lubricant (see Chapter 1).

19 Install the brake components (see Chapter 9).

20 Install the wheels and lug nuts.

21 Lower the vehicle and tighten the lug nuts to the torque listed in the Chapter 1 Specifications.

20   Front axle assembly (4WD models) - removal and installation

1991 through 1994 models

Removal

Refer to illustration 20.7

1 Raise the front of the vehicle and install jackstands under the radius arm brackets.

2 Disconnect the driveshaft from the front axle yoke (see Section 14).

3 Remove the front brake calipers (see Chapter 9). **Caution:** *Tie the calipers up with*

wire to keep any strain off the flexible brake lines.

4 Detach the tie-rod ends from the steering linkage (see Chapter 10).

5 Remove the stabilizer bar and its connecting links (see Chapter 10).

6 Position a floor jack under the axle arm and slightly compress the front coil spring. Remove the nut securing the lower portion of the spring to the axle beam.

7 Carefully lower the jack and remove the coil spring, spacer, seat and stud **(see illustration)**. **Caution:** *The axle arm assembly must be supported on the jack throughout spring removal and installation. Do not let the arm assembly hang suspended by the brake hose.*

8 Detach the shock absorber from the radius arm bracket.

9 Detach the radius arm and bracket from the axle arm and remove them from the vehicle.

10 Remove the pivot bolt securing the right axle arm assembly to the frame crossmember. Remove the clamps securing the axleshaft boot to the axleshaft slip yoke and axleshaft. Slide the rubber boot over the stub shaft.

11 Once the right driveaxle is disconnected from the slip yoke, lower the floor jack and remove the right axle arm assembly.

12 Position the jack under the differential housing and remove the bolt securing the left axle arm to the frame crossmember. Lower and remove the left axle arm assembly.

Installation

13 Installation is the reverse of the removal procedure.

14 Tighten all fasteners to the torque listed in this Chapter's Specifications and related Chapters.

1995 and later models

Removal

Refer to illustrations 20.21, 20.22 and 20.23

15 Loosen the wheel lug nuts, raise the

vehicle and support it securely on jackstands. Remove the wheel(s). **Caution:** *If the vehicle is equipped with Automatic Ride Control (ARC), make sure the air suspension switch is turned to the OFF position before the vehicle is raised to prevent damage to the system components (see Chapter 10).*

16 Remove the caliper and brake disc as outlined in Chapter 9, then disconnect the ABS sensor and secure the sensor and wire harness aside.

17 Remove the hub driveaxle/nut. Place a prybar between two of the wheel studs to prevent the hub from turning while loosening the nut. **Warning:** *Discard the wheel hub nut and washer assembly and replace it with a new one, this nut is designed for a single torque sequence and cannot be re-used.*

18 Remove the brake hose support bracket-to-strut bolt.

19 Support the lower control arm with a floor jack, remove the lower control arm balljoint nut and separate the control arm from the steering knuckle (see Chapter 10). **Warning:** *The jack must remain in this position throughout the entire procedure.*

20 Remove the stabilizer bar (see Chapter 10).

21 Using a prybar or slide hammer with CV joint adapter, pry the inner CV joint assembly from the front differential **(see illustration)**. Be careful not to damage the front differential. Suspend the axle with a piece of wire - don't let it hang, or damage to the outer CV joint may occur.

22 Press the driveaxle out of the hub with a two-jaw puller, using a prybar or large screwdriver to keep the hub from turning **(see illustration)**. **Caution:** *Never use a hammer to remove the driveaxle from the hub or damage to the end of the driveaxle will occur.*

23 Once the driveaxle is loose from the hub splines, pull out on the strut/knuckle assembly and guide the outer CV joint out of the hub. Remove the support wire and carefully detach the driveaxle from the vehicle **(see illustration)**.

8

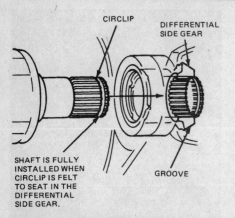

20.30 The inner CV joint stub shaft is completely seated when the circlip on the shaft snaps into the groove in the differential side gear

24   Refer to Section 14 and disconnect the front driveshaft from the front differential.
25   Disconnect the front axle vent hose. You may have to cut off the original clamp and install a screw-type clamp during reinstallation.
26   Support the differential with a floorjack, strapping or chaining the differential to the jack head.
27   Remove the three mounting bolts and remove the axle from the vehicle with the jack. One bolt holds the right axle tube to the frame rail, one mounts the left side of the differential to the frame, and the top bolt (just above the pinion and facing the engine) secures the top of the differential to the crossmember.

Installation

Refer to illustration 20.30
28   Installation of the differential housing is the reverse of the disassembly procedure.

Tighten the carrier mounting bolts to the torque listed in this Chapter's Specifications.
29   Install a **new** circlip on the inner stub shaft splines.
30   Coat the differential seal lips with multi-purpose grease and insert the stub shaft into the differential side gear until the shaft is seated and the circlip snaps into place **(see illustration)**.
31   Pull out on the strut/knuckle assembly and insert the outer CV joint stub shaft into the hub (make sure the splines are aligned). Push the shaft as far into the hub as possible by hand.
32   Support the outer CV joint housing and carefully tap on the hub, using a soft-faced hammer, until enough threads on the stub shaft are exposed to thread the **old** drive-axle/hub nut on. **Caution:** *Don't allow any force to be transmitted to the inner portion of the CV joint.*
33   Tighten the nut until the stub shaft is

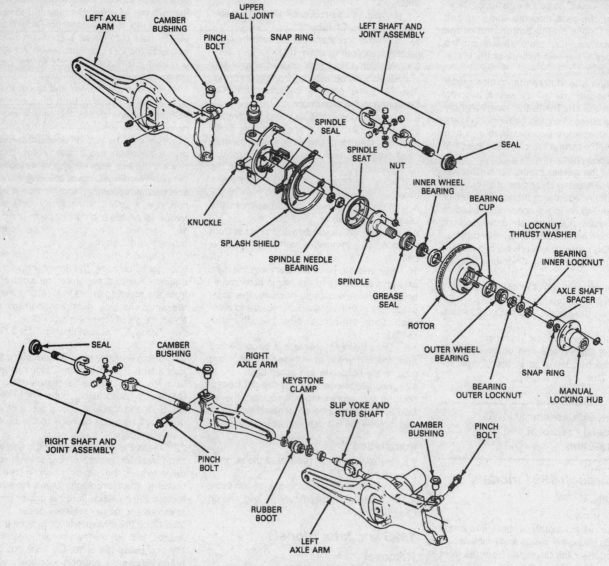

21.4 An exploded view of the 4WD front axle (1991 through 1994 models)

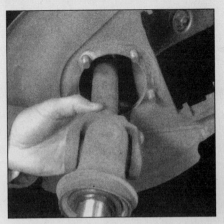

21.5 Remove the spindle securing nuts (arrows), then tap the spindle loose with a soft face hammer and remove it

21.6 Remove the spindle seat

21.7a Withdraw the left shaft and joint assembly

21.7b If necessary, tap the slinger off with a hammer - a press is needed to reinstall it

pulled completely into the hub, then remove the nut and discard it.

34 Install the driveaxle/hub washer and a **new** nut. Tighten the nut to the torque listed in this Chapter's Specifications while preventing the hub from turning by placing a screwdriver between two wheel studs.

35 Insert the balljoint stud into the steering knuckle. Install a **new** nut and tighten it to the torque listed in the Chapter 10 Specifications.

36 Install the stabilizer bar (see Chapter 10).

37 Install the brake disc and caliper (see Chapter 9).

38 Install the brake hose support bracket bolt and ABS sensor.

39 Install the wheel and lug nuts and lower the vehicle. Tighten the lug nuts to the torque listed in the Chapter 1 Specifications.

40 Check the front differential lubricant level and add, if necessary (see Chapter 1).

21 Front axleshaft and joint assembly - removal, component replacement and installation

1991 through 1994 models
Removal
Refer to illustrations 21.4, 21.5, 21.6, 21.7a, 21.7b and 21.8

1 Loosen the lug nuts on the front wheels.
2 Raise the front of the vehicle, support it

securely on jackstands and remove the front wheels.

3 Remove the front brake calipers (see Chapter 9). Hang the calipers out of the way with pieces of wire. Don't disconnect the brake hoses.

4 Remove the locking hubs, wheel bearings and locknuts **(see illustration)**. Refer to Chapter 1, Section 28, for the hub/brake disc and wheel bearing removal procedure.

5 Remove the nuts securing the spindle to the steering knuckle. Tap the spindle with a plastic or soft faced hammer to jar it free, then remove it **(see illustration)**.

6 Remove the spindle seat **(see illustration)**.

7 On the left side of the vehicle, pull the shaft and joint assembly out of the carrier **(see illustration)**. If necessary, tap the slinger off the shaft with a hammer **(see illustration)**. **Note:** *Don't remove the slinger unless necessary. You'll need a press to install it.*

8 Working on the right side, remove the metal clamps from the shaft and joint assembly and the stub shaft. Slide the rubber boot onto the stub shaft and pull the shaft and joint assembly from the splines of the stub shaft **(see illustration)**.

Spindle bearing and oil seal replacement
Refer to illustrations 21.12 and 21.13

9 Place the spindle in a padded vise so the vise grips the spindle's second step.
10 Remove the bearing and oil seal with a slide hammer and puller attachment.

11 Clean all dirt and grease from the spindle bore. Make sure there are no nicks or burrs in the bearing bore.
12 Install the new bearing in the bore with the manufacturer's mark out. Install the bearing with a bearing/seal installer **(see illustration)**.

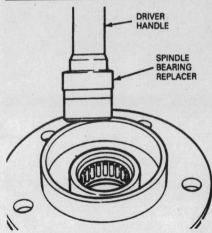

DRIVER HANDLE

SPINDLE BEARING REPLACER

21.12 Drive the spindle bearing out with a bearing driver

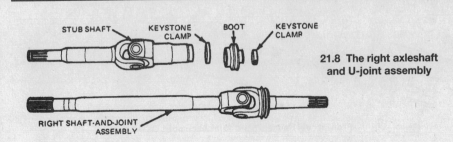

STUB SHAFT   KEYSTONE CLAMP   BOOT   KEYSTONE CLAMP

21.8 The right axleshaft and U-joint assembly

RIGHT SHAFT-AND-JOINT ASSEMBLY

8

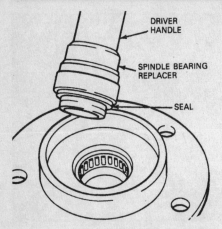

21.13 Drive in a new seal with its lip toward the driver

21.14a Mark the shaft and yoke so they can be reassembled in the same relative position

21.14b Remove the snap-rings from the joint with a screwdriver

13   Use the same tools to install the seal **(see illustration)**. The lip of the seal faces the installer. After installation, coat the lip of the seal with high-temperature lubricant.

Axleshaft assembly universal joint replacement

Refer to illustrations 21.14a, 21.14b and 21.15

14   Mark the orientation of the shaft and

21.15 Position the joint and sockets in a vise to press the bearings out - the small socket is smaller than the bearing cup, the large socket is large enough to receive the bearing cup

21.24   1995 and later driveaxle assembly - exploded view

1   Nut and washer assembly
2   Front wheel hub
3   Excluder seal
4   Driveaxle and outer joint assembly
5   Boot clamp
6   Boot
7   Boot clamp
8   Driveshaft
9   Stop ring
10   Circlip
11   Spider assembly
12   Tri-lobe insert
13   Housing
14   Inner CV joint
15   Boot clamp
16   Inner CV joint
17   Circlip
18   Housing
19   Front wheel hub

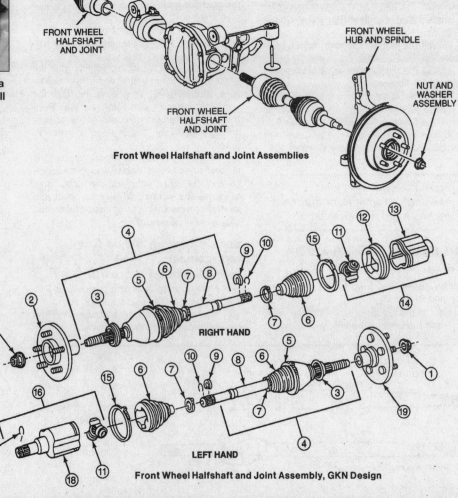

Front Wheel Halfshaft and Joint Assemblies

Front Wheel Halfshaft and Joint Assembly, GKN Design

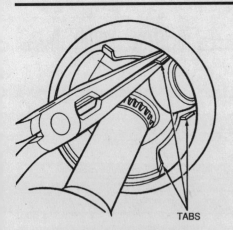

21.28 Bend the retaining tabs down slightly . . .

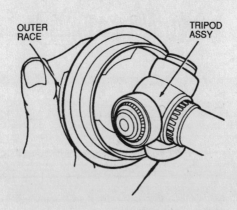

21.29 . . . and remove the outer race from the tripod assembly

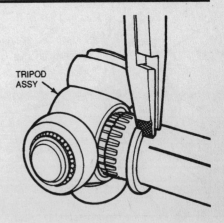

21.30 Move the stop ring down the axle shaft

yoke, then remove the internal snap-rings from the U-joint with a screwdriver **(see illustrations)**.

15   The remainder of the procedure is the same as for driveshaft universal joints, which is described in Section 15 **(see illustration)**.

Installation

16   If the slinger was removed from the spindle, have it pressed on by a machine shop.

17   On the right side of the carrier, install the rubber boot and new metal clamps on the stub shaft slip yoke. There is no blind spline to ensure correct alignment, so pay special attention to the yoke ears and make sure they are in phase. Be sure the assembly is fully seated and crimp the metal clamp.

18   On the left side of the carrier, slide the shaft and joint assembly through the knuckle and engage the splines on the shaft in the carrier.

19   Install the splash shield. Install the spindle on the steering knuckle, then install and tighten the spindle nuts to the torque listed in Chapter 10.

20   Install the front wheel bearings and locking hubs (see Chapter 1).

21   Install the brake caliper (see Chapter 9).

22   Install the front wheels and lower the vehicle.

23   Tighten the lug nuts to the torque listed in Chapter 1.

1995 and later models
Inner CV joint and boot replacement

Disassembly
Refer to illustrations 21.24, 21.28, 21.29, 21.30 and 21.31

24   1995 and later models are equipped with Constant Velocity (CV) joint type driveaxle assemblies **(see illustration)**. The inboard tripod design CV joint and driveaxle boot are serviceable units. The outer CV joint is not serviceable and must be replaced as an assembly. However, the outer driveaxle boot is serviceable with a similar procedure as the inner boot.

25   Remove the driveaxle from the vehicle (see Section 20).

26   Mount the driveaxle in a vise. The jaws of the vise should be lined with wood or rags to prevent damage to the axleshaft.

27   Cut the boot clamps from the boot and discard them.

28   Bend the retaining tabs slightly to allow for tripod removal **(see illustration)**.

29   Remove tripod assembly from outer race **(see illustration)**.

30   Move the inner (exposed) stop ring down the shaft about 1/2-inch **(see illustration)**.

31   Move the tripod down the shaft towards the inner snap-ring until the circlip is visible on the end of the driveaxle. Remove the circlip and remove the tripod assembly from the driveaxle **(see illustration)**.

32   No further disassembly of the tripod is possible. Inspect the tripod rollers, roller bearings and races carefully for damage, worn spots and smooth operation. Damaged or worn tripods cannot be rebuilt and must be replaced.

33   Remove the inner stop ring from the axleshaft and remove the old boot.

Reassembly
Refer to illustrations 21.38a, 21.38b and 21.40

34   Slide a new clamp and the inner CV joint boot on the axleshaft.

35   Install a new inner stop ring past the second ring groove about 1/2-inch.

36   Install the tripod assembly on the driveaxle with the chamfered side inward.

37   Push the tripod assembly down the axle far enough to allow circlip installation. Install the new circlip and push the tripod assembly towards the axle end until the tripod seats on

8

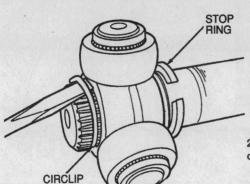

21.31 Remove the circlip and pull the tripod assembly off the axleshaft

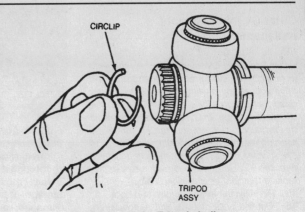

31.38a Install the Tripod circlip

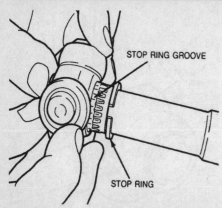

21.38b Push the tripod assembly toward the axle end and move the stop ring into the groove

STOP RING GROOVE

STOP RING

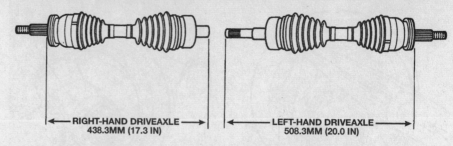

RIGHT-HAND DRIVEAXLE
438.3MM (17.3 IN)

LEFT-HAND DRIVEAXLE
508.3MM (20.0 IN)

21.40 Driveaxle assembly length specifications - 1995 and 1996 models

the circlip and the inner stop ring groove is exposed.

38   Move the inner stop ring to its groove to secure the tripod assembly **(see illustrations)**.

39   Fill the outer race with CV joint grease and spread some on the inside of the boot as well. The left axle tripods use about 6.5 ounces of grease and the right axles use about 7 ounces. Push the tripod assembly into outer race and bend the six retaining tabs back to their original shape.

40   Wipe any excess grease from the axle boot groove on the outer race. Seat the small diameter of the boot in the recessed area on the axleshaft and install the clamp. Push the other end of the boot onto the outer race and move the race in-or-out to adjust the axle to the proper length **(see illustration)**. This applies only to 1995 and 1996 models. The 1997 and later models have grooves in the axleshafts that locate the boot clamps.

41   With the axle set to the proper length, equalize the pressure in the boot by inserting a dull screwdriver between the boot and the outer race. Don't damage the boot with the tool.

42   Install the boot clamp. A pair of special clamp-crimping pliers are required.

43   Install a new clip on the stub axle.

44   Install the driveaxle as described in Section 20.

Outer CV joint and boot replacement

Refer to illustration 21.50

Note: *The outer CV joint is a non-serviceable item and is permanently retained to the driveaxle. If any damage or excessive wear occurs to the axle or the outer CV joint, the driveaxle and outer CV joint assembly must be replaced. Complete rebuilt driveaxle assemblies are be available at most auto parts stores. If after disassembly and inspection the CV joint is deemed serviceable, replace the boot as described below.*

45   Remove the inner CV joint and boot.

46   Cut the boot clamps and remove them from the outer CV joint.

47   Detach the boot from the CV joint and slide it off the axleshaft.

48   Check the grease for contamination by rubbing a small amount between your fingers. If the grease feels gritty, the CV joint is contaminated and must be replaced.

49   If the grease is not contaminated, remove all traces of grease from the CV joint using solvent and a soft bristle brush. Thoroughly dry the CV joint using compressed air, if available. All traces of solvent must be removed or damage to the CV joint may occur. **Warning:** *Wear eye protection when using compressed air!*

50   Inspect the cage, balls and races for pitting, score marks, cracks and other signs of wear or damage **(see illustration)**. Shiny, polished spots are normal and will not adversely affect CV joint performance.

52   Install the new outer boot and clamps onto the driveaxle shaft. Completely pack the CV joint with the grease provided in the boot kit. Force the grease into the joint and around the balls and races. Spread the remaining grease evenly inside the boot. Total grease fill is approximately 6.0 ounces.

53   Position the outer boot on the CV joint and using the appropriate boot clamp pliers, tighten the clamps.

21.50 After the old grease has been completely removed and the solvent has dried, rotate the outer joint housing through its full range of motion and inspect the bearing surfaces for wear or damage - if the balls, race or cage appear damaged, replace the driveaxle shaft and outer joint assembly

54   Assemble the inner CV joint and boot and install the driveaxle.

22   Right stub shaft assembly, carrier oil seals and bearings (1994 and earlier models) - removal and installation

Removal

Refer to illustrations 22.7, 22.9 and 22.10

1   Loosen the lug nuts on the front wheels.

2   Raise the front of the vehicle, support it securely on jackstands and remove the front wheels.

3   Detach the front driveshaft from the axle yoke (see Section 14). Secure the driveshaft aside.

4   Remove the spindles and the shaft and joint assemblies from both sides (see Section 19).

5   Place a transmission jack under the carrier and unbolt the carrier from the left axle arm. Separate the carrier from the arm. After the lubricant has drained, remove the carrier.

6   Place the carrier in a vise or on a workbench.

7   Turn the shaft assembly so the open side of the C-clip is exposed, then push the C-clip off with a pair of screwdrivers **(see illustration)**.

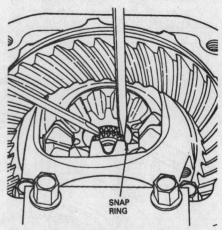

SNAP RING

22.7 Push the C-clip off the shaft with a pair of screwdrivers

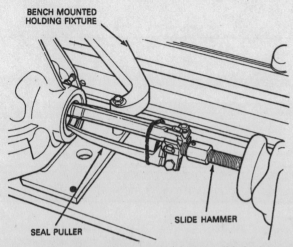

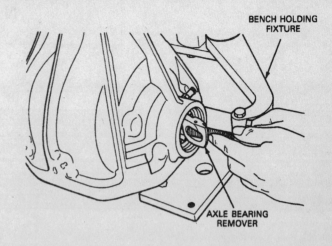

22.9 Remove the passenger side oil seal with a slide hammer and a puller that hooks to the inside of the seal

22.10 Remove the caged needle bearings with a slide hammer and puller

8   Pull the slip yoke and shaft assembly out of the carrier.

9   Remove the oil seals with a slide hammer and puller **(see illustration)**. The driver's side can be removed with an attachment that grips the outer edge of the seal. A hammer and chisel can also be used, being careful not to gouge the housing or the seal bore.

10   Remove the caged needle bearings with a slide hammer and a puller attachment **(see illustration)**.

Installation

Refer to illustrations 22.11 and 22.14

11   Clean and inspect the bearing bore for nicks and burrs before installing a new caged bearing. Position the bearing with the manufacturer's marks facing out, then drive the bearing in until it is fully seated in the bore **(see illustration)**.

12   Coat the oil seal with high-temperature lubricant and drive it into the carrier housing.

13   Install the slip yoke and shaft assembly into the carrier so the C-clip groove in the shaft is visible.

14   Push the C-clip into the groove with a screwdriver **(see illustration)**. Make sure it is completely seated in the groove. **Caution:** *Don't push on the center of the C-clip with the screwdriver, as it may be damaged. Push in the notches of the clip.*

15   Clean all traces of gasket sealant from the mating surfaces with lacquer thinner or gasket remover. Apply RTV sealant in an unbroken 1/4-inch wide bead, inboard of the bolt holes, and install the differential cover.

16   Position the carrier on the transmission jack and install it in position on the support arm. Use the guide pins for alignment. Install and tighten the bolts in a clockwise or counterclockwise pattern to the torque listed in this Chapter's Specifications.

17   Install the shear bolt securing the carrier to the axle arm and tighten to the torque listed in this Chapter's Specifications.

18   Install the shaft and joint assemblies and spindles.

19   Connect the driveshaft to the yoke (see Section 14).

20   Fill the carrier with lubricant (see Chapter 1).

21   Install the front wheels and lower the vehicle.

22   Tighten the lug nuts to the torque listed in Chapter 1.

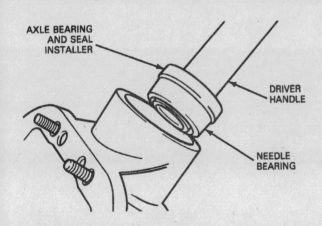

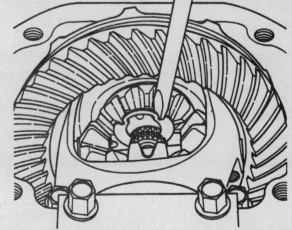

22.11 Drive the bearings in with the manufacturer's marking facing out

22.14 Push the C-clip into the groove with a screwdriver - don't press on the center of the C-clip or it may be damaged

8

Notes

Chapter 9 Brakes

Contents

Specifications

Brake fluid type	DOT 3 brake fluid

Drum brakes

Drum wear limit	Specified on drum
Minimum lining thickness	1/16 inch above rivet heads

Disc brakes

Pad lining minimum thickness	1/8 inch
Front brake disc	
Minimum thickness*	
1994 and earlier	0.810 inch
1995 and 1996	
2WD	0.964 inch
4WD	0.810 inch
1997 and later	0.98 inch
Runout limit	
1996 and earlier	0.003 inch
1997 and later	0.0005 inch
Thickness variation limit (parallelism)	
1996 and earlier	0.002 inch
1997 and later	0.0002 inch
Rear brake disc	
Minimum thickness*	0.409 inch
Runout limit	
1996 and earlier	0.003 inch
1997 and later	0.0002 inch
Thickness variation (parallelism)	0.0005 inch

*Refer to minimum thickness cast into the disc (it supersedes the information printed here)

Torque specifications

	Ft-lbs (unless otherwise indicated)
Backing plate-to-axle housing nuts	25 to 35
Brake booster-to-firewall nuts	15 to 20
Brake hose -to-caliper banjo bolt	23 to 29
Caliper mounting bolts	
1996 and earlier	
Front	38 to 48
Rear	20
1997 and later	
Front	21 to 26
Rear	20
Caliper mounting bracket-to-spindle bolts	73 to 97
Caliper adapter-to-rear axle flange nuts	80
Master cylinder-to-brake booster nuts	13 to 25
RABS valve mounting screw	11 to 14
RABS sensor mounting bolt	25 to 30

9

1 General information

Refer to illustration 1.3

General description

All models covered by this manual are equipped with hydraulically-operated, power-assisted brake systems. All front brake systems are disc type, while the rear brakes are either disc or drum type. Some models are equipped with an Anti-lock Brake System (ABS), which is described in Section 2.

All brakes are self-adjusting. The front and rear disc brakes automatically compensate for pad wear, while the rear drum brakes incorporate an adjustment mechanism which is activated as the brakes are applied.

The hydraulic system is a split design, meaning there are separate circuits for the front and rear brakes **(see illustration)**. If one circuit fails, the other circuit will remain functional and a warning indicator will light up on the dashboard, showing that a failure has occurred.

Master cylinder

The master cylinder is located under the hood, mounted to the power brake booster, and is best recognized by the large fluid reservoir on top. The removable plastic reservoir is partitioned to prevent total fluid loss in the event of a front or rear brake hydraulic system failure.

The master cylinder is designed for the "split system" mentioned earlier and has separate primary and secondary piston assemblies, the piston nearest the firewall being the primary piston, which applies hydraulic pressure to the front brakes.

Power brake booster

The power brake booster, utilizing engine manifold vacuum and atmospheric pressure to provide assistance to the hydraulically operated brakes, is mounted on the firewall in the engine compartment.

Parking brake

The parking brake mechanically operates the rear brakes only.

On drum brake models the parking brake cables pull on a lever attached to the brake shoe assembly, causing the shoes to expand against the drum. On models with rear disc brakes, the cables operate small parking brake shoes inside the brake disc hub.

Precautions

There are some general cautions and warnings involving the brake system on this vehicle:

a) *Use only brake fluid conforming to DOT 3 specifications.*
b) *The brake pads and linings may contain asbestos fibers which are hazardous to your health if inhaled. Whenever you work on brake system components,* clean all parts with brake system cleaner. Do not allow the fine dust to become airborne. Also, wear an approved filtering mask.
c) *Safety should be paramount whenever any servicing of the brake components is performed. Do not use parts or fasteners which are not in perfect condition, and be sure that all clearances and torque specifications are adhered to. If you are at all unsure about a certain procedure, seek professional advice. Upon completion of any brake system work, test the brakes carefully in a controlled area before putting the vehicle into normal service.*
d) **Warning:** *Some models covered by this manual are equipped with an air suspen-sion system. Always disconnect the electrical power to the suspension system before lifting or towing (see Chapter 10). Failure to perform this procedure may result in unexpected shifting or movement of the vehicle, which could cause personal injury.*

If a problem is suspected in the brake system, don't drive the vehicle until it's fixed.

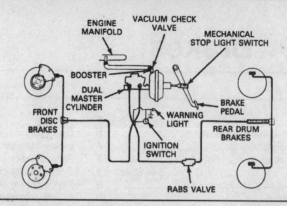

1.3 Typical dual master cylinder brake system (non-ABS models)

2 Anti-lock Brake System (ABS) - general information

1   Some models are equipped with an Anti-lock Brake System (ABS). Early models are equipped with rear wheel anti-lock brake

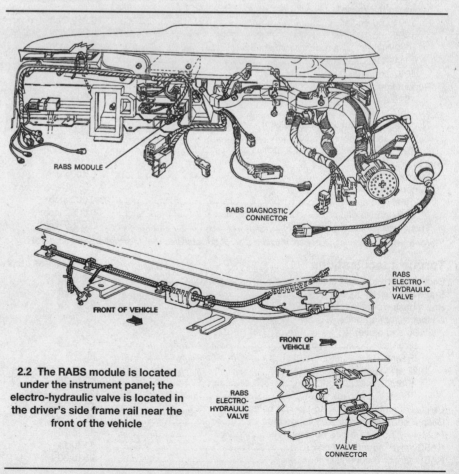

2.2 The RABS module is located under the instrument panel; the electro-hydraulic valve is located in the driver's side frame rail near the front of the vehicle

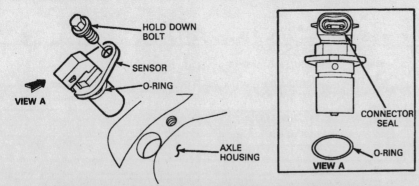

2.16 The rear wheel speed sensor mounts on the rear axle housing

systems while later models are equipped with 4-wheel anti-lock brake systems. The ABS system is designed to maintain vehicle steerability, directional stability and optimum deceleration under severe braking conditions and on most road surfaces. It does so by monitoring the rotational speed of each wheel and controlling the brake line pressure to each wheel during braking. This prevents the wheel from locking-up and provides maximum vehicle controllability.

Rear wheel Anti-lock Brake System (RABS)

Refer to illustration 2.2

2   A Rear Anti-lock Brake System (RABS) is used on early models. The system consists of a computer module, anti-lock brake valve, an exciter ring and a sensor **(see illustration)**.

3   Disconnect the cable from the negative battery terminal.

RABS module

4   Working under the instrument panel, press on the plastic tab on the electrical connector and disconnect it from the RABS module **(see illustration 2.2)**.

5   Remove the screws securing the RABS module to the instrument panel bracket and remove it.

6   Installation is the reverse of the removal Steps.

RABS valve

7   Disconnect the two brake lines from the RABS valve **(see illustration 2.2)**. Plug the ends of the lines to prevent the loss of hydraulic fluid and the entry of foreign matter and moisture.

8   Disconnect the electrical connector from the RABS valve.

9   Remove the screw securing the RABS valve to the frame rail and take it out.

10   To install, position the RABS valve on the frame rail. Install the screw and tighten it to the torque listed in this Chapter's Specifications.

11   Connect the electrical connector to the valve.

12   Install the brake lines and tighten to the torque listed in this Chapter's Specifications. **Caution:** *Do not overtighten the fittings.*

13   Bleed the hydraulic system.

Speed sensor

Refer to illustration 2.16

14   Disconnect the electrical connector from the sensor located on the rear axle housing.

15   Prior to removing the sensor, thoroughly clean the rear axle housing surrounding the sensor.

16   Remove the bolt securing the sensor and remove it from the rear axle housing **(see illustration)**. Do not allow any dirt to fall into the interior of the rear axle housing.

17   Thoroughly clean the sensor mounting surface on the rear axle housing.

18   Inspect and clean the magnetized sensor pole piece. Remove any small metal particles that could cause erratic system operation.

19   If installing a new sensor, lightly lubricate the O-ring seal with clean engine oil.

20   If installing an old sensor, remove the old O-ring seal, install a new one and lubricate it with clean engine oil.

21   Firmly grasp the sides of the sensor (do not apply pressure to the electrical connector) and push it into the mounting hole.

22   Align the mounting flange bolt hole with that on the rear axle housing and install the hold-down bolt. Tighten the bolt to the torque listed in this Chapter's Specifications.

23   Connect the electrical connector to the sensor.

Exciter ring

24   To service the exciter ring the rear axle must be disassembled and the ring gear removed with a press. It is recommended that this procedure be performed by a dealer service department or other repair shop.

Four wheel Anti-lock Brake System

Hydraulic control unit

Refer to illustrations 2.25a, 2.25b and 2.25c

25   The hydraulic control unit is located in the left (driver's side) front corner of the engine compartment. It consists of a brake pressure control valve block, a pump motor and a hydraulic control unit reservoir with a fluid level indicator assembly **(see illustrations)**.

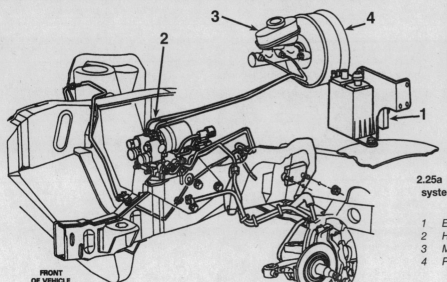

2.25a Four wheel anti-lock brake system components - 1994 and earlier models

1   *Electronic control module*
2   *Hydraulic control unit*
3   *Master cylinder*
4   *Power brake booster*

9

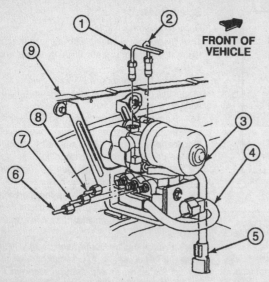

FRONT OF VEHICLE

2.25b Four wheel anti-lock brake system hydraulic control unit details - 1995 and 1996 models

1   Primary brake tube
2   Secondary brake tube
3   Pump motor
4   Solenoid control wiring
5   Pump motor wiring
6   Front brake tube (right side)
7   Front brake tube (left side)
8   Rear brake tube
9   Inner fender panel (left side)

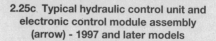

2.25c Typical hydraulic control unit and electronic control module assembly (arrow) - 1997 and later models

26   During normal braking conditions, brake hydraulic fluid from the master cylinder enters the hydraulic control unit through two inlet ports and passes through four normally-open inlet valves, one to each wheel.

27   When the anti-lock brake control module senses that a wheel is about to lock up, the anti-lock brake control module closes the appropriate inlet. This prevents any more fluid from entering the affected brake. If the module determines that the wheel is still decelerating, the module opens the outlet valve, which bleeds off pressure in the affected brake.

Wheel sensors

Refer to illustration 2.29

28   The ABS system uses four "variable-reluctance" sensors to monitor wheel speed ("reluctance" is a term used to indicate the amount of resistance to the passage of flux lines - lines of force in a magnetic field - through a given material). Each sensor contains a small inductive coil that generates an electromagnetic field. When paired with a toothed sensor ring which interrupts this field as the wheels turn, each sensor generates a low-voltage analog signal. This voltage signal, which rises and falls in proportion to wheel rotation speed, is continuously sampled by the control module, converted into digital data inside the module and processed.

29   The front wheel sensors **(see illustration)** are mounted in the steering knuckle in close proximity to the toothed indicator rings, which are pressed onto the wheel hubs on 2WD models. On 4WD models, the front wheel indicator rings are an integral part of the front wheel bearings and are not serviceable separately. The rear wheel sensor is mounted in the rear axle housing and the sensor ring is mounted along with the ring gear inside the rear axle.

Brake control module

Refer to illustration 2.30

30   The brake control module is the "brain"

of the ABS system. The module constantly monitors the incoming analog voltage signals from the four ABS wheel sensors, converts these signals to digital form, processes this digital data by comparing it to the map (program), makes decisions, converts these (digital) decisions to analog form and sends them to the hydraulic control unit, which opens and closes the front and/or rear circuits as necessary. On 1994 and earlier models, the module is mounted on a bracket behind the EVAP canister **(see illustration 2.25a)**. On 1995 and 1996 models, the module is mounted in the front left fenderwell behind the plastic inner liner **(see illustration)**. On 1997 and later models, the control unit is mounted to the hydraulic control unit **(see illustration 2.25c)**.

31   On 1997 and later models with 4WD, there is also a G-switch, which detects vehicle movement and prevents all four wheels from sliding if one wheel locks up (which would otherwise signal the module that the vehicle is stopped, due to the ABS speed sensor).

32   The module has a self-diagnostic capability which operates during both normal driving as well as ABS system operation. During normal operation, a yellow ABS warning light on the dash should glow for three seconds

whenever the vehicle is started. This is the internal self-test the system performs at every startup. You may also hear a slight noise when you first reach 12 mph after a startup. This is part of the self-check in which the pump is turned on for a half-second and is normal.

33   If a malfunction occurs during driving, a red "BRAKE" warning indicator or an amber (yellow on 1997 and later models) "CHECK ANTI-LOCK BRAKES" warning indicator will light up on the dash. On 1997 and later models, the warning light simply says ABS.

a) *If the red BRAKE light glows, the brake fluid level in the master cylinder reservoir has fallen below the level established by the fluid level switch. Top up the reservoir and verify that the light goes out.*

b) *If the amber or yellow CHECK ANTI-LOCK BRAKES light glows, the ABS has been turned off because of a symptom detected by the module. Normal power-assisted braking is still operational, but the wheels can now lock up if you're involved in a panic-stop situation. A diagnostic code is also stored in the module when a warning indicator light comes on; when retrieved by a service*

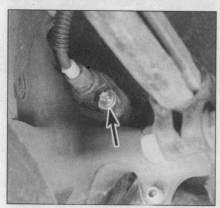

2.29 ABS front wheel sensor (arrow)

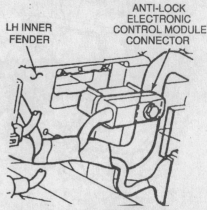

2.30 ABS control module location - 1995 and 1996 models

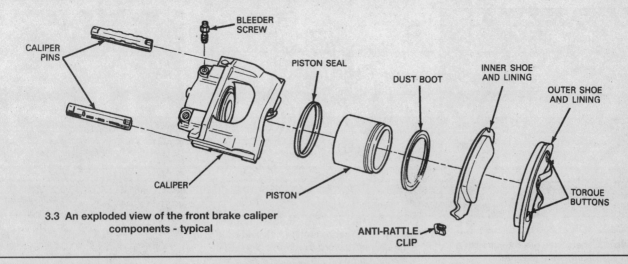

CALIPER PINS

BLEEDER SCREW

PISTON SEAL

DUST BOOT

INNER SHOE AND LINING

OUTER SHOE AND LINING

CALIPER

PISTON

TORQUE BUTTONS

ANTI-RATTLE CLIP

3.3 An exploded view of the front brake caliper components - typical

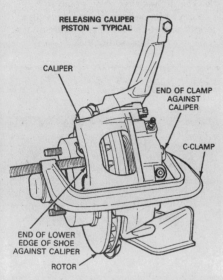

RELEASING CALIPER PISTON – TYPICAL

CALIPER

END OF CLAMP AGAINST CALIPER

C-CLAMP

END OF LOWER EDGE OF SHOE AGAINST CALIPER

ROTOR

3.6a Use a large C-clamp and push the piston back into the bore just enough to allow the caliper to slide off the brake disc easily - note the one end of the C-clamp is on the flat area of the inner side of the caliper and the other end (screw end) is pressing on the outer pad (1994 and earlier)

technician, the code indicates the area or component where the problem is located. Once the problem is fixed, the code is cleared. These procedures, however, are beyond the scope of the home mechanic.

Relays

34   The main relay and pump motor relay are located in a relay box along with three other relays mounted near the battery on the left side of the engine compartment.

Diagnosis and repair

Warning: *If a dashboard warning light comes on and stays on while the vehicle is in operation, the ABS system requires immediate attention!*

35   Although a special electronic ABS diagnostic tester is necessary to properly diagnose the system, the home mechanic can perform a few preliminary checks before taking the vehicle to a dealer who is equipped with this tester.

a) *Check the brake fluid level in the reservoir.*
b) *Verify that the control module electrical connector is securely connected.*
c) *Check the electrical connectors at the hydraulic control unit.*
d) *Check the fuses.*
e) *Follow the wiring harness to each wheel and check that all connections are secure and that the wiring is not damaged.*

If the above preliminary checks do not rectify the problem, the vehicle should be diagnosed by a dealer service department or other qualified repair shop. Due to the rather complex nature of this system, all actual repair work must be done by the dealer service department or repair shop.

3   Brake pads - replacement

Refer to illustrations 3.3, 3.6a through 3.6m and 3.6n through 3.6t

Warning 1: *Disc brake pads must be replaced on both front wheels or both rear wheels at the same time - never replace the pads on only one wheel. Also, the dust created by the brake system may contain asbestos, which is harmful to your health. Never blow it out with compressed air and don't inhale any of it. An approved filtering mask should be worn when working on the brakes. Do not, under any circumstances, use petroleum-based solvents to clean brake parts. Use brake system cleaner only!*

Warning 2: *1995 and later models use a special lubricating gel on certain parts of the brake caliper. This gel is non-toxic under most conditions. If it is heated to a certain point, however, it can give off toxic fumes which can cause health problems or even*

death. Never apply heat to these calipers and don't smoke around them, either (the gel might get on your cigarette). Always wash your hands thoroughly after working on the brakes.

1   Remove the cover from the brake fluid reservoir and siphon out about 1/2 of the brake fluid.
2   Loosen the wheel lug nuts, raise the vehicle and support it securely on jackstands.
3   Remove the wheels. Work on one brake assembly at a time, using the assembled brake for reference if necessary **(see illustration)**.
4   Inspect the brake disc carefully as outlined in Section 5.
5   If machining is necessary, follow the information in that Section to remove the disc, at which time the pads can be removed from the caliper as well.
6   Follow the accompanying photos, beginning with illustration 3.6a, for the front brake pad replacement procedure. Be sure to stay in order and read the caption under each illustration. If you're replacing the rear brake pads, follow the photos beginning with **illustration 3.6n**.

3.6b On 1994 and earlier models, use pliers and squeeze the caliper retaining pin while prying the other end until the tabs on the pin enter the spindle groove . . .

9

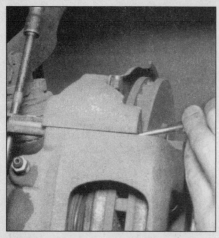

3.6c   . . . then drive out the pin with a punch and hammer

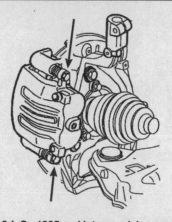

3.6d   On 1995 and later models, remove the lower caliper mounting bolt and rotate the caliper upward to remove the pads. **Note:** *Don't wipe the lubricating gel from these bolts or the bolt boots. At the time of publication there was no substitute for this gel (also read the* **Warning** *at the beginning of Section 3)*

3.6e   If the caliper isn't going to be removed for service, suspend it with a length of wire to relieve any strain on the brake hose

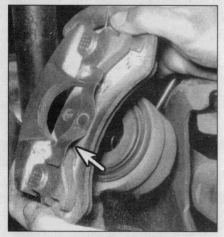

3.6f   This spring clip (arrow) holds the outside brake pad to the caliper on 1994 and earlier models - push down on the pad, releasing the locking tabs and slide the pad out of the caliper, then remove the inner pad and the anti-rattle clips. On 1995 and later models, detach the pads from the caliper anchor plate

3.6g   On 1997 and later models, with the caliper rotated upward out of the way, remove the pads (A) from the caliper anchor plate (B), and remove the stainless-steel slippers (C)

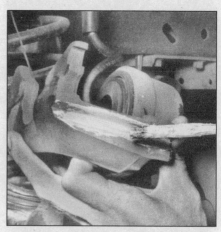

3.6h   Prior to installing the caliper, lightly lubricate the V-grooves where the caliper slides into the anchor plate with disc brake caliper slide rail grease (1994 and earlier models)

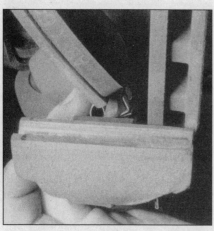

3.6i   Make sure the tabs on the spring clip are positioned correctly and the clip is fully seated

3.6j   Insert a new anti-rattle clip on the lower end of the inner pad

3.6k   Compress the anti-rattle clip and slide the upper end of the pad into position

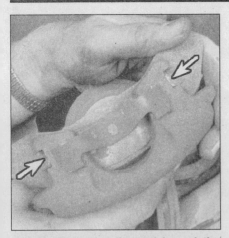

3.6l On 1994 and earlier models, push the outer pad into position on the caliper ears, making sure the torque buttons on the pad seat fully in the retention notches (arrows)

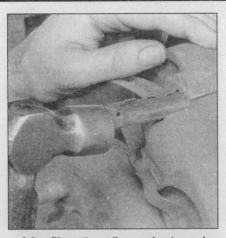

3.6m Place the caliper and outer pad assembly over the disc and inner pad, then drive the caliper pins into their grooves or, on 1995 and later models, lower the caliper and install the lower mounting bolt and tighten to the torque listed in this Chapter's Specifications

3.6n Remove the two rear caliper bolts (upper and lower arrows) with a Torx Drive bit - DO NOT remove the brake hose banjo bolt (middle arrow) unless you are planning to overhaul the caliper

3.6o Depress the caliper piston by squeezing the old pad with a pair of large adjustable pliers

3.6p Pry off the outer brake pad with a screwdriver

3.6q Remove the inner brake pad by pulling the retaining clips out of the piston

3.6r To install the inner brake pad, place it in position like this, then push it all the way in until the retaining clips are fully seated

3.6s To install the outer brake pad, slide it down like this until the locating pins (arrow) are fully engaged with the holes in the caliper housing

3.6t Remove the slippers (arrows), clean the mounting surfaces and reinstall the slippers, then install the caliper and pads over the disc, install the caliper bolts and tighten them to the torque listed in this Chapter's Specifications

9

7   When reinstalling the caliper on 1995 and later models, be sure to tighten the caliper bolts to the torque listed in this Chapter's Specifications. **Caution:** *On 1995 and later models, the caliper mounting bolts are "lubricated for life" with a special lubricant inside the caliper bolt boots. If the boots are to be replaced, use only silicone brake caliper grease/dielectric compound to lubricate them. Regular hi-temp brake grease will cause the boots to swell.*

8   After the job has been completed, firmly depress the brake pedal a few times to bring the pads into contact with the disc.

9   Check the brake fluid level and add some, if necessary, to bring it to the appropriate level (see Chapter 1).

5.5a To check disc runout, mount a dial indicator as shown and rotate the disc

5.5b Using a swirling motion, remove the glaze from the disc with sandpaper or emery cloth

4   Brake caliper - removal and installation

Warning 1: *Dust created by the brake system may contain asbestos, which is harmful to your health. Never blow it out with compressed air and don't inhale any of it. An approved filtering mask should be worn when working on the brakes. Do not, under any circumstances, use petroleum-based solvents to clean brake parts. Use brake system cleaner only!*

Warning 2: *1995 and later models use a special lubricating gel on certain parts of the brake caliper. This gel is non-toxic under most conditions. If it is heated to a certain point, however, it can give off toxic fumes which can cause health problems or even death. Never apply heat to these calipers and don't smoke around them, either (the gel might get on your cigarette). Always wash your hands thoroughly after working on the brakes.*

Warning 3: *If the vehicle is equipped with ABS, make sure you plug the brake hose immediately after disconnecting it from the brake caliper, to prevent the fluid from draining out of the line and air entering the Hydraulic Control Unit (HCU). The HCU on an ABS system cannot be bled without a very expensive tool.*

Removal

1   Apply the parking brake and block the wheels opposite the end being worked on. Loosen the wheel lug nuts, raise the vehicle and support it securely on jackstands. Remove the wheel.

2   Unscrew the brake hose banjo bolt and detach the hose from the caliper. **Caution:** *On ABS-equipped models, plug the brake hose immediately to prevent air from getting into the Hydraulic Control Unit (HCU). If air gets into the HCU, you may not be able to bleed the brakes properly at home. On non-ABS models, wrap a plastic bag around the end of the hose to prevent fluid loss and contamination. Note: If the caliper will not be completely removed from the vehicle - as for pad inspection or disc removal - leave the*

hose connected and suspend the caliper with a length of wire. This will save the trouble of bleeding the brake system.

3   Refer to the first six steps in Section 3 to separate the caliper from the steering knuckle (front) or the torque plate (rear) - it's part of the brake pad replacement procedure.

Installation

4   Refer to Section 3 for the caliper installation procedure, as it is part of the brake pad replacement procedure.

5   Connect the brake hose to the caliper, using new sealing washers. Tighten the banjo bolt to the torque listed in this Chapter's Specifications.

6   Bleed the brakes as outlined in Section 10. This is not necessary if the banjo bolt was not loosened or removed (if the caliper was removed for access to other components, for example).

7   Install the wheel and lower the vehicle. Tighten the lug nuts to the torque listed in the Chapter 1 Specifications. Pump the brake pedal several times to bring the pads into contact with the disc.

8   Test the operation of the brakes before placing the vehicle into normal service.

5   Brake disc - inspection, removal and installation

Inspection

Refer to illustrations 5.5a, 5.5b, 5.6a and 5.6b
Note: *This procedure applies to front and rear disc brake assemblies.*

1   Loosen the wheel lug nuts, raise the vehicle and support it securely on jackstands. Remove the wheel.

2   Remove the brake caliper as outlined in Section 4. It's not necessary to disconnect the brake hose for this procedure. After removing the caliper bolts, suspend the caliper out of the way with a piece of wire.

Don't let the caliper hang by the hose and don't stretch or twist the hose.

3   If you're working on a 1995 or later 4WD model, reinstall three lug nuts (inverted) to hold the disc against the hub. It may be necessary to install washers between the disc and the lug nuts to take up space.

4   Visually check the disc surface for score marks and other damage. Light scratches and shallow grooves are normal after use and may not always be detrimental to brake operation, but deep score marks - over 0.015-inch - require disc removal and refinishing by an automotive machine shop. Be sure to check both sides of the disc. If pulsating has been noticed during application of the brakes, suspect disc runout.

5   To check disc runout, place a dial indicator at a point about 1/2-inch from the outer edge of the disc **(see illustration)**. Set the indicator to zero and turn the disc. The indicator reading should not exceed the specified allowable runout limit. If it does, the disc should be refinished by an automotive machine shop. **Note:** *Professionals recommend resurfacing of brake discs regardless of the dial indicator reading (to produce a smooth, flat surface that will eliminate brake pedal pulsations and other undesirable symptoms related to questionable discs). At the very least, if you elect not to have the discs resurfaced, deglaze the surface with emery cloth or sandpaper (use a swirling motion to ensure a non-directional finish)* **(see illustration).**

6   The disc must not be machined to a thickness less than the specified minimum refinish thickness. The minimum wear (or discard) thickness is cast into the inside of the disc **(see illustration)**. The disc thickness can be checked with a micrometer **(see illustration).**

Removal

7   If you are working on a 2WD model, refer to Chapter 1, *Front wheel bearing*

check, repack, and adjustment for the disc removal procedure.

8   If you are working on a 1994 or earlier 4WD model, refer to Chapter 1, *Front hub lock, spindle bearing, and wheel bearing maintenance* for front disc removal.

9   If you are working on a 1995 or later 4WD model, remove the caliper mounting bracket (front only), and slide the rotor off the hub assembly.

10   If you are working on rear wheel disc brake systems, remove the rear caliper (see Section 4). Remove the keeper nuts (if equipped) from the wheel studs. If the disc binds on the parking brake shoes, remove the plug from the backing plate inspection hole. With a suitable tool, back-off the parking brake shoes by turning the parking brake adjuster clockwise for the driver's side and counter clockwise for the passenger's side.

Installation

11   Install the disc by reversing the removal procedure.

12   Install the caliper and brake pad assembly over the disc and position it on the steering knuckle or mounting bracket (front), or on the torque plate (rear) (see Section 4). Install the caliper bolts and tighten them to the torque listed in this Chapter's Specifications.

13   Install the wheel, then lower the vehicle to the ground. Depress the brake pedal a few times to bring the brake pads into contact with the rotor. Bleeding of the system will not be necessary unless the brake hose was disconnected from the caliper. Check the opera-

5.6a The minimum thickness limit is cast into the inside of the disc (typical marking shown, look at your disc for exact measurement)

tion of the brakes carefully before placing the vehicle into normal service.

---

6   Brake shoes (rear) - replacement

Refer to illustrations 6.4a through 6.4y and 6.5
Warning: *Drum brake shoes must be replaced on both wheels at the same time - never replace the shoes on only one wheel. Also, the dust created by the brake system may contain asbestos, which is harmful to your health. Never blow it out with com-*

5.6b Use a micrometer to measure disc thickness at several points, about 1/2-inch from the edge

pressed air and don't inhale any of it. An approved filtering mask should be worn when working on the brakes. Do not, under any circumstances, use petroleum-based solvents to clean brake parts. Use brake system cleaner only!
Caution: *Whenever the brake shoes are replaced, the retractor and hold-down springs should also be replaced. Due to the continuous heating/cooling cycle that the springs are subjected to, they lose their tension over a period of time and may allow the shoes to drag on the drum and wear at a much faster rate than normal.*

1   Loosen the wheel lug nuts, raise the rear of the vehicle and support it securely on jackstands. Block the front wheels to keep the vehicle from rolling.

2   Release the parking brake.

3   Remove the wheel. **Note:** *All four rear brake shoes must be replaced at the same time, but to avoid mixing up parts, work on only one brake assembly at a time.*

4   Follow the accompanying photos **(illustrations 6.4a through 6.4y)** for the inspec-

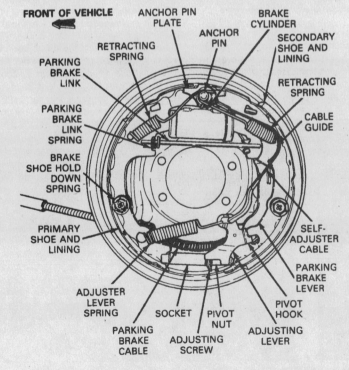

6.4a Components of the rear drum brake assembly (left side shown)

6.4b If the drum is difficult to remove, you may have to retract the brake shoes: Remove the rubber plug from the backing plate . . .

9

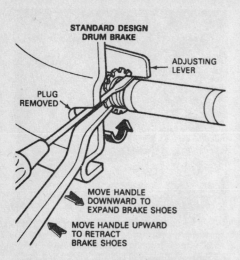

STANDARD DESIGN
DRUM BRAKE

ADJUSTING
LEVER

PLUG
REMOVED

MOVE HANDLE
DOWNWARD TO
EXPAND BRAKE SHOES

MOVE HANDLE UPWARD
TO RETRACT
BRAKE SHOES

6.4c ... insert a small screwdriver through the hole, push the adjusting lever off the star wheel and rotate the star adjuster with a brake adjustment tool or another screwdriver as shown

6.4f Pull up on the adjusting cable and disconnect the cable eye from the anchor pin

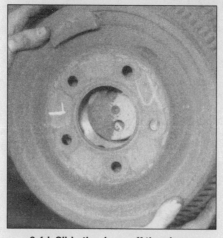

6.4d Slide the drum off the shoes

6.4g Remove the anchor pin plate

tion and replacement of the brake shoes. Be sure to stay in order and read the caption under each illustration. **Note:** *If the brake drum cannot be easily pulled off, pry the rubber plug from the backing plate inspection hole and insert a screwdriver and a brake*

6.4e Use a spring removal tool to remove the shoe retracting springs

6.4h Remove the shoe retaining springs and pins (one on each shoe)

adjusting tool to lift the adjusting lever and rotate the adjusting screw. This will cause the brake shoes to pull together. Spray the adjuster assembly with penetrating oil and allow the oil to soak in if the mechanism is difficult to turn. The drum should now come off.

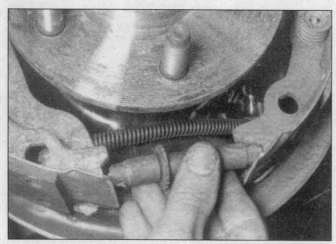

6.4i Pull the shoes apart and remove the adjusting screw

6.4j Remove the primary shoe, then slide out the parking brake strut and spring

6.4k Remove the adjuster pawl

6.4l Pull the secondary shoe away from the backing plate

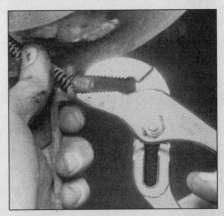

6.4m Separate the parking brake cable and spring from the actuating lever

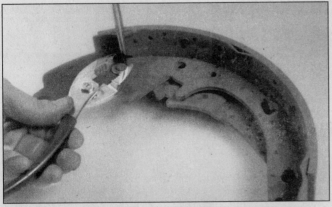

6.4n Remove the retaining clip which holds the parking brake actuating lever to the brake shoes

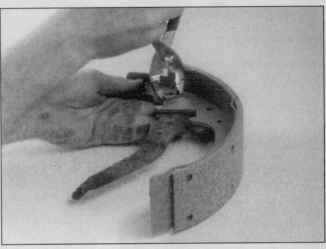

6.4o Install the parking brake actuator lever on the new brake shoe and install the retaining clip

6.4p Lubricate the threads and end of the adjusting screw assembly with high-temperature brake grease

6.4q Lightly coat the shoe guide pads on the backing plate with high-temperature brake grease

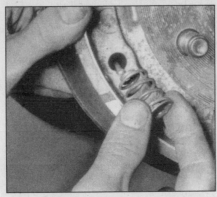

6.4r Position the shoes on the backing plate, insert the retaining pins through the backing plate and shoes and put the springs over them

6.4s Install the retaining spring caps

9

6.4t Make sure the slots in the wheel cylinder pushrods and the parking brake strut properly engage the brake shoes

6.4u Install the adjusting screw with the long end pointing towards the front of the vehicle

6.4v Install the adjusting pawl

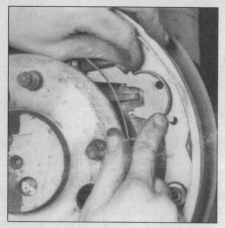

6.4w Install the anchor pin plate, cable guide and cable

6.4x Install the shoe guide and adjusting cable eye to the anchor pin, then install the shoe retracting springs

6.4y Connect the cable and spring to the lever, then install the drum and adjust the brake shoe-to-drum clearance (adjust the shoes so they rub slightly as the drum is turned, then back-off the adjuster until they don't rub)

5   Before reinstalling the drum it should be checked for cracks, score marks, deep scratches and hard spots, which will appear as small discolored areas. If the hard spots cannot be removed with fine emery cloth or if any of the other conditions listed above exist, the drum must be taken to an automotive machine shop to have it turned. **Note:** *Professionals recommend resurfacing the drums*

6.5 The maximum permissible diameter specification is cast into the brake drum

whenever a brake job is performed. Resurfacing will eliminate the possibility of out-of-round drums. If the drums are worn so much that they can't be resurfaced without exceeding the maximum allowable diameter, which is stamped into the drum **(see illustration)**, *then new ones will be required. At the very least, if you elect not to have the drums resurfaced, remove the glazing from the surface with medium-grit emery cloth using a swirling motion.*

6   Install the brake drum on the hub.

7   Mount the wheel, install the lug nuts, then lower the vehicle.

8   Make a number of forward and reverse stops to adjust the brakes until satisfactory pedal action is obtained.

7   Wheel cylinder - removal and installation

Note: *The time for wheel cylinder replacement is usually indicated by fluid leakage or sticky operation. New wheel cylinders are not very expensive at your local auto parts store,*

and their purchase makes more sense than trying to rebuild one at home. Always replace both rear wheel cylinders at the same time, never just one side.

Removal

Refer to illustration 7.4

1   Raise the rear of the vehicle and support it securely on jackstands. Block the front wheels to keep the vehicle from rolling.

2   Remove the brake shoe assembly (see Section 6).

3   Remove all dirt and foreign material from around the wheel cylinder. **Warning:** *Use brake cleaner spray to ensure that there is no airborne dust that could contain asbestos.*

4   Disconnect the brake line **(see illustration)**. Don't pull the brake line away from the wheel cylinder.

5   Remove the wheel cylinder mounting bolts.

6   Detach the wheel cylinder from the brake backing plate and immediately plug the brake line to prevent fluid loss and contamination.

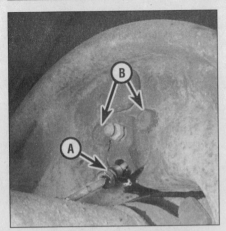

7.4 To remove the wheel cylinder, disconnect the brake line fitting (A) and remove the two mounting bolts (B)

8.4a At the master cylinder, disconnect the electrical connectors from the Cruise switch (A), and the low-fluid switch (B)

8.4b Remove the two master cylinder mounting nuts (arrows) at the brake booster studs

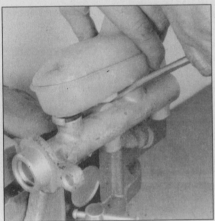

8.6 If you must remove the fluid reservoir to replace leaking seals or a broken reservoir, gently pry it off with a screwdriver or small prybar

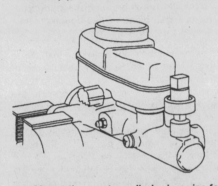

8.9 Mount the master cylinder in a vise for the bench bleeding procedure

8.10 To bench bleed the master cylinder, attach a pair of bleeder tubes to the outlet ports and submerge the ends in the reservoir

Installation

7   Place the wheel cylinder in position, install the mounting bolts and tighten them to the torque listed in this Chapter's Specifications.

8   Connect the brake line and install the brake shoe assembly.

9   Bleed the brakes (see Section 10).

8   Master cylinder - removal and installation

Warning: *On 1996 and earlier models equipped with four-wheel Anti-lock Brake System (ABS), do not attempt to remove the master cylinder. Have the master cylinder replaced at a dealer service department or other qualified repair shop. Removing a master cylinder from an ABS system can allow air to get into the ABS hydraulic control unit, which requires a special bleeding procedure impossible to perform at home.*

Removal

Refer to illustrations 8.4a, 8.4b and 8.6

1   Place rags under the brake line fittings and prepare caps or plastic bags to cover the ends of the lines once they are disconnected. **Caution:** *Brake fluid will damage paint. Cover all body parts and be careful not to spill fluid during this procedure.*

2   Unscrew the tube nuts at the ends of the brake lines where they enter the master cylinder. To prevent rounding off the flats on these nuts, a flare-nut wrench, which wraps around the fitting, should be used.

3   Pull the brake lines away from the master cylinder slightly and plug the ends to prevent contamination and entry of air.

4   Disconnect the cruise switch at the front of the master cylinder and the brake fluid level warning light switch, if equipped **(see illustration)**. Remove the two master cylinder mounting nuts, and detach the master cylinder from the vehicle **(see illustration)**.

5   Remove the reservoir cap, then discard any fluid remaining in the reservoir.

6   If the master cylinder is being replaced, but the old reservoir is being reused, pry the reservoir from the old master cylinder **(see illustration)**. It is a friction-fit within the grommets. Lubricate the new reservoir grommets with silicone grease and press them into the master cylinder body. Make sure they're

properly seated. **Note:** *If silicone grease is not available, use clean brake fluid.*

7   Lay the reservoir on a hard surface and press the master cylinder body onto the reservoir, using a rocking motion.

8   Inspect the reservoir cap and diaphragm for cracks and deformation. Replace any damaged parts with new ones and attach the diaphragm to the cap.

Bench-bleeding

Refer to illustrations 8.9 and 8.10

Note: *Whenever the master cylinder is removed, the complete hydraulic system must be bled. The time required to bleed the system can be reduced if the master cylinder is filled with fluid and bench bled before the master cylinder is installed on the vehicle.*

9   The master cylinder should be supported in such a manner that the piston can be depressed and brake fluid will not spill during the bench-bleeding procedure. Mount the master cylinder in a vise with the vise jaws clamping on the mounting flange **(see illustration)**.

10   Attach a pair of master cylinder bleeder tubes (available at auto parts stores) to the outlet ports **(see illustration)**. Position the ends of the tubes into the master cylinder reservoir.

9

11   Fill the reservoirs with brake fluid.
12   Using a rod, large Phillips screwdriver or drift punch. push the piston assembly into the bore to force air from the master cylinder. Air will be expelled through the submerged ends of the tubes. Stroke the piston until no air bubbles can be seen leaving the tubes.
13   Remove the bleeder tubes, refill the master cylinder reservoirs and install the diaphragm and cap assembly.

Installation

14   Carefully install the master cylinder by reversing the removal steps, then bleed the brakes (see Section 10).

---

9   Brake hoses and lines - inspection and replacement

Caution: *If the vehicle is equipped with ABS, make sure you plug the brake line immediately after disconnecting it from the brake hose, to prevent the fluid from draining out of the line and air entering the HCU. The HCU on an ABS system cannot be bled without a very expensive tool.*

Inspection

1   About every six months, with the vehicle raised and supported securely on jackstands, the rubber hoses which connect the steel brake lines with the front and rear brake assemblies should be inspected for cracks, chafing of the outer cover, leaks, blisters and other damage. These are important and vulnerable parts of the brake system and inspection should be complete. A light and mirror will be helpful for a thorough check. If a hose exhibits any of the above conditions, replace it with a new one.

Replacement
Flexible hose
Refer to illustrations 9.2 and 9.3
2   Using a flare nut wrench, disconnect the brake line from the hose fitting, being careful not to bend the frame bracket or brake line. Hold the fitting on the hose with a wrench to prevent the metal line from twisting and the frame bracket from bending **(see illustration)**.
3   Remove the large retaining clip **(see illustration)** and detach the hose from the bracket and the body. **Caution:** *Plug the metal brake line immediately to prevent air from getting into the hydraulic control unit (HCU). If air gets into the HCU, you will not be able to bleed the brakes properly at home.*
4   Remove the banjo bolt from the caliper and discard the sealing washers.
5   Connect the hose to the caliper, using new sealing washers. Tighten the banjo bolt to the torque listed in this Chapter's Specifications.
6   Without twisting the hose, connect the other end of the line to the bracket on the chassis.

9.2 To disconnect the fitting that attaches the flexible brake hose to the metal brake line at the bracket in the wheel well, use a backup wrench on the hose fitting to ensure that the metal line doesn't twist

7   Connect the metal brake line to the hose fitting by hand, then, using a flare nut wrench, tighten the fitting securely. Be sure to use a wrench on the hose fitting to prevent the bracket from bending or the metal line from twisting.
8   When the brake hose installation is complete, there should be no kinks in the hose. Make sure the hose doesn't contact any part of the suspension. Check this by turning the wheels to the extreme left and right positions. If the hose makes contact, remove it and correct the installation as necessary.

Metal brake line
9   When replacing brake lines be sure to use the correct parts. Don't use copper tubing for any brake system components. Purchase steel brake lines from a dealer or auto parts store.
10   Prefabricated brake line, with the tube ends already flared and fittings installed, is available at auto parts stores and dealers. These lines may be available pre-bent to the proper shapes. If pre-bent lines are not available, purchase a straight section the same diameter and length as the original section and use the proper bending tools to shape the new line to match the original.
11   When installing the new line make sure it's securely supported in the brackets and has plenty of clearance between moving or hot components.
12   After installation, check the master cylinder fluid level and add fluid as necessary. Bleed the brake system as outlined in the next Section and test the brakes carefully before driving the vehicle in traffic.

---

10   Brake hydraulic system - bleeding

Refer to illustration 10.8
Warning: *Wear eye protection when bleeding*

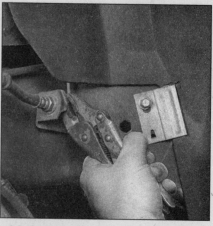

9.3 Once the fitting has been loosened sufficiently, remove this large retainer clip and separate the hose from the bracket

the brake system. If the fluid comes in contact with your eyes, immediately rinse them with water and seek medical attention.
Note: *Bleeding the hydraulic system is necessary to remove any air that manages to find its way into the system when it's been opened during removal and installation of a hose, line, caliper or master cylinder.*

Conventional brakes (non-ABS)

1   It will probably be necessary to bleed the system at all four brakes if air has entered the system due to low fluid level, or if the brake lines have been disconnected at the master cylinder.
2   If a brake line was disconnected only at a wheel, then only that caliper or wheel cylinder must be bled.
3   If a brake line is disconnected at a fitting located between the master cylinder and any of the brakes, that part of the system served by the disconnected line must be bled.
4   Remove any residual vacuum from the brake power booster by applying the brake several times with the engine off.
5   Remove the master cylinder reservoir cover and fill the reservoir with brake fluid. Reinstall the cover. **Note:** *Check the fluid level often during the bleeding operation and add fluid as necessary to prevent the fluid level from falling low enough to allow air bubbles into the master cylinder.*
6   Have an assistant on hand, as well as a supply of new brake fluid, an empty clear plastic container, a length of 3/16-inch plastic, rubber or vinyl tubing to fit over the bleeder valve and a wrench to open and close the bleeder valve.
7   Beginning at the right rear wheel, loosen the bleeder valve slightly, then tighten it to a point where it is snug but can still be loosened quickly and easily.
8   Place one end of the tubing over the bleeder valve and submerge the other end in brake fluid in the container **(see illustration)**.
9   Have the assistant pump the brakes

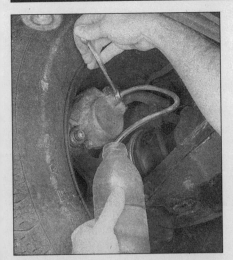

10.8 When bleeding the brakes, a hose is connected to the bleed screw at the caliper or wheel cylinder and then submerged in brake fluid - air will be seen as bubbles in the tube and container (all air must be expelled before moving to the next wheel)

slowly a few times to get pressure in the system, then hold the pedal firmly depressed.

10   While the pedal is held depressed, open the bleeder valve just enough to allow a flow of fluid to leave the valve. Watch for air bubbles to exit the submerged end of the tube. When the fluid flow slows after a couple of seconds, close the valve and have your assistant release the pedal.

11   Repeat Steps 9 and 10 until no more air is seen leaving the tube, then tighten the bleeder valve and proceed to the left rear wheel, the right front wheel and the left front wheel, in that order, and perform the same procedure. Be sure to check the fluid in the master cylinder reservoir frequently.

12   Never use old brake fluid. It contains moisture which will deteriorate the brake system components.

13   Refill the master cylinder with fluid at the end of the operation.

14   Check the operation of the brakes. The pedal should feel solid when depressed, with no sponginess. If necessary, repeat the entire process. **Warning:** *Do not operate the vehicle if you are in doubt about the effectiveness of the brake system.*

Anti-lock brake system (ABS)

15   Brake systems on 1996 and earlier models with the four wheel anti-lock brake system cannot be bled at home if air gets into the master cylinder and/or the hydraulic control unit (HCU). The first step in the bleeding procedure for these two components requires a special anti-lock test adapter which must be plugged into the control module. Any attempt to bleed the master cylinder and HCU without this special device will trap air in the HCU, which will result in a spongy brake pedal. All other models, earlier and later, can be bled conventionally.

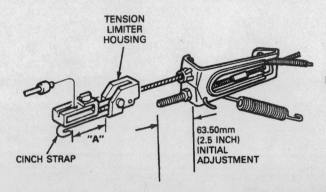

TENSION LIMITER HOUSING

11.4 Hold the tension limiter and tighten the equalizer nut up the rod to adjust the parking brake

CINCH STRAP   "A"

63.50mm (2.5 INCH) INITIAL ADJUSTMENT

16   As long as no air has gotten into the master cylinder or the HCU, the brake lines and the calipers can be bled in the conventional manner. Refer to Steps 1 through 14 above. However, if the pedal feels spongy after a bleeding procedure, or the ABS warning light on the instrument panel stays lit or blinks, have the vehicle towed to a dealer service department or other properly equipped repair facility for further testing with the proper tools.

11   Parking brake - adjustment

1991 and 1992 models

Note: *Adjust the rear brakes prior to adjusting the parking brake cable.*

1   The parking brake is a cable system that incorporates a tension limiter. If the parking brake system is in normal operating condition, depressing the parking brake pedal to the floor will automatically set the proper tension. **Note:** *The brake drums must be cold for this adjustment to be correct.*

Initial adjustment

Refer to illustration 11.4

2   Perform this procedure only if a new tension limiter is installed.

3   Fully apply the parking brake.

4   Hold the threaded rod on the end of the right-hand brake cable from turning and turn the nut 2-1/2-inches up the rod **(see illustration)**. As you turn the nut, keep an eye on the cinch strap. It should slip.

5   After the nut is tightened, dimension A **(see illustration 11.4)** should be less than 1-3/8-inch.

Field adjustment

Refer to illustration 11.8

6   This procedure is used to correct looseness in the system when a new tension limiter isn't installed.

7   Fully apply the parking brake.

8   Scribe a mark on the threaded rod at the equalizer nut to note the original position **(see illustration)**.

9   Grip the threaded rod to prevent it from turning and tighten the equalizer nut six full turns past its original position.

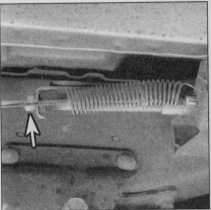

11.8 Scribe a mark on the threaded rod to indicate the position of the equalizer nut

10   Release the parking brake and check for rear wheel drag. The cables should be tight enough to provide full application of the rear brake shoes with the pedal fully applied. The cable should be loose enough to ensure complete release of the brake shoes when the lever is in the released position.

1993 and 1994 models

11   These models use a parking brake control assembly with an automatic tensioning device. No adjustment of the cable is necessary.

1995 and later models

12   The adjustment on the rear disc brake systems depends on the inner diameter of the rotor/parking brake drum housing and the thickness of the parking brake lining. Refer to the parking brake shoes in Section 13 for the check and replacement procedure.

12   Parking brake cables - replacement

1991 and 1992 models

Front cable

Refer to illustrations 12.1 and 12.11

1   Raise the vehicle and support it securely on jackstands. Release the parking brake completely, then loosen the equalizer nut and

9

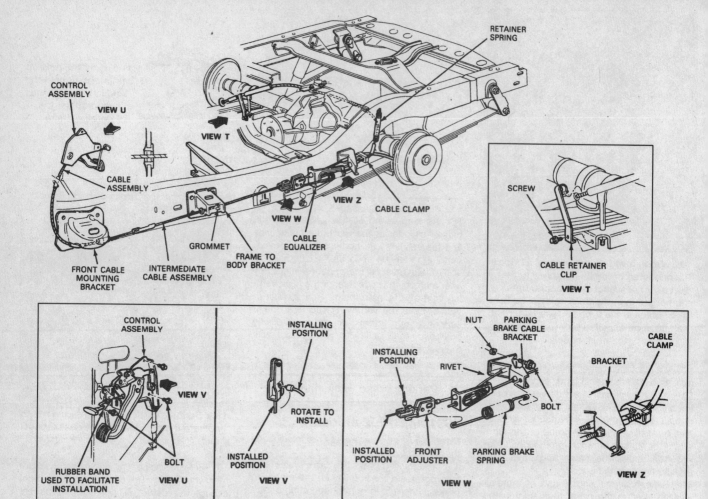

12.1 Parking brake details on 1991 and 1992 models

remove the cable end from the tension limiter (see illustration). Refer to illustration 11.4 for additional views.

2   Remove the intermediate cable from the front cable.

3   From inside the vehicle, disconnect the cable from the pedal assembly.

4   Working in the passenger compartment, remove the cable from the vehicle pulling it into the passenger compartment.

5   To install, pull the assembly through the cable housing from the passenger compartment to the cable equalizer.

6   From underneath the vehicle, fasten the cable to the adjuster bracket.

Rear cable

7   Raise the vehicle and support it securely on jackstands. Release the parking brake completely, then remove the hubcap, the wheel and tire, and brake drum.

8   Remove the locknut on the threaded rod at the equalizer.

9   Disconnect the cable end from the equalizer.

10   Depress the prongs that retain the cable

housing to the frame bracket and pull out the cable and housing from the crossmember.

11   On the wheel side of the backing plate, compress the retainer fingers so the retainer passes through the hole in the backing plate (see illustration). This can be done with pliers, but a small hose clamp tightened around the prongs will evenly compress them. Once the cable is started through the backing plate, remove the clamp.

12   Lift the cable out of the slot in the park-

ing brake lever attached to the secondary brake shoe and remove the cable through the backing plate hole.

13   Installation is the reverse of removal.

1993 and 1994 models

Front and intermediate cable

Refer to illustration 12.14

14   Raise the vehicle and support it securely on jackstands. Relieve the tension on the

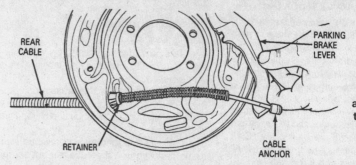

12.11 Press the retainer fingers and pass the assembly through the backing plate

12.14 Parking brake details on 1993 and 1994 models

1   Parking brake control assembly
2   Bolt
3   Front parking brake cable
4   Parking brake intermediate cable
5   Grommet
6   Bracket
7   Nut
8   Bolt
9   Rear guide bracket
10 Rear parking brake cable retainer
11 Parking brake retainer clip
12 Screw
13 Clip
14 Screw
15 Right side cable assembly
16 Left side cable assembly
17 Parking brake cable equalizer
A   Cable end assembly
B   Rotate cable down
C   Pull pin to adjust cable

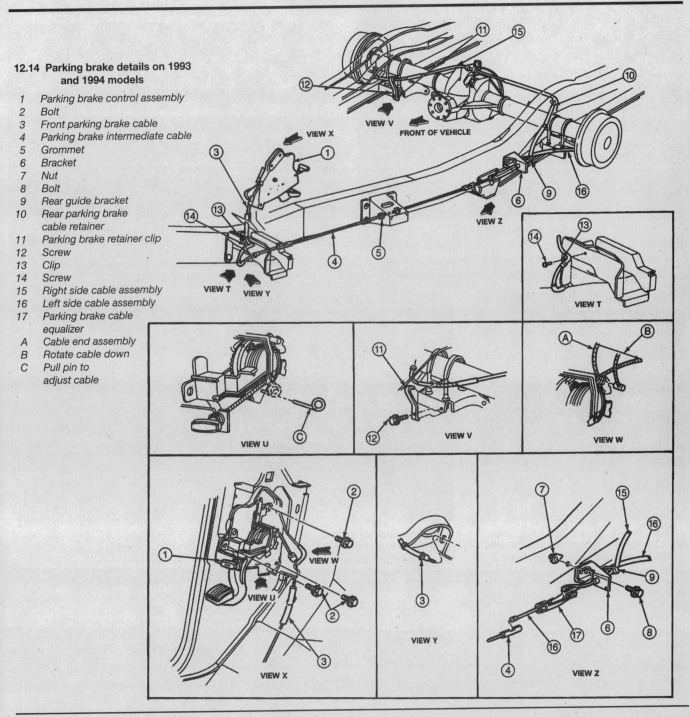

parking brake cable by pulling on the front cable and inserting 5/32 drill bit or pin into the hole provided in the parking brake control assembly **(see illustration).**
15 Disconnect the front cable from the intermediate and then the rear cable and conduit from the intermediate cable.
16 From inside the vehicle, disconnect the cable from the pedal assembly.
17 Working in the passenger compartment, remove the cable from the vehicle pulling it into the passenger compartment.
18 To install, pull the assembly through the cable housing from the passenger compartment to the cable equalizer.

19 From underneath the vehicle, fasten the cable to the adjuster bracket.

Rear cable

20 Raise the vehicle and support it securely on jackstands. Release the parking brake completely, then remove the hubcap, the wheel and tire, and brake drum.
21 Disconnect the rear cable from the intermediate cable and separate the cable and housing from the bracket.
22 Disconnect the rear cable from the rear bracket.
23 On the wheel side of the backing plate, compress the retainer fingers so the retainer

passes through the hole in the backing plate.
24 Lift the cable out of the slot in the parking brake lever attached to the secondary brake shoe and remove the cable through the backing plate hole.
25 Installation is the reverse of removal.

1995 and later models

Front cable

Refer to illustrations 12.26 and 12.28

26 Release the parking brake mechanism, and while an assistant pulls on the front cable (away from the control assembly), insert a 5/32-inch pin (such as a drill bit) into the ser-

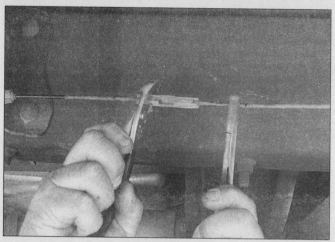

12.26 Release the parking brake, pull on the cable, and insert a 5/32 inch pin or drill bit through the service hole (arrow)

12.28 Release the front cable from the connector at the intermediate cable

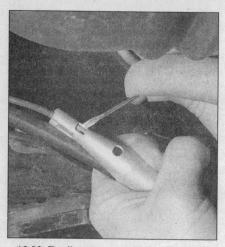

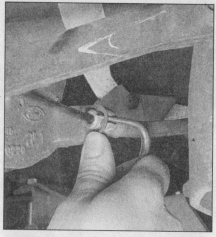

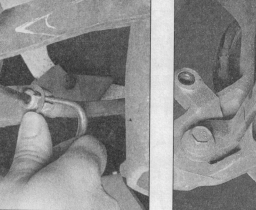

12.30 To disconnect the left rear cable from the intermediate cable, pry open this locking tang with a small screwdriver

12.31a On models with rear disc brakes, remove this retaining clip from the bracket on the torque plate, then pull the cable forward (toward the front of the vehicle) and guide it out of the bracket . . .

12.31b . . . then disconnect the cable from the parking brake lever

vice hole to hold the ratchet mechanism tension (**see illustration**).

27   Remove the bolt and clamp holding the front cable (refer to Chapter 11 for removal of the inner fender splash shield for access).

28   Release the rear of the front cable from the front of the intermediate cable connector (**see illustration**).

Intermediate and rear cables

Refer to illustrations 12.30, 12.31a, 12.31b and 12.32

Note: *The following procedure applies to the intermediate cable (the short cable between the front cable and the two rear cables) and to either rear cable.*

29   Disconnect the front cable from the connector to the intermediate cable (**see illustration 12.28**).

30   Disconnect the rear of the intermediate cable from the connector to the rear cables (**see illustration**).

31   Disconnect the rear cable. On models with rear drum brakes, you'll need to remove the brake drum and disassemble the brake (see Section 6). Disconnect the parking brake cable from the actuating lever and backing plate (see Step 11). On models with rear disc brakes, remove the cable end from the lever and remove the cable from the bracket (**see illustrations**).

32   Installation is the reverse of removal. When you reattach the two rear cables to the intermediate cable, make sure that the left cable is on top (**see illustration**).

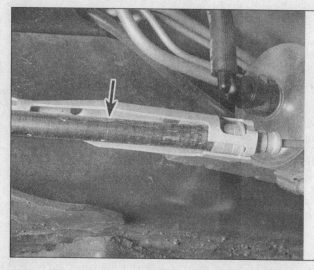

12.32 When you reattach the two rear parking brake cables to the intermediate cable, make sure the left rear cable (arrow) is on top, and attached to the forward end of the cable connector; the right rear cable should be under the left rear cable, and should be attached to the rear end of the connector

13.5a   Remove the parking brake shoe hold-down springs and pins from both the leading and trailing shoes

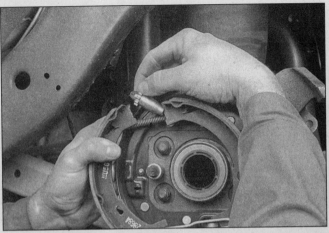

13.5b   Rotate the adjuster screw to its shortest length, then spread the parking brake shoes apart and remove the adjuster

13   Parking brake shoes (rear disc brakes only) - inspection and replacement

Refer to illustrations 13.5a through 13.5g
Warning: *Dust created by the brake system may contain asbestos, which is hazardous to your health. Never blow it out with compressed air and don't inhale any of it. An approved filtering mask should be worn when working on the brakes. Do not, under any circumstances, use petroleum-based solvents to clean brake parts. Use brake system cleaner only!*

1   Loosen the rear wheel lug nuts. Raise the rear of the vehicle and place it securely on jackstands. Remove the rear wheels.
2   Remove the caliper (see Section 4) and the brake disc (see Section 5). It's not necessary to disconnect the brake hose from the caliper; but hang the caliper out of the way with a piece of wire to prevent damage to the hose.
3   Inspect the thickness of the lining material on the shoes. If the lining has worn down

13.5c   Detach the adjuster return spring from the parking brake shoes

to 3/64-inch or less, the shoes must be replaced.
4   Disconnect the rear parking brake cable from the parking brake lever (see Section 12).
5   Follow the accompanying photos **(see illustrations 13.5a through 13.5g)** for the

13.5d   Remove the parking brake shoes, the return springs and the parking brake lever as an assembly; make sure you don't damage the rubber grommet in the backing plate for the parking brake lever

parking brake shoe replacement procedure. Be sure to stay in order and read the caption under each illustration.

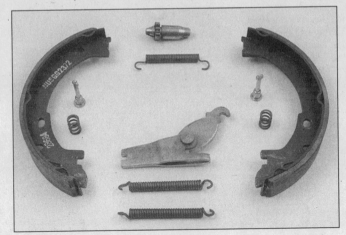

13.5e   Disassemble the parking brake assembly on the bench

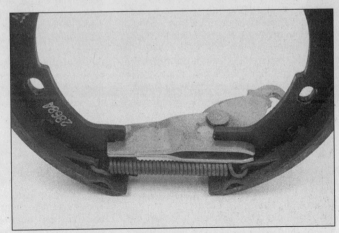

13.5f   Assemble the parking brake lever and return springs onto the shoes

9

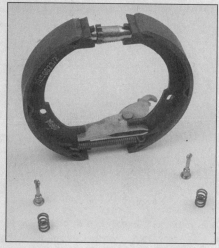

13.5g Install the assembly onto the backing plate, attach the hold down pins, hold down springs, adjuster screw and return spring (shown assembled on the bench for clarity)

6   Install the brake disc. Temporarily thread three of the wheel lug nuts onto the studs to hold the disc in place.

7   Remove the rubber access plug from the backside of the brake torque plate. Adjust the parking brake shoe clearance by turning the adjuster star wheel with a brake adjusting tool or screwdriver until the shoes contact the disc and the disc can't be turned. Back-off the adjuster eight notches, then install the hole plug.

8   Install the brake caliper. Be sure to tighten the bolts to the torque listed in this Chapter's Specifications.

9   Install the wheel and tighten the lug nuts to the torque listed in the Chapter 1 Specifications.

---

14   Power brake booster - check, removal, installation and adjustment

---

1   The power brake booster unit requires no special maintenance apart from periodic inspection of the vacuum hose and the case.

2   Dismantling of the brake booster requires special tools and is not ordinarily done by the home mechanic. If a problem develops, install a new or factory rebuilt unit.

Check

3   Begin the power booster check by depressing the brake pedal several times with the engine off and make sure that there is no change in the pedal reserve distance. The reserve distance is the distance between the pedal and the floor when the pedal is fully depressed.

4   Now, depress the pedal and start the engine. If the pedal goes down slightly, operation is normal. Release the brake pedal and let the engine run for a couple of minutes.

14.10 Working under the instrument panel, disconnect the brake light switch and booster pushrod from the brake pedal

5   Turn off the engine and depress the brake pedal several times slowly. If the pedal goes down farther the first time but gradually rises after the second or third depression, the booster is airtight.

6   Start the engine and depress the brake pedal, then stop the engine with the pedal still depressed. If there is no change in the reserve distance after holding the pedal for about 30-seconds, the booster is airtight.

7   If the pedal feels "hard" when the engine is running, the booster isn't operating properly or there is a vacuum leak in the hose to the booster.

Removal

Refer to illustrations 14.10 and 14.11

8   Remove the nuts attaching the master cylinder to the booster (see Section 8) and carefully pull the master cylinder forward until it clears the mounting studs. Use caution so as not to bend or kink the brake lines.

9   Detach the manifold vacuum hose from the booster check valve.

10   Working in the passenger compartment under the steering column, unplug the electrical connector from the brake light switch **(see illustration)**, then remove the pushrod retaining clip and nylon washer from the brake pedal pin. Slide the pushrod off the pin.

11   Remove the nuts attaching the brake booster to the firewall **(see illustration)**.

12   Carefully detach the booster from the firewall and lift it out of the engine compartment.

Installation

13   Place the booster into position on the firewall and tighten the mounting nuts to the torque listed in this Chapter's Specifications. Connect the pushrod and brake light switch to the brake pedal. Install the retaining clip in the brake pedal pin.

14   Install the master cylinder to the booster, tightening the nuts to the torque listed in this Chapter's Specifications.

15   Carefully check the operation of the brakes before driving the vehicle in traffic.

14.11 Remove the four booster mounting nuts

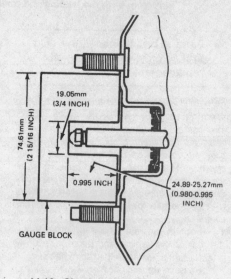

14.16 Check the master cylinder-to-booster pushrod clearance with a gauge fabricated from sheet metal

Adjustment

Refer to illustration 14.16

16   Some boosters feature an adjustable pushrod. They are matched to the booster at the factory and most likely will not require adjustment, but if a misadjusted pushrod is suspected, a gauge can be fabricated out of heavy gauge sheet metal **(see illustration)**.

17   Some common symptoms caused by a misadjusted pushrod include dragging brakes (if the pushrod is too long) or excessive brake pedal travel accompanied by a groaning sound from the brake booster (if the pushrod is too short).

18   To check the pushrod length, unbolt the master cylinder from the booster and position it to one side. It isn't necessary to disconnect the hydraulic lines, but be careful not to bend them.

19   Apply vacuum to the booster with a hand-held vacuum pump, then place the pushrod gauge against the end of the

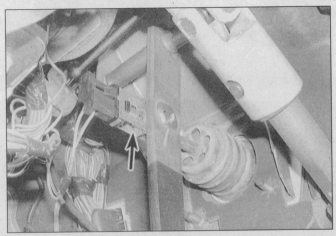

15.2 The brake light switch (arrow) is attached to the brake pedal

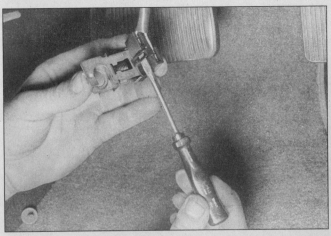

15.3 Use a small screwdriver to disengage the locking tab from the electrical connector

pushrod, exerting a force of approximately five pounds to seat the pushrod in the power unit. The rod measurement should fall somewhere between the minimum and maximum specifications on the gauge. If it doesn't, adjust it by holding the knurled portion of the pushrod with a pair of pliers and turning the end with a wrench.

20   When the adjustment is complete, reinstall the master cylinder and check for proper brake operation before driving the vehicle in traffic.

15   Brake light switch - removal and installation

Removal

Refer to illustrations 15.2 and 15.3

1   Remove the under dash panel.

2   Locate the brake light switch assembly near the top of the brake pedal and disconnect the switch from the brake pedal by removing the retaining clip **(see illustration)**.

3   Use a small screwdriver to unlock the electrical connector, then unplug the connector from the brake light switch **(see illustration)**.

Installation

Refer to illustration 15.5

4   Install the switch to the electrical connector by snapping the clip into place.

5   Reconnect the assembly to the brake pedal **(see illustration)**.

6   Install the under dash panel.

7   Check the brake lights for proper operation.

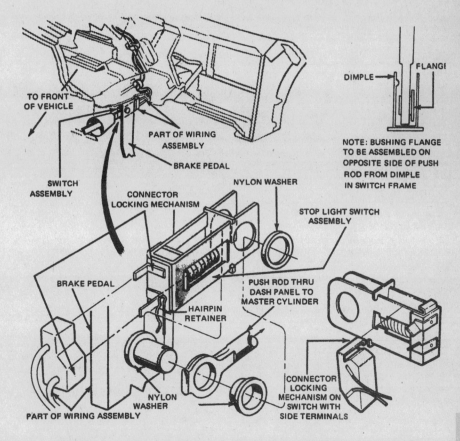

15.5 Installation details of a typical earlier model brake light switch assembly

9

Notes

Chapter 10
Suspension and steering systems

Contents

Specifications

General

Front suspension	
1991 through 1994 models	Twin I-beam with coil springs
1995 and later models	Upper and lower control arms with torsion bars
Rear suspension	Leaf spring
Steering	
1991 through 1994 models	Ford integral power steering
1995 and later models	Rack and pinion steering
Power steering pump	Ford Model C-II
Power steering gear	
Gear ratio	17:1
Fluid type and capacity	See Chapter 1

Torque specifications

Ft-lbs (unless otherwise indicated)

Front suspension

1991 through 1994

2WD models

Shock absorber-to-radius arm nut	39 to 53
Shock absorber upper nut	25 to 35
Spring retainer bolt	191 to 231
Stabilizer bar bolts	16 to 24
Stabilizer bar end nuts	108 to 144 in-lbs
Radius arm-to-frame nut	81 to 120
Radius arm bracket-to-frame bolt	77 to 110
Radius arm front bracket-to-axle arm bolt	191 to 231
Axle pivot bolt	120 to 150
Axle pivot bracket nuts	77 to 90
Jounce bumper bolt	18 to 26
Balljoint pinch bolt	48 to 65

Torque specifications (continued)   **Ft-lbs** (unless otherwise indicated)

Front suspension

1991 through 1994 (continued)

4WD models

Spindle-to-knuckle nuts	40 to 50
Shock absorber-to-radius arm nut	39 to 53
Shock absorber upper nut	25 to 35
Spring retainer nut	70 to 100
Stabilizer bar bolts	35 to 50
Stabilizer bar link nuts	30 to 44
Radius arm-to-rear bracket nut	80 to 120
Radius arm front bracket front bolts	27 to 37
Radius arm front bracket lower bolt	190 to 230
Radius arm-to-front bracket upper stud	190 to 230
Axle pivot bolt	120 to 150
Axle pivot bracket bolts	
Left bracket (right axle arm)	155
Right bracket (left axle arm)	77 to 110
Jounce bumper bolt	18 to 26
Balljoint pinch bolt	65 to 85

1995 and later models (2WD and 4WD)

Shock absorber-to-lower control arm

Without ARC	15 to 21
With ARC	18 to 24

Shock absorber upper nut

Without ARC	30 to 40
With ARC	25 to 34
Upper spindle pinch bolt/nut	30 to 40
Lower balljoint nut	83 to 113
Wheel hub nut	157 to 213
Lower control arm pivot bolts	111 to 148
Upper control arm pivot bolts	83 to 112
Front stabilizer bar link nuts	15 to 21
Front stabilizer bar bushing clamp bolts	25 to 34
Hub and bearing assembly-to-steering knuckle bolts (4WD)	70 to 80

Rear suspension

1991 through 1994

Rear leaf spring U-bolt nut	88 to 108
Shock-to-lower bracket nut	39 to 53
Shock-to-upper bracket nut	39 to 53
Shackle-to-spring nut	74 to 115
Spring-to-front bracket bolt	64 to 91
Spring shackle-to-rear bracket bolt	74 to 114
Stabilizer-to-mounting bracket bolt	30 to 42
Stabilizer-to-link nut	50 to 68
Stabilizer link-to-frame bolt	50 to 68

1995 and later

Rear leaf spring U-bolt nut	65 to 87
Shock absorber-to-lower bracket nut	39 to 53

Shock absorber-to-upper bracket nut

Without ARC	39 to 53
With ARC	14 to 19
Rear shackle-to-spring nut	74 to 115
Spring-to-front bracket bolt	50 to 68
Spring shackle-to-rear bracket bolt	74 to 114
Stabilizer-to-mounting bracket bolt	30 to 40
Stabilizer-to-link nut	50 to 68
Stabilizer link-to-frame bolt	44 to 59

Steering system

1991 through 1994

Drag link-to-Pitman arm nut	51 to 73
Drag link-to-steering connecting rod nut	51 to 73
Tie-rod adjusting sleeve nut	30 to 42
Tie-rod to spindle nut	51 to 73
Pitman arm-to-steering gear nut	170 to 228
Flex coupling-to-steering gear input shaft bolt	25 to 34
Steering gear-to-flex coupling bolt	25 to 34

Power steering gear hoses to gear	20 to 30
Power steering gear-to-frame bolts	50 to 61
Power steering pump-to-bracket bolts	35 to 47
Steering wheel bolt...	23 to 33
1995 and later	
Power steering gear pressure line fittings..	20 to 25
Power steering pressure line-to-pump..	40 to 54
Power steering gear mounting bolts/nuts	94 to 127
Power steering pump mounting bolts	
4.0L pushrod V6...	30 to 40
4.0L SOHC V6 and 5.0L V8	15 to 20
Power steering pump pulley bolts...	15 to 20
Steering wheel bolt...	24 to 34
Tie-rod to spindle nut	45 to 60
Tie-rod jam nut	50 to 68

1   General information

Front suspension

1991 through 1994 models

The front suspension on 1991 through 1994 2WD models is a twin I-beam type, which is composed of coil springs, I-beam axle arms, radius arms, upper and lower balljoints and spindles, tie-rods, shock absorbers and an optional stabilizer bar. The 4WD model is basically the same except the front driveline system is composed of a two-piece driveaxle assembly.

The front suspension consists of two independent axle arm assemblies **(see illustration 25.9a in Chapter 1)**. One end of the assembly is anchored to the frame and the other is supported by the coil spring and radius arm. The spindle is connected to the axle by upper and lower balljoints. The balljoints are constructed of a lubricated-for-life special bearing material. Lubrication points are found on the tie-rods and steering linkage. Movement of the spindles is controlled by the tie-rods and the steering linkage.

Two adjustments can be performed on the axle assembly. Camber is adjusted by removing and replacing an adapter between the upper balljoint stud and the spindle on 2WD models. 4WD models require replacing the camber adapter on the upper balljoint stud. Adapters are available in 0-degree, 1/2-degree, 1-degree and 1-1/2-degree increments. Toe-in adjustment is accomplished on both models by turning the tie-rod adjusting sleeve.

1995 and later models

The front suspension on 1995 and later models consists of upper and lower control arms, shock absorbers, torsion bars and a front stabilizer bar **(see illustrations 12.6 and 13.4)**. The inner ends of the control arms are attached to the frame; the outer ends are attached to the spindle. The upper control

arms pivot on eccentric bolts. The lower control arms pivot on two separate, non-adjustable bolts. The shocks and springs are mounted between the lower control arms and the frame. The front stabilizer bar is attached to the frame with mounting clamps.

The hydraulic shock absorbers are of the direct, double acting type, with later models having low pressure gas shocks. Both shock absorbers are of the telescoping design and come equipped with rubber grommets at the mounting points for quiet operation. The low pressure gas shock absorbers are sealed and charged with nitrogen gas to reduce shock absorber fade and improve ride. The shock absorbers are non-adjustable. The shock absorbers are not

rebuildable and must be replaced as complete assemblies.

Rear suspension
Refer to illustrations 1.6 and 1.7

The rear suspension uses shock absorbers and semi-elliptical leaf springs. The forward end of each spring is attached to the bracket on the frame side rail **(see illustration)**. The rear of each spring is shackled to a bracket on the frame rail. The rear shock absorbers are direct, double acting units, mounted behind the axle. On four-door models, there is also a side-action shock absorber mounted ahead of the rear axle on the right side.

Some later 4WD models are equipped

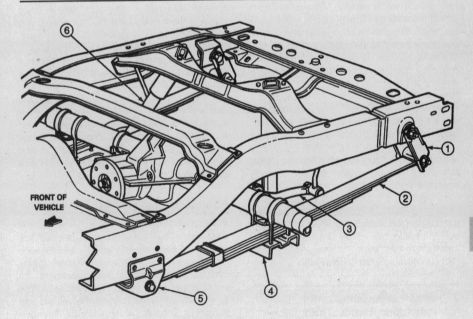

FRONT OF VEHICLE

1.6 Rear suspension components

1	*Shackle*	*3*	*Stabilizer bar*	*5*	*Spring hanger*
2	*Leaf spring*	*4*	*Spring plate*	*6*	*Shock absorber*

10

**1.7 Automatic Ride Control (ARC)
system components**

1   Front gate solenoid
2   Front shock absorber
3   Generic Electronic Module
4   Powertrain Control Module
5   Automatic Ride Control
6   Message Center Indicator
7   Vehicle Speed Sensor
8   Steering sensor
9   Ignition switch
10 Rear height sensor
11 Rear fill solenoid
12 Air compressor
13 ARC service switch
14 Rear gate solenoid
15 Rear shock absorber
16 Air line
17 Front fill solenoid
18 Stoplight switch
19 Front height sensor
20 Compressor relay

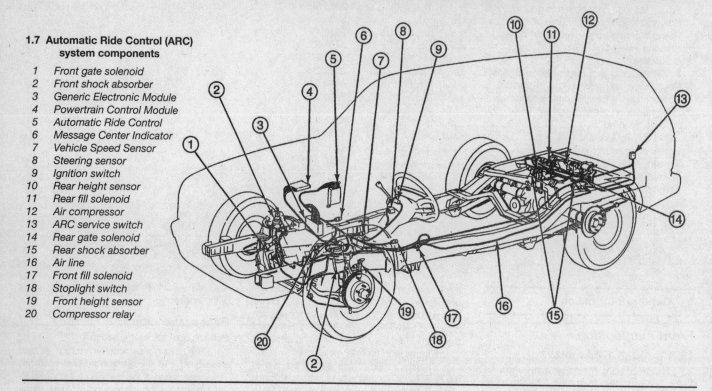

with the Automatic Ride Control (ARC) **(see illustration)**. This system allows the driver to control the ride of the suspension (hard or soft) depending upon the road conditions and the load of the vehicle. See Section 20 for more information.

Since most procedures that are dealt with in this chapter involve jacking up the vehicle and working underneath it, a good pair of jackstands will be needed. A hydraulic floor jack is the preferred type of jack to lift the vehicle, and it can also be used to support certain components during various operations.
Warning 1: *Never, under any circumstances, rely on a jack to support the vehicle while working under it.*
Warning 2: *Whenever any of the suspension or steering fasteners are loosened or removed they must be inspected and, if necessary, replaced with new ones of the same part number or of original equipment quality and design. Torque specifications must be followed for proper reassembly and component retention.*
Caution: *If the vehicle is equipped with Automatic Ride Control (ARC), make sure the air suspension switch is turned to the OFF position before the vehicle is raised, towed or jump started to prevent damage to the system components (see Section 20).*

2   Shock absorbers, front - inspection, removal and installation

Inspection

1   The common test of shock damping is simply to bounce the corners of the vehicle several times and observe whether or not the vehicle stops bouncing once you let go if it. A slight rebound and settling indicates good damping, but if the vehicle continues to bounce several times, the shock absorbers must be replaced.
2   If the shock absorbers stand up to the bounce test, visually inspect the shock body for signs of fluid leakage, punctures or deep dents in the metal of the body. Replace any shock absorber that is leaking or damaged, even if it passed the bounce test in Step 1.
3   After removing the shock absorber, pull the piston rod out and push it back in several times to check for smooth operation throughout the travel of the piston rod. Replace the shock absorber if it gives any signs of hard or soft spots in the piston rod travel.
4   Prior to installing the new shock absorbers, pump the piston rod fully in and out several times to lubricate the seals and fill the hydraulic sections of the unit.

Removal

Refer to illustrations 2.8, 2.9, 2.10a and 2.10b
Caution: *The low-pressure gas shock absorbers are pressurized to 135 PSI with nitrogen gas. Do not attempt to open, puncture or apply heat to the shock absorbers.*
5   Loosen the front wheel lug nuts on the side to be dismantled. Raise the vehicle and support it securely on jackstands. Remove the front wheel. **Caution:** *If the vehicle is equipped with Automatic Ride Control (ARC), make sure the air suspension switch is turned to the OFF position before the vehicle is raised to prevent damage to the system com-*

ponents (see Section 20).
6   On vehicles equipped with the ARC system, disconnect the electrical connectors from the shock absorbers and pry the harness retainers from their grommets in the frame.
7   On vehicles equipped with the ARC system, disconnect the airline from the shock absorber. Push IN and hold the plastic ring on the shock absorber. While holding the ring, pull out on the airline.
8   Remove the top nut with a deep socket while holding the shaft with an open-end wrench **(see illustration)**. Lift off the washer.
9   On ARC systems, remove the front left height sensor from the upper bracket.

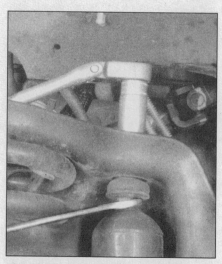

**2.8 Hold the shock absorber shaft while
turning the top retainer nut**

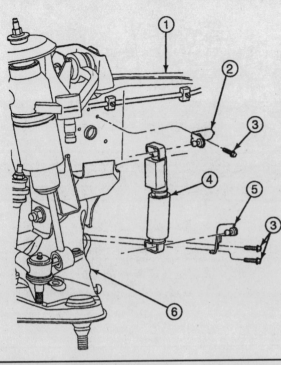

2.9 On ARC systems, remove the height sensor from the chassis

1   Frame
2   Front height sensor upper bracket
3   Bolt
4   Front height sensor
5   Front height sensor lower bracket
6   Front suspension lower arm

2.10a The shock absorber lower mounting bracket is attached to the radius arm on 1994 and earlier models

Release the spring clip and pull the sensor from the lower ball stud **(see illustration)**.
10   On 1991 through 1994 models, remove the bolt and nut securing the shock absorber to the radius arm **(see illustration)**. On 1995 and later models, remove the nuts from the lower control arm **(see illustration)**.
11   To remove the shock absorber, slightly compress the shock and remove it from its brackets.

Installation

12   Installation is the reverse of the removal steps with the following additions.

a) Install and tighten the new nuts and bolts to the torque listed in this Chapter's Specifications.

b) Install the wheel and lug nuts, lower the vehicle and tighten the lug nuts to the torque listed in the Chapter 1 Specifications.

---

3   Front wheel spindle - removal and installation

1991 through 1994
2WD models

Removal

Refer to illustrations 3.3 and 3.6
1   Remove the front wheel and brake disc (see Chapter 1).
2   Remove the brake dust shield.
3   Remove the cotter pin and nut and disconnect the tie-rod end from the spindle **(see illustration)**.

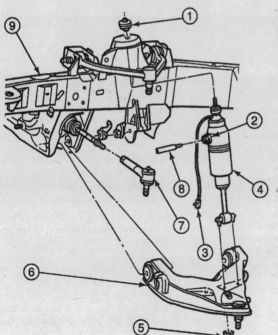

2.10b Front shock absorber installation details - ARC system (non-ARC shock mounting is similar)

1   Nut, washer and insulator assembly
2   Plastic release ring
3   Electrical connector
4   Front shock absorber
5   Nut
6   Front suspension lower arm
7   Tie-rod end
8   Air line
9   Frame

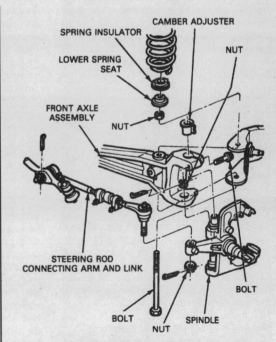

3.3 Front spindle installation details - 1991 through 1994 2WD models

10

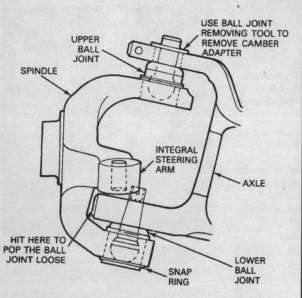

3.6 Separate the upper balljoint with a special balljoint tool, then strike the edge of the front spindle (arrows) with a hammer to break it loose from the lower balljoint stud

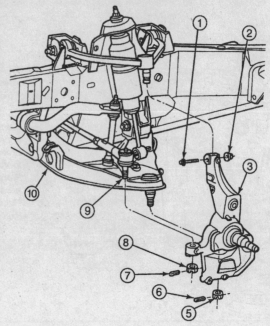

3.26 Front spindle installation details - 1995 and later 2WD models

1   Upper spindle bolt
2   Nut
3   Spindle
4   Not used
5   Nut
6   Cotter pin
7   Cotter pin
8   Nut
9   Tie rod end
10 Front suspension lower arm

4   Remove the cotter pin from the lower balljoint stud nut, then remove the nut.
5   Remove the clamp bolt from the upper balljoint.
6   Remove the camber adjuster from the upper balljoint **(see illustration)**. The upper balljoint must be separated with a balljoint removal tool, which clamps on and has a center bolt that presses the stud out without damage to any of the components. These are available at most auto parts stores.
7   Strike the inside of the spindle near the lower balljoint to break the spindle loose from the balljoint stud **(see illustration 3.6)**. **Caution:** *Don't use a fork-type separator to detach the balljoint. This will damage the balljoint seal.*
8   Remove the spindle together with the balljoints.

Installation

9   Before installation, check that the upper and lower balljoint seals were not damaged during spindle removal and that they are positioned correctly. Replace if necessary.
10 Position the spindle and balljoints in the axle arm.
11 Install the camber adjuster on the upper balljoint, making sure it's aligned correctly.
12 Tighten the lower balljoint nut to the torque listed in this Chapter's Specifications, then tighten further until a cotter pin hole lines up. Install a new cotter pin and bend it to secure the nut.
13 Install the clamp bolt on the upper balljoint and tighten it to the torque listed in this Chapter's Specifications.
14 Install the dust shield. Install the brake disc and the front wheel (see Chapter 1).
15 Have the front-end alignment checked by a dealer service department or an alignment shop.

4WD models

16 Refer to Chapter 8 for the axleshaft removal procedure, then refer to the 2WD spindle removal and installation procedure described above.

1995 and later
2WD models

Refer to illustration 3.26

17 Place the steering wheel in the center position.
18 Loosen the wheel lug nuts. Raise the vehicle and support it securely on jackstands. Remove the wheel. **Caution:** *If the vehicle is equipped with Automatic Ride Control (ARC), make sure the air suspension switch is turned to the OFF position before the vehicle is raised to prevent damage to the system components (see Section 20).*
19 Remove the disc brake caliper (see Chapter 9). Wire the caliper to the underbody to prevent damage to the brake hose.
20 Remove the brake disc (see Chapter 9).
21 Remove the ABS sensor, if equipped, then remove the dust shield from the spindle.
22 Disconnect the tie-rod end from the spindle (see Section 25).
23 Place a jack under the lower arm of the balljoint area. Raise the jack until it supports the spring load on the lower arm. The jack must remain in this position throughout the remainder of this procedure. **Warning:** *Failure to perform this step could result in serious injury.*
24 Remove the lower balljoint nut and the upper balljoint pinch bolt from the spindle (see Section 5).
25 Unload the torsion bar using a special tool designed for this purpose.
26 Remove the spindle **(see illustration)**.
27 Installation is the reverse of removal with

the following additions:
a) *Be sure to tighten all fasteners to the torque values listed in this Chapter's Specifications.*
b) *Have the front-end alignment checked by a dealer service department or an alignment shop.*

4WD models

Note: *On 4WD models the spindle is referred to as the steering knuckle.*

28 Refer to Chapter 8 for removal of the driveaxle hub nut and driveaxle, and Section 4 for removal of the hub and bearing assembly.
29 Follow the Steps above for the 2WD models to remove the steering knuckle (spindle).
30 Installation is the reverse of removal. Be sure to tighten all fasteners to the torque values listed in this Chapter's Specifications and have the front-end alignment checked by a dealer service department or an alignment shop.

---

4   Hub and bearing assembly, front (1995 and later 4WD models) - removal and installation

Note: *The hub and bearing assembly is a sealed unit and isn't serviceable. If it's defective, it must be replaced.*

Removal

Refer to illustrations 4.5 and 4.6

1   Put the vehicle in gear, apply the parking brake and break loose the driveaxle/hub nut with a socket and large breaker bar (see Chapter 8).
2   Loosen the wheel lug nuts, raise the vehicle and support it securely on jackstands.

4.5 To detach the splash shield from the steering knuckle on 1995 and later 4WD models, remove these three bolts (arrows)

4.6 To detach the hub and bearing assembly from the steering knuckle, remove the three bolts (arrows)

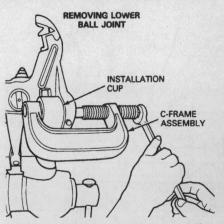

5.3 Using the special tools to remove the balljoints (2WD models)

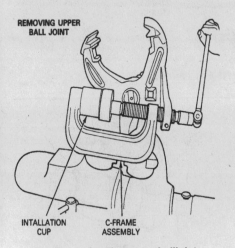

5.4 Removing the upper balljoint (early 2WD models)

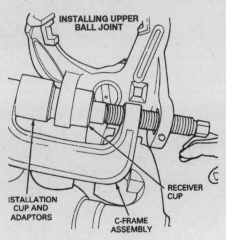

5.5 Note the additional receiver cup required to install the upper balljoint - install the upper balljoint first (2WD models)

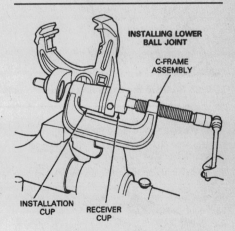

5.7 Installing the lower balljoint (2WD models)

Caution: *If the vehicle is equipped with Automatic Ride Control (ARC), make sure the air suspension switch is turned to the OFF position before the vehicle is raised to prevent damage to the system components (see Section 20).*

3   Remove the wheel. Remove the drive-axle/hub nut.

4   Unbolt the brake caliper and hang it out of the way with a piece of wire, then remove the caliper anchor bracket and the brake disc (see Chapter 9).

5   On models with 4-wheel ABS (4WABS), remove the disc splash shield from the steering knuckle **(see illustration)**, then remove the speed sensor retaining bolt, remove the sensor from the knuckle and set the sensor and wire harness safely aside.

6   Remove the hub assembly-to-steering knuckle bolts **(see illustration)**.

7   Tap the hub assembly from side-to-side to break it loose from the steering knuckle. Pull the hub assembly off the end of the driveaxle. Wrap the end of the driveaxle with a rag to prevent damaging it. If the hub is stuck on the splines on the end of the driveaxle, use a puller to free it.

Installation

8   Installation is the reverse of the removal procedure. Be sure to lubricate the driveaxle splines with multi-purpose grease, and tighten all of the fasteners to the torque listed in this Chapter's Specifications.

5   Balljoints - removal and installation

Warning: *Do not heat the axle or balljoint to aid removal since the temper may be removed from the component(s), leading to premature failure.*

1991 through 1994
2WD models
Refer to illustrations 5.3, 5.4, 5.5 and 5.7

1   Remove the spindle (see Section 3).

2   Remove the snap-ring from the lower balljoint. **Note:** *Remove the lower balljoint first.*

3   Install a special balljoint assembly tool and the appropriate size receiving cup (available from most auto parts stores) on the lower balljoint **(see illustration)**. Tighten the special tool and press the balljoint out of the spindle. **Note:** *If the C-frame assembly tool and receiver cup are not available or will not remove the balljoint, take the spindle assembly to a dealer service department or automotive machine shop and have the balljoints pressed out.*

4   Use the same tool setup used in Step 3 on the upper balljoint and press the balljoint out **(see illustration)**.

5   Install the upper balljoint with the C-frame assembly, balljoint receiver cup and installation cup **(see illustration)**. **Note:** *Always install the upper balljoint first since the special tool must pass through the lower balljoint receptacle in the axle.*

6   Turn the screw in the C-frame clockwise and press the balljoint into the axle until it is completely seated. **Caution:** *Don't heat the axle or balljoint to aid in installation since the temper may be removed from the component(s) leading to premature failure.*

7   Install the lower balljoint with the C-frame assembly, balljoint receiver cup and installation cup **(see illustration)**.

10

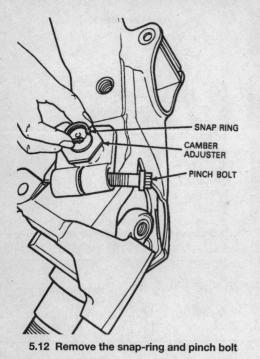

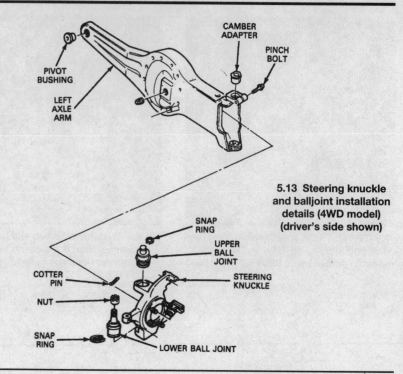

5.13 Steering knuckle and balljoint installation details (4WD model) (driver's side shown)

5.12 Remove the snap-ring and pinch bolt

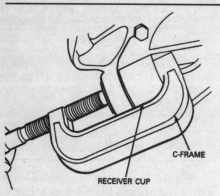

5.18 Using the special tools to remove the balljoints (4WD models)

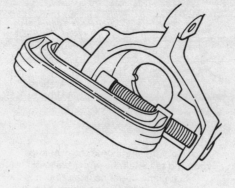

5.19 Removing the upper balljoint (4WD models)

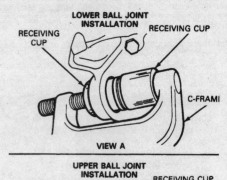

VIEW A

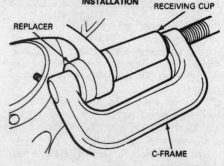

VIEW B

5.22 Note the additional receiver cup required to install the upper balljoint; install the lower balljoint first (4WD models)

8   Turn the screw in the C-frame clockwise and press the balljoint into the axle until it is completely seated. **Caution:** *Don't heat the axle or balljoint to aid in installation since the temper may be removed from the component(s) leading to premature failure.* Install the lower balljoint snap-ring.

9   Have the front-end alignment checked by a dealer service department or an alignment shop.

4WD models

Refer to illustrations 5.12, 5.13, 5.18, 5.19 and 5.22

10   Remove the spindle and shaft and joint assembly (see Chapter 8).

11   Remove the cotter pin and nut securing the tie-rod (see Section 20).

12   Remove the snap-ring from the upper balljoint, then remove the pinch bolt **(see illustration)**.

13   Remove the cotter pin from the lower balljoint, then loosen the nut to the end of the

stud (but no farther) **(see illustration)**.

14   Hit the inside of the steering knuckle near each balljoint with a hammer. This will break the axle arm loose from the balljoint studs. Remove the steering knuckle from the axle arm.

15   Remove the camber adjuster sleeve **(see illustration 5.12)**. Mark the position of the slot in the camber adjuster. It should be installed in its original position to maintain the correct alignment. **Note:** *If the camber adjuster is difficult to remove, use a Pitman arm puller (available at most auto parts stores).*

16   Remove the nut from the lower balljoint.

17   Place the steering knuckle in a vise. Remove the snap ring (if equipped) from the lower balljoint socket. **Note:** *Always remove the lower balljoint first.*

18   Install a special C-frame balljoint assembly tool and receiver cup (available at auto parts stores) on the lower balljoint **(see illustration)**. Tighten the special tool and press

the lower balljoint out of the steering knuckle. **Note:** *If the C-frame assembly tool and receiver tool are not available or won't remove the balljoint, take the knuckle assembly to a dealer service department or automotive machine shop and have the balljoints pressed out.*

19   Use the same tool setup used in Step 18

6.2 With the vehicle supported on jackstands, use a floor jack to raise and lower the front axle

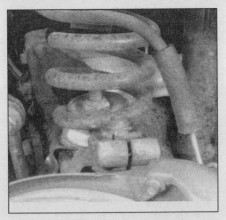

6.5a Remove the lower spring retainer nut

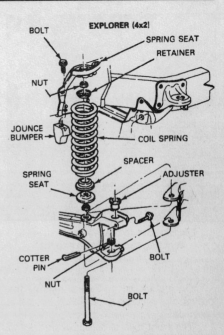

6.5b Front coil spring installation details (2WD models)

on the upper balljoint and press the balljoint out (see illustration).
20 Clean the steering knuckle balljoint bores.
21 Insert the lower balljoint into the knuckle as straight as possible. Note: *Always install the lower balljoint first since the special tool must pass through the upper balljoint receptacle in the axle.*
22 Position the C-frame assembly, balljoint receiver cup and installation cup on the lower balljoint (see illustration).
23 Turn the screw in the C-frame clockwise and press the balljoint into the steering knuckle until it is completely seated. Caution: *Don't heat the axle or balljoint to ease installation since the temper may be removed from the component(s) leading to premature failure.* If the balljoint won't go in all the way, realign the C-frame and receiver cup.
24 Install the lower balljoint snap-ring (if equipped).
25 Position the C-frame assembly, balljoint receiver cup and installation cup on the upper balljoint (see illustration 5.22).
26 Turn the screw in the C-frame clockwise and press the balljoint into the axle until it is completely seated. Caution: *Don't heat the axle or balljoint to aid in installation since the temper may be removed from the component(s) leading to premature failure.*
27 Install the camber adjuster into the axle arm. Place the slot in the original position noted during removal. Install the camber adjuster with the arrow pointing toward the outside for positive camber or with the arrow pointing toward the inside of the vehicle for negative camber. Zero camber bushings do not have an arrow and may be rotated in either direction as long as the lugs on the yoke engage the slots in the bushing. Caution: *The following tightening sequence must be followed exactly when securing the steering knuckle. Excessive spindle turning effort may result in reduced steering returnability if this procedure is not followed.*
28 Install the steering knuckle in the axle arm. Do not disrupt the camber adjuster during installation.

29 Install a new nut on the lower balljoint stud and tighten it to the torque listed in this Chapter's Specifications, then advance the nut until a cotter pin slot lines up. Install a new cotter pin and bend the ends over completely.
30 Install the snap-ring on the upper balljoint stud. Install the pinch bolt and tighten it to the torque listed in this Chapter's Specifications. Note: *The camber adjuster will position itself in the knuckle during adjustment. DO NOT try to change its position.*
31 Attach the tie-rod end to the knuckle. Tighten the nut to the torque listed in this Chapter's Specifications, install a new cotter pin and bend the ends over completely.
32 The remainder of installation is the reverse of the removal steps.
33 Have the front-end alignment checked by a dealer service department or an alignment shop.

1995 and later models

34 The balljoints on upper and lower control arms are not removable or serviceable. If a balljoint is damaged or worn, replace the control arm.

6 Coil spring (1991 through 1994 models) - removal and installation

Note: *1995 and later models are equipped with torsion bars instead of coil springs at the front, while the rear suspension uses leaf springs on all models.*

Removal

Refer to illustrations 6.2, 6.5a, 6.5b and 6.5c
1 Loosen the front wheel lug nuts on the side to be dismantled. Raise the front of the vehicle, support it securely on jackstands, block the rear wheels and set the parking brake. Remove the front wheel.
2 Place a floor jack under the axle (see illustration).
3 Remove the bolt and nut securing the shock absorber to the radius arm (see Section 2).

4 Remove the brake caliper assembly and suspend it with a length of wire to relieve any strain on the brake hose (see Chapter 9).
5 At the lower end of the spring, remove the retaining nut and the retainer securing the spring to the front axle (see illustration). Note: *The nut is attached to a stud on 4WD models and on 2WD models it is attached to a bolt that runs through the axle (see illustrations).*

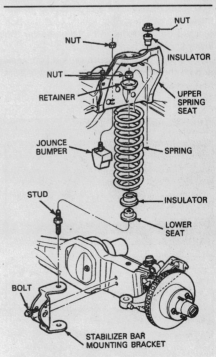

6.5c Front coil spring installation details (4WD models)

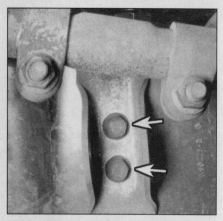

7.2 Remove the two bolts (arrows) securing the radius arm to the front bracket

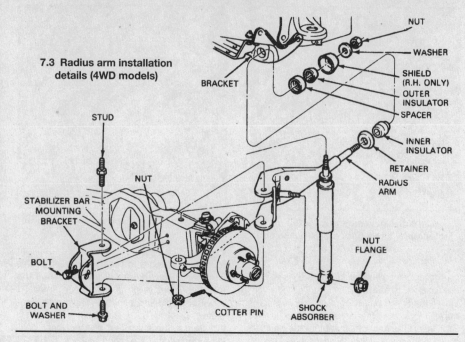

7.3 Radius arm installation details (4WD models)

6   On models so equipped, remove the through bolt securing the stabilizer bar to the front axle.
7   The front axle should now be free to allow spring removal.
8   If necessary, use a pry bar and lift the spring up and over the bolt or stud that passes through the lower spring seat. Rotate the spring so the upper built-in spring seat retainer is cleared, then remove the spring.

Installation

9   Install the spring lower seat and insulator onto the front axle.
10   Push the front axle down to allow installation of the spring. Install the upper end of the spring into the upper seat and rotate the spring into place.
11   If necessary, use a pry bar and lift the lower end of the spring up and over the bolt or stud on the axle and into place on the lower spring seat and insulator.
12   Apply slight pressure on the floor jack and lift the front axle until the spring is correctly seated.
13   Install the lower spring retainer with a new nut. Tighten the nut to the torque listed in this Chapter's Specifications.

7.4 Remove the nut, washer and insulator and remove the rear radius arm support

14   Install the brake caliper assembly (see Chapter 9).
15   Install the bolt and new nut securing the shock absorber to the radius arm. Tighten the nut to the torque listed in this Chapter's Specifications.
16   Install the wheel and lug nuts, lower the vehicle and tighten the lug nuts to the torque listed in Chapter 1.
17   Have the front-end alignment checked by a dealer service department or an alignment shop.

7   Radius arm (1991 through 1994 models) - removal and installation

Removal

Refer to illustrations 7.2, 7.3 and 7.4
1   Remove the spring assembly (see Section 6).
2   On 2WD models, perform the following:
  a) *Remove the spring lower seat from the radius arm.*
  b) *Remove the bolt and nut securing the radius arm to the front axle and front bracket* **(see illustration).**
3   On 4WD models, perform the following:
  a) *Remove the spring lower seat and stud (see Section 6).*
  b) *Remove the bolts securing the radius arm to the front axle bracket* **(see illustration).**
4   From the rear side of the radius arm bracket, remove the nut, rear washer, shield (passenger side only) and insulator **(see illustration).**
5   Remove the radius arm and remove the inner insulator and retainer from the radius arm threaded end **(see illustration 7.3).**

Installation

6   Install the front-end of the radius arm onto the front axle.
7   On 2WD models, from underneath the axle, install the attaching bolt and a new nut. Tighten the nut only finger tight at this time.
8   On 4WD models, position the front-end of the radius arm onto the bracket and axle. Install the bolts and stud (and washer on the left axle only) in the bracket. Tighten the bolts and stud only finger tight at this time.
9   Install the rear retainer and insulator onto the threaded end of the radius arm.
10   Install the radius arm into the rear bracket and install the rear insulator, washer and new nut. Tighten the nut to the torque listed in this Chapter's Specifications.
11   Tighten the bolts and nuts installed in Step 7 and Step 8 to the torque listed in this Chapter's Specifications.
12   On 2WD models, install the spring lower seat onto the radius arm.
13   On 4WD models, install the spring lower seat and insulator onto the left radius arm.
14   Install the spring assembly (see Section 6).
15   Have the front-end alignment checked by a dealer service department or an alignment shop.

8   Radius arm insulators (1991 through 1994 models) - replacement

1   Remove the front spring (see Section 6).
2   Loosen the axle arm pivot bolt.
3   Remove the shock absorber upper nut and compress the shock.
4   From the rear side of the radius arm bracket, remove the nut, rear washer, insulator, shield (right side only) and spacer **(see**

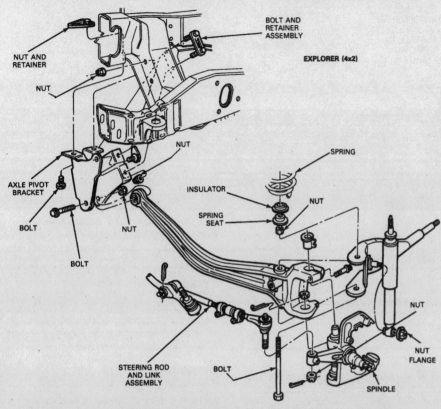

BOLT AND
RETAINER
ASSEMBLY

EXPLORER (4x2)

NUT AND
RETAINER

NUT

NUT

SPRING

AXLE PIVOT
BRACKET

INSULATOR

BOLT

SPRING
SEAT

NUT

NUT

BOLT

BOLT

NUT

STEERING ROD
AND LINK
ASSEMBLY

BOLT

NUT
FLANGE

SPINDLE

9.5 Front axle arm installation details - 2WD models (driver's side shown)

illustration or 7.3).

5   Raise the front axle arm with a floor jack until the radius arm is level.

6   Push the radius arm forward until it is free of the radius arm bracket. **Note:** *If necessary on 4WD models, detach the driveshaft from the front axle flange.*

7   Remove the front insulator from the radius arm threaded end.

8   Installation is the reverse of the removal steps with the following additions:

a) *Install new insulators and nut.*

b) *Tighten the bolts and nut to the torque listed in this Chapter's Specifications.*

9   Front axle arm (1991 through 1994 2WD models) - removal and installation

Note: *Front axle arm removal and installation for 4WD models is covered in Chapter 8.*

Removal

Refer to illustration 9.5

1   Raise the front of the vehicle, support it securely on jackstands, block the rear wheels and set the parking brake. Position the front wheels in the straight ahead position.

2   Remove the coil spring (see Section 6) and front spindle (see Section 3).

3   Remove the radius arm (see Section 7).

4   Remove the front stabilizer bar (see Section 15).

5   Remove the bolt and nut securing the axle arm to the frame pivot bracket **(see illustration)**. Remove the axle arm.

Installation

6   Install the axle arm to the frame pivot bracket with a new retaining bolt and nut. Tighten the nut finger-tight.

7   Install the front stabilizer bar (see Section 15).

8   Install the front axle arm onto the radius arm (see Section 7).

9   Install the front spindle (see Section 3) and the coil spring (see Section 6).

10   Install the axle arm retaining bolt and nut and tighten them slightly. Don't torque them to specifications yet.

11   Install the wheel and lug nuts. Lower the vehicle and tighten the lug nuts to torque listed in Chapter 1.

12   Tighten the axle arm retaining bolt and nut to the torque listed in this Chapter's Specifications.

13   Have the front-end alignment checked by a dealer service department or an alignment shop.

10   Axle pivot bushing (1991 through 1994 models) - removal and installation

2WD models

1   Remove the front spring (see Section 6).

2   To remove the left axle pivot bushing, remove the retaining bolt and nut, then pull the pivot end of the axle down until the bushing is exposed.

3   To remove the right axle pivot bushing, remove the axle arm from the frame (see Section 9).

4   Specialized tools are required to replace the bushings. Have the old bushings pressed out and new ones pressed in by a dealer or suspension shop.

4WD models

5   Remove the front axle assembly (see Chapter 8).

6   Specialized tools are required to replace the bushings. Have the old bushings pressed out and new ones pressed in by a dealer or suspension shop.

7   The remainder of installation is the reverse of the removal Steps.

11   Axle pivot bracket (1991 through 1994 models) - removal and installation

2WD models

Removal

1   Loosen the front wheel lug nuts on the side to be dismantled. Raise the front of the vehicle, support it securely on jackstands, block the rear wheels and set the parking brake. Remove the front wheel.

2   Remove the front spring (see Section 6), the radius arm (see Section 7), the wheel spindle (see Section 3) and the front axle arm (see Section 9).

3   Remove the bolts and nuts securing the axle pivot bracket to the frame and remove the bracket from the frame crossmember **(see illustration 9.5)**.

Installation

4   Position the front axle bracket onto the frame crossmember. Install the new bolts from within the frame crossmember and out through the bracket, then install the new nuts. After all of the bolts and nuts have been installed, tighten them to the torque listed in this Chapter's Specifications.

5   The remainder of installation is the reverse of the removal Steps. Have the front-end alignment checked by a dealer service department or an alignment shop.

4WD models

Removal

Refer to illustrations 11.9 and 11.10

6   Loosen the wheel lug nuts on the side to be dismantled, raise the front of the vehicle and support it securely on jackstands. Block the rear wheels and apply the parking brake. Remove the wheel.

7   Remove the front spring (see Section 6).

8   Place a jack beneath the axle arm near the pivot bracket to support it.

10

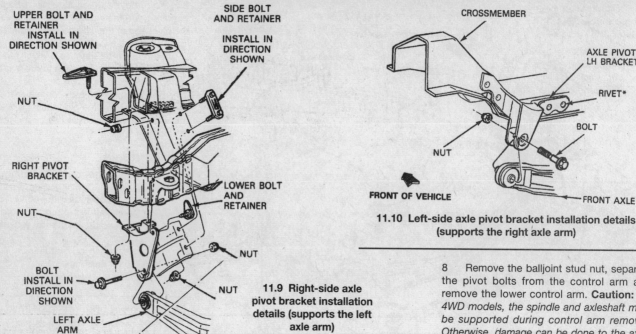

11.10 Left-side axle pivot bracket installation details
(supports the right axle arm)

11.9 Right-side axle pivot bracket installation details (supports the left axle arm)

9   On the right-side axle pivot bracket (which supports the left axle arm), remove the nuts and the upper and side bolts and retainers. Discard the lower bolt and retainer. Remove the axle pivot bracket from the frame crossmember **(see illustration)**.

10   On the left-side axle pivot bracket (which supports the right axle arm), use a 9/16-inch drill bit and drill out the rivets securing the bracket to the frame. Remove the axle bracket from the frame crossmember **(see illustration)**.

Installation

11   Install the left-side axle pivot bracket onto the frame crossmember and align the 9/16 inch holes of the bracket and frame. Install 9/16-12x1-1/2-inch Grade 8 bolts **(see illustration 11.10)**. **Caution:** *Be sure to use Grade 8 bolts, not a lesser grade (see the fastener grading chart in the front of this book). Install the washer and nuts onto the bolts and tighten to the torque listed in this Chapter's specifications.*

12   To install the right-side axle pivot bracket, perform the following:

a) *Install the right-side axle pivot bracket onto the frame crossmember (see illustration 11.9). Install all bolts in the direction shown in the illustration. The side bolt heads must face the engine oil pan to give maximum clearance. Install new nuts and tighten to the torque listed in this Chapter's specifications.*

b) *Use a 9/16-inch drill bit and drill out the lower mounting hole bracket and frame crossmember.*

c) *Install a 9/16-inch Grade 8 replacement bolt with two flat washers and a new retaining nut. Tighten the nut to the torque listed in this Chapter's Specifications.*

13   The remainder of installation is the reverse of the removal Steps. Have the front-end alignment checked by a dealer service department or an alignment shop.

12   Lower control arm (1995 and later models) - removal, inspection and installation

Removal

Refer to illustration 12.6

1   Loosen the wheel lug nuts. Raise the front of the vehicle and support it securely on jackstands. Remove the front wheel. **Caution:** *If the vehicle is equipped with Automatic Ride Control (ARC), make sure the air suspension switch is turned to the OFF position before the vehicle is raised to prevent damage to the system components (see Section 20).*

2   Remove the shock absorber (see Section 2).

3   Disconnect the stabilizer bar (see Section 15).

4   Position the floor jack directly under the lower control arm to give support during removal.

5   Remove the torsion bar (see Section 14).

6   Remove and discard the cotter pin from the balljoint stud **(see illustration)**. Loosen the castle nut on the stud one or two turns. Rap the spindle sharply in the immediate vicinity of the stud to relieve stud pressure and loosen the stud in the knuckle. If the stud won't come loose, you might have to resort to a "picklefork" type of balljoint stud separator. **Caution:** *The use of a picklefork balljoint separator will usually result in balljoint boot damage. It may be necessary to purchase a special tool to remove the balljoint.*

7   Loosen the two control arm pivot bolts.

8   Remove the balljoint stud nut, separate the pivot bolts from the control arm and remove the lower control arm. **Caution:** *On 4WD models, the spindle and axleshaft must be supported during control arm removal. Otherwise, damage can be done to the axleshaft joints by too much angular movement.*

Inspection

9   Inspect the bushings for cracks and tears. If they're damaged or worn, take the control arm to an automotive machine shop to have new bushings installed. This procedure requires a number of specialized tools, so it's not worth tackling at home. If the balljoint is worn out the entire control arm must be replaced.

Installation

10   Install the lower control arm pivot bolts and nuts. Don't fully tighten the nuts at this time.

11   Install the torsion bar (see Section 15).

12   Place a floor jack under the lower control arm, raise the lower control arm and guide it into position.

13   Insert the balljoint stud into the spindle, install the castle nut and tighten it to the torque listed in this Chapter's Specifications. Continue to tighten the nut, if necessary, until the hole in the stud is in line with the slot in the nut. Install a new cotter pin, bending the ends of the cotter pin away from the wheel.

14   Install the shock absorber (see Section 2).

15   Attach the stabilizer bar (see Section 15).

16   Install the wheel, remove the jackstands, lower the vehicle and tighten the wheel lug nuts to the torque listed in this Chapter's Specifications.

17   Tighten the lower control arm pivot bolt nuts to the torque listed in this Chapter's Specifications.

18   Adjust the vehicle ride height by turning the torsion bar bolt (see Section 14).

19   Have the front-end alignment checked by a dealer service department or an alignment shop.

12.6 Front suspension components - 1995 and later 2WD models

1   Stabilizer bar stud and bushing
2   Nut, washer and insulator assembly
3   Bolt
4   Frame
5   Torsion bar (left side)
6   Upper control arm
7   Nut
8   Nut and washer
9   Lower control arm
10 Not used
11 Tie-rod end
12 Nut
13 Nut
14 Stabilizer bar link
15 Stabilizer bar
16 Stabilizer bar bushing
17 Stabilizer bar bracket
18 Bolt

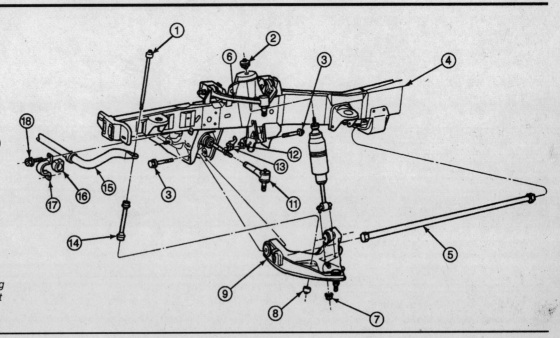

13 Upper control arm (1995 and later models) - removal, inspection and installation

Note: The upper control arm and balljoint cannot be separated and must be replaced as a single unit.

Removal

Refer to illustration 13.4

1   Loosen the wheel lug nuts, raise the vehicle, support it securely on jackstands and remove the wheel. **Caution:** If the vehicle is equipped with Automatic Ride Control (ARC), make sure the air suspension switch is turned to the OFF position before the vehicle is raised to prevent damage to the system components (see Section 20).

2   Place a floor jack under the lower control arm and raise it slightly. The jack must remain in this position during the entire operation.

3   Remove the pinch bolt that secures the upper balljoint stud to the spindle/steering knuckle.

4   Remove the upper control arm pivot bolts **(see illustration)** and detach the arm from the frame.

Inspection

5   Inspect the control arm bushings for cracks and tears. If they're damaged or worn, take the control arm to an automotive machine shop to have new bushings installed. This procedure requires a number of specialized tools, so it's not worth tackling at home. If the balljoint is worn out, the control arm must be replaced.

Installation

6   Position the control arm on the frame and install the two pivot bolts, cams and nuts. Tighten the bolts to the torque listed in this Chapter's Specifications.

7   Insert the upper balljoint stud into the spindle and install the pinch bolt and nut. Tighten the nut to the torque listed in this Chapter's Specifications.

8   Install the wheel and tighten the lug nuts to the torque listed in this Chapter's Specifications.

9   Have the front-end alignment checked by a dealer service department or an alignment shop.

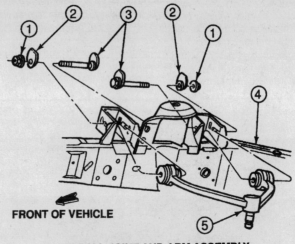

FRONT OF VEHICLE

LH BALL JOINT AND ARM ASSEMBLY

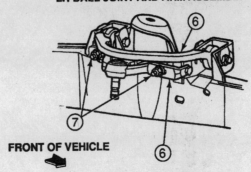

FRONT OF VEHICLE

RH BALL JOINT AND ARM ASSEMBLY (TWO-PIECE DESIGN)

13.4 Upper control arm installation details - 1995 and later models

1   Nut
2   Front suspension upper adjuster cam
3   Front suspension upper adjuster cam bolt
4   Frame
5   Upper control arm ball joint
6   Upper control arm
7   Nut

10

14.4 Raise the adjuster assembly using a heavy-duty, two-jaw puller - the puller must have a bridge that keeps the jaws from separating

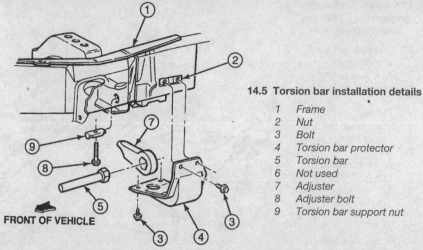

FRONT OF VEHICLE

14.5 Torsion bar installation details

1   *Frame*
2   *Nut*
3   *Bolt*
4   *Torsion bar protector*
5   *Torsion bar*
6   *Not used*
7   *Adjuster*
8   *Adjuster bolt*
9   *Torsion bar support nut*

14   Torsion bar (1995 and later models) - removal and installation

Note: *Before removing or adjusting the torsion bar(s), measure the existing ride height (the difference in measurement to the ground from the center of the lower arm's inner mounting bolt and the lowest part of the spindle) of the vehicle.*

Removal

Refer to illustrations 14.4 and 14.5

1   Position the front wheels in the straight ahead position. Raise the vehicle and support it securely on jackstands. **Caution:** *If the vehicle is equipped with Automatic Ride Con-trol (ARC), make sure the air suspension switch is turned to the OFF position before the vehicle is raised to prevent damage to the system components (see Section 20).*
2   Disconnect the torsion bar protector from the frame.
3   Unscrew the bolt from the torsion bar nut while counting the number of turns to remove the bolt. Record this number on a piece of paper for installation purposes.
4   Use a heavy-duty two-jaw puller to raise the adjuster **(see illustration)**. **Warning:** *The puller must have an adjustable bridge that keeps the jaws from spreading apart under pressure. Once the puller is in place, tighten the nuts on the bridge.*
5   Remove the torsion bar nut **(see illustration)**. Lower the adjuster.
6   Slide the torsion bar into the lower con-

trol arm and remove the adjuster.
7   Remove the torsion bar from the lower control arm.

Installation

8   Installation is the reverse of removal, with the following points:

a) *Be sure to lubricate the threads of the adjuster nut and use a new adjuster bolt (factory-coated with a special adhesive).*

b) *When installing the adjuster bolt, screw it in the same number of turns it took to remove the old one, plus two more turns. If necessary, turn the torsion bar adjuster bolt in or out to adjust the ride height so the vehicle is level from side-to-side, with the vehicle sitting on its suspension, not a jack or jackstands. If the vehicle is equipped with Automatic Ride Control (ARC), disconnect the air lines from the shock absorbers before attempting adjustment. Also, roll the vehicle back and forth and jounce the suspension before adjusting and in between adjustments. Have the front-end alignment checked by a dealer service department or an alignment shop.*

15   Stabilizer bar, front - removal and installation

Removal

Refer to illustration 15.2a and 15.2b

1   Position the front wheels in the straight ahead position. Raise the vehicle and support it securely on jackstands. **Caution:** *If the vehicle is equipped with Automatic Ride Con-trol (ARC), make sure the air suspension switch is turned to the OFF position before the vehicle is raised to prevent damage to the system components (see Section 20).*
2   Detach the stabilizer bar from the axle arms (1994 and earlier models) or from the link attached to the lower control arm (1995 and later models), then from the frame **(see illustrations)**.
3   Remove the stabilizer bar assembly.

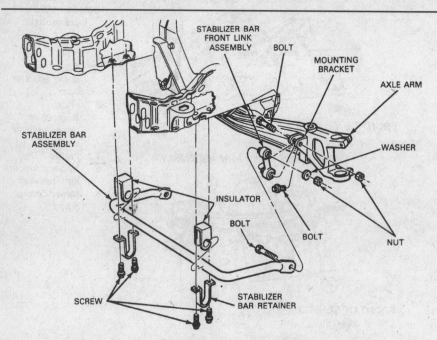

STABILIZER BAR FRONT LINK ASSEMBLY
BOLT
MOUNTING BRACKET
AXLE ARM
WASHER
STABILIZER BAR ASSEMBLY
INSULATOR
BOLT
BOLT
NUT
SCREW
STABILIZER BAR RETAINER

15.2a Stabilizer bar installation details - 1991 through 1994 models (2WD model shown, 4WD similar)

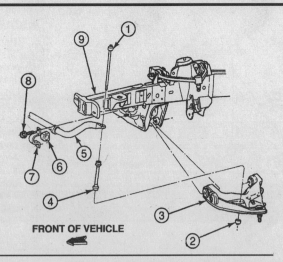

15.2b Stabilizer bar installation details on (1995 and later models)

1   Front stabilizer bar bolt
2   Nut
3   Lower control arm
4   Front stabilizer bar link
5   Front stabilizer bar
6   Front stabilizer bar bushing
7   Front stabilizer bar bracket
8   Bolt
9   Frame

FRONT OF VEHICLE

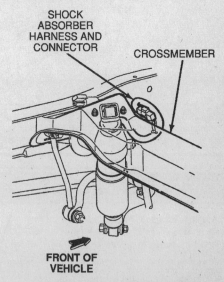

16.3 Details of the rear shock absorber - models equipped with Automatic Ride Control

FRONT OF VEHICLE

Installation

4   Installation is the reverse of the removal steps. Tighten the bolts and nuts to the torque listed in this Chapter's Specifications. **Note:** *On 1997 and 1998 models, the stabilizer-to-frame bracket bolts are self-tapping type. If they will not hold in the frame anymore, contact a dealer for a kit that contains special fasteners to fix the stripped holes.*

16 Shock absorbers, rear - inspection, removal and installation

Inspection

1   Refer to Section 3 for shock absorber inspection procedures.

Removal

Refer to illustration 16.3

2   Place a floor jack under the axle adjacent to shock absorber being removed. Raise the vehicle just enough to take the load off the shock absorber. **Caution:** *If the vehicle is equipped with Automatic Ride Control (ARC), make sure the air suspension switch is turned to the OFF position before the vehicle is raised to prevent damage to the system components (see Section 20).*

3   Disconnect the ARC electrical connector from the shock absorber harness **(see illustration).**

4   If equipped with ARC, disconnect the air line from the shock. Push in on the plastic ring on the shock absorber and while holding the ring in, pull out the air line.

5   Remove the nut and bolt securing the lower end of the shock absorber to the spring plate **(see illustration 17.4).** Remove the nut securing the top of the shock absorber to the upper mounting bracket on the frame.

Installation

7   Installation is the reverse of the removal steps. Tighten the nuts and bolts to the torque listed in this Chapter's Specifications.

17 Rear leaf spring - removal and installation

Removal

Refer to illustrations 17.4, 17.5 and 17.6

1   Raise the rear of the vehicle until the weight is off the rear springs but the tires are still touching the ground. **Caution:** *If the vehicle is equipped with Automatic Ride Control (ARC), make sure the air suspension switch is turned to the OFF position before the vehicle is raised to prevent damage to the system components.*

2   Support the vehicle securely on jackstands and support the rear axle with a jack. DO NOT get under a vehicle that's supported only by a jack, even if the tires are still installed.

3   Remove the nut and bolt securing the lower end of the shock absorber to the spring plate. Compress the shock up and out of the way.

4   Remove the nuts from the U-bolts and remove the U-bolts and the spring plate from the spring **(see illustration).**

5   Remove the bolts and nuts securing the shackle assembly at the rear of the spring **(see illustration).** Let the spring pivot down and rest on the floor.

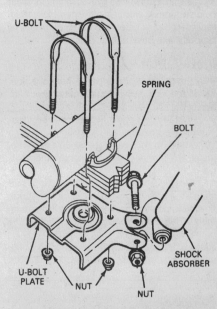

17.4 Rear leaf spring U-bolt and spring plate installation details

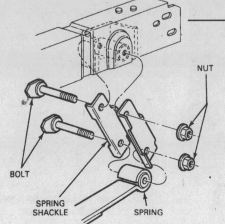

17.5 Rear shackle installation details

10

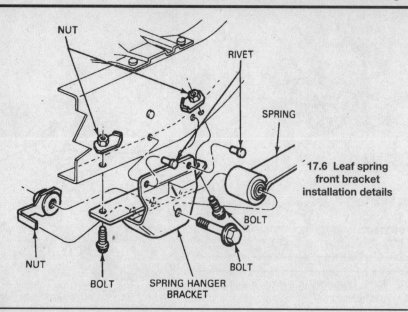

17.6 Leaf spring front bracket installation details

6   Remove the spring hanger bolt and nut at the front of the spring **(see illustration)**. Remove the spring.

7   Inspect the spring eye bushings for wear or distortion. If worn or damaged have them replaced by a dealer service department or properly-equipped shop.

Installation

8   Place the spring in the rear shackle. Install the bolt and nut and tighten them finger-tight.

9   Install the spring in the front bracket. Tighten the bolt and nut finger-tight.

10   Position the rear shackle on the frame. Install the bolt and nut and tighten them finger-tight.

11   Position the axle on the spring and install the spring seat. Make sure the spring tie-bolt is positioned in the hole in the spring, then install the U-bolts and nuts. Tighten the nuts finger-tight.

12   Lower the vehicle to the ground so its weight is resting on the tires. Tighten the spring bracket bolt and nut, shackle bolts and nuts and U-bolt nuts to the torque values listed in this Chapter's Specifications.

18   Stabilizer bar, rear - removal and installation

Removal

Refer to illustrations 18.2a and 18.2b

1   Raise the vehicle and support it securely on jackstands. **Caution:** *If the vehicle is equipped with Automatic Ride Control (ARC), make sure the air suspension switch is turned to the OFF position before the vehicle is raised to prevent damage to the system components (see Section 20).*

2   Remove the nut and washer securing the rear stabilizer bar ends to the link at each end **(see illustration)**. Unbolt the retainers from the rear axle housing **(see illustration)**.

3   Remove the mounting brackets, retainer and the stabilizer bar from the vehicle.

4   Inspect the rubber isolators on the stabilizer bar and replace if necessary.

Installation

5   Position the stabilizer bar onto the rear axle assembly. Position the retainer with the UP mark facing up.

6   Install the stabilizer bar and retainer onto the mounting brackets. Make sure the UP mark on the retainer is facing up toward the floor pan.

7   Install the U-bolts and nuts. Tighten the nuts finger-tight.

8   Move the stabilizer bar ends up into position and connect them to the links. Install the bolt, washer and nut on each end. Tighten the bolts and nuts to the torque listed in this Chapter's Specifications.

9   Tighten the retainer bolts and link nuts to the torque listed in this Chapter's Specifications.

19   Rear anti-windup bar (5.0L V8 models only) - removal and installation

Removal

Refer to illustrations 19.3 and 19.4

1   Raise the vehicle and support it securely on jackstands. **Caution:** *If the vehicle is equipped with Automatic Ride Control (ARC), make sure the air suspension switch is turned to the OFF position before the vehicle is raised to prevent damage to the system components (see Section 20).*

2   Remove the rear wheels.

3   Unbolt the rear end of the bar from the axle bracket **(see illustration)**.

4   Unbolt the front-end of the bar from the frame bracket **(see illustration)**.

6   Remove the track bar.

Installation

7   Installation is the reverse of removal. Be sure to tighten the track bar fasteners to the torque listed in this Chapter's Specifications. Activate the air suspension system, if equipped.

18.2a Remove the link nut . . .

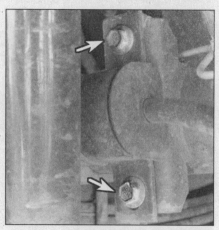

18.2b . . . and retainer bolts, then remove the stabilizer bar

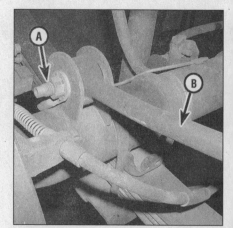

19.3 Remove the bolt (A) and nut holding the anti-windup bar (B) to the rear axle - 5.0L models

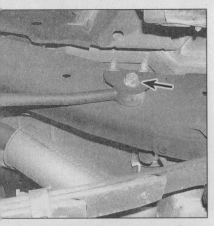

19.4 Anti-windup bar front mounting bolt (arrow)

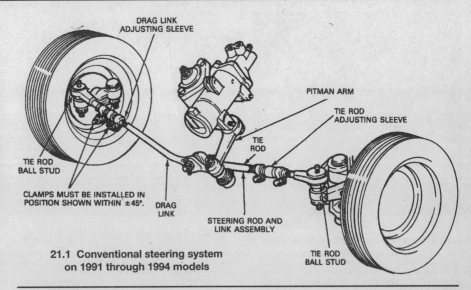

21.1 Conventional steering system on 1991 through 1994 models

20 Automatic Ride Control system - general information

The ARC system is a computer-controlled suspension that uses unique components to provide a smooth ride for normal driving conditions without sacrificing handling performance. An air spring integral with each shock absorber, provides automatic load-leveling and allows the vehicle height to be adjusted over a span of 2 inches.

The ARC system adjusts the vehicle height on the front and rear axles separately through the use of solenoid valves, an air compressor and air lines. Four system-specific shock absorbers are used to move the height setting of the chassis using hydraulic pistons and valves. The shocks not only control vehicle height, but the computer can change the shock valving, firm or soft, to suit driver and sensor input.

Ride quality is superior to non-ARC models, because ARC-specific torsion bars and other components are used on these models that are softer than conventional parts. Under normal circumstances, ride quality is softer, until the computer determines that firmer suspension action is needed. Whenever the transfer case is shifted into 4WH, the ARC system switches to firmer shock valving and raises the vehicle one inch.

It also uses two height sensors, a steering sensor and several other sensors on the transfer case and elsewhere on the vehicle to monitor driver and road inputs. Whenever the Door-Ajar light comes on in the vehicle, the ARC computer stores a vehicle height reading. Then as passengers or cargo are loaded, when the door closes the computer signals the ARC shocks to maintain the stored ride height.

Disabling the system

The air suspension switch is located behind the jack stowage door in the luggage compartment area. To disable the system, turn off the switch. This cuts the power to the air suspension control module, which deactivates the system. The system should be turned off anytime the vehicle is going to be raised off the ground so that it won't react to the higher vehicle height.

21 Steering system - general information

Refer to illustrations 21.1 and 21.2

The steering system on 1991 through 1994 models consists of a Pitman arm, drag link, steering connecting rod and tie-rods (see illustration). The Pitman arm transfers the steering gear movements through the drag link and steering connecting rod to the tie-rods at each end. The tie-rods move the spindles (or knuckle) and front wheels to the desired steering movement. The tie-rods are equipped with an adjusting sleeve for setting the toe-in. All models are equipped with the Ford Integral Power Steering gear that consists of a belt-driven Ford C-II pump and associated lines and hoses. The power steering pump reservoir fluid level should be checked periodically (see Chapter 1). The steering wheel operates the steering shaft, which actuates the steering gear through universal joints and the intermediate shaft. Looseness in the steering can be caused by wear in the steering shaft universal joints, the steering gear, the tie-rod ends and loose retaining bolts.

The steering system on 1995 and later models consists of integral power steering rack-and-pinion gear that incorporates a constant-diameter rack (see illustration). The steering gear is mounted within the crossmember. The steering gear is a hydraulic/mechanical unit which uses an integral piston and rack design to provide power assisted steering.

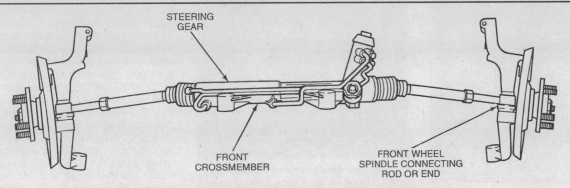

21.2 Rack-and-pinion steering system on 1995 and later models

10

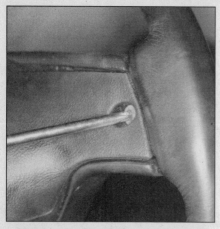

22.2 Remove the horn pad screws and remove the horn pad

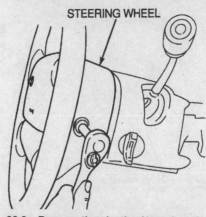

22.3a Remove the plastic plugs, then the screw from each side of the steering wheel

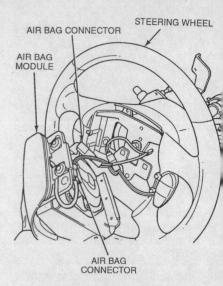

22.3b Disconnect the airbag electrical connector

22.5 Lift the damper off the steering wheel (if equipped)

22.6 There should be factory alignment marks on the steering wheel and shaft; if not, make your own

22.7 Thread the steering wheel bolt partway in so the puller will have something to push against, then remove the steering wheel with a puller like the one shown here - DO NOT strike the wheel to remove it

22   Steering wheel - removal and installation

Warning: *Some models have airbags. Always disconnect the negative battery cable, then the positive battery cable and wait two min-* utes before working in the vicinity of the impact sensors, steering column or instrument panel to avoid the possibility of accidental deployment of the airbag, which could cause personal injury (see Chapter 12).

22.8 Make sure the horn spring is in position before installing the wheel

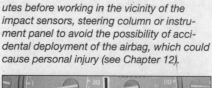

22.9 On airbag-equipped models, the mark (arrow) on the sliding contact should be aligned at the top of the steering column before installing the steering wheel

Removal

Refer to illustrations 22.2, 22.3a, 22.3b, 22.5, 22.6 and 22.7

1   Disconnect the negative cable from the battery (and from the positive terminal if equipped with an airbag).

2   On non-airbag systems, remove the screws securing the horn pad to the steering wheel **(see illustration)**, then partially pull the horn pad away from the steering wheel, unplug the electrical connector then remove the horn pad.

3   On models with airbags, pry out the two plugs covering the driver's side airbag module screws on the sides of the steering wheel, remove the two screws, lift the airbag off and detach the electrical connector **(see illustrations)**. Store the airbag in a safe place until needed. **Warning:** *Whenever handling an airbag module, always keep the trim-side pointed away from your body. Never place*

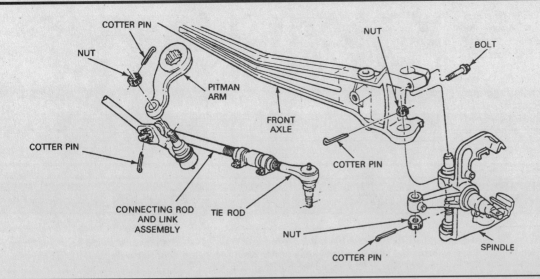

23.2a Steering linkage components (1994 and earlier 2WD models)

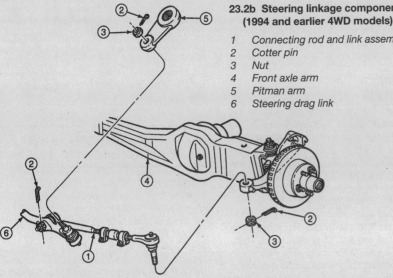

23.2b Steering linkage components (1994 and earlier 4WD models)

1   Connecting rod and link assembly
2   Cotter pin
3   Nut
4   Front axle arm
5   Pitman arm
6   Steering drag link

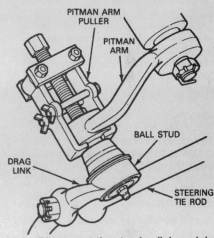

23.2c Disconnect the steering linkage joints with a Pitman arm puller (available at most auto parts stores)

the airbag module on a bench of other surface with the trim-side down. Always place the airbag module in a safe location with the trim-side facing up.

4   Remove the steering wheel bolt.

5   Lift off the damper (if equipped) **(see illustration).**

6   Look for factory alignment marks on the steering wheel and column **(see illustration).** If they aren't visible, make your own marks. These marks will be used during installation to align the wheel and column.

7   Use a puller to remove the steering wheel **(see illustration). Caution:** *Don't hammer on the shaft to remove the steering wheel. Discard the original steering wheel bolt.*

Installation

Refer to illustrations 22.8 and 22.9

8   Make sure the horn spring is in position **(see illustration).** Align the mark on the steering wheel hub with the mark on the steering shaft and slip the wheel onto the shaft. Install the new mounting bolt and

tighten it to the torque listed in this Chapter's Specifications.

9   The remainder of installation is the reverse of the removal Steps. **Caution:** *On airbag systems, make sure the airbag wiring does not get pinched between the airbag sliding contact and the steering wheel. Also, the airbag sliding contact must be properly aligned before installing the steering wheel* **(see illustration).**

23   Drag link (1991 through 1994 models)- removal and installation

Removal

Refer to illustrations 23.2a, 23.2b and 23.2c

1   Raise the front of the vehicle, support it securely on jackstands, block the rear wheels and set the parking brake. Position the front wheels in the straight ahead position.

2   Remove the cotter pins and nuts securing the drag link to the connecting rod and to the Pitman arm **(see illustrations).** Use a Pit-

man arm puller or balljoint separator to separate the drag link from the connecting rod and the Pitman arm **(see illustration).**

3   Loosen the clamp bolts on the right tie-rod adjusting sleeve.

4   Apply a paint mark on the drag link threads adjacent to the right tie-rod adjusting sleeve.

5   Unscrew the drag link from the adjusting sleeve and remove the drag link.

Installation

6   Install the drag link into the right tie-rod adjusting sleeve and thread it in the sleeve until the marks you made in Step 4 align.

7   Install the drag link ball stud into the Pitman arm. Position the connecting rod arm ball end into the drag link. Make sure the front wheels and steering wheel are in the straight ahead position. Make sure the ball ends are seated in the taper to prevent them from rotating while tightening the nuts.

8   Install new nuts on the studs of the Pitman arm and connecting rod and tighten them to the torque listed in this Chapter's

10

Specifications. Install new cotter pins and bend the ends over completely. Tighten the right tie-rod adjusting sleeve clamp bolts to the torque listed in this Chapter's Specifications.

9   Install the wheel and lug nuts. Lower the vehicle and tighten the lug nuts to torque listed in the Chapter 1 Specifications.

10   Have the front-end alignment checked by a dealer service department or an alignment shop.

24 Steering connecting rod (1991 through 1994 models)- removal and installation

Removal

1   Raise the front of the vehicle, support it securely on jackstands, block the rear wheels and set the parking brake. Position the front wheels in the straight ahead position.

2   Remove the cotter pin and nut from the ball end of the steering connecting rod (see illustration 23.2a or 23.2b).

3   Use a Pitman arm puller and remove the ball end from the drag link (see illustration 23.2c).

4   Loosen the clamp bolts on the tie-rod adjusting sleeve.

5   Apply a paint mark on the connecting rod threads adjacent to the adjusting sleeve. Unscrew the connecting rod from the tie-rod adjusting sleeve.

6   Remove the steering connecting rod.

Installation

7   Install the steering connecting rod onto the tie-rod adjusting sleeve and thread it in until the marks made in Step 5 align.

8   Install the steering connecting rod end into the drag link. Make sure the front wheels and steering wheel are in the straight ahead position. Make sure the ball end is seated in the taper to prevent them from rotating while tightening the nut.

9   Install a new nut on the stud and tighten to the torque listed in this Chapter's Specifications. Install a new cotter pin and bend the ends over completely. Tighten the tie-rod adjusting sleeve clamp bolts to the torque listed in this Chapter's Specifications.

10   Install the wheel and lug nuts. Lower the vehicle and tighten the lug nuts to torque listed in the Chapter 1 Specifications.

11   Have the alignment checked by a dealer service department or an alignment shop.

25 Tie-rod ends - removal and installation

Warning: *Whenever any of the suspension or steering fasteners are loosened or removed they must be replaced with new ones - discard the originals and don't re-use them. They must be replaced with new ones of the same part number or of original equipment*

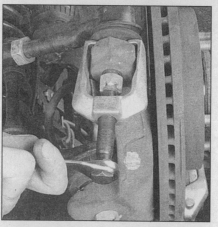

25.4 To separate the outer tie-rod end from the spindle, remove the cotter pin, back off but don't remove the nut, install a small puller and force the ballstud out of the spindle arm

quality and design. Torque specifications must be followed for proper reassembly and component retention.

Removal

Refer to illustration 25.6

1   Loosen the front wheel lug nuts on the side to be dismantled. Raise the vehicle and support it securely on jackstands. **Caution:** *If the vehicle is equipped with Automatic Ride Control (ARC), make sure the air suspension switch is turned to the OFF position before the vehicle is raised to prevent damage to the system components (see Section 20).*

2   Remove the front wheel.

3   Remove the cotter pin and loosen the nut on the tie-rod end stud. Discard the cotter pin.

4   Disconnect the tie-rod from the steering spindle with a Pitman arm puller (see illustration).

5   On 1994 and earlier models, loosen the clamp bolts on the tie-rod adjusting sleeve. On 1995 and later models, loosen the tie-rod end jam nut and back it off several turns.

6   Apply a paint mark on the threads adjacent to the adjusting sleeve or tie-rod end. Unscrew the tie-rod end from the adjusting sleeve or connecting rod.

Installation

7   Install the tie-rod end into the adjusting sleeve or connecting rod. Thread the tie-rod in until the marks made in Step 6 align.

8   Install the tie-rod onto the steering spindle. Make sure the front wheels and steering wheel are in the straight ahead position. Make sure the tie-rod stud is seated in the taper to prevent it from rotating while tightening the nut.

9   Install a new nut on the stud and tighten it to the torque listed in this Chapter's Specifications. Install a new cotter pin and bend the ends over completely.

10   Tighten the tie-rod adjusting sleeve clamp bolts or jam nut to the torque listed in

26.4 The power steering gear line fittings are accessible through the wheel well

this Chapter's Specifications. Make sure the tie-rod is positioned correctly in the same position as when it was removed.

11   Install the wheel and lug nuts. Lower the vehicle and tighten the lug nuts to the torque listed in Chapter 1.

12   Have the front-end alignment checked by a dealer service department or an alignment shop.

26 Steering gear - removal and installation

Warning 1: *On models equipped with airbags, make sure the steering shaft is not turned while the steering gear or box is removed or you could damage the sliding contact for the airbag system. To prevent the shaft from turning, turn the ignition key to the lock position before beginning work or run the seat belt through the steering wheel and clip the seat belt into place.*

Warning 2: *Whenever any of the suspension or steering fasteners are loosened or removed they must be replaced with new ones - discard the originals and don't re-use them. They must be replaced with new ones of the same part number or of original equipment quality and design. Torque specifications must be followed for proper reassembly and component retention.*

1991 through 1994 models

Removal

Refer to illustrations 26.4, 26.6, 26.9 and 26.10

1   Set the front wheels to the straight-ahead position.

2   Disconnect the cable from the negative battery terminal.

3   Turn the ignition key to the Run position to unlock the steering wheel.

4   Place a pan under the steering gear. Remove the power steering hoses/lines and cap the ends and the ports to prevent excessive fluid loss and contamination (see illustration and Sections 27 and 28).

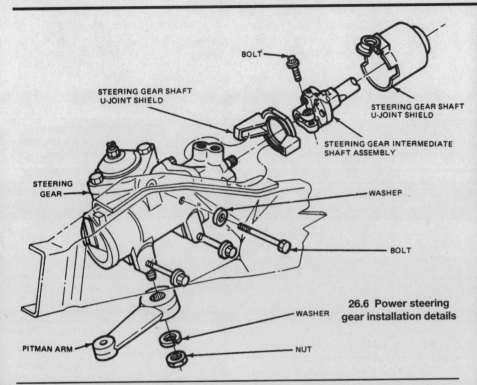

26.6 Power steering gear installation details

26.9 Remove the Pitman arm from the gear with a Pitman arm puller (available at auto parts stores)

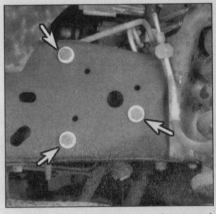

26.10 The power steering gear is secured to the frame member by three bolts

5   Remove the upper and lower U-joint shaft coupling shield from the steering gear.

6   Remove the bolt securing the flex coupling to the steering gear (see illustration).

7   Loosen the front wheel lug nuts. Raise the front of the vehicle, support it securely on jackstands, block the rear wheels and set the parking brake. Remove both front wheels.

8   Remove the large nut and washer securing the Pitman arm to the sector shaft.

9   Use a Pitman arm puller and remove the Pitman arm from the sector shaft (see illustration). Caution: Don't hammer on the end of the puller as the steering gear will be damaged.

10   Loosen all three bolts securing the steering gear box to the frame (see illustration). Hold onto the steering gear and remove the bolts and washers and remove the steering gear.

Installation

11   Make sure the steering gear is centered. Turn the gear input shaft to full lock in one direction, then count and record the number of turns required to rotate it to the opposite full lock position. Turn the pinion shaft back through one half of the number of turns just counted to center the assembly.

12   Check that the front wheels are in the straight ahead position.

13   Install the steering gear input shaft lower shield onto the steering gear box. Install the upper shield onto the intermediate shaft.

14   Position the flat on the gear input shaft so it is facing down. Position the flex coupling on the steering gear input shaft, aligning the flat with the flat on the input shaft.

15   Install all three bolts securing the steering gear box to the frame. Tighten the bolts to the torque listed in this Chapter's Specifications.

16   Align the two blocked teeth on the Pitman arm with the four missing teeth on the steering gear sector shaft and install the Pitman arm. Install the washer and new nut and tighten the nut to the torque listed in this Chapter's Specifications.

17   Install the bolt securing the flex coupling to the steering gear input shaft. Tighten the bolt to the torque listed in this Chapter's Specifications.

18   Move the flex coupling shield into place on the steering gear input shield.

19   Install the front wheel(s) and lug nuts.

20   Lower the vehicle and remove the floor jack. Tighten the lug nuts to the torque listed in the Chapter 1 Specifications.

21   Connect the pressure and return lines to the steering gear assembly and tighten to the torque listed in this Chapter's Specifications (see Sections 27 and 28).

22   Turn the ignition key Off and connect the negative battery cable.

23   Fill the fluid reservoir with the specified fluid and refer to Section 29 for the power steering bleeding procedure.

24   Have the front-end alignment checked by a dealer or front-end alignment shop.

1995 and later models

Removal

Refer to illustrations 26.28, 26.29, 26.30, 26.31 and 26.33

25   Set the front wheels to the straight-ahead position. Raise and securely support the vehicle on jackstands. Caution: If the vehicle is equipped with Automatic Ride Control (ARC), make sure the air suspension switch is turned to the OFF position before the vehicle is raised to prevent damage to the system components (see Section 20).

26   Disconnect the cable from the negative battery terminal.

27   Secure the steering wheel in the straight-ahead position, but do not lock the column.

28   Remove the bolt retaining the lower steering column shaft to the steering gear input shaft (see illustration). Note: Do not rotate the steering wheel when the intermedi-

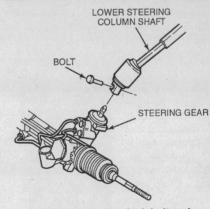

26.28 Remove the pinch bolt and separate the lower steering column shaft from the steering gear

10

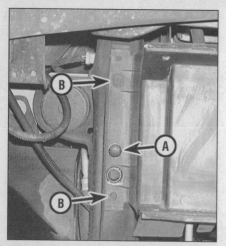

26.29 Remove the plastic pushpins (A) and bolts (B) from each side and remove the plastic air deflector panel

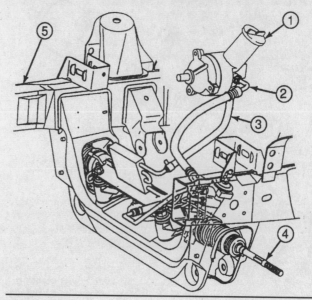

26.30 Power steering system components - 1995 and later models

1   Power steering pump
2   Power steering pressure hose
3   Power steering cooler and hose assembly
4   Steering gear
5   Frame

ate shaft is disconnected from the steering column or damage to the airbag sliding contact will result.

29   Remove the plastic air deflector panel under the radiator area **(see illustration)**. Remove the stabilizer bar from the chassis (see Section 15).

30   Unscrew the disconnect fittings from the power steering hoses at the power steering gear **(see illustration)**. Plug all the lines and fittings to avoid any entry of dirt or contamination.

31   Remove the power steering cooler from the engine compartment **(see illustration)**.

32   Separate the tie-rod ends from the wheel spindle using a Pitman arm puller (see Section 25).

33   Remove the nuts and bolts that retain the steering gear to the front crossmember **(see illustration)**.

34   Remove the steering gear from the vehicle. Begin by rotating the top of the steering rack towards the front of the vehicle, then slide the gear as far to the right as possible within the crossmember. Now angle the left side of the rack ahead of the crossmember and pull the whole assembly out to the left.

35   Check all the mounting bracket insulators and grommets and replace any that are worn or damaged.

Installation

36   Install the steering gear mounting insulators into the steering gear housing.

37   Position the steering gear into the front crossmember (reverse procedure of Step 34) and install the nuts, bolts and washers that retain the unit. Tighten the nuts/bolts to the torque listed in this Chapter's specifications.

38   Install the power steering cooler. Connect the hoses to the power steering gear.

39   Install the tie rod ends onto the steering arms.

40   Verify that the steering wheel has not rotated from its original position and install the steering column lower intermediate shaft over the steering gear input shaft.

41   Replace the front wheel and tires and lower the vehicle.

42   The remaining installation is the reverse of removal.

43   Have the vehicle's alignment checked.

27   Power steering pump - removal and installation

Note: *Refer to Section 28 for information on quick disconnect fittings used on the power steering system.*

Removal

Refer to illustrations 27.4a, 27.4b, 27.4c, 27.5a, 27.5b, 27.6a, 27.6b, 27.7a, 27.7b and 27.7c

1   Disconnect the cable from the negative battery terminal.

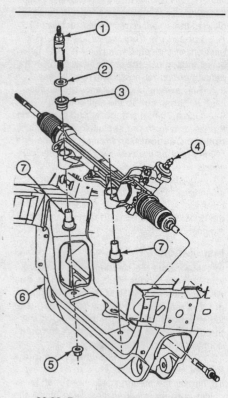

26.33 Power steering rack (gear) assembly installation details

1   Bolt
2   Washer
3   Steering gear insulator
4   Steering gear
5   Nut
6   Front crossmember
7   Steering gear insulator

26.31 Remove the power steering cooler (arrow) - 5.0L V8 model shown

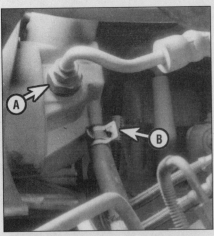

27.4a Power steering pump pressure line (A) and return hose (B) - 4.0L pushrod V6 (viewed through the wheelwell)

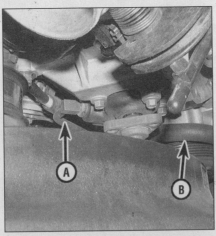

27.4b Power steering pump pressure line (A) and return hose (B) - 4.0L SOHC V6

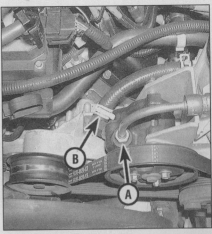

27.4c Power steering pump pressure line (A) and return hose (B) - 5.0L V8

27.5a On SOHC V6 engines, the power steering fluid reservoir must be removed (arrows indicate mounting bolts) before removing the belt guard

27.5b Remove the mounting bolts (arrows) and the belt guard - SOHC V6

2   Remove the drivebelt (see Chapter 1).

3   Remove the engine cooling fan, if necessary (see Chapter 3). Remove the engine oil dipstick tube, if necessary.

4   Disconnect the power steering return hose and drain the fluid into a suitable container (see illustrations). Disconnect the pressure hose from the pump.

5   On SOHC V6 models, remove the power steering fluid reservoir and the belt guard (see illustrations).

6   On either V6 model, remove the drivebelt pulley as follows:

a) On SOHC V6 models, the pulley can be removed by unbolting it from the pump (see illustrations).

b) On pushrod V6 models, a pump pulley removal tool is required (see illustration). Position the tool onto the pump hub, hold the hub with a wrench and rotate the tool shaft clockwise to pull the pulley from the pump. Do not apply excessive force on the pulley shaft as it may damage the internal parts of the pump.

27.6a On SOHC V6 models, secure the pulley with a prybar and remove the pulley bolts

27.6b On 4.0L pushrod V6 models, remove the power steering pump pulley with a special pulley removal tool (available at auto parts stores)

10

27.7b Power steering pump mounting bolts (arrows) - 4.0L SOHC V6

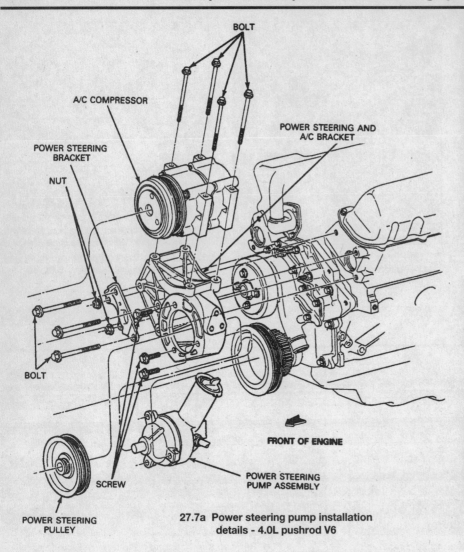

BOLT

A/C COMPRESSOR

POWER STEERING AND A/C BRACKET

POWER STEERING BRACKET

NUT

BOLT

SCREW

POWER STEERING PULLEY

FRONT OF ENGINE

POWER STEERING PUMP ASSEMBLY

27.7a Power steering pump installation details - 4.0L pushrod V6

27.7c On 5.0L V8 models, the front power steering pump mounting bolts (A) can be accessed through the holes in the pulley; remove the rear mounting bolt (B)

27.10 On 4.0L pushrod V6 models, a special tool is used to install the power steering pump pulley

7   On pushrod V6 models, remove the mounting bracket, if necessary. Remove the bolts securing the pump to the mounting bracket **(see illustrations)**. Note: *On 5.0L V8 models, it isn't necessary to remove the pul-* ley. *There are holes in the pulley that allow removal of the three front mounting bolts* **(see illustration)**.
8   Remove the pump from the mounting bracket.

Installation
Refer to illustration 27.10
9   Place the pump in the bracket and install the mounting bolts. Tighten the bolts to the torque listed in this Chapter's Specifications.
10   Install the pulley on V6 models. On 4.0L pushrod V6 models, install the pulley onto the pump using a special pulley installer **(see illustration)**. Be sure the pulley removal groove is toward the front of the vehicle and is flush with the end of the shaft (within 0.010-inch).
11   Install the pressure and return hoses to the proper fittings on the pump (see Section 26).
12   The remainder of installation is the reverse of removal.
13   Fill the pump reservoir with the specified fluid and bleed the system (see Section 29).

28   Power steering line quick-disconnect fittings

Refer to illustrations 28.1 and 28.3
1   If a fitting leaks, note whether the leak is between the tubing and the flare nut or between the flare nut and the pump **(see illustration)**. If it's between the tubing and the flare nut, replace the hose as an assembly. If it's between the flare nut and the pump, repair the leak (see Steps 2 and 3 below).
2   Make sure the nut is tightened to the torque listed in this Chapter's Specifications.
3   If the fitting still leaks, unscrew the nut and check the Teflon washer. Replace the washer if its condition is in doubt. **Note:** *It may be necessary to stretch the washer with a center punch or similar tapered tool so it will fit over the flare nut* **(see illustration)**. The washer will slowly return to its original size after being stretched.
4   The O-ring in the fitting can't be replaced. If it's causing the leak, the hose must be replaced as an assembly.

NOTE: IF LEAK OCCURS HERE, REPLACE HOSE ASSEMBLY

TUBE NUT

O-RING

NOTE: IF LEAK OCCURS HERE, TIGHTEN NUT TO SPECIFICATION. REPLACE TEFLON SEAL IF NECESSARY.

TEFLON SEAL

SNAP RING

PUMP OUTLET

NOTE: ALWAYS REPLACE THIS SEAL WHEN A LINE IS REMOVED

28.1 Quick-disconnect fitting installation details

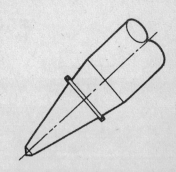

28.3 To install a new Teflon washer, stretch it with a center punch or similar tool until it fits over the flare nut - it will slowly shrink to its original size

5   If a quick-disconnect fitting disconnects while the pump is in operation, replace the hose as an assembly.

29 Power steering system bleeding

1   The power steering system must be bled whenever a line is disconnected. Bubbles can be seen in power steering fluid which has air in it and the fluid will often have a tan or milky appearance. On later models, low fluid level can cause air to mix with the fluid, resulting in a noisy pump as well as foaming of the fluid.
2   Open the hood and check the fluid level in the reservoir, adding the specified fluid necessary to bring it up to the proper level (see Chapter 1).
3   Start the engine and slowly turn the steering wheel several times from left-to-right and back again. Do not turn the wheel completely from lock-to-lock. Check the fluid level, topping it up as necessary until it remains steady and no more bubbles are visible.

30 Wheel alignment - general information

Refer to illustration 30.1

A wheel alignment refers to the adjustments made to the wheels so they're in proper angular relationship to the suspension and the ground. Wheels that are out of proper alignment not only affect steering control, but also increase tire wear. The adjustment most commonly required is the toe-in adjustment but camber adjustment is also possible **(see illustration)**.

Getting the proper wheel alignment is a very exacting process, one in which complicated and expensive machines are necessary to perform the job properly. Because of this, you should have a technician with the proper equipment perform these tasks. We will, however, attempt to give you a basic idea of what's involved with wheel alignment so you

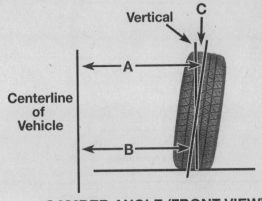

CAMBER ANGLE (FRONT VIEW)

Vertical

C

A

Centerline of Vehicle

B

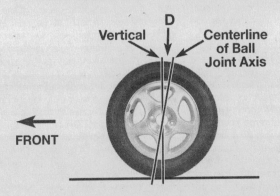

Vertical

D

Centerline of Ball Joint Axis

FRONT

30.1 Front-end alignment details

CASTER ANGLE (SIDE VIEW)

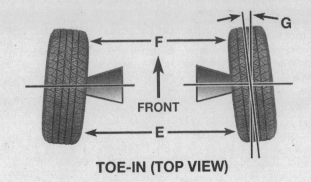

G

F

FRONT

E

TOE-IN (TOP VIEW)

10

METRIC TIRE SIZES
P 185 / 80 R 13

TIRE TYPE
P-PASSENGER
T-TEMPORARY
C-COMMERCIAL

ASPECT RATIO
(SECTION HEIGHT)
————————————
(SECTION WIDTH)
70
75
80

RIM DIAMETER
(INCHES)
13
14
15

SECTION WIDTH
(MILLIMETERS)
185
195
205
ETC

CONSTRUCTION TYPE
R-RADIAL
B-BIAS - BELTED
D-DIAGONAL (BIAS)

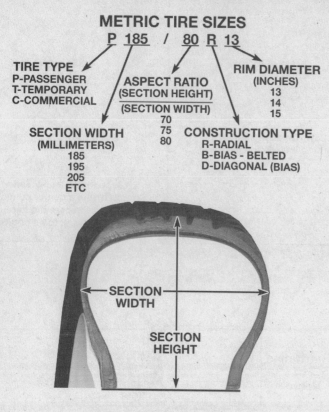

SECTION WIDTH

SECTION HEIGHT

31.1 Metric tire size code

can better understand the process and deal intelligently with the shop that does the work.

Toe-in is the turning in of the wheels. The purpose of a toe specification is to ensure parallel rolling of the wheels. In a vehicle with zero toe-in, the distance between the front edges of the wheels will be the same as the distance between the rear edges of the wheels. The actual amount of toe-in is normally only a fraction of an inch. Incorrect toe-in will cause the tires to wear improperly by making them scrub against the road surface. On the front-end, toe adjustment is controlled by the tie-rod end position on the tie-rod.

Camber is the tilting of the wheels from the vertical when viewed from the front or rear of the vehicle. When the wheels tilt out at the top, the camber is said to be positive (+). When the wheels tilt in at the top the camber is negative (-). The amount of tilt is measured in degrees from the vertical and this measurement is called the camber angle. This angle affects the amount of tire tread which contacts the road and compensates for changes in the suspension geometry when the vehicle is cornering or traveling over an

undulating surface. On the front-end, camber can be adjusted within a certain range by either a camber sleeve on top of the spindle (twin I-beam models) or at the inner eccentric mounts for the upper A-arms on later models. Camber adjustment must be done with the vehicle on the alignment rack, by an experienced technician.

Caster and camber adjustments on the front suspension can only be done by an alignment shop, while the vehicle is on the alignment machine. On twin I-beam models, worn I-beam-to-crossmember bushings can cause the camber to be beyond the range of the camber adjusters, and the bushings will have to be replaced (requires a press). If the caster/camber can't be corrected within this range of adjustment, the suspension components must be checked for damage.

On 1994 and earlier models the camber can be adjusted by replacing the camber adjuster in the upper balljoint socket. On 1995 and later models the camber (as well as caster) is adjusted by repositioning the cams on the upper control arm pivots. Due to the requirement for special alignment tools, this

job beyond the scope of the home mechanic. Have this work performed by a Ford dealer or a front-end alignment specialist.

31 Wheels and tires - general information

Refer to illustration 31.1

1   All vehicles covered by this manual are equipped with metric-size fiberglass or steel belted radial tires **(see illustration)**. The use of other size or type tires may affect the ride and handling of the vehicle. Don't mix different types of tires, such as radials and bias belted, on the same vehicle, since handling may be seriously affected. Tires should be replaced in pairs on the same axle, but if only one tire is being replaced, be sure it's the same size, structure and tread design as the other.

2   Because tire pressure affects handling and wear, the tire pressures should be checked at least once a month or before any extended trips (see Chapter 1).

Wheels must be replaced if they're bent, dented, leak air, have elongated bolt holes, are heavily rusted, out of vertical symmetry or if the lug nuts won't stay tight. Wheel repairs by welding or peening aren't recommended.

3   Tire and wheel balance is important to the overall handling, braking and performance of the vehicle. Unbalanced wheels can adversely affect handling and ride characteristics as well as tire life. Whenever a tire is installed on a wheel, the tire and wheel should be balanced by a shop with the proper equipment.

Inspection

4   Check the tire pressures (cold) weekly (see Chapter 1).

5   Inspect the sidewalls and treads periodically for damage and signs of abnormal or uneven wear.

6   Make sure the wheel lug nuts are properly tightened.

7   Periodically inspect the wheels for elongated or damaged lug holes, distortion and nicks in the rim. Replace damaged wheels.

8   Clean the wheels inside and out and check for rust and corrosion, which could lead to wheel failure.

9   If the wheel and tire are balanced on the vehicle, one wheel stud and lug hole should be marked whenever the wheel is removed so it can be reinstalled in the original position. If balanced on the vehicle, the wheel should not be moved to a different axle position.

10   To remove alloy wheels, carefully pry off the center cap to expose the lug nuts. Do not use impact wrenches on alloy wheels.

Chapter 11 Body

Contents

1 General information

These models feature a welded body that is attached to a separate frame. Certain components are particularly vulnerable to accident damage and can be unbolted and repaired or replaced. Among these parts are the body moldings, bumpers, hood, doors, liftgate and all glass.

Only general body maintenance procedures and body panel repair procedures within the scope of the do-it-yourselfer are included in this Chapter.

2 Body - maintenance

1   The condition of your vehicle's body is very important, because the resale value depends a great deal on it. It's much more difficult to repair a damaged body than it is to repair mechanical components. The hidden areas of the body, such as the fenderwells, the frame and the engine compartment, are equally important, although they don't require as frequent attention as the rest of the body.

2   Once a year, or every 12,000 miles, it's a good idea to have the underside of the body steam cleaned. All traces of dirt and oil will be removed and the area can then be inspected carefully for rust, damaged brake lines, frayed electrical wires, damaged cables and other problems. The front suspension components should be greased after completion of this job.

3   At the same time, clean the engine and the engine compartment with a steam cleaner or water-soluble degreaser.

4   The fenderwells should be given close attention, since undercoating can peel away and stones and dirt thrown up by the tires can cause the paint to chip and flake, allow-

11

ing rust to set in. If rust is found, clean down to the bare metal and apply an anti-rust paint.

5   The body should be washed about once a week (or when dirty). Wet the vehicle thoroughly to soften the dirt, then wash it down with a soft sponge and plenty of clean soapy water. If the surplus dirt is not washed off very carefully, it can wear down the paint.

6   Spots of tar or asphalt thrown up from the road should be removed with a cloth soaked in kerosene or lamp oil.

7   Once every six months, wax the body and chrome trim. If a chrome cleaner is used to remove rust from any of the vehicle's plated parts, remember that the cleaner also removes part of the chrome, so use it sparingly and wax over the cleaned area for protection.

3   Vinyl trim - maintenance

Don't clean vinyl trim with detergents, caustic soap or petroleum-based cleaners. Plain soap and water works just fine, with a soft brush to clean dirt that may be ingrained. Wash the vinyl as frequently as the rest of the vehicle.

After cleaning, application of a high quality rubber and vinyl protectant will help prevent oxidation and cracks. The protectant can also be applied to weatherstripping, vacuum lines and rubber hoses, which often fail as a result of chemical degradation, and to the tires. **Caution:** Do not use vinyl/rubber protectants on the steering wheel, pedal pads or driver's seat.

4   Upholstery and carpets - maintenance

1   Every three months remove the carpets or mats and clean the interior of the vehicle (more frequently if necessary). Vacuum the upholstery and carpets to remove loose dirt and dust.

2   Leather upholstery requires special care. Stains should be removed with warm water and a very mild soap solution. Use a clean, damp cloth to remove the soap, then wipe again with a dry cloth. Never use alcohol, gasoline, nail polish remover or thinner to clean leather upholstery.

3   After cleaning, regularly treat leather upholstery with a leather wax. Never use car wax on leather upholstery.

4   In areas where the interior of the vehicle is subject to bright sunlight, cover leather seats with a sheet if the vehicle is to be left out for any length of time.

5   Body repair - minor damage

See photo sequence

Repair of minor scratches

1   If the scratch is superficial and does not penetrate to the metal of the body, repair is very simple. Lightly rub the scratched area

with a fine rubbing compound to remove loose paint and built up wax. Rinse the area with clean water.

2   Apply touch-up paint to the scratch, using a small brush. Continue to apply thin layers of paint until the surface of the paint in the scratch is level with the surrounding paint. Allow the new paint at least two weeks to harden, then blend it into the surrounding paint by rubbing with a very fine rubbing compound. Finally, apply a coat of wax to the scratch area.

3   If the scratch has penetrated the paint and exposed the metal of the body, causing the metal to rust, a different repair technique is required. Remove all loose rust from the bottom of the scratch with a pocket knife, then apply rust inhibiting paint to prevent the formation of rust in the future. Using a rubber or nylon applicator, coat the scratched area with glaze-type filler. If required, the filler can be mixed with thinner to provide a very thin paste, which is ideal for filling narrow scratches. Before the glaze filler in the scratch hardens, wrap a piece of smooth cotton cloth around the tip of a finger. Dip the cloth in thinner and then quickly wipe it along the surface of the scratch. This will ensure that the surface of the filler is slightly hollow. The scratch can now be painted over as described earlier in this Section.

Repair of dents

Warning: *On models equipped with airbags, always disconnect the negative battery cable, then the positive battery cable and wait two minutes before working in the vicinity of the impact sensors, steering column or instrument panel to avoid the possibility of accidental deployment of the airbag, which could cause personal injury (see Chapter 12).*

4   When repairing dents, the first job is to pull the dent out until the affected area is as close as possible to its original shape. There is no point in trying to restore the original shape completely as the metal in the damaged area will have stretched on impact and cannot be restored to its original contours. It is better to bring the level of the dent up to a point about 1/8-inch below the level of the surrounding metal. In cases where the dent is very shallow, it is not worth trying to pull it back out at all.

5   If the back side of the dent is accessible, it can be hammered out gently from behind using a soft-face hammer. While doing this, hold a block of wood firmly against the opposite side of the metal to absorb the hammer blows and prevent the metal from being stretched.

6   If the dent is in a section of the body which has double layers, or some other factor makes it inaccessible from behind, a different technique is required. Drill several small holes through the metal inside the damaged area, particularly in the deeper sections. Screw long, self tapping screws into the holes just enough for them to get a good grip in the metal. Now the dent can be pulled out by pulling on the protruding heads of the screws with locking pliers.

7   The next stage of repair is the removal of the paint from the damaged area and from an inch or so of the surrounding metal. This is easily done with a wire brush or sanding disk in a drill motor, although it can be done just as effectively by hand with sandpaper. To complete the preparation for filling, score the surface of the bare metal with a screwdriver or the tang of a file or drill small holes in the affected area. This will provide a good grip for the filler material. To complete the repair, see the Section on filling and painting.

Repair of rust holes or gashes

8   Remove all paint from the affected area and from an inch or so of the surrounding metal using a sanding disk or wire brush mounted in a drill motor. If these are not available, a few sheets of sandpaper will do the job just as effectively.

9   With the paint removed, you will be able to determine the severity of the corrosion and decide whether to replace the whole panel, if possible, or repair the affected area. New body panels are not as expensive as most people think and it is often quicker to install a new panel than to repair large areas of rust.

10   Remove all trim pieces from the affected area except those which will act as a guide to the original shape of the damaged body, such as headlight shells, etc. Using metal snips or a hacksaw blade, remove all loose metal and any other metal that is badly affected by rust. Hammer the edges of the hole inward to create a slight depression for the filler material.

11   Wire brush the affected area to remove the powdery rust from the surface of the metal. If the back of the rusted area is accessible, treat it with rust inhibiting paint.

12   Before filling is done, block the hole in some way. This can be done with sheet metal riveted or screwed into place, or by stuffing the hole with wire mesh.

13   Once the hole is blocked off, the affected area can be filled and painted. See the following subsection on filling and painting.

Filling and painting

14   Many types of body fillers are available, but generally speaking, body repair kits which contain filler paste and a tube of resin hardener are best for this type of repair work. A wide, flexible plastic or nylon applicator will be necessary for imparting a smooth and contoured finish to the surface of the filler material. Mix up a small amount of filler on a clean piece of wood or cardboard (use the hardener sparingly). Follow the manufacturer's instructions on the package, otherwise the filler will set incorrectly.

15   Using the applicator, apply the filler paste to the prepared area. Draw the applicator across the surface of the filler to achieve the desired contour and to level the filler surface. As soon as a contour that approximates the original one is achieved, stop working the paste. If you continue, the paste will begin to stick to the applicator. Continue to add thin layers of paste at 20-minute intervals until the

8.3 Scribe or draw alignment marks on the hood to ensure proper alignment on reinstallation

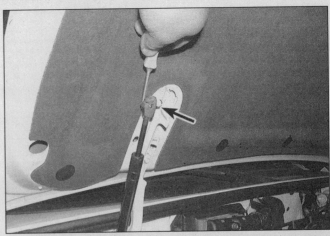

8.5 On later models, pry the upper end of the hood support cylinder from the ball stud on the hood hinge (arrow) - make sure the hood is supported by a helper while you do this

level of the filler is just above the surrounding metal.

16   Once the filler has hardened, the excess can be removed with a body file. From then on, progressively finer grades of sandpaper should be used, starting with a 180-grit paper and finishing with 600-grit wet-or-dry paper. Always wrap the sandpaper around a flat rubber or wooden block, otherwise the surface of the filler will not be completely flat. During the sanding of the filler surface, the wet-or-dry paper should be periodically rinsed in water. This will ensure that a very smooth finish is produced in the final stage.

17   At this point, the repair area should be surrounded by a ring of bare metal, which in turn should be encircled by the finely feathered edge of good paint. Rinse the repair area with clean water until all of the dust produced by the sanding operation is gone.

18   Spray the entire area with a light coat of primer. This will reveal any imperfections in the surface of the filler. Repair the imperfections with fresh filler paste or glaze filler and once more smooth the surface with sandpaper. Repeat this spray-and-repair procedure until you are satisfied that the surface of the filler and the feathered edge of the paint are perfect. Rinse the area with clean water and allow it to dry completely.

19   The repair area is now ready for painting. Spray painting must be carried out in a warm, dry, windless and dust free atmosphere. These conditions can be created if you have access to a large indoor work area, but if you are forced to work in the open, you will have to pick the day very carefully. If you are working indoors, dousing the floor in the work area with water will help settle the dust which would otherwise be in the air. If the repair area is confined to one body panel, mask off the surrounding panels. This will help minimize the effects of a slight mismatch in paint color. Trim pieces such as chrome strips, door handles, etc. will also need to be masked off or removed. Use masking tape and several thickness of newspaper for the masking operations.

20   Before spraying, shake the paint can thoroughly, then spray a test area until the spray painting technique is mastered. Cover the repair area with a thick coat of primer. The thickness should be built up using several thin layers of primer rather than one thick one. Using 600-grit wet-or-dry sandpaper, rub down the surface of the primer until it is very smooth. While doing this, the work area should be thoroughly rinsed with water and the wet-or-dry sandpaper periodically rinsed as well. Allow the primer to dry before spraying additional coats.

21   Spray on the top coat, again building up the thickness by using several layers of paint. Begin spraying in the center of the repair area and then, using a circular motion, work outward until the whole repair area and about two inches of the surrounding original paint is covered. Remove all masking material 10 to 15 minutes after spraying on the final coat of paint. Allow the new paint at least two weeks to harden, then use a very fine rubbing compound to blend the edges of the new paint into the existing paint, Finally, apply a coat of wax.

6   Body repair - major damage

1   Major damage must be repaired by an auto body shop. These shops have the specialized equipment required to do the job properly.

2   If the damage is extensive, the frame must be checked for proper alignment or the vehicle's handling characteristics may be adversely affected and other components may wear at an accelerated rate.

3   Due to the fact that all of the major body components (hood, fenders, etc.) are separate and replaceable units, any seriously damaged components should be replaced rather than repaired. Sometimes the components can be found in a wrecking yard that specializes in used vehicle components, often at a considerable savings over the cost of new parts.

7   Hinges and locks - maintenance

Once every 3,000 miles, or every three months, the hinges and latch assemblies on the doors, hood and the tailgate or the liftgate should be given a few drops of light oil or lock lubricant. The door latch strikers should also be lubricated with a thin coat of grease to reduce wear and ensure free movement. Lubricate the door and the liftgate locks with spray-on graphite lubricant.

8   Hood - removal, installation and adjustment

Warning: *On models equipped with airbags, always disconnect the negative battery cable, then the positive battery cable and wait two minutes before working in the vicinity of the impact sensors, steering column or instrument panel to avoid the possibility of accidental deployment of the airbag, which could cause personal injury (see Chapter 12).*
Note: *The hood is heavy and somewhat awkward to remove and install - at least two people should perform this procedure.*

Removal and installation
Refer to illustrations 8.3 and 8.5

1   Open the hood and support it in the open position.

2   Cover the fenders and cowl with blankets or heavy cloths to protect the paint.

3   Scribe or draw alignment marks around the bolt heads to ensure proper alignment on reinstallation **(see illustration)**.

4   Disconnect the underhood light connector on the driver's side and the ground strap on the passenger side of the hood.

5   Have an assistant support the hood while you pry off the hood support cylinder (if equipped) from each hood hinge **(see illustration)**.

6   Have an assistant support the hood while you remove the hood-to-hinge assembly bolts.

11

These photos illustrate a method of repairing simple dents. They are intended to supplement *Body repair - minor damage* in this Chapter and should not be used as the sole instructions for body repair on these vehicles.

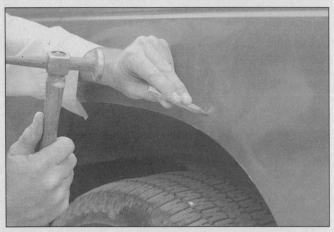

1   If you can't access the backside of the body panel to hammer out the dent, pull it out with a slide-hammer-type dent puller. In the deepest portion of the dent or along the crease line, drill or punch hole(s) at least one inch apart . . .

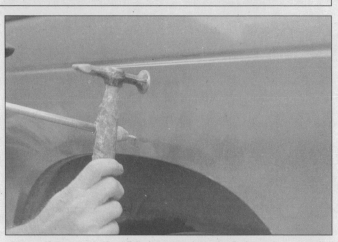

2   . . . then screw the slide-hammer into the hole and operate it. Tap with a hammer near the edge of the dent to help 'pop' the metal back to its original shape. When you're finished, the dent area should be close to its original contour and about 1/8-inch below the surface of the surrounding metal

3   Using coarse-grit sandpaper, remove the paint down to the bare metal. Hand sanding works fine, but the disc sander shown here makes the job faster. Use finer (about 320-grit) sandpaper to feather-edge the paint at least one inch around the dent area

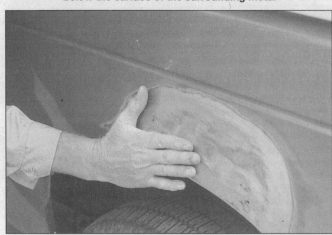

4   When the paint is removed, touch will probably be more helpful than sight for telling if the metal is straight. Hammer down the high spots or raise the low spots as necessary. Clean the repair area with wax/silicone remover

5   Following label instructions, mix up a batch of plastic filler and hardener. The ratio of filler to hardener is critical, and, if you mix it incorrectly, it will either not cure properly or cure too quickly (you won't have time to file and sand it into shape)

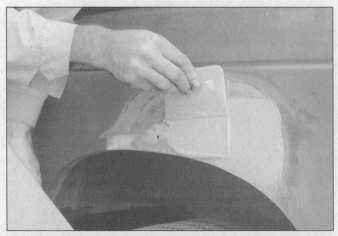

6   Working quickly so the filler doesn't harden, use a plastic applicator to press the body filler firmly into the metal, assuring it bonds completely. Work the filler until it matches the original contour and is slightly above the surrounding metal

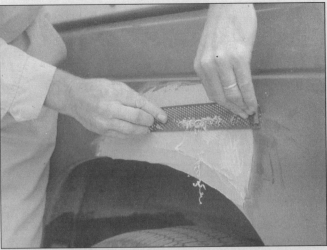

7   Let the filler harden until you can just dent it with your fingernail. Use a body file or Surform tool (shown here) to rough-shape the filler

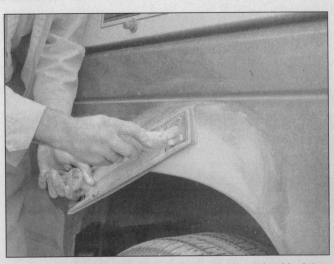

8   Use coarse-grit sandpaper and a sanding board or block to work the filler down until it's smooth and even. Work down to finer grits of sandpaper - always using a board or block - ending up with 360 or 400 grit

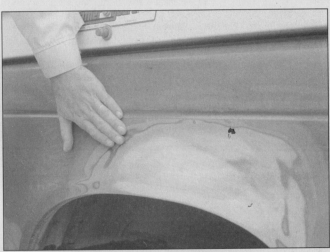

9   You shouldn't be able to feel any ridge at the transition from the filler to the bare metal or from the bare metal to the old paint. As soon as the repair is flat and uniform, remove the dust and mask off the adjacent panels or trim pieces

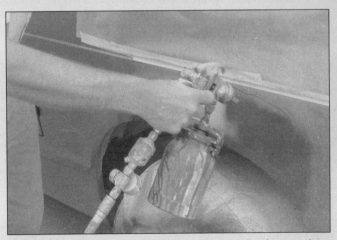

10   Apply several layers of primer to the area. Don't spray the primer on too heavy, so it sags or runs, and make sure each coat is dry before you spray on the next one. A professional-type spray gun is being used here, but aerosol spray primer is available inexpensively from auto parts stores

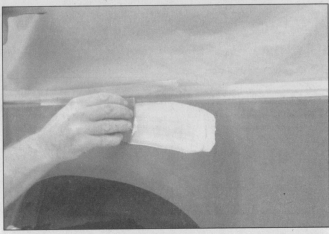

11   The primer will help reveal imperfections or scratches. Fill these with glazing compound. Follow the label instructions and sand it with 360 or 400-grit sandpaper until it's smooth. Repeat the glazing, sanding and respraying until the primer reveals a perfectly smooth surface

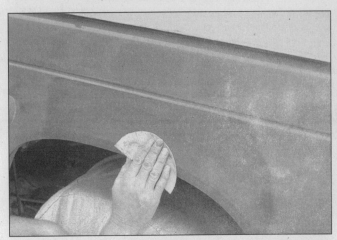

12   Finish sand the primer with very fine sandpaper (400 or 600-grit) to remove the primer overspray. Clean the area with water and allow it to dry. Use a tack rag to remove any dust, then apply the finish coat. Don't attempt to rub out or wax the repair area until the paint has dried completely (at least two weeks)

9.1 Remove the plastic rivets and take off the shield (arrow) to access the hood latch on later models

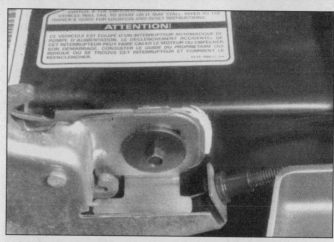

9.2 Mark the position of the hood latch for proper installation

7   With the help of an assistant, lift off the hood.

8   Installation is the reverse of removal with the following additions:

a) *Align the hood and hinges using the alignment marks made in Step 3.*

b) *Be sure to tighten the bolts securely.*

Adjustment

9   The hood can be adjusted to obtain a flush fit between the hood and fenders.

10   Loosen the hood retaining bolts.

11   Move the hood from side-to-side or front-to-rear until the hood is properly aligned with the fenders at the front. Tighten the bolts securely.

12   Loosen the bolts securing the hood latch assembly.

13   Move the latch until alignment is correct with the hood latch striker. Tighten the latch bolt securely.

9   Hood latch control cable - removal and installation

Warning: *On models equipped with airbags, always disconnect the negative battery cable, then the positive battery cable and wait two minutes before working in the vicinity of the impact sensors, steering column or instrument panel to avoid the possibility of accidental deployment of the airbag, which could cause personal injury (see Chapter 12).*

Removal

Refer to illustrations 9.1, 9.2, 9.3 and 9.4

1   Open the hood and support it in the open position. On 1996 and later models, remove the large plastic shield that covers the latch area **(see illustration)**.

2   Scribe or draw alignment marks around the hood latch to ensure proper alignment on reinstallation **(see illustration)**.

3   Remove the bolts securing the hood latch and take the latch out **(see illustration)**.

4   Detach the cable bushing and lift the cable eye off the anchor post **(see illustration)**.

5   Remove the cable clips from the radiator support bracket and apron.

6   Remove the screws that secure the hood release handle to the instrument panel.

7   Remove the tie wrap that secures the cable to the steering column bracket.

9.3 The hood latch is secured by two bolts (arrows)

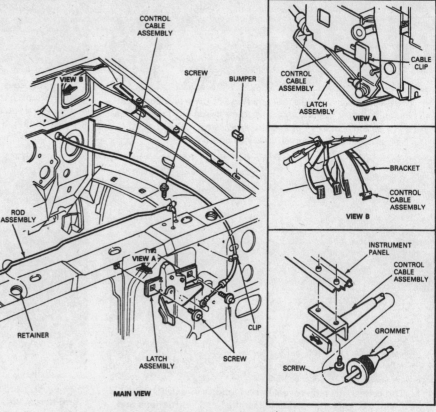

9.4 Hood release cable installation and routing details

8   Working in the passenger compartment, remove the cable grommet from the firewall, then pull the cable through.

Installation

9   Installation is the reverse of removal with the following additions:

a) *Insert the cable through the grommet in the firewall and make sure the grommet is properly seated within the firewall hole.*

b) *Before closing the hood, operate the control cable and make sure the latch control operates correctly.*

10   Door - removal, installation and alignment

Removal and installation

Refer to illustration 10.4

1   With the door in the open position, place a jack under the door or have an assistant hold the door while the hinge bolts are removed. **Note:** *Place thick padding on top of the jack to protect the door's painted finish.*

2   On models so equipped, remove the upper and lower hinge access cover plates. Disconnect the electrical connectors to the door wiring harness.

3   Scribe or paint lines around the door hinges to ensure proper alignment on reinstallation.

4   Remove the hinge-to-door bolts **(see illustration)** and carefully lift off the door.

5   Installation is the reverse of the removal steps. Align the door.

Alignment

6   Check the alignment of the door all around the edge. The door should be evenly spaced in the opening and should be flush with the body.

7   Adjust if necessary as follows:

a) *Note how the door is misaligned and determine which bolts need to be loosened to correct it.*

b) *Loosen the bolts just enough so the door can be moved with a padded pry bar. Reposition the door as necessary, then tighten the bolts securely.*

c) *The door lock striker can also be adjusted both up-and-down and sideways to provide positive engagement with the lock mechanism (see Section 12).*

8   Tighten the hinge-to-body bolts securely.

11   Door hinge - removal and installation

Note: *If both hinges are to be replaced, it is easier to leave the door in place and replace one hinge at a time.*

Removal

Refer to illustration 11.4

1   With the door in the open position, place a jack under the door or have an assistant hold the door while the door hinge bolts are removed. **Note:** *If a jack is used, place thick padding on top of the jack to protect the door's painted finish.*

10.4 Remove the door hinge bolts and lift the door off

2   On models so equipped, remove the upper and lower hinge access cover plates.

3   Scribe or draw lines around the door hinges to ensure proper alignment on reinstallation.

4   Remove the hinge-to-door bolts, then the hinge-to-body bolts **(see illustration)**. Carefully remove the hinge(s) or lift off the door.

Installation

5   Installation is the reverse of removal with the following additions:

a) *Apply a body sealant to the hinge before installing it onto the body panel.*

b) *Adjust the door if necessary (see Section 10).*

12   Door latch striker - removal, installation and adjustment

Removal and installation

Refer to illustration 12.2

1   Disconnect the negative battery cable (see Chapter 1).

2   Use a Torx bit to unscrew the door latch striker stud from the door jamb **(see illustration)**. Installation is the reverse of removal. Adjust if necessary.

12.2 Use a Torx bit to unscrew the door latch striker stud from the door jamb

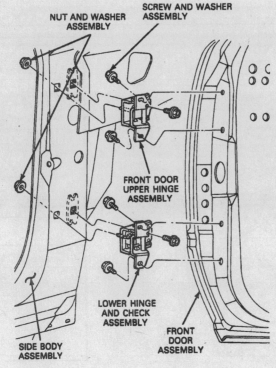

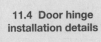

11.4 Door hinge installation details

NUT AND WASHER ASSEMBLY

SCREW AND WASHER ASSEMBLY

FRONT DOOR UPPER HINGE ASSEMBLY

LOWER HINGE AND CHECK ASSEMBLY

FRONT DOOR ASSEMBLY

SIDE BODY ASSEMBLY

11

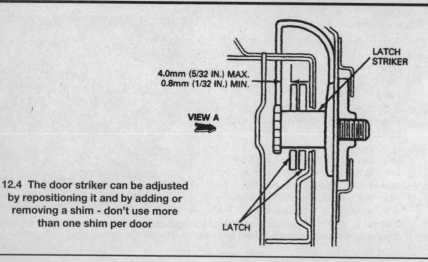

4.0mm (5/32 IN.) MAX.
0.8mm (1/32 IN.) MIN.

LATCH
STRIKER

VIEW A

LATCH

12.4 The door striker can be adjusted by repositioning it and by adding or removing a shim - don't use more than one shim per door

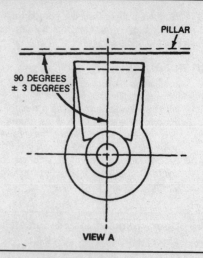

PILLAR

90 DEGREES
± 3 DEGREES

VIEW A

Adjustment

Refer to illustration12.4

3   The door latch striker can be adjusted vertically and laterally as well as fore-and-aft. **Note:** *Don't use the door latch striker to compensate for door misalignment.*

4   The door latch striker can be shimmed to obtain the correct clearance between the

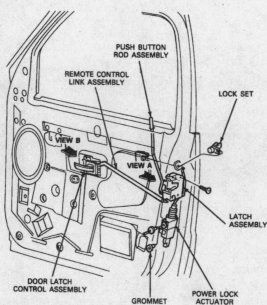

13.2a Remove the two screws (arrows) and the interior door handle

13.2b Front door latch and inside handle (four-door model shown, rear door latch similar)

latch and the striker **(see illustration)**. Don't use more than one shim per door.

5   To check the clearance between the latch jamb and the striker area, spread a layer of dark grease on the striker.

6   Open and close the door several times and note the pattern in the grease.

7   Move the door striker assembly laterally to provide a flush fit at the door and pillar or at the quarter panel.

8   Securely tighten the door latch striker after adjustment is complete.

13   Inner door handle and latch assembly - removal and installation

Removal

Refer to illustrations 13.2a, 13.2b and 13.4

1   Remove the door trim panel and water-

shield (see Section 16).

2   Remove the screws securing the inner door handle assembly to the door **(see illustrations)**.

3   Disconnect the handle link from the handle and lock cylinder.

4   Detach the rods from the latch, then remove the latch screws and remove the latch assembly **(see illustration)**.

Installation

5   Installation is the reverse of the removal steps, plus the following additions:

a) If a new latch is being installed, install the rod retaining clips in it.

b) Attach the control rod and lock cylinder rod to the latch before installing the latch.

c) Tighten the latch screws securely.

d) Check operation of the handle and lock before installing the watershield and door panel.

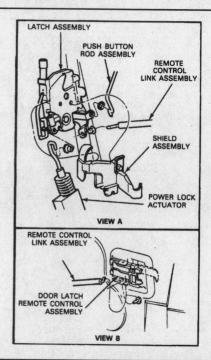

LATCH ASSEMBLY

PUSH BUTTON
ROD ASSEMBLY

REMOTE
CONTROL
LINK ASSEMBLY

SHIELD
ASSEMBLY

POWER LOCK
ACTUATOR

VIEW A

REMOTE CONTROL
LINK ASSEMBLY

DOOR LATCH
REMOTE CONTROL
ASSEMBLY

VIEW B

PUSH BUTTON
ROD ASSEMBLY

REMOTE CONTROL
LINK ASSEMBLY

LOCK SET

VIEW B

VIEW A

LATCH
ASSEMBLY

DOOR LATCH
CONTROL ASSEMBLY

GROMMET

POWER LOCK
ACTUATOR

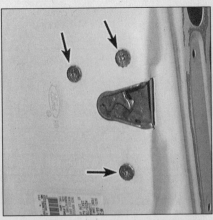

13.4 Remove the Torx screws (arrows) retaining the latch assembly to the door

14   Outer door handle - removal and installation

Refer to illustration 14.2

1   Remove the door trim panel and pull away the watershield (see Section 16). **Note:** *The window glass must be in the full-raised position*.

2   Disconnect the door latch actuator rod from the latch **(see illustration)**.

3   Support the door handle in the open position.

4   Drill out both rivets that secure the handle to the door and remove the handle.

5   Installation is the reverse of the removal steps. Position the handle in the door and install the two pop-rivets.

15   Door lock cylinder - removal and installation

Refer to illustration 15.3

1   Raise the window all the way.

2   Remove the door trim panel and watershield (see Section 16).

3   Disconnect the door latch control-to-cylinder rod **(see illustration)**.

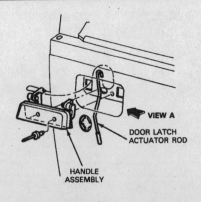

14.2 Outer door handle assembly details

4   Use a pair of pliers and slide the cylinder retainer clip away from the lock cylinder and door.

5   Remove the lock cylinder from the door.

6   Install the lock cylinder into the door opening from the outside and push the retainer clip into place. Make sure it's seated correctly.

7   Reconnect the lock cylinder rod-to-door latch control.

8   Check for proper lock operation, then install the watershield and the door trim panel.

16   Door trim panel and watershield - removal and installation

Removal

1996 and earlier models

Refer to illustrations 16.2a, 16.2b, 16.3, 16.4, 16.5, 16.6, 16.7, 16.9 and 16.10

1   Place the window in the full-down position. Disconnect the negative battery cable.

2   Remove two screws above the door handle and the one screw at the upper front part of the trim panel **(see illustrations)**.

3   Carefully pry out the door handle trim with a screwdriver **(see illustration)**. On base

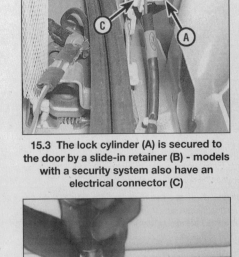

15.3 The lock cylinder (A) is secured to the door by a slide-in retainer (B) - models with a security system also have an electrical connector (C)

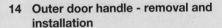

16.2a On early models, remove two screws above the door handle . . .

models without electric windows, remove the screw in the window handle and remove the handle and escutcheon, if equipped.

4   Insert a screwdriver into the pry notch in the window switch panel **(see illustration)**. Pry the switch panel up and pull it out.

5   Remove the screws and separate the window and door lock switches from the switch panel **(see illustration)**.

16.2b . . . and one screw that secures the upper front corner of the panel

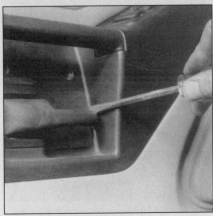

16.3 Carefully pry the door handle trim piece out

16.4 Carefully pry out the switch panel, using the notch made for this purpose

11

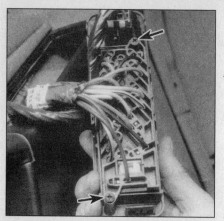

16.5 Separate the window and door lock switches from the plate after removing the screws (arrows)

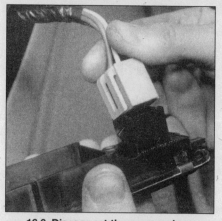

16.6 Disconnect the power mirror connector from the switch and remove the plate

16.7 Carefully pry the trim clips out of the door with a wide-bladed screwdriver or panel removal tool - don't pull on the panel

6   Release the locking finger on the power mirror connector and disconnect the connector **(see illustration)**.

7   Insert a wide-bladed screwdriver or trim removal tool between the door panel and door **(see illustration)**. Carefully pry the trim clips out of the door.

8   Lift the panel straight up to disengage its top edge from the door. On front door panels, lift the panel up at least 1-1/2 inches to disengage the hidden hooks from the door frame. Remove the door panel.

9   Detach the door light socket, if equipped **(see illustration)**.

10   If necessary, carefully peel the watershield from the door **(see illustration)**. Be careful not to tear it.

1997 and later models

Refer to illustrations 16.13, 16.15, 16.16 and 16.17

11   Place the window in the full-down position. Disconnect the negative battery cable.

12   Remove two screws above the door handle.

13   Carefully pry out the door handle trim with a screwdriver **(see illustration)**. On base models without electric windows, remove the screw in the window handle and remove the

16.9 Pull the panel out and detach the door light socket

handle and escutcheon, if equipped.

14   Insert a wide-bladed screwdriver or trim removal tool between the door panel and door. Carefully pry the trim clips out of the door. Lift the panel straight up to disengage its top edge from the door.

15   Pull the door panel away from the door and detach the switch panel from the door trim panel **(see illustration)**.

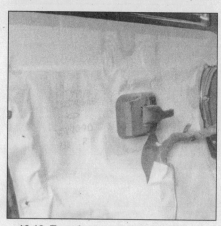

16.10 To gain access to the door lock mechanism and window glass, peel the watershield from the door, being careful not to tear it

16   Turn the switch panel over, remove the screws and separate the window and door lock switches from the switch panel **(see illustration)**.

17   If necessary, carefully peel the watershield from the door **(see illustration)**. Be careful not to tear it.

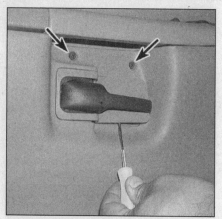

16.13 On 1997 and later models, remove the two screws (arrows) above the handle and pry off the door handle trim

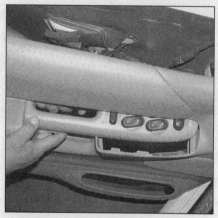

16.15 Detach the switch plate from the door panel and maneuver it through the opening

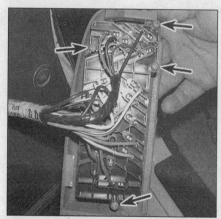

16.16 Turn the switch plate over, remove the screws (arrows) and disconnect the electrical connector from the switches

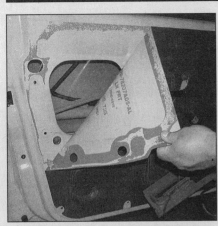

16.17 Carefully peel back the watershield

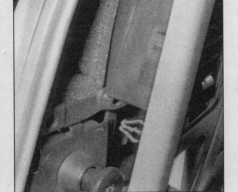

16.19 To install the panel, align the trim clips with their holes and push in on the panel

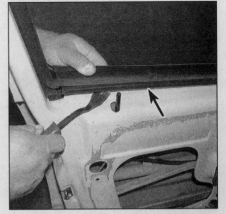

17.4 Pry up and remove the inside weatherstrip (arrow)

Installation

Refer to illustration 16.19

18   Apply adhesive to the watershield, if necessary, and press the watershield into position.

19   To install the door panel, engage the top edge with the top of the door and press the panel down. Align the trim clips with their holes and push the door panel in until the clips are seated **(see illustration)**.

20   The remainder of installation is the reverse of removal.

17 Front door window glass - removal, installation and adjustment

Removal

Refer to illustrations 17.4, 17.6a and 17.6b

1   Remove the door trim panel and watershield (see Section 16).

2   Remove the door speakers (if equipped) (see Chapter 12).

3   Raise the window all the way to the top.

4   Remove the inside belt weatherstrip from the door **(see illustration)**.

5   On early models, detach the rear glass run retainer from the door and take it out.

6   Lower the window enough to provide access to the glass bracket and retention rivets **(see illustrations)**. Using the appropriate size drill bit, drill out the center of the rivets and remove the rivets.

7   Lift the glass out through the top of the door, leaning it to the outside.

Installation

8   After reinserting the glass and installing two new pop-rivets, installation is the reverse of removal, with the following additions:

a) Lubricate the regulator rollers, shafts and tracks with multi-purpose grease.

b) Raise and lower the window to check its operation before installing the speakers (if equipped), watershield and door trim panel.

Adjustment

Refer to illustration 17.10

9   Lower the glass two-to-three inches from the full-up position.

10   Loosen the guide assembly nut and washer **(see illustration)**.

11   Push the glass toward the rear until it bottoms out within the door frame.

12   Move the window guide post toward the rear within the retention slot in the door inner panel. Tighten the guide assembly nuts securely.

13   Raise and lower the window several times and check for proper fit.

14   Install the watershield and door trim panel.

18 Front door window regulator - removal and installation

Removal

Refer to illustrations 18.3, 18.4 and 18.6

1   Remove the door trim panel and watershield (see Section 16).

2   Remove the window glass (see Section 17). On models with power windows, disconnect the electrical connector from the window motor.

3   Using the appropriate size drill bit, drill out the center of the rivets and remove the rivets that secure the equalizer bracket **(see illustration)**.

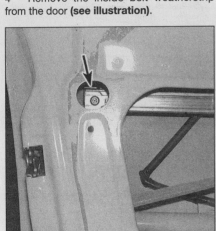

17.6a Lower the window until the two glass channel rivets align with the holes in the door - arrow indicates rivet at rear of door

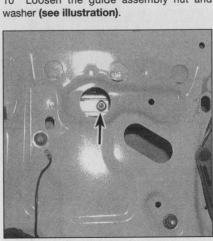

17.6b Location of glass channel rivet (arrow) at front of door

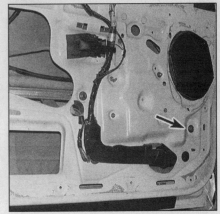

17.10 Loosen the glass guide fastener (arrow) to adjust the door glass

11

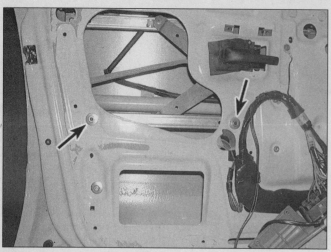

18.3 Remove the rivets (arrows) holding the equalizer bar to the door

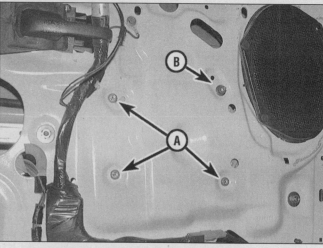

18.4 Drill-out and remove the window regulator rivets (A) and remove the one nut (B)

4   Carefully drill out and remove the rivets that secure the regulator base plate to the inner door panel and remove the one nut **(see illustration)**.

5   Remove the regulator and glass bracket.

Installation

6   If the regulator guides must be replaced, flatten the tab in the end of the channel, then remove the arm slides from the channel **(see illustration)**. Install the new guides in the

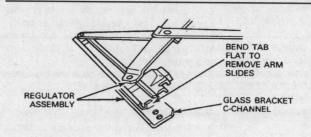

BEND TAB FLAT TO REMOVE ARM SLIDES

REGULATOR ASSEMBLY

GLASS BRACKET C-CHANNEL

18.6 Bend the tab down and back up carefully - if it's cracked or broken, the bracket assembly must be replaced

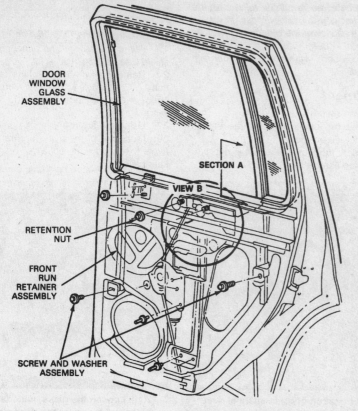

DOOR WINDOW GLASS ASSEMBLY

SECTION A

VIEW B

RETENTION NUT

FRONT RUN RETAINER ASSEMBLY

SCREW AND WASHER ASSEMBLY

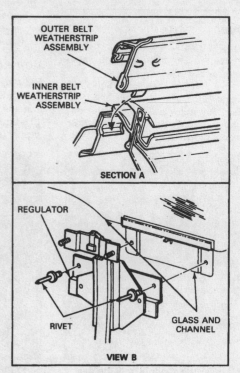

OUTER BELT WEATHERSTRIP ASSEMBLY

INNER BELT WEATHERSTRIP ASSEMBLY

SECTION A

REGULATOR

RIVET

GLASS AND CHANNEL

VIEW B

19.2 Rear door window glass installation details

channel and bend the tab back to its original position. **Note:** *If the tab is cracked or broken during this procedure, the bracket assembly must be replaced with a new one.*

7   If a new bracket is installed, be sure the rubber bumper is installed on it in the correct position. **Warning:** *DO NOT remove the regulator counterbalance spring unless the regulator arms are secured so they can't move. The spring may unwind suddenly and cause injury.*

8   Install the regulator and glass bracket in the door. Use the base plate locator tab to position the regulator base plate.

9   Attach the regulator to the inner door panel. This can be done with rivets, or with a pair of 1/4-20 x 1-inch bolts with nuts and washers. Metric bolts, nuts and washers of the equivalent size can be substituted for the 1/4-20 fasteners. Tighten the screws securely.

10   The remainder of installation is the reverse of removal.

---

19   Rear door window glass - removal and installation

Removal

Refer to illustration 19.2

1   Remove the door trim panel and watershield (see Section 16).

2   Pull the inner and outer belt weatherstrips off the door **(see illustration)**. On 1996 and later models, remove only the inside weatherstrip.

3   Remove the screw that secures the front retainer. Pull the retainer down and remove it through the access hole in the door.

4   Raise the window until it's four inches from the top of its travel.

5   Using the appropriate size drill bit, drill out the center of the rivets, remove the rivets and lift the window up and out of the door, angling it toward the outside of the door.

Installation

6   Installation is the reverse of removal, with the following additions:
   a) *Insert the rear edge of the window fully into the rear glass run.*
   b) *Raise and lower the window several times to check operation before installing the trim panel and weathershield.*

---

20   Rear door window regulator - removal and installation

Removal

1   Remove the trim panel and watershield (see Section 16).

2   Disconnect the power window electrical connector from the regulator (if equipped).

3   Remove the door window (see Section 19).

4   Remove the nuts and washers that secure the window guide (if equipped) and lift the guide out of the door through the top.

5   If equipped with power windows, using the appropriate size drill bit, drill out the center of the rivets and remove the motor bracket rivets.

6   Remove the rivets and nuts that secure the regulator and remove the regulator.

Installation

7   Installation is the reverse of removal, with the following additions:
   a) *The regulator rivets can be replaced with 1/4-20 x 1/2-inch bolts, nuts and washers or metric equivalents.*
   b) *Raise and lower the window several times to check operation before installing the watershield and trim panel.*

---

21   Liftgate - removal and installation

Refer to illustrations 21.2 and 21.4

1   Open the liftgate and support it in the open position. **Warning:** *The liftgate is heavy and may fall if it isn't securely supported.*

21.2 Pry out the spring clip from the ball socket and detach the support cylinder from the liftgate

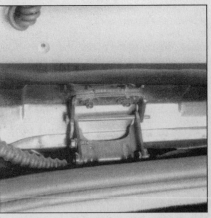

21.4 Index the hinge for proper installation before removing the two bolts securing the liftgate

2   Use a screwdriver and pry out the spring clip from the ball socket at the end of the support cylinder and detach the support cylinder from the liftgate **(see illustration)**.

3   Disconnect the electrical connector at the top of the liftgate.

4   Scribe or draw lines around the liftgate hinges and bolts to ensure proper alignment on reinstallation **(see illustration)**.

5   Remove the liftgate-to-hinge nuts or screws and carefully remove the liftgate.

6   Installation is the reverse of removal. Tighten the hinge nuts or screws securely.

---

22   Liftgate panel, lock cylinder and latches - removal and installation

Liftgate panel

Refer to illustrations 22.1, 22.2, 22.3 and 22.4

1   Remove the screws securing the panel to the liftgate assembly **(see illustration)**. Some models use plastic rivets instead of screws. Use a trim tool to pry the plastic rivets out.

2   Carefully pry up the liftgate lock knob bezel and remove the one screw accessible through the lock knob hole **(see illustration)**.

22.1 Remove the screws or plastic rivets that secure the upper part of the trim panel

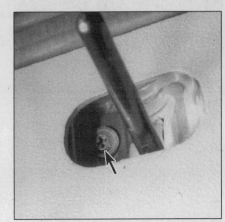

22.2 Remove one screw accessible through the lock knob hole

11

3   Remove the liftgate inside handle screws **(see illustration)**.

4   Use a trim tool to pry out the plastic rivets at the top of the panel, then use a flat-blade trim tool all around the panel to release the hidden clips **(see illustration). Caution:** *Don't pull on the panel as this will rip the plastic retainer clips out of the panel.*

5   Installation is the reverse of removal.

Liftgate latch control and lock actuator

Refer to illustrations 22.8, 22.10, 22.11 and 22.12

6   On models equipped with an electric liftgate lock actuator, disconnect the negative battery cable from the battery.

7   Refer to Step 1 through 4 and remove the interior panel and watershield.

8   Disconnect the liftgate lock actuator electrical connector **(see illustration)**. Using the appropriate size drill bit, drill out the center of the mounting rivet and remove the rivet.

9   Disconnect the rod connecting the actuator to the latch mechanism. On some models, the rod has a bent S-shape at the end, which fits into a hole in the latch control assembly. This type may require you to remove the latch control assembly to angle the rod end out to disengage the actuator.

10   Apply alignment marks around the window latch bolts, then disconnect the electrical connector and remove the window latch **(see illustration)**.

11   Remove the vent panel by prying out the plastic rivets **(see illustration)**.

12   Disconnect the liftgate latch rods and liftgate handle rod from the control assembly **(see illustration)**. Remove the control assembly mounting bolts and remove the control assembly.

13   Installation is the reverse of removal.

Outside lock cylinder

Refer to illustration 22.14

14   On earlier models, use a pair of pliers and slide the cylinder retainer clip away from the lock cylinder and liftgate, much like a door lock. On 1997 and later models, remove the latch control assembly and actuator (see Steps 6 through 12) and the two nuts holding the lock cylinder on the inside of the liftgate

22.3 Remove the liftgate handle screws

(see illustration).

15   Remove the lock cylinder from the liftgate.

16   Install the lock cylinder into the liftgate opening from the outside and push the retaining clip into place. Make sure it's seated completely. On later models, reinstall the two nuts.

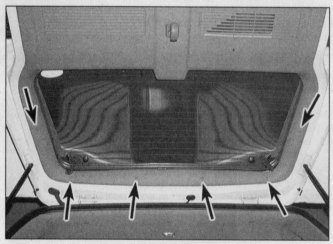

22.4 Remove the six plastic rivets (arrows) before prying the panel clips off

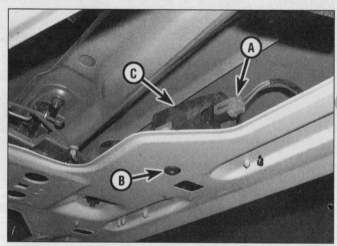

22.8 Disconnect the electrical connector (A), drill out the rivet (B), and remove the actuator (C)

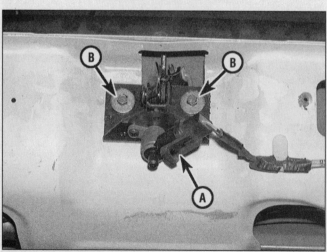

22.10 Disconnect the electrical connector (A) and remove the nuts (B) to remove the window latch

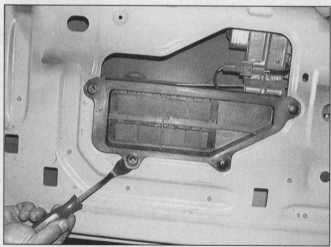

22.11 Remove the plastic rivets and remove the vent panel for access to the latch rods

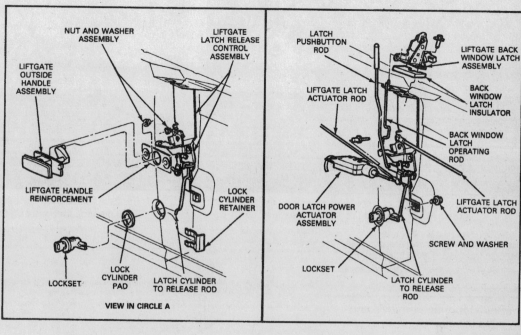

22.12 Liftgate latch control assembly installation details

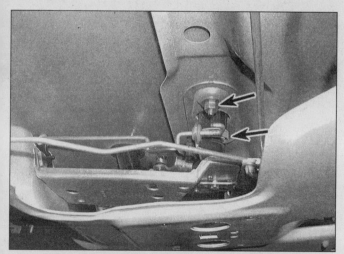

22.14 On later models, remove the two nuts (arrows) and the lock cylinder

22.17 Open the two clips (arrow indicates one) and release the latch rods

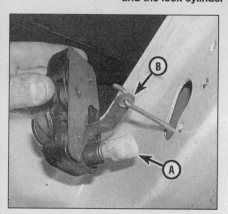

22.18 Remove the three bolts, pull the latch away and disconnect the electrical connector (A) and rod (B)

Liftgate latch

Refer to illustrations 22.17 and 22.18

17   Disconnect the actuator rods from the latch **(see illustration)**.

18   Remove the screws securing the latch to the liftgate and pull the rod and latch out as an assembly. On models so equipped, pull the latch assembly out and disconnect the electrical connector from the back **(see illustration)**.

19   Installation is the reverse of removal. **Note:** *The control rods should be snapped into their clips when the rods have no tension on them.*

23   Liftgate glass - removal and installation

Refer to illustration 23.4

1   Open the liftgate window and support it in the open position.

2   On early models, remove the retainer rings and clevis pins from both support cylinders. On later models, pry the support cylinder sockets from their ball-studs **(see illustration 21.2)**.

3   Disconnect the electrical connectors for the rear glass defogger (at the top of the glass on each side).

4   Remove the screws securing the hinges and remove the glass panel assembly. On

11

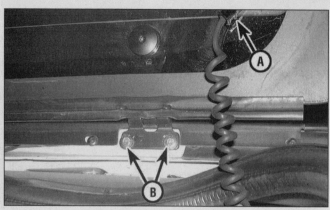

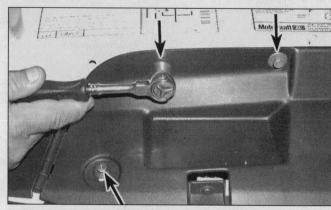

23.4 Disconnect the defogger wire (A) and remove the two hinge screws (B) on each side of the liftgate glass

24.4 Remove the screws (arrows) and plastic rivets (on end pieces), then take off the plastic shield

some models, the screws are concealed by plastic access covers **(see illustration)**.

5   Installation is the reverse of removal.

24   Radiator grille - removal and installation

Warning: *On models equipped with airbags, always disconnect the negative battery cable, then the positive battery cable and wait two minutes before working in the vicinity of the impact sensors, steering column or instrument panel to avoid the possibility of accidental deployment of the airbag, which could cause personal injury (see Chapter 12).*

1996 and earlier models

Refer to illustrations 24.2a and 24.2b

1   Open the hood.

2   Remove the screws and detach the grille from the vehicle **(see illustrations)**.

3   Installation is the reverse of the removal procedure. Be careful not to overtighten the screws.

1997 through 2001 models

Refer to illustrations 24.4, 24.6a and 24.6b

4   Open the hood and remove the screws and plastic rivets holding the plastic shield over the top of the radiator **(see illustration)**.

5   Refer to Chapter 12 and remove the headlights.

6   Remove the grille mounting screws, and disengage the grille from the four clips along the bottom of the grille opening **(see illustrations)**.

7   Installation is the reverse of the removal procedure. **Note:** *Make sure the grille locating tabs are aligned before installing the screws.*

2001 models

8   Remove the front bumper cover (see Section 26).

9   Remove the four pin-type retainers at the base of the grille.

10   Release the six clips around the perimeter of the grille and remove the grille.

11   Installation is the reverse of the removal procedure.

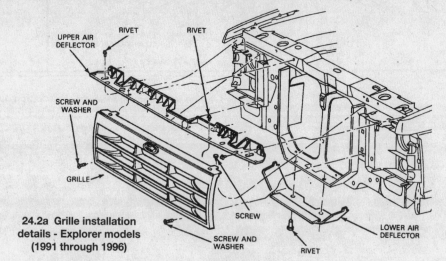

24.2a Grille installation details - Explorer models (1991 through 1996)

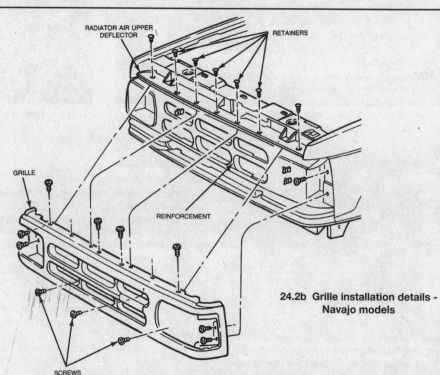

24.2b Grille installation details - Navajo models

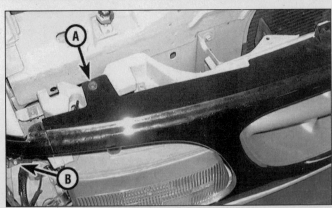

24.6a Grille mounting screws on 1997 and later models - on each side there is one screw at the top (A) and one in the marker light area (B)

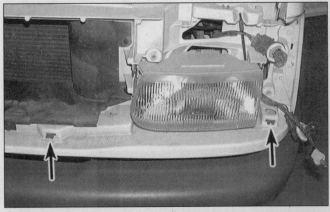

24.6b Release the grille from the clips (arrows) on each side and along the bottom - grille removed for clarity

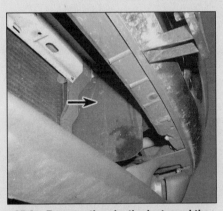

25.2a Remove the plastic rivets and the air deflector (arrow indicates one) on each side of the radiator on 1997 and later models (Explorer shown, Mountaineer similar, but with bolted support brackets where the deflectors are shown here)

25.2b Remove the bolt (arrow) holding the grille reinforcement panel to the hood latch support

25.2c Remove the grille opening reinforcement panel by removing the nut (arrow) from underneath . . .

25 Front fender - removal and installation

Warning: *On models equipped with airbags, always disconnect the negative battery cable, then the positive battery cable and wait two minutes before working in the vicinity of the impact sensors, steering column or instrument panel to avoid the possibility of accidental deployment of the airbag, which could cause personal injury (see Chapter 12).*

Removal

Refer to illustrations 25.2a, 25.2b, 25.2c, 25.2d, 25.2e, 25.3 and 25.7

1   Remove the grille (see Section 24) and the headlight trim and housing.

2   On 1997 and later models, remove the grille opening reinforcement panel **(see illustrations)** and the front bumper (refer to Section 26). Also remove the running board (if so equipped) and the front section of the rocker panel trim.

3   Open the front door and then remove the single fender bolt **(see illustration)**. On 1997 and later models remove, fender-to-radiator core support bolt. Remove the antenna if it interferes.

4   Open the hood and remove the four bolts across the top of the fender. **Note:** *On models with speed control, unbolt the speed control from the fender when removing the right-side fender.*

5   Raise the vehicle and support it securely on jackstands.

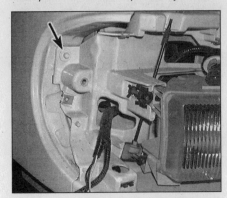

25.2d . . . and the bolt (arrow) on the front side retaining the panel to the fenders

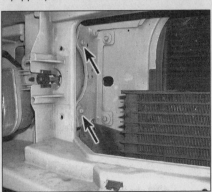

25.2e Remove the support bracket bolts (arrows) on each side of the radiator and remove the panel

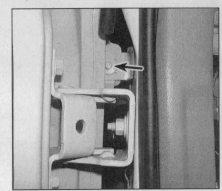

25.3 Remove the front fender bolt (arrow) through the door opening

11

6   Remove the screws or plastic rivets and detach the inner fender panel. Removing the front wheel/tire makes the job easier.

7   Remove the bolt at the lower front edge of the fender (see illustration).

8   Remove the two bolts attaching the fender to the rocker panel. Carefully detach the fender from the vehicle. **Note:** *On models so equipped, remove the running board or front section of the lower rocker panel trim to access the two lower-rear fender bolts.*

Installation

9   Installation is the reverse of the removal procedure, but don't fully tighten the bolts.

10   Check the alignment of the fender to the door edge and the hood (close the hood to check for an even gap), then tighten the bolts securely.

26   Front bumper and valance - removal and installation

Warning: *On models equipped with airbags, always disconnect the negative battery cable, then the positive battery cable and wait two minutes before working in the vicinity of the impact sensors, steering column or instrument panel to avoid the possibility of accidental deployment of the airbag, which could cause personal injury (see Chapter 12).*

Note: *The bumper is heavy and somewhat awkward to remove and install; at least two people should perform this procedure.*

Valance

1   Remove the screws and plastic rivets securing the valance to the bumper, pull the valance back and remove it.

2   Installation is the reverse of removal.

Front bumper (1991 through 2000 models)

Refer to illustrations 26.4 and 26.6

3   Raise the front of the vehicle, support it securely on jackstands, block the rear wheels and set the parking brake.

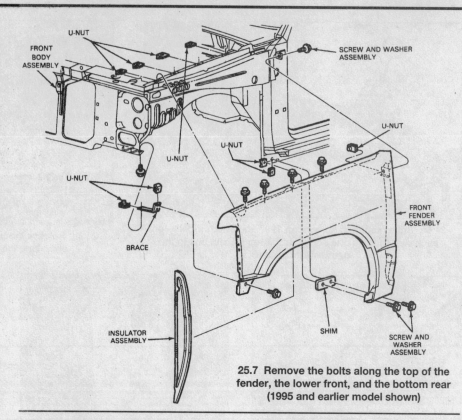

25.7 Remove the bolts along the top of the fender, the lower front, and the bottom rear (1995 and earlier model shown)

4   On 1997 through 2000 models, remove the plastic air deflector panel below the radiator (see illustration).

5   On models equipped with optional fog lights, disconnect the electrical connectors from the fog lights, then unbolt and remove the fog lights.

6   Have an assistant hold onto the bumper assembly, then remove the bolts/nuts on each side securing the bumper to the frame (see illustration). On some models, there are studs on the bumper and nuts to hold them to the chassis.

7   Installation is the reverse of the removal steps with the following additions:

   a) *If you're installing a new bumper, transfer the license plate holder and any additional trim items to the new bumper.*

   b) *The bumper can be adjusted by repositioning the fasteners in the brackets. Tighten the fastener nuts securely.*

Front bumper (2001 models)

Note: *The grille opening is an integral part of the front bumper.*

8   Open the hood.

9   Remove the bolts and retaining clips and remove the radiator grille sight shield.

10   Remove both headlight and parking lamp assemblies (see Chapter 12).

11   Remove the three bolts and remove the lower radiator air deflector.

12   On each side, remove the three fender splash shield bolts.

13   Within the wheel well, remove the front bumper cover bolt on each side.

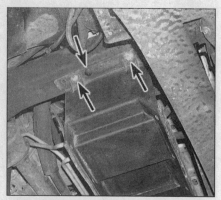

26.4 Where applicable, remove the air deflector panel below the radiator (arrows indicate bolts and plastic rivet)

26.6 Remove the two nuts (arrows) on each side to remove the front bumper from the frame

27.2 Rear bumper mounting bolts are welded to a strap (arrow) - remove the two nuts on the opposite side and pull the bolts/strap out

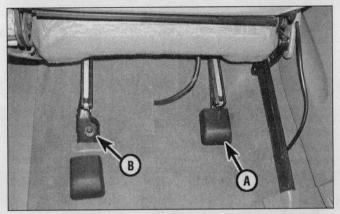

29.5 Pry off the plastic covers (A indicates one in place) and remove the rear bolts (B indicates one bolt with cover removed)

29.6 Remove the plastic cover (A) and the bolt (B) at the slider bar

14   Release the six locking tabs and remove the front bumper cover assembly.

15   Installation is the reverse of the removal procedure.

---

27   Rear bumper - removal and installation

Note: *The bumper is heavy and somewhat awkward to remove and in-stall - at least two people should perform this procedure.*

Removal

Refer to illustration 27.2

1   Remove the license plate light socket from the bumper (see Chapter 12).

2   Have an assistant hold onto the bumper assembly and remove the nuts and retainer on each side securing the bumper to the frame **(see illustration). Note:** *If there were any spacers between the bumper brackets and the frame, save them for use on reassembly.*

3   If necessary, remove the bumper brackets, stone deflector and step plate.

Installation

4   Installation is the reverse of removal with the following additions:

 a) If you're installing a new bumper, transfer the brackets, stone deflector and any additional items to the new bumper.

 b) When installing the step pad, start at the locator hole at the center of the bumper.

 c) Loosely install the bumper fasteners, then adjust the bumper position so there is a 3/4-inch horizontal gap between the bumper and body.

 d) Tighten the fasteners securely after adjustment.

---

28   Windshield glass - replacement

Replacement of the windshield and fixed glass requires the use of special fast-setting adhesive/caulk materials and some specialized tools and techniques. These operations should be left to a dealer service

29.7 Front seat-to-floor bolts (arrows)

department or a shop specializing in automotive glass work.

---

29   Seats - removal and installation

Warning: *On models equipped with air bags, always disconnect the negative battery cable, then the positive cable, then wait two minutes before working in the vicinity of the impact sensors, steering column, instrument panel or front seats (2001 models so equipped) to avoid the possibility of accidental deployment of the airbag, which could cause personal injury (see Chapter 12).*

Front seat

1996 and earlier models

1   Remove the pin securing the insulators at the front of the seat track, then remove both seat track insulators.

2   Remove the four bolts securing the seat track to the floorpan and lift the seat from the vehicle.

3   Installation is the reverse of the removal steps. Tighten the retaining bolts securely.

1997 through 2001 models (except 2001 models with side airbags)

Refer to illustrations 29.5, 29.6 and 29.7

4   On models with power seats, position

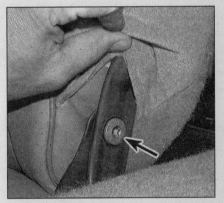

29.10 The seat cushion frame is secured to the seat back frame with Torx screws

the seat all the way up and disconnect the battery (see Chapter 1).

5   At the rear of the seat, remove the seat-to-floor bolts **(see illustration).**

6   Remove the bolt from the slider bar **(see illustration).**

7   Remove the front seat-to-floor bolts and remove the seat **(see illustration).** On power models, disconnect the electrical connectors underneath before removing the seat from the vehicle.

2001 models with side air bags

On models equipped with side air bags, do not attempt to remove the front seats, since a special scan-type tool is required. Take the vehicle to a dealer service department.

Rear seat cushion - 60/40 split seat

Refer to illustrations 29.10 and 29.11

8   Remove the rear seat latch handle.

9   Remove the outboard cushion latch cover.

10   Remove the Torx screws that secure the seat back to the cushion frame **(see illustration).** Separate the cushion from the seat back by pulling it forward while holding the seat back in place.

11

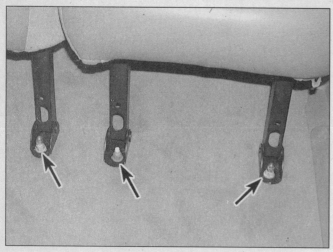

29.11 The seat cushion frame is secured to the floor with nuts (arrows)

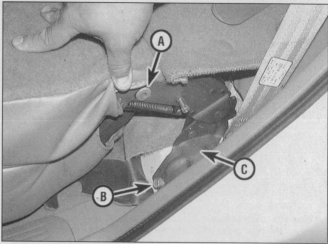

29.14 Remove the Torx bolt (A) on each side of the seat - if necessary, remove the floor bolt (B) and seat belt bolt/nut (C)

11   Remove the nuts that secure the seat cushion and frame **(see illustration)**. Take the seat cushion and frame out.

12   Installation is the reverse of removal. Tighten the nuts and Torx screws securely.

Rear seat back - 60/40 split seat

Refer to illustration 29.14

13   Remove the seat latch handle and cushion latch cover.

14   Remove the Torx screws that secure the seat back to the cushion frame **(see illustration)**.

15   Pull the seat back away from the cushion and remove it.

16   Remove the nut and bolt that secure the seat back and seat belt to the floor (on the side nearest the door).

17   Pull the seat back toward the side of the vehicle to detach it from the center pivot bracket, then lift it out.

18   If necessary, remove the nut and bolt that secure the buckle and seat back pivot bracket **(see illustration 29.14)**.

19   Installation is the reverse of removal. Tighten the nuts and Torx screws securely.

Rear seat cushion - 50/50 split seat

20   Remove the nut and detach the cushion link arm from the seat cushion.

21   Tilt the cushion forward, remove the nuts that secure it to the floor and take it out.

22   Installation is the reverse of removal. Tighten the cushion pivot and link arm nuts securely.

Rear seat back - 50/50 split seat

23   Remove the nut and detach the cushion link arm from the seat cushion.

24   Remove the nut and bolt that secure the seat back latch and seat belt to the floor (on the side nearest the door). **Note:** *A special seat-belt bit, available at your auto parts store, is required to remove seat belt bolts.*

25   Tilt the seat back forward, pull it toward the door to disengage it from the center pivot

31.2a Push down on the ash tray retainer and pull the ash tray out

and take it out of the vehicle.

26   If necessary, remove the link arm cover (if equipped) and link arm from the seat back.

27   If necessary, remove the pivot bracket and seat buckle from the floor.

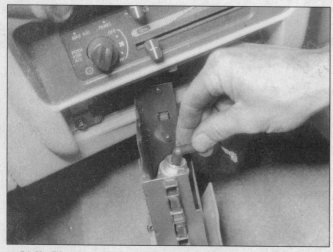

31.2b Disconnect the electrical connector from the cigarette lighter

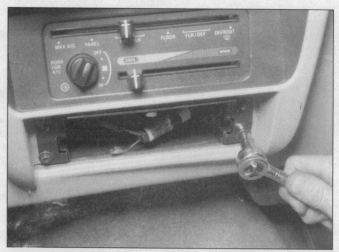

31.3a Remove the ash tray bracket screws

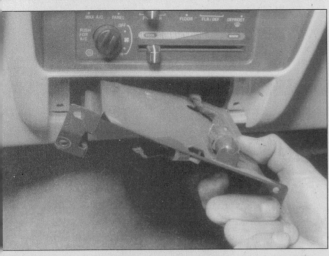

31.3b Pull the bracket out, disconnect the electrical connector and remove the bracket

31.4a Pull back sharply on the edge of the instrument cluster bezel . . .

28   Installation is the reverse of the removal steps. Tighten the retaining bolts and nuts securely.

30   Seat belt check

1   Check the seat belts, buckles, latch plates and guide loops for obvious damage and signs of wear.
2   Check that the seat belt reminder light comes on when the ignition key is turned to the Run or Start position.
3   The seat belts are designed to lock up during a sudden stop or impact, yet allow free movement during normal driving. Check that the retractors return the belt against your chest while driving and rewind the belt fully when the buckle is unlatched.
4   If any of the above checks reveal problems with the seat belt system, replace parts as necessary. **Note:** *A special seat-belt bit, available at your auto parts store, is required to remove seat belt bolts.*

31   Instrument cluster bezel - removal and installation

Warning: *On models equipped with airbags, always disconnect the negative battery cable, then the positive battery cable and wait two minutes before working in the vicinity of the impact sensors, steering column or instrument panel to avoid the possibility of accidental deployment of the airbag, which could cause personal injury (see Chapter 12).*

1994 and earlier models

Refer to illustrations 31.2a, 31.2b, 31.3a, 31.3b, 31.4a and 31.4b
1   Disconnect the negative battery cable from the battery.
2   Squeeze the ash tray retaining clip and pull the ash tray out. Disconnect the cigarette lighter electrical connector and remove the ash tray **(see illustrations)**.
3   Remove the ashtray bracket screws and pull the bracket out **(see illustrations)**. Dis-

connect the electrical connector at the rear of the bracket and remove it.
4   Carefully pull back the edge of the instrument cluster bezel near the ashtray and disengage the clips from the instrument panel **(see illustrations)**. Work around the perimeter of the bezel, detaching each clip. Depress the hazard warning switch on the steering column to provide additional clearance, disconnect the 4WD electronic shift electrical connector (if equipped) and remove the bezel.
5   Installation is the reverse of the removal procedure.

1995 and later models

Refer to illustrations 31.7, 31.8a, 31.8b, 31.9a and 31.9b
6   Disconnect the negative battery cable, then the positive battery cable and wait two minutes before proceeding any further.
7   Unbolt both the parking brake release handle bracket from the dash, and the hood release handle bracket **(see illustration)**.

31.4b . . . to disengage the clips from the instrument panel

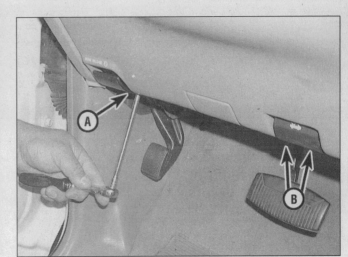

31.7 Remove the bolts from the parking brake release (A) and the hood release (B)

11

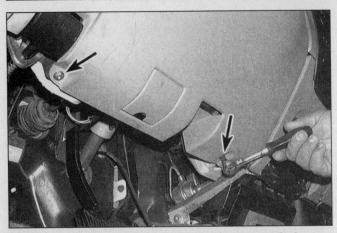

31.8a Remove the screws (arrows) at the bottom of the lower cover and pull the clips out at the top to remove the lower instrument panel cover

31.8b Remove the screws (arrows) and the steel reinforcement panel

8   Remove the lower instrument panel cover below the steering column, then remove the four screws and the steel reinforcement panel **(see illustrations)**.

9   Remove the bezel retaining screws **(see illustrations)**.

10   Tilt the steering wheel down to the lowest position. Pull the top of the instrument cluster bezel outward while disengaging the lower half. Remove the bezel from the vehicle.

11   Installation is the reverse of removal.

32   Instrument panel - removal and installation

Warning: *On models equipped with airbags, always disconnect the negative battery cable, then the positive battery cable and wait two minutes before working in the vicinity of the impact sensors, steering column or instrument panel to avoid the possibility of accidental deployment of the airbag, which could cause personal injury (see Chapter 12).*

Note: *The instrument panel is heavy and somewhat awkward to remove and install - at least two people should perform this procedure.*

1994 and earlier models

Refer to illustration 32.9

1   Disconnect the cable from the negative battery terminal.

2   Remove the screws retaining the hood release cable to the lower instrument panel cover. Remove the screws at the bottom of the lower cover and unsnap the top of the cover from the instrument panel. Remove the cover.

3   Remove the lower instrument panel cover reinforcement brace.

4   Remove the ash tray and the ashtray retainer.

5   Remove the steering column covers (see Section 33).

6   Remove the instrument cluster bezel (see Section 31) and the instrument cluster (see Chapter 12).

7   Remove the inside A-pillar moldings and the kick-panels from each side. Remove the passenger-side lower insulation panel.

8   Remove the nut retaining the instrument panel to the brake pedal support bracket and disconnect the wiring harness from the steering column.

9   Remove the bolts retaining each side of the instrument panel to the cowl sides and

the screws retaining the instrument panel to the cowl top **(see illustration)**. Carefully pull the instrument panel back slightly and disconnect all the electrical connectors from the lights, switches, radio and heater/air conditioning control assembly. Disconnect the antenna from the radio and the vacuum lines and control cables from the heater/air conditioning control assembly.

10   Carefully remove the instrument panel from the vehicle.

11   Installation is the reverse of removal.

1995 and later models

Refer to illustrations 32.16, 32.19, 32.20, 32.21, 32.23, 32.27a, 32.27b, 32.27c and 32.27d

12   Disconnect the negative battery cable, then the positive battery cable and wait two minutes before proceeding any further.

13   Position the steering wheel in the lock position. **Note:** *Make sure the steering wheel remains in the lock position during the entire procedure or damage to the airbag sliding contact will occur.*

14   Remove the center console, if equipped (see Section 36).

15   Remove the lower instrument panel

31.9a Remove the screws (arrows) at the top of the bezel . . .

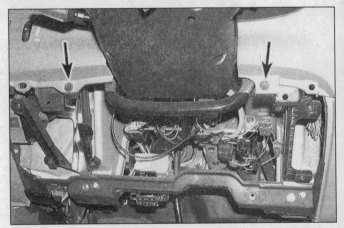

31.9b . . . and the two below the bezel (arrows) and remove the instrument cluster bezel

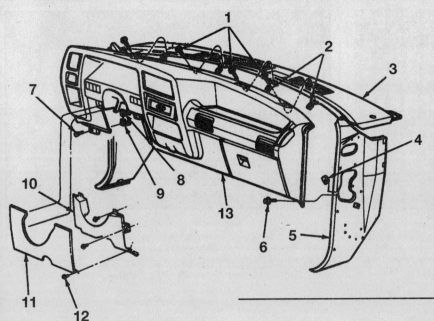

**32.9 Instrument panel installation details –
1994 and earlier models**

1   Instrument panel-to-cowl top screws
2   Nut inserts
3   Cowl top
4   U-nut
5   Cowl side
6   Instrument panel mounting bolt (right side)
7   Instrument panel mounting bolt (left side)
8   U-nut
9   Parking brake release bracket
10 Lower instrument panel cover reinforcement
11 Lower instrument panel cover
12 Screw
13 Instrument panel

cover, the reinforcement panel and the
instrument cluster bezel (see Section 31).
Remove the instrument cluster (see Chapter
12).

16   Remove the lower insulation panel on
the passenger side, by removing the plastic
pushpins and rivets **(see illustration)**.

17   Remove the A-pillar moldings from both
sides. Remove both kick panels by removing
the plastic rivets.

**32.16 Instrument panel installation details –
1995 and later models**

1   Defroster grille
2   Passenger airbag cover
3   Screw
4   Screw
5   Instrument panel
6   Glove box
7   Pushpin
8   Rivet
9   Sound insulator
10 Screw
11 Center bracket
12 Screw
13 Nut
14 Brace
15 Screw
16 Ash tray
17 Screw
18 Center trim panel
19 Radio opening panel
20 Access cover
21 Lower cover
22 Reinforcement panel
23 Nut
24 Steering column reinforcement
25 Screw
26 Shift indicator
27 Instrument cluster
28 Fuse panel cover
29 Instrument cluster bezel

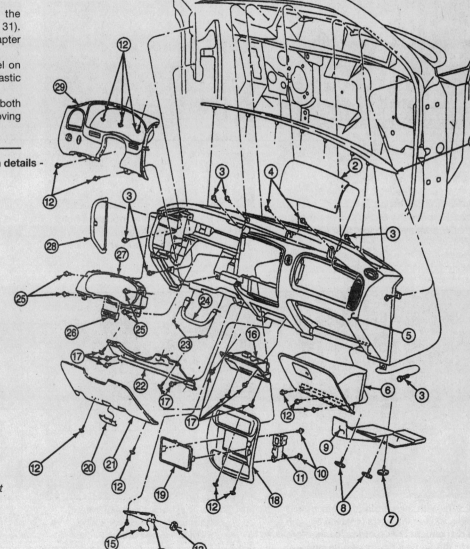

11

32.19 Remove the two bolts (arrows) inside the glove box area, pull out the passenger air bag enough to disconnect its electrical connector, then remove the airbag

32.20 After removing the lighter and the knob for the rear washer switch, remove the center trim panel screws (arrows)

32.21 Remove the bolts (arrows) and the instrument panel brace

32.23 Remove the two mounting bolts (arrows) behind the fuse panel cover at the left of the instrument panel

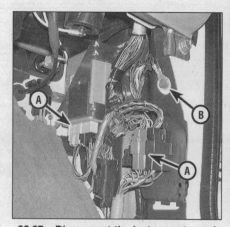

32.27a Disconnect the instrument panel electrical connectors (A) and ground wires (B) in the right kick-panel area

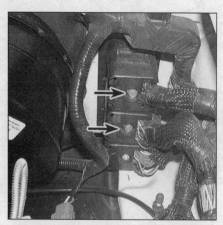

32.27b With the distribution box pulled aside, loosen the bolts (arrows) enough to disconnect the electrical connectors at the firewall on the engine side - the bolts do not come out

32.27c Pull back on the bulkhead connector on the driver's side of the lower firewall

32.27d With the instrument panel pulled away, remove the bolt (arrow) in the center of the main electrical connector

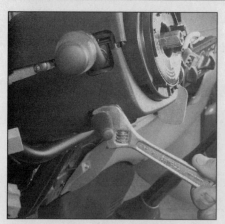

33.3 Use a small wrench to remove the tilt wheel lever (if equipped)

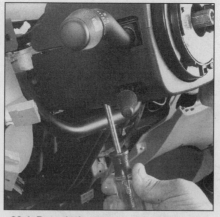

33.4 Detach the screws - then separate the steering column cover halves to remove them from the column

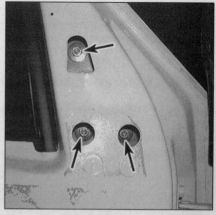

34.3 Remove the three nuts (arrows) and detach the mirror

18 Open the glove box, press the release tabs in and lower the glove box. Remove the hinge retaining screws and remove the glove box.

19 Remove the two mounting screws for the passenger airbag, pull the unit away from the instrument panel and disconnect the electrical connector **(see illustration)**. *Warning: Handle the airbag module with care! Always carry the module with the trim side facing away from your body and place the module in a secure location with the trim side facing up.*

20 Remove the two screws and carefully pull the center trim panel off the instrument panel **(see illustration)**. Remove the radio (see Chapter 12) and the heater/air conditioning control assembly (see Chapter 3).

21 At the right of the throttle pedal, remove the bolts and the lower instrument panel brace **(see illustration)**.

22 Remove the steering column mounting bolts, lower and support the steering column.

23 Pry off the fuse panel cover at the left edge of the instrument panel and remove the two bolts **(see illustration)**.

24 Remove the one bolt on the lower right side of the instrument panel.

25 Pry out the defroster grilles and remove the screws securing the upper edge of the instrument panel.

26 Carefully pull the instrument panel back slightly and disconnect all the electrical connectors and wiring harness retainers attached to the instrument panel.

27 On 1997 and later models, it is possible to leave the harness attached to the instrument panel. Disconnect the wiring harness from the following connectors:

a) *Disconnect the electrical connector at the brake pedal position switch (on the brake pedal).*

b) *At the lower right side of the cowl, disconnect all the electrical harness connectors and ground wires routed to the instrument panel* **(see illustration)**.

c) *In the left side of the engine compartment pull the power distribution box out of its clips and move it aside enough to loosen the bolts on the bulkhead connectors* **(see illustration)**. *Disconnect the connectors.*

d) *Use the handle to pull out the electrical connector on the driver's side* **(see illustration)**.

e) *Pull the instrument panel back and loosen the bolt at the large bulkhead connector until the connector can be disconnected* **(see illustration)**.

28 Installation is the reverse of the removal procedure. *Warning: Do not connect the battery until all electrical connections have been secured, especially the passenger airbag.*

33 Steering column cover - removal and installation

Refer to illustrations 33.3 and 33.4

Warning: *On models equipped with airbags, always disconnect the negative battery cable, then the positive battery cable and wait two minutes before working in the vicinity of the impact sensors, steering column or instrument panel to avoid the possibility of accidental deployment of the airbag, which could cause personal injury (see Chapter 12).*

1 Disconnect the negative battery cable.

2 Remove the key lock cylinder (see

Chapter 12). **Note:** *This is only necessary for removal of the upper column cover.*

3 Remove the tilt wheel lever **(see illustration)**.

4 Remove the screws from the lower steering column cover **(see illustration)**.

5 Separate the cover halves and detach them from the steering column.

6 Installation is the reverse of removal.

34 Sideview mirrors - removal and installation

Refer to illustration 34.3

1 Remove the door trim panel and the plastic watershield (see Section 16).

2 Disconnect the electrical connector from the mirror. On most models, this will require removal of the radio speaker for that door to access the mirror connector (see Chapter 12 for speaker removal).

3 Peel back the insulator pad, then remove the three mirror retaining nuts and detach the mirror from the vehicle **(see illustration)**.

4 Installation is the reverse of removal.

35 Cowl cover - removal and installation

Refer to illustration 35.2

1 Remove the windshield wiper arms (see Chapter 12).

2 Remove the screw securing the left hand and right hand cowl covers **(see illustration)**, then pull up both cowl panels, which are held to the body with clips.

3 If the panels are being removed for replacement (not just for access to something under the covers), disconnect the windshield washer hoses from the nozzles on the panels. **Caution:** *The plastic washer jets are extremely fragile. Exercise great care when removing the hoses. It is safer to simply cut the hoses and splice them later using a small piece of plastic or copper tubing.*

4 Installation is the reverse of removal.

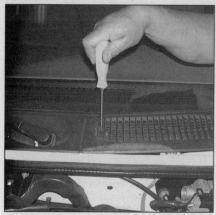

35.2 Remove the center screw, then pry up the left and right cowl panels, held by clips

11

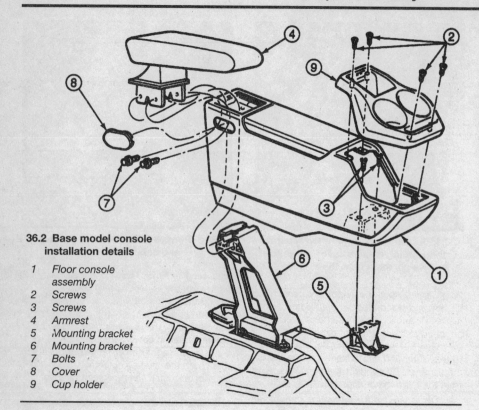

36.2 Base model console installation details

1   Floor console assembly
2   Screws
3   Screws
4   Armrest
5   Mounting bracket
6   Mounting bracket
7   Bolts
8   Cover
9   Cup holder

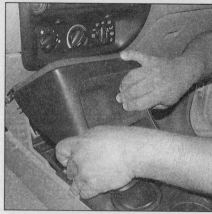

36.6 Remove the front trim panel/message center from the console

36   Center console - removal and installation

Warning: *On models equipped with airbags, always disconnect the negative battery cable, then the positive battery cable and wait two minutes before working in the vicinity of the impact sensors, steering column or instrument panel to avoid the possibility of accidental deployment of the airbag, which could cause personal injury (see Chapter 12).*
1   Disconnect the negative battery cable, then the positive battery cable and wait two minutes before proceeding any further.

Base model

Refer to illustration 36.2
2   Remove the two plastic screw covers on each side of the console **(see illustration)**. Remove the four bolts and the armrest.
3   Remove the screw in the armrest opening.
4   Remove the screws retaining the cup holder and remove the cup holder. Remove the two screws securing the front half of the console and remove the console.
5   Installation is the reverse of removal.

High line model

Refer to illustrations 36.6, 36.7 and 36.8
6   Using your hands, press in on the sides of the front trim panel/message center unsnapping the retaining clips **(see illustration)**. Disconnect the electrical connectors and remove the panel.
7   Using the same procedure, remove the rear trim panel/cup holder from the console **(see illustration)**.
8   At the front of the console, loosen the bolt in the center of the main electrical connector, and separate the connector halves. Remove the two console mounting bolts **(see illustration)**.
9   At the center of the console, below the cup holder, remove the two mounting bolts, then slide the console rearward to free it from the rear floor mount.
10   Installation is the reverse of removal.

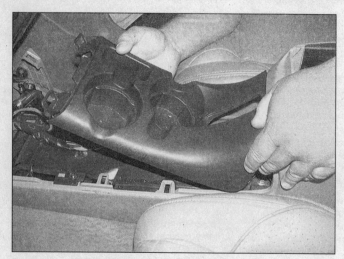

36.7 Remove the trim panel/cup holder

36.8 Loosen the bolt (A) and disconnect the electrical connector at the front of the console, then remove the two console mounting bolts (B)

Chapter 12
Chassis electrical system

Contents

1 General information

Warning: *To prevent electrical shorts, fires and injury, always disconnect the cable from the negative terminal of the battery before checking, repairing or replacing electrical components.*

The chassis electrical system of this vehicle is a 12-volt, negative ground type. Power for the lights and all electrical accessories is supplied by a lead/acid-type battery which is charged by the alternator.

This chapter covers repair and service procedures for various chassis (non-engine related) electrical components. For information regarding the engine electrical system components (battery, alternator, distributor and starter motor), see Chapter 5.

It should be noted that when portions of the electrical system are serviced, the negative battery cable should be disconnected from the battery to prevent electrical shorts and/or fires.

2 Electrical troubleshooting - general information

A typical electrical circuit consists of an electrical component, any switches, relays, motors, fuses, fusible links or circuit breakers, etc. related to that component and the wiring and connectors that link the components to both the battery and the chassis. To help you pinpoint an electrical circuit problem, wiring diagrams are included at the end of this manual.

Before tackling any troublesome electrical circuit, first study the appropriate wiring diagrams to get a complete understanding of what makes up that individual circuit. Trouble spots, for instance, can often be isolated by noting if other components related to that circuit are often routed through the same fuse and ground connections.

Electrical problems usually stem from simple causes such as loose or corroded connectors, a blown fuse, a melted fusible link or a bad relay. Visually inspect the condition of all fuses, wires and connectors in a problem circuit before troubleshooting it.

The basic tools needed for electrical troubleshooting include a circuit tester, a high impedance (10 M-ohm) digital voltmeter, a continuity tester and a jumper wire with an inline circuit breaker for bypassing electrical components. Before attempting to locate or define a problem with electrical test instruments, use the wiring diagrams to decide where to make the necessary connections.

Voltage checks

Perform a voltage check first when a circuit is not functioning properly. Connect one lead of a circuit tester to either the negative battery terminal or a known good ground.

Connect the other lead to a connector in the circuit being tested, preferably nearest to the battery or fuse. If the bulb of the tester lights up, voltage is present, which means that the part of the circuit between the con- nector and the battery is problem free. Continue checking the rest of the circuit in the same fashion.

When you reach a point at which no voltage is present, the problem lies between that point and the last test point with voltage. Most of the time the problem can be traced to a loose connection. **Note:** *Keep in mind that some circuits receive voltage only when the ignition key is in the Accessory or Run position.*

Finding a short circuit

One method of finding shorts in a circuit is to remove the fuse and connect a test light or voltmeter in its place. There should be no voltage present in the circuit. Move the electrical connectors from side-to-side while watching the test light. If the bulb goes on, there is a short to ground somewhere in that area, probably where the insulation has been rubbed through. The same test can be performed on each component in a circuit, even a switch.

Ground check

Perform a ground test to check whether a component is properly grounded. Disconnect the battery and connect one lead of a self-powered test light, known as a continuity tester, to a known good ground. Connect the other lead to the wire or ground connection being tested. If the bulb goes on, the ground is good.

If the bulb does not go on, the ground is not good.

Continuity check

A continuity check determines if there are any breaks in a circuit - if it is conducting electricity properly. With the circuit off (no power in the circuit), a self-powered continuity tester can be used to check the circuit. Connect the test leads to both ends of the circuit, and if the test light comes on the circuit is passing current properly. If the light doesn't come on, there is a break somewhere in the circuit. The same procedure can be used to test a switch, by connecting the continuity tester to the power in and power out sides of the switch. With the switch turned on, the test light should come on.

Finding an open circuit

When diagnosing for possible open circuits it is often difficult to locate them by sight because oxidation or terminal misalignment are hidden by the connectors. Merely wiggling a connector on a sensor or in the electrical connector may correct the open circuit condition. Remember this if an open circuit is indicated when troubleshooting a circuit. Intermittent problems may also be caused by oxidized or loose connections.

Electrical troubleshooting is simple if you keep in mind that all electrical circuits are basically electricity running from the battery, through the wires, switches, relays, fuses and fusible links to each electrical component (light bulb, motor, etc.) and then to ground, from which it is passed back to the battery. Any electrical problem is an interruption in the flow of electricity to and from the battery.

3   Connectors - general information

Refer to illustration 3.3

Always release the lock lever(s) before attempting to unplug inline type connectors. There are a variety of lock lever configurations. Although nothing more than a finger is usually necessary to pry lock levers open, a small pocket screwdriver is effective for hard-to-release levers. Once the lock levers are released, try to pull on the connectors them-

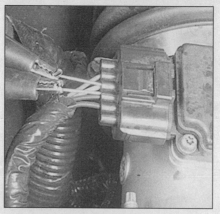

3.3 To backprobe a connector, insert a small, sharp probe (such as a straight pin) into the back of the connector, along side the wire, until it contacts the terminal inside; connect your meter leads to the probes to test a functioning circuit

selves, not the wires, when unplugging two connector halves (there are times, however, when this is not possible - use good judgment).

It is usually necessary to know which side, male or female, of the connector you're checking. Male connectors are easily distinguished from females by the shape of their internal pins.

When checking continuity or voltage with a circuit tester, insertion of the test probe into the receptacle may open the fitting to the connector and result in poor contact. Instead, insert the test probe from the wire harness side of the connector, known as "backprobing" **(see illustration).**

4   Fuses - general information

Refer to illustrations 4.1a, 4.1b, 4.1c, 4.2 and 4.7

The electrical circuits are protected by a combination of fuses, cartridge-type fusible links and circuit breakers. There are two fuse panels; one on the power distribution box in the engine compartment and one in the inte-

4.1a The power distribution box is located in the engine compartment - fuses are located under the cover

rior (see illustrations).

The fuse panels are equipped with miniaturized fuses because their compact dimensions and convenient blade-type terminal design allow fingertip removal and installation. If an electrical component fails, always check the fuse first. The best way to check the fuses is with a test light. Check for power at the exposed terminal tips of each fuse. If power is present at one side of the fuse but not the other, the fuse is blown. A blown fuse can also be identified by visually inspecting it **(see illustration).**

Each fuse protects one or more circuits. Consult your owner's manual - it will have the most accurate guide for your vehicle.

The cover of the interior fuse panel contains a fuse puller tool and spare fuses. **Note:** *On engine compartment and interior fuse boxes, there is a fuse identification guide on the inside of the cover, either stamped into the plastic or on a printed sheet.*

To open the engine compartment fuse panel (also called a power distribution box), press on the tab above the Ford emblem, then slide the cover out of its tabs and lift it off. To open the footwell fuse panel on 1991

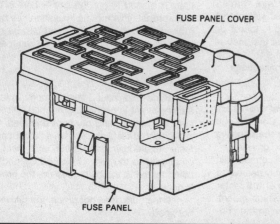

FUSE PANEL COVER

FUSE PANEL

4.1b On 1991 through 1994 models, the interior fuse panel is located above the driver's footwell - lift the handle and let the panel swing down for access

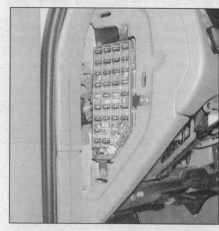

4.1c The interior fuse panel on 1995 and later models is at the left end of the instrument panel (shown with plastic cover removed)

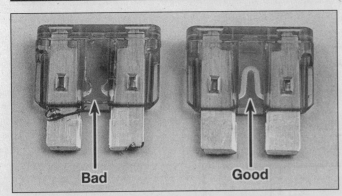

4.2 The fuses can easily be checked visually to see if they are blown (the fuse on the left is blown)

Fuse Value Amps	Color Code
4	Pink
5	Tan
10	Red
15	Light Blue
20	Yellow
25	Natural
30	Light Green

4.7 Each fuse amp value has a corresponding color code

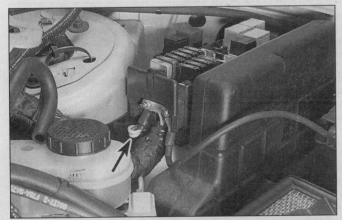

5.2a Conventional type fusible links (arrow) can be found exiting the power distribution box in the engine compartment

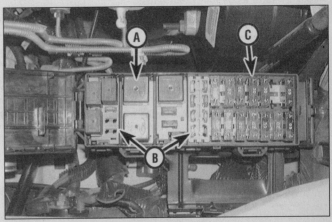

5.2b On later models the power distribution box contains relays, fusible links and fuses

A   Relays
B   Mini-fuses

C   Cartridge-type fusible links (maxi-fuses)

through 1994 models, lift the handle and let the fuse panel swing down. On 1995 and later models, pry open the plastic cover at the left end of the instrument panel (access with the driver's door open). Use the plastic fuse puller to remove and insert fuses. Don't twist the fuses during removal or installation. Twisting could force the terminal open too far, resulting in a bad connection.

Be sure to replace blown fuses with the correct type and amp rating. Fuses of different ratings are physically interchangeable, but replacing a fuse with one of a higher or lower value than specified is not recommended. Each electrical circuit needs a specific amount of protection. The amperage value of each fuse is usually molded into the fuse body. Different colors are also used to denote fuses of different amperage types. The accompanying color code **(see illustration)** shows common amperage values and their corresponding colors. **Caution:** *Always turn off all electrical components and the ignition switch before replacing a fuse. Never bypass a fuse with pieces of metal or foil. Serious damage to the electrical system could result.*

If the replacement fuse immediately fails, do not replace it again until the cause of the problem is isolated and corrected. In most cases, this will be a short circuit in the wiring caused by a broken or deteriorated wire.

5   Fusible links - general information

Refer to illustrations 5.2a and 5.2b

Some circuits are protected by fusible links. These links are used in circuits which are not ordinarily fused, such as the ignition circuit.

In addition to the conventional type of fusible link (described below) which is located in the wiring harness **(see illustration)**, there is also cartridge type fusible links located in the engine compartment fuse block that are similar to a large fuses and, after disconnecting the negative battery cable, are simply unplugged and replaced by a unit of the same amperage **(see illustration)**. Some fusible links are held in place by a screw which must be loosened before removing the link.

Conventional type fusible links cannot be repaired, a new link of the same size wire should be installed in its place. The procedure is as follows:

a) *Disconnect the cable from the negative battery terminal.*
b) *Disconnect the fusible link from the wiring harness.*
c) *Cut the damaged fusible link out of the wiring just behind the connector.*

d) *Strip the insulation back approximately 1/2-inch.*
e) *Position the connector on the new fusible link and crimp it into place.*
f) *Use rosin core solder at each end of the new link to obtain a good solder joint.*
g) *Use plenty of electrical tape around the soldered joint. No wires should be exposed.*
h) *Connect the battery ground cable. Test the circuit for proper operation.*

6   Circuit breakers - general information

Refer to illustration 6.3

Circuit breakers protect accessories such as the windshield wipers, windshield washer pump, interval wiper, low washer fluid, etc. Circuit breakers are located in the power distribution box or interior fuse box. Refer to the cover of your fuse box or the fuse panel guide in your owner's manual for the location of the circuit breakers used in your vehicle.

Because a circuit breaker resets itself automatically, an electrical overload in a circuit breaker protected system will cause the circuit to fail momentarily, then come back on. If the circuit does not come back on, check it immediately.

12

6.3 Perform a continuity test with an ohmmeter to check a circuit breaker - infinite resistance indicates a bad circuit breaker

a) *Remove the circuit breaker from the fuse panel.* **Note:** *The circuit breaker for the rear wiper is mounted behind the glove compartment.*

b) *Using an ohmmeter, verify that there is continuity between both terminals of the circuit breaker* **(see illustration).** *If there is no continuity, replace the circuit breaker.*

c) *Install the old or new circuit breaker. If it continues to cut out, a short circuit is indicated. Troubleshoot the appropriate circuit (see the wiring diagrams at the back of this book) or have the system checked by a professional mechanic.*

7   Relays - general information and testing

General information

1   Several electrical accessories in the vehicle, such as the fuel injection system, horns, starter, and fog lamps use relays to transmit the electrical signal to the component. Relays use a low-current circuit (the control circuit) to open and close a high-current circuit (the power circuit). If the relay is

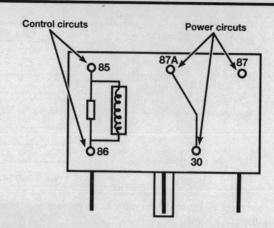

7.4 Most relays are marked on the outside to easily identify the control and power circuits

defective, that component will not operate properly. The various relays are mounted in engine compartment **(see illustration 5.2b)** and several locations throughout the vehicle (see Chapter 5). If a faulty relay is suspected, it can be removed and tested using the procedure below or by a dealer service department or other repair shop. Defective relays must be replaced as a unit.

Testing

Refer to illustration 7.4

2   It's best to refer to the wiring diagram for the circuit to determine the proper hook-ups for the relay you're testing. However, if you're not able to determine the correct hook-up from the wiring diagrams, you may be able to determine the test hook-ups from the information that follows.

3   On most relays, two of the terminals are the relay's control circuit (they connect to the relay coil which, when energized, closes the large contacts to complete the circuit). The other terminals are the power circuit (they are connected together within the relay when the control circuit coil is energized).

4   The relays are marked as an aid to help you determine which terminals are the control circuit and which are the power circuit **(see illustration).**

5   Remove the relay from the vehicle and check for continuity between the relay power circuit terminals. In the example illustrated here, there should be no continuity between terminal 30 and 87.

6   Connect a fused jumper wire between one of the two control circuit terminals and the positive battery terminal. Connect another jumper wire between the other control circuit terminal and ground. When the connections are made, the relay should click. On some relays, polarity may be critical, so, if the relay doesn't click, try swapping the jumper wires on the control circuit terminals.

7   With the jumper wires connected, check for continuity between the power circuit terminals. Now, there should be continuity between terminals 30 and 87.

8   If the relay fails any of the above tests, replace it.

8   Multi-function switch - check and replacement

Warning: *On models equipped with airbags, disconnect the negative battery cable, then the positive battery cable and wait two minutes before working in the vicinity of the*

8.2a Multi-function switch terminal identification guide

Switch positions	Hazard warning	Continuity between
Neutral	OFF ON	1 and 6; 1 and 8 3 and 5; 3 and 8 3 and 2; 3 and 6
Left	OFF	4 and 6; 4 and 2
Right	OFF	4 and 8; 4 and 5

8.2b Turn signal and hazard switch continuity chart

Switch positions	Continuity between
Flash to Pass hold lever in this position	12 and 11 9 and 10
High beam	9 and 11
Low beam	9 and 10

8.2c Headlight dimmer switch continuity chart

Switch positions	Test terminals	Ohmmeter readings (+ or – 10%)
Wiper OFF	13 and 14	47.6 K-ohms
Wash OFF	14 and 15	103.3 K-ohms
Intermitent	13 and 14	11.3 K-ohms
Low	13 and 14	4.08 K-ohms
High	13 and 14	0-ohms
Wash ON Wiper OFF	14 and 15	0-ohms
Delay control knob @ maximum position	14 and 15	103.3 K-ohms
Delay control knob @ minimum position	14 and 15	3.3 K-ohms

8.2d Windshield wiper switch continuity chart

8.5 Squeeze the upper extension shroud at the top and bottom and take it off the steering column

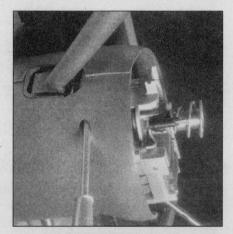

8.6a Remove the shroud attachment screws

impact sensors, steering column or instrument panel to avoid the possibility of accidental deployment of the airbag, which could cause personal injury (see Section 30).
Note: *The multi-function switch is located on the left side of the steering column. It incorporates the turn signal, headlight dimmer and windshield wiper/washer functions into one switch.*

Check

Refer to illustrations 8.2a, 8.2b, 8.2c and 8.2d

1   Remove the multi-function switch (see below).

2   Using an ohmmeter or self-powered test light and the accompanying diagrams, check for continuity between the indicated switch terminals with the switch in each of the indi-cated positions **(see illustrations)**. If the continuity isn't as specified, replace the switch.

Replacement

1991 through 1994 models

Refer to illustrations 8.5, 8.6a, 8.6b and 8.7

3   Remove the steering wheel (see Chapter 10).

4   Remove the screws that secure the instrument panel trim piece below the steering column and take the trim piece off (see Chapter 11).

5   On models equipped with a tilt steering wheel, squeeze the upper extension shroud at the six and twelve o'clock positions and release it from the retaining plate **(see illustration)**.

6   Remove the two screws securing the steering column upper and lower shrouds and remove both shrouds **(see illustrations)**.

7   Remove the two screws securing the switch assembly to the steering column **(see illustration)**.

8   Use a small screwdriver and release the tangs, then pull the electrical connector from the switch.

9   Remove the switch assembly.

10   Installation is the reverse of the removal steps. Be sure the horn harness grommet is positioned correctly. Check the steering column and switch for proper operation.

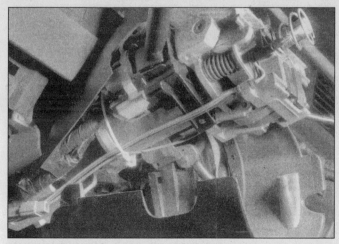

8.6b Remove the lower shroud, then the upper shroud - note the position of the grommet for the horn electrical connector

8.7 Remove the multi-function switch retaining screws (upper screw shown)

12

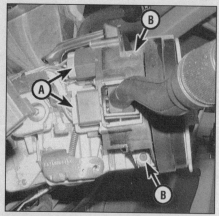

8.12 Multi-function switch electrical connectors (A) and mounting screws (B) - 1995 and later models

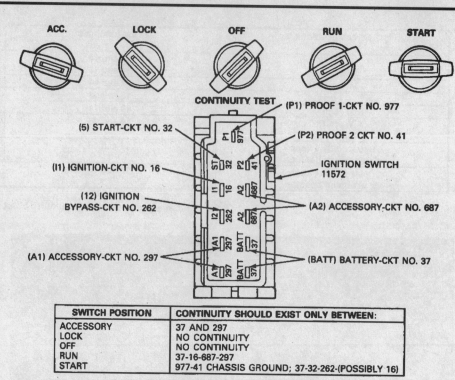

SWITCH POSITION	CONTINUITY SHOULD EXIST ONLY BETWEEN:
ACCESSORY	37 AND 297
LOCK	NO CONTINUITY
OFF	NO CONTINUITY
RUN	37-16-687-297
START	977-41 CHASSIS GROUND; 37-32-262-(POSSIBLY 16)

NOTE: THE FOLLOWING CIRCUITS ARE CONNECTED IN PAIRS INTERNALLY IN THE SWITCH: 37, 687, AND 297.

10.2 Ignition switch terminal guide and continuity chart

9.1 On 1995 and later models, the flasher relay is located under the instrument panel to the right of the steering column

1995 and later models

Refer to illustration 8.12

11   Refer to Chapter 11 and remove the upper and lower steering column covers. **Note:** *It isn't necessary to remove the steering wheel on 1995 and later models.*
12   Disconnect the electrical connectors, then remove the two mounting screws and remove the switch **(see illustration)**.

9   Turn signal/hazard flasher relay - check and replacement

Warning: *On models equipped with airbags, disconnect the negative battery cable, then the positive battery cable and wait two minutes before working in the vicinity of the impact sensors, steering column or instrument panel to avoid the possibility of accidental deployment of the airbag, which could cause personal injury (see Section 30).*

Check

Refer to illustration 9.1

1   The turn signal and hazard flashers are controlled by a flasher relay. The flasher relay is located in the fuse box on 1994 and earlier models. On 1995 and later models it is located under the instrument panel, near the steering column **(see illustration)**.
2   When the flasher relay is functioning properly, an audible click can be heard during its operation. If the turn signals fail to operate on one side or the other and the flasher relay does not make its characteristic clicking sound, a faulty turn signal bulb is indicated.
3   If both turn signals fail to blink, the problem may be due to a blown fuse, a faulty flasher relay, a defective switch or a bad connection. if a quick check of the fuse box indicates that the turn signal fuse has blown, check the wiring for a short before installing a new fuse.

Replacement

4   Disconnect the negative cable from the battery.
5   Remove the flasher by pulling it straight out of the fuse box or electrical connector.
6   Make sure the new flasher is identical to the original and install the new flasher relay into the fuse box or instrument panel connector.

10   Ignition switch - check and replacement

Warning: *On models equipped with airbags, disconnect the negative battery cable, then the positive battery cable and wait two minutes before working in the vicinity of the impact sensors, steering column or instrument panel to avoid the possibility of accidental deployment of the airbag, which could cause personal injury (see Section 30).*

Check

Refer to illustration 10.2

1   The ignition switch is mounted on the steering column behind the instrument panel. To check the switch, remove the lower instrument panel cover and disconnect the electrical connector from the switch (see below).
2   Using an ohmmeter or self-powered test light and the accompanying diagrams, check for continuity between the indicated switch terminals with the switch in each of the indicated positions **(see illustrations)**. If the continuity isn't as specified, replace the switch.

Replacement

Refer to illustration 10.6

3   Disconnect the negative battery cable (and the positive battery cable if equipped with airbags).
4   Turn the ignition key lock cylinder to the Off position.
5   Remove the driver side lower instrument panel cover and the reinforcement panel behind it (see Chapter 11).
6   Disconnect the ignition switch electrical connector and remove the switch retaining screws **(see illustration)**.
7   Disengage the ignition switch from the actuator rod and remove the switch from the vehicle.
8   Make sure the new ignition switch is in

10.6 Unplug the electrical connector and remove the ignition switch retaining screws (arrows)

the Off position.

9   Place the new switch in position and engage the actuator rod. Install the retaining screws and tighten them securely.

10   The remainder of the installation is the reverse of removal. Check for proper operation of the ignition switch in the lock, start and accessory positions.

11 Ignition lock cylinder - replacement

Refer to illustration 11.3
Warning: *On models equipped with airbags, disconnect the negative battery cable, then* the positive battery cable and wait two minutes before working in the vicinity of the impact sensors, steering column or instrument panel to avoid the possibility of accidental deployment of the airbag, which could cause personal injury (see Section 30).

1   Disconnect the negative cable of the battery (and the positive battery cable if equipped with airbags).

2   Insert the key in the lock cylinder and turn the ignition key lock cylinder to the first (Run) position.

3   Insert a small scribe or punch through the hole in the lower steering column cover and depress the retaining pin while pulling out on the lock cylinder **(see illustration)**.

4   To install the lock cylinder, press the retaining pin in and place the lock cylinder into the steering column. Make' sure the lock cylinder is fully seated and the retaining pin is engaged before turning the key to the Lock position. Check the lock cylinder operation in each key position.

12 Windshield wiper motor - removal and installation

Front wiper motor

Refer to illustrations 12.7, 12.8 and 12.9

1   Turn the windshield wiper switch On.

2   Turn the ignition switch on and keep your hand on the key. When the wiper arms move to the straight up position, turn the ignition switch Off.

3   Disconnect the negative cable from the battery.

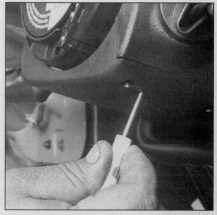

11.3 To remove the lock cylinder, place the key in the "RUN" position and depress the retaining pin with a small punch while pulling the cylinder straight out

4   Remove the right side wiper arm and blade assembly (see Section 13).

5   On 1991 through 1996 models, remove the pivot nut from the right side linkage post. Allow the linkage to drop down into the cowl. On 1997 and later models, refer to Chapter 11 for removal of the cowl vent panels.

6   On 1991 through 1996 models, remove the wiper linkage access cover. Reach through the access opening and unsnap the wiper motor clip.

7   Push the clip away from the linkage until it clears the crank pin nib, then push the clip off the linkage **(see illustration)**.

8   Remove the wiper linkage from the motor crankpin. On 1997 and later models,

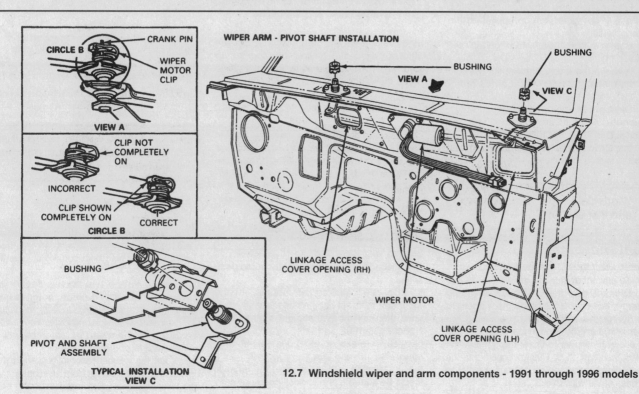

12.7 Windshield wiper and arm components - 1991 through 1996 models

12

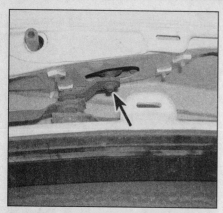

12.8 On 1997 and later models, remove the cowl vent panels and disconnect the linkage from the motor (arrow)

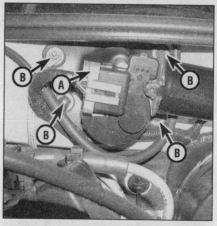

12.9 Disconnect the electrical connector (A) and remove the mounting bolts (B)

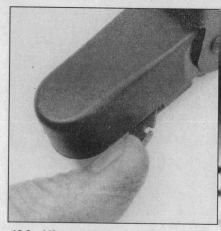

13.2a Lift up on the arm, pull on the slide latch and pull the arm off the spindle

disconnect the linkage arms by pulling back the clip at the end of each arm **(see illustration)**.

9   Disconnect the electrical connector from the motor **(see illustration)**.

10   Remove the screws securing the motor and remove the motor. **Note:** *Some fasteners are stud/bolts. First remove the nut and the electrical ground strap or harness from the stud, then remove the bolt.*

11   Installation is the reverse of the removal steps. Make sure the wiper blades are in the parked position before attaching the linkage to the motor.

Rear wiper motor

12   Disconnect the negative cable from the battery.

13   Lift the wiper arm off the window. Grasp the base of the wiper arm and pull it off the motor spindle.

14   Remove the liftgate trim panel (see Chapter 11).

15   Remove the motor bracket screws and pull the motor and bracket out of the grommet.

16   Disconnect the electrical connector, detach the wiring locator pins and take the motor out.

17   Installation is the reverse of the removal steps.

13   Windshield wiper arm - removal and installation

Front wiper arm

Refer to illustrations 13.2a and 13.2b

1   To prevent damage to the windshield and also to make sure the wipers are operating under normal conditions, keep the windshield wet during this step. Prior to removing the wiper arm, turn the windshield wipers switch On, allow the wipers to travel through several cycles, then turn it Off. This will ensure that the wiper arm is in the parked position parallel to the base of the windshield.

2   Swing the wiper arm away from the

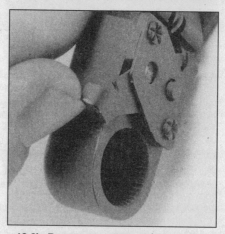

13.2b Be sure to pull the latch back far enough to clear the spindle

windshield. Slide the latch away from the wiper spindle and then allow the wiper arm to go down. It will be held away from the glass. Grasp the wiper arm, push down gently and wiggle it to remove it.

3   Installation is the reverse of the removal steps with the following additions.

a)   *Position the arm back onto the post in the parked position. Do not position the arm too low as it will hit the base of the windshield during normal wiper operation.*

b)   *Make sure the latch is correctly seated in the groove in the wiper spindle.*

Rear wiper arm

Refer to illustration 13.5

4   Place the rear wipers in the parked position.

5   On 1997 and earlier models, lift the wiper arm to its service position **(see illustration)**, then pull the arm straight off the motor spindle. On 1998 and later models, lift the pivot cover to expose the mounting nut, then remove the nut and arm.

6   Installation is the reverse of the removal steps. Be sure the arm is positioned correctly on the motor spindle.

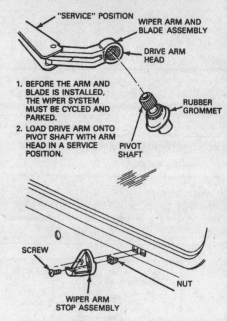

13.5 Rear wiper arm installation details - 1997 and earlier models

14   Windshield washer reservoir and pump assembly - removal and installation

Front washer

Refer to illustrations 14.1, 14.6 and 14.9

1   Use a small screwdriver to unlock the electrical connector tabs, then unplug the washer pump electrical connector **(see illustration)**.

2   Remove the two screws securing the washer reservoir and pump motor assembly to the fenderwell. **Note:** *On later models, the windshield washer reservoir and coolant recovery tank are part of the same unit, see Chapter 3 for mounting details.*

3   Lift the washer reservoir up and disconnect the small hose from the base of the reservoir. Place your finger over the end of the small hose fitting on the reservoir to pre-

Chapter 12   Chassis electrical system

12-9

14.1 Disconnect the electrical connector for the pump motor (arrow)

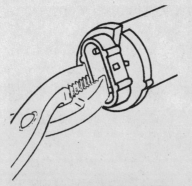

14.6 Grip the wall around the electrical terminal and pull the motor/pump assembly from the reservoir

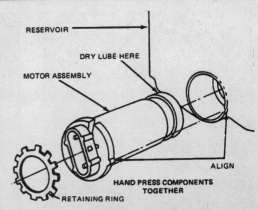

14.9 Windshield washer motor/pump installation details

vent spilling the washer fluid in the engine compartment. Remove the reservoir.

4   Drain the washer fluid from the reservoir into a clean container. If the fluid is kept clean it can be reused.

5   Use a small screwdriver and carefully pry out the retaining ring securing the pump motor in the reservoir receptacle.

6   Grasp one wall surrounding the electrical terminal with a pair of pliers, then pull the motor, seal and impeller assembly out of the reservoir **(see illustration). Note:** *If the impeller and seal separate from the motor they can be reassembled after removal.*

7   Flush out the reservoir with clean water to remove any residue. Inspect it for any foreign matter.

8   Inspect the reservoir pump chamber prior to installing an old motor into a new reservoir. Clean it if necessary.

9   Lubricate the outer surface of the seal with powdered graphite to make installation easier **(see illustration).**

10   Align the small projection on the motor end cap with the slot in the reservoir and push it in until the seal seats against the bottom of the motor receptacle in the reservoir.

11   Use a 1-inch 12-point socket and hand press the retaining ring securely against the motor and plate.

12   Connect the hose to the fitting on the

base of the reservoir.

13   Install the reservoir in the engine compartment and secure with the two screws.

14   Connect the electrical connector.

15   **Caution:** *Do not operate the pump without fluid in the reservoir, as it would be damaged. Fill the reservoir with fluid, operate the pump and check for leaks.*

Rear washer

Refer to illustration 14.17

Note: *This procedure applies only to 1991 through 1996 models, the later models use the front washer reservoir for both front and rear washers.*

16   Remove the quarter trim panel from the driver's side of the cargo area.

17   Disconnect the filler hose from the reservoir tank **(see illustration).**

18   Disconnect the electrical connector from the pump motor.

19   Pour the fluid into a container, then disconnect the discharge hose from the reservoir and take the reservoir out.

20   Carefully pry the pump out of the reservoir with a small screwdriver.

21   Remove the seal and filter (a one-piece assembly).

22   Installation is the reverse of the removal steps, with the following additions:

a) *Lubricate the inside of the seal with*

soapy water, then push the pump in and make sure it's firmly seated.

b) *Fill the reservoir slowly so air won't be trapped in the system.* **Caution:** *Don't operate the washers without fluid in the reservoir.*

15   Headlight switch - replacement

Warning: *On models equipped with airbags, disconnect the negative battery cable, then the positive battery cable and wait two minutes before working in the vicinity of the impact sensors, steering column or instrument panel to avoid the possibility of accidental deployment of the airbag, which could cause personal injury (see Section 30).*

1   Disconnect the negative battery cable from the battery (and the positive cable if equipped with airbags).

2   Remove the switches for the heated rear window and rear windshield washer (see Section 16).

1991 through 1994 models

Refer to illustrations 15.4, 15.5, 15.7 and 15.8

3   Pull the headlight switch knob out as far as it will go.

4   Reach up along the side of the switch (on the side nearest the driver's door) and

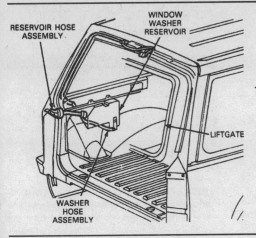

14.17 Details of the rear windshield washer reservoir and hoses on 1991 through 1996 models

15.4 Depress the shaft release button (arrow) on the switch assembly and pull the shaft out (switch removed for clarity)

15.5 Unscrew the plastic retaining nut (upper arrow) with your fingers only, then remove the two screws

15.7 Lift the tangs and unplug the connector from the switch

15.8 Align the switch tab with the notch in the instrument panel

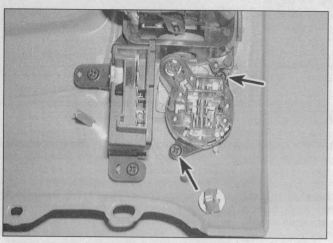

15.11 The headlight switch is mounted behind the instrument cluster trim panel with two screws (arrows)

16.3 Remove the switch assembly from the instrument panel and disconnect the electrical connectors

push the knob release button **(see illustration)**. While holding the button down, pull the knob and shaft out of the switch.

5 Unscrew the switch retaining nut with your fingers **(see illustration)**. **Caution:** *The nut is plastic. Don't use tools on it.*

6 Remove two screws that secure the bottom end of the switch bracket **(see illustration 15.5)**. Pull the bracket and switch out of the instrument panel.

7 Disconnect the electrical connector and remove the switch **(see illustration)**.

8 Installation is the reverse of the removal Steps. Align the switch tab with the notch in the instrument panel **(see illustration)**.

1995 and later models

Refer to illustration 15.11

9 Pull off the headlight switch knob.

10 Remove the instrument cluster trim panel (see Chapter 11).

11 Disconnect the electrical connector, remove the two screws and remove the switch **(see illustration)**.

12 Installation is the reverse of removal.

16   Heated rear window and rear wiper switch - removal and installation

Refer to illustrations 16.3 and 16.4

Warning: *On models equipped with airbags, disconnect the negative battery cable, then the positive battery cable and wait two minutes before working in the vicinity of the impact sensors, steering column or instrument panel to avoid the possibility of accidental deployment of the airbag, which could cause personal injury (see Section 30).*

1 Disconnect the negative cable from the battery (and the positive cable if equipped with airbags).

2 Remove the instrument cluster trim panel (see Chapter 11).

3 Pry the switch assembly out of the instrument panel **(see illustration)**. Disconnect the electrical connectors and remove the switch.

4 To remove an individual switch, squeeze the retainer tabs and press the switch out of the bracket **(see illustration)**. On 1995 and

later models, the rear window wiper/washer switch is on the center dash finish panel.

5 Installation is the reverse of the removal Steps.

16.4 To replace an individual switch, squeeze the retainer tabs and press the switch out of the bracket

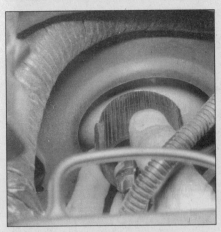

17.3 Unplug the electrical connector from the headlight

17.4 Rotate the retaining ring about 1/8-turn counterclockwise (viewed from the rear) and slide it off the base

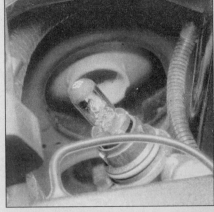

17.5 Pull the bulb straight out of the socket and push the new one in - don't touch the bulb glass with your bare fingers

17 Headlight bulb - replacement

Refer to illustrations 17.3, 17.4 and 17.5
Warning: *Halogen gas filled bulbs are under pressure and may shatter if the surface is scratched or the bulb is dropped. Wear eye protection and handle the bulbs carefully, grasping only the base whenever possible. Do not touch the surface of the bulb with your fingers because the oil from your skin could cause the bulb to overheat and fail prematurely. If you do touch the bulb surface, clean it with rubbing alcohol.*

1   These models have a headlight with a separate halogen bulb. When the light is burned out, only the bulb needs to be replaced, while the housing is replaced only if it is damaged (see Section 18).
2   Disconnect the negative cable from the battery.
3   Disconnect the electrical connector from the back of the headlight **(see illustration)**.
4   On 1997 and later models, remove the headlight housing assembly (see Section 19) Rotate the retaining ring about 1/8-turn counterclockwise (viewed from the rear) and slide it off the base **(see illustration)**.
5   Carefully pull the bulb straight out of the socket **(see illustration)**. Do not rotate the

bulb during removal.
6   Without touching the glass with bare fingers, insert the bulb into the socket. Position the flat on the plastic base up.
7   Align the socket locating tabs with the grooves in the forward part of the plastic base, then push the socket firmly into the base. Make sure the base mounting flange contacts the socket.
8   Slide the retaining ring on. Turn it clockwise until it hits the stop. On 1997 and later models, install the headlight housing.
9   Connect the electrical connector and test the headlight operation.
10   Have the headlight adjustment checked and, if necessary, adjusted by a dealer service department or service station at the earliest opportunity.

18 Headlight housing - removal and installation

Warning: *Halogen gas filled bulbs are under pressure and may shatter if the surface is scratched or the bulb is dropped. Wear eye protection and handle the bulbs carefully, grasping only the base whenever possible. Do not touch the surface of the bulb with your fingers because the oil from your skin could cause the bulb to overheat and fail prema-*

turely. If you do touch the bulb surface, clean it with rubbing alcohol.
1   Disconnect the negative battery cable.
2   On 2001 models, open the hood and remove the plastic headlight cover. Pry up and remove the two retaining pins on the headlight housing and release it from the body. Disconnect the electrical connector and remove the housing.
3   Remove the plastic panel across the top of the radiator. On early models its is attached with screws and plastic rivets, while on 1997 and later models, each side of the panel is hinged. Release a catch and flip the hinged cover up for access to the rear of the headlight assembly. Refer to Section 17 and remove the bulb.
4   Remove the screws and the turn/park light assembly (see Section 20).
5   On 1996 and earlier models, remove the grille (see Chapter 11).
6   Using needle-nose pliers, squeeze the plastic clips on the headlight housing to release it from the body.
7   Installation is the reverse of removal.

19 Headlights - adjusting

Refer to illustrations 19.1 and 19.3
Note: *The headlights must be aimed correctly. If adjusted incorrectly they could momentarily blind the driver of an oncoming vehicle and cause a serious accident or seriously reduce your ability to see the road. The headlights should be checked for proper aim every 12 months and any time a new headlight is installed or front end body work is performed. It should be emphasized that the following procedure is only an interim step which will provide temporary adjustment until the headlights can be adjusted by a properly equipped shop.*
1   Headlights have two adjusting screws, one near the fender controlling up-and-down movement **(see illustration)** and one behind the inboard side of the headlight controlling left-and-right movement.

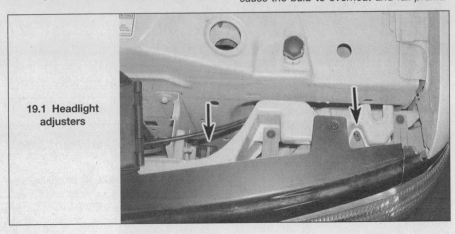

19.1 Headlight adjusters

12

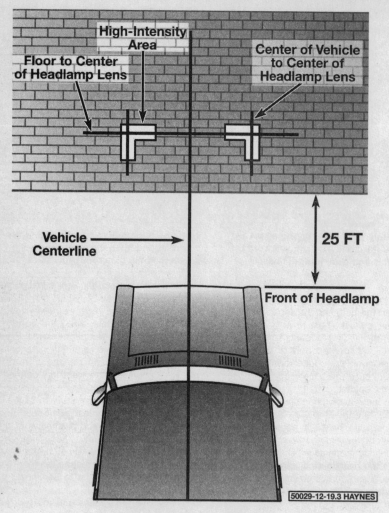

19.3 Headlight aiming details

High-Intensity Area

Floor to Center of Headlamp Lens

Center of Vehicle to Center of Headlamp Lens

Vehicle Centerline

25 FT

Front of Headlamp

50029-12-19.3 HAYNES

20.2a Remove two screws that secure the top of the lamp housing (the screw indicated by the arrow is the headlight adjustment screw that affects the vertical position of the headlight beam) - 1991 through 1996 models

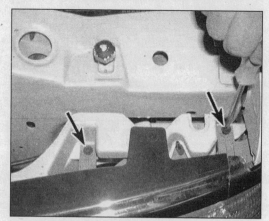

20.2b Parking light housing screws (arrows) - 1997 and later models (Explorer shown, Mountaineer has only one screw)

2   There are several methods of adjusting the headlights. The simplest method requires a blank wall and a level floor.

3   Position masking tape vertically on the wall in reference to the vehicle centerline and the centerline of both headlights (see illustration).

4   Position a horizontal line in reference to the centerline of all headlights. **Note:** *It may be easier to position the tape on the wall with the vehicle parked only a few inches away.*

5   Adjustment should be made with the vehicle parked 25 feet from the wall, sitting level, the gas tank half-full and no unusually heavy load in the vehicle.

6   Starting with the low beam adjustment, position the high intensity zone so it is two inches below the horizontal line and two inches to the right of the headlight vertical line. Adjustment is made by turning the adjusting screw closest to the fender as necessary to raise or lower the beam (see illustration 18.1). The other adjusting screw should be used in the same manner to move the beam left or right.

7   With the high beams on, the high intensity zone should be vertically centered with the exact center just below the horizontal line. **Note:** *It may not be possible to position the headlight aim exactly for both high and low beams. If a compromise must be made, keep in mind that the low beams are the most used and have the greatest effect on driver safety.*

8   Have the headlights adjusted by a dealer service department or service station at the earliest opportunity.

20.3a Pull the housing straight out . . .

20 Bulb replacement

Parking/turn signal/front side marker lamp

Refer to illustrations 20.2a, 20.2b, 20.3a, 20.3b, 20.4 and 20.5

1   Open the hood.

20.3b . . . to expose the connectors - 1991 through 1996 models shown

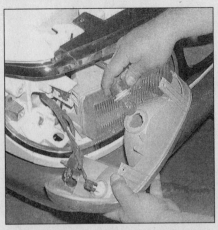

20.4 Rotate the bulb socket to align the lugs and take the socket out - 1997 model shown

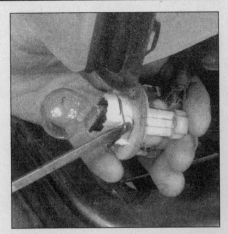

20.5 Pry the retaining clip off the bulb and pull the bulb out of the socket

20.8 Remove two screws that secure the top of the lamp housing

20.9 Pull the housing away from the body

20.10 Lift the retaining tabs and unplug the electrical connectors

2   Remove the two screws that secure the top of the lamp assembly **(see illustrations)**.

3   Pull the assembly straight out to expose the bulb sockets **(see illustrations)**. **Note:** *On 1997 and later models, the park/turn/side light assembly is separate from the headlight assembly.*

4   Turn the socket in either direction to align the lugs on the socket with the notches in the housing and pull the socket out **(see illustration)**.

5   To replace the parking/turn signal bulb, remove the clip from the socket **(see illustration)**.

6   Pull the bulb straight out of the socket.

7   Installation is the reverse of the removal Steps.

Rear combination lights (stop, tail, turn and backup)

Refer to illustrations 20.8, 20.9, 20.10, 20.11 and 20.12

8   On 2001 Sport Trac models, lower the lift gate. Detach the three rubber plugs and remove the three screws securing the combination light housing. On all other models, remove the two screws securing the top of

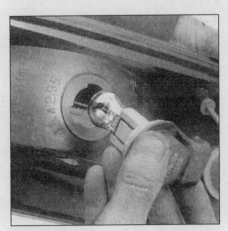

20.11 Rotate the bulb socket and remove it from the housing

the rear combination light assembly **(see illustration)**.

9   Pull the housing straight out from the body **(see illustration)**.

10   Spread the connector prongs and separate the connector from the bulb socket **(see illustration)**.

20.12 Spread the retainer prongs and pull the bulb out of the socket

11   Rotate the bulb socket to align the lugs and take the socket out of the housing **(see illustration)**.

12   Carefully pry apart the retaining clips and pull the bulb out of the socket **(see illustration)**.

13   Installation is the reverse of the removal Steps.

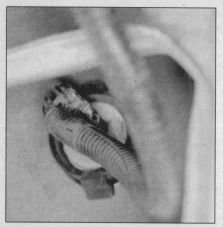

20.14 Reach up under the bumper and remove the bulb socket

20.19 Pull the lamp down and disconnect the electrical connector

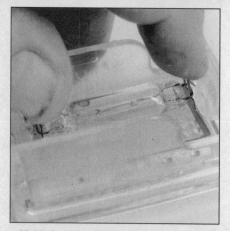

20.20 Squeeze the retainer prongs and remove the bulb

20.23 Carefully pry the lens loose and detach it from the headliner, 1991 through 1996 model shown, later models similar but larger

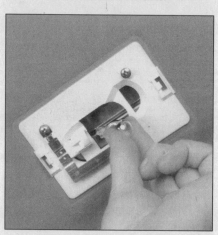
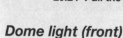

20.24 Pull the bulb out

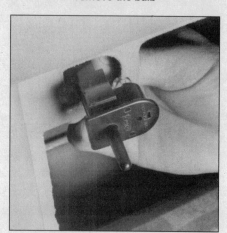

20.28a Squeeze the retainer tabs on the light assembly and pull it out

License plate bulb

Refer to illustration 20.14

14   Reach up under the bumper and grasp the socket **(see illustration)**.

15   Twist the socket counterclockwise and remove it from the housing.

16   Pull the bulb out of the socket.

17   Installation is the reverse of the removal Steps.

Cargo light

Refer to illustrations 20.19 and 20.20

18   Carefully pry the cargo light out of the headliner. **Caution:** *Only one side of the light can be pried down. If it won't go easily, try prying the other side. Don't use excessive force.*

19   Detach the light from the connector **(see illustration)**.

20   Squeeze the bulb retainer prongs together **(see illustration)**.

21   Slip the bulb off the prongs and install a new one.

22   The remainder of installation is the reverse of the removal Steps.

Dome light (front)

Refer to illustrations 20.23 and 20.24

23   Use a small screwdriver and carefully pry the dome light lens away from the base **(see illustration)**. Remove the lens. **Note:** *1997 and later models incorporate two small map lights in the dome light housing.*

24   Pull the bulb straight out of the socket **(see illustration)**.

25   Install a new bulb and press it in all the way, then install the dome light lens assembly.

Glove compartment light

Refer to illustrations 20.28a, 20.28b and 20.30

26   Open the glove compartment and remove the contents.

27   Squeeze the sides of the glove compartment together and swing it toward you, out of the opening in the dash panel. Let the glove compartment hang down.

28   Reach into the glove compartment. Squeeze the prongs on the bulb housing together and remove it from the opening **(see illustrations)**.

29   Lift the retaining tab on the electrical connector and pull the housing out of the connector.

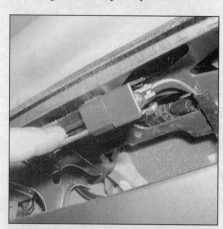

20.28b On later models the bulb housing is part of the glove compartment latch assembly (1997 model shown)

30   Carefully pull the bulb out of the housing **(see illustration)**.

31   Installation is the reverse of the removal Steps.

Instrument cluster bulbs

Refer to illustration 20.33

32   Remove the instrument cluster to gain

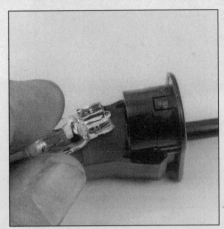

20.30 Pull the bulb out of the socket

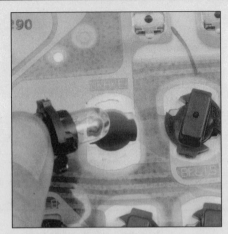

20.33 Rotate the connector counterclockwise and remove it from the cluster

20.35 Remove the screws and lift the lamp body off without damaging the paint - early model shown

20.36 Disconnect the electrical connector, work the socket free of the lamp body and pull the bulb out

20.40 Pry the lens off

20.41 Pry the bulb out of the clips and push a new one in

access to the bulbs (see Section 24).

33 Turn the bulb/socket assembly 1/4-turn counterclockwise. Remove the bulb from the socket and install a new bulb **(see illustration)**. **Note:** *The instrument cluster printed circuit board is fragile. Be careful when removing the bulb assemblies.*

High mount brake light

Refer to illustrations 20.35 and 20.36
Warning: *On 1997 and later models, the high mount brake light assembly contains high-voltage components. The replacement procedure is difficult, and potentially dangerous. Replacement on these models is best left to a dealer service department.*

34 Remove the screws that secure the lamp assembly to the body.
35 Lift the lamp assembly off, taking care not to damage the paint **(see illustration)**.
36 Disconnect the electrical connector **(see illustration)**.
37 Work the bulb socket(s) out of the lamp body.
38 Pull the bulb out of the socket.
39 Installation is the reverse of the removal Steps.

Vanity mirror light

Refer to illustrations 20.40 and 20.41
40 Pry the lens off the light **(see illustration)**.
41 Pry the bulb out of its clips **(see illustration)** and push a new one in.
42 Push the lens back into position.

21 Radio - removal and installation

Refer to illustration 21.2
Warning: *On models equipped with airbags, disconnect the negative battery cable, then the positive battery cable and wait two minutes before working in the vicinity of the impact sensors, steering column or instrument panel to avoid the possibility of accidental deployment of the airbag, which could cause personal injury (see Section 30).*

1 On 1994 and earlier models, remove the instrument cluster trim panel (see Chapter 11).
2 Insert a special radio removal tool (available at most auto parts stores) into the radio face plate **(see illustration)**. Push the tool in about 1-inch to release the retaining clips on each side.

3 Using the special tool, pull the radio partially out of the instrument panel, then disconnect the electrical connectors and antenna lead from the radio.
4 Pull the radio assembly out of the instrument panel.

21.2 Insert the two radio removal tools (arrows) to release the clips, then pull the radio out with the tools (push the tools toward each other to grip the radio)

12

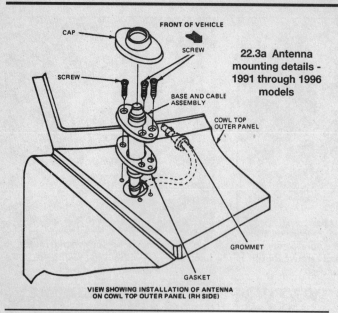

22.3a Antenna mounting details - 1991 through 1996 models

VIEW SHOWING INSTALLATION OF ANTENNA ON COWL TOP OUTER PANEL (RH SIDE)

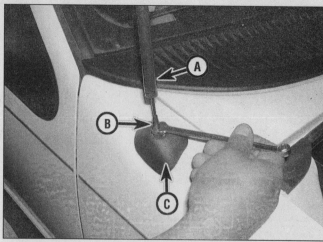

22.3b Antenna mounting, 1997 and later models - pull up the sleeve (A), then unscrew the mast (B), remove the cap (C) and the mounting screws

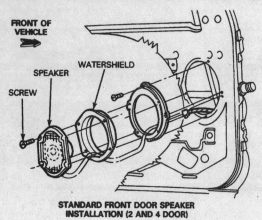

STANDARD FRONT DOOR SPEAKER INSTALLATION (2 AND 4 DOOR)

23.2 Radio speaker installation details (typical)

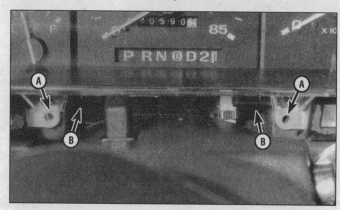

24.3 Remove the cluster securing screws - there's a screw at each upper corner and two at the bottom of the cluster (A) - on automatic transmission models, remove the two screws (B) and detach the gearshift indicator from the cluster

5   Installation is the reverse of the removal steps.

22 Radio antenna - removal and installation

Refer to illustrations 22.3a and 22.3b

1   Remove the radio from the instrument panel and disconnect the antenna from the rear of the radio.
2   Disconnect the antenna cable from the plastic clips along the top of the defroster nozzle.
3   Use a small screwdriver and carefully pry the antenna cap from the base and remove the cap (see illustrations). Note: *On 1997 and later models, remove the mast first, then the cap and the base screws.*
4   Tie a piece of string onto the radio end of the antenna cable and the other end to the instrument panel. This will be used to pull the new antenna cable back through the same path as the old one.
5   Remove the screws securing the antenna base to the body and slowly pull the

antenna cable out through the body opening. Remove the antenna and gasket.
6   Installation is the reverse of the removal steps with the following additions.
   a) *Be sure to install a new gasket.*
   b) *Untie the string and attach it to the new antenna cable. Slowly pull the string and antenna cable back through the body opening. Discard the string.*

23 Speakers - removal and installation

Refer to illustration 23.2

1   Remove the door trim panel or rear quarter panel (see Chapter 11).
2   Remove the screws securing the speaker to the trim panel or speaker bracket (see illustration).
3   Lift out the speaker and disconnect the electrical connector at the speaker. Caution: *Do not operate the radio with the speakers disconnected.*
4   Installation is the reverse of the removal steps.

24 Instrument cluster - removal and installation

Warning: *On models equipped with airbags, disconnect the negative battery cable, then the positive battery cable and wait two minutes before working in the vicinity of the impact sensors, steering column or instrument panel to avoid the possibility of accidental deployment of the airbag, which could cause personal injury (see Section 30).*

Removal

1991 through 1996 models

Refer to illustrations 24.3, 24.6a and 24.6b

1   Disconnect the negative battery cable from the battery.
2   Refer to Chapter 11 for removal of the instrument cluster trim bezel.
3   Remove the four screws securing the instrument cluster (see illustration).
4   On automatic transmission models, remove the screws securing the gear indicator to the instrument cluster and slide the

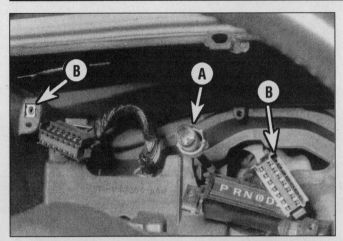

24.6a Squeeze the flat on the speedometer cable (A) to disconnect it; the connectors (B) plug into sockets in the printed circuit board (instrument cluster removed for clarity)

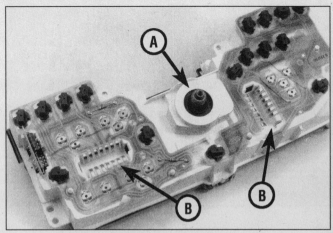

24.6b Rear view of the instrument cluster

A   *Speedometer cable socket*
B   *Electrical connector sockets*

24.10 Remove the four instrument panel screws (arrows)

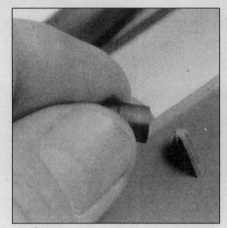

24.12 If any of the anti-rattle insulators fall off during cluster removal, find them and reinstall them on their posts

indicator down and out of the cluster **(see illustration 24.3)**. It is not necessary to disconnect the indicator.

5   Carefully pull the instrument cluster assembly partially out from the instrument panel.

6   Reach under the instrument panel, press on the flat portion of the quick-disconnect plastic connector and detach the speedometer cable from the instrument cluster **(see illustrations)**. **Note:** *If there isn't enough slack in the speedometer cable to pull the cluster out far enough, disconnect the cable at the transmission or transfer case under the vehicle, then push it inside the vehicle. Once there is enough slack in the cable, reach behind the cluster and disconnect it.*

7   Squeeze the locking tabs on the printed circuit electrical connectors, then disconnect them from the printed circuit board **(see illustrations 24.6a and 24.6b)**.

8   Remove the instrument cluster assembly.

1997 and later models
Refer to illustration 24.10

9   Refer to Chapter 11 and remove the instrument cluster bezel.

10   Remove the screws and pull the cluster

forward **(see illustration)**.

11   Disconnect the electrical connectors at the rear of the cluster and the shift indicator cable (See Chapter 7B).

Installation
Refer to illustration 24.12

12   Installation is the reverse of the removal steps, with the following additions:

a) *Apply a 3/16-inch diameter ball of silicone dielectric compound to the drive hole of the speedometer head prior to installing the cable (if equipped).*

b) *There are several vibration dampers fitted over prongs on the cluster (see illustration). These tend to fall off when the cluster is pulled out. Make sure they're all reinstalled.*

25   Gauges and speedometer cable - removal and installation

Speedometer cable and casing

1   Perform Steps 1 through 11 of Section 25. **Note:** *Later models do not have a*

speedometer cable. The speedometer is electronic and responds to signals from the vehicle speed sensor.

2   To remove the cable only, pull out the cable core from the casing and install a new one by reversing the removal steps.

3   Remove the casing as follows:

a) *Remove the screw and clamp securing the speedometer cable to the firewall.*

b) *Working under the vehicle, remove the clamp screw and remove the cable and casing from the transmission or transfer case.*

4   Connect the speedometer casing to the head and (if removed) install the casing to the transmission using a new O-ring.

5   Install the instrument cluster (see Section 24).

Gauges
Refer to illustration 25.7, 25.8, and 25.9
Note: *US Federal law requires that the odometer in any replacement speedometer must register the same mileage as that on the removed speedometer.*

12

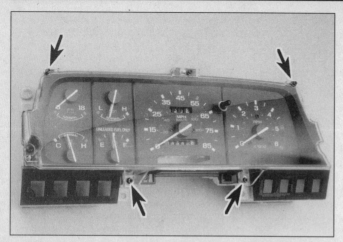

25.7 Remove the screws and take the lens off the cluster

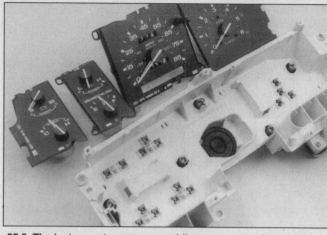

25.8 The instrument gauge assemblies are removed separately - early model shown

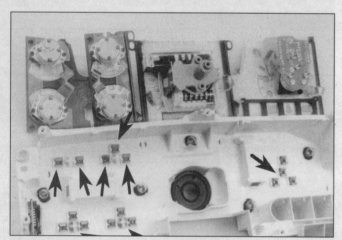

25.9 Each gauge assembly plugs into prongs on the back of the cluster - early model shown

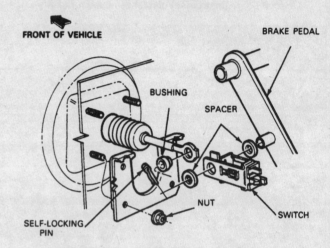

26.1 Typical brake light switch installation details

6   Remove the instrument cluster (see Section 24).

7   Remove the lens and mask from the instrument cluster **(see illustration)**.

8   Gently push on the back of the speedometer to loosen it from the retainer prongs in the cluster housing **(see illustration)**.

9   Lift the gauges with a gentle rocking motion to separate them from the retainer prongs in the cluster housing **(see illustration)**. Note: *The side gauge assemblies must be removed before taking out the center gauge assembly on 1997 and later models.*

10   Installation is the reverse of the removal steps. Apply a 3/16-inch diameter ball of silicone dielectric compound to the drive hole of the speedometer head prior to installing the cable (if equipped).

26 Brake light switch - replacement

Refer to illustration 26.1

1   The brake light switch is located on a flange or a bracket protruding from the brake pedal support **(see illustration)**.

2   Disconnect the electrical connector from the switch.

3   Remove the hairpin retainer and spacer securing the switch to the brake pedal arm.

4   Slide the switch, pushrod, nylon washers and bushing away from the pedal, then remove the switch.

5   To install, place the switch with the U-shaped side nearest the pedal and directly over or under the pin. The bushing must be in the pushrod eyelet with the washer face on the side closes to the retaining pin.

6   Slide the switch up or down so the master cylinder pushrod and bushing are trapped, then push the switch firmly down.

7   Install the plastic washer and hairpin retainer.

8   Connect the electrical connector to the switch. Make sure the electrical connector is routed correctly and will travel the full swing of the pedal without binding.

27 Power window system - general information

The power window system operates the

electric motors mounted in the doors which lower and raise the windows. The system consists of the control switches, the motors (regulators), glass mechanisms and the associated wiring.

Diagnosis can usually be limited to simple checks of the wiring connections and motors for minor faults which can be easily repaired. These are:

a) *Inspect the power window actuating switches for broken wires and loose connections.*

b) *Check the power window fuse/and or circuit breaker.*

c) *Remove the door panel(s) and check the power window motor wires to see if they're loose or damaged. Inspect the glass mechanisms for damage which could cause binding.*

28 Power mirrors - removal and installation

Removal of power mirrors is similar to manual mirrors. Refer to procedure in Chapter 11.

29 Power door lock system - general information

The power door lock system operates the door lock actuators mounted in each door. The system consists of the switches, actuators and associated electrical wiring.

Diagnosis can usually be limited to simple checks of the wiring connectors and actuators for minor faults which can be easily repaired. These include:

a) *Check the system fuse and circuit breaker.*

b) *Check the switch wiring for damage or loose connections.*

c) *Check the switch for continuity in the closed position. Also check the switches for sticking. If a switch sticks closed the actuator internal circuit breaker will trip and stay tripped until voltage is removed.*

d) *Remove the door panel(s) and check the actuator electrical connections for looseness or damage. Inspect the actuator rods and door lock linkage to make sure they are not bent, damaged or binding.*

e) *Remove the electrical connector at the actuator and with a test light or voltmeter, check for available voltage to the actuator as you cycle the switch. If voltage is present, but the actuator doesn't operate when connected, the actuator is probably defective. If no voltage is present at the connector, the switch, relay (if equipped) or wiring is probably defective.*

30 Airbag - general information

Later models are equipped with a Supplemental Inflatable Restraint System (SRS), more commonly known as an airbag. This system is designed to protect the driver and front seat passenger from serious injury in the event of a head-on or frontal collision. It consists of and airbag module in the center of the steering wheel, the right side of the instrument panel and on 2001 models (if equipped), a side airbag mounted on the outer side of both front seats. Crash sensors are mounted on the front of the vehicle and a diagnostic monitor which also contains a backup power supply is located in the passenger compartment.

Airbag module

Steering wheel-mounted

The airbag inflator module contains a housing incorporating the cushion (airbag) and inflator unit, mounted in the center of the steering wheel. The inflator assembly is mounted on the back of the housing over a hole through which gas is expelled, inflating the bag almost instantaneously when an electrical signal is sent from the system. A coil assembly on the steering column under the module carries this signal to the module.

This coil assembly can transmit an electrical signal regardless of steering wheel position. The igniter in the airbag converts the electrical signal to heat and ignites the sodium aside/copper oxide powder, producing nitrogen gas, which inflates the bag.

Instrument panel-mounted

The airbag is mounted above the glove compartment and designated by the letters SRS (Supplemental Restraint System). It consists of an inflator containing an igniter, a bag assembly, a reaction housing and a trim cover.

This airbag is considerably larger that the steering wheel-mounted unit and is supported by the steel reaction housing. The trim cover is textured and painted to match the instrument panel and has a molded seam which splits when the bag inflates. As with the steering-wheel-mounted airbag, the igniter electrical signal converts to heat, converting sodium aside/iron oxide powder to nitrogen gas, inflating the bag.

Side air bag (some 2001 models)

The optional side air bag is mounted on the outboard side of the seatbacks of both front seats. It consists of an inflator containing an igniter, a bag assembly and a reaction housing.

The special seat cover material is color matched with seat fabric and is designed to allow airbag deployment. As with the other air bags assemblies, the igniter electrical signal converts to heat, converting the sodium aside/iron oxide powder to nitrogen gas, inflating the bag.

Sensors

The steering wheel and instrument panel-mounted system has three sensors; two forward sensors at the front of the vehicle and a safing sensor inside the electronic diagnostic monitor.

The optional side air bag system has two sensors located on the "B" pillars at the trailing edge of the front door.

The forward and passenger compartment sensors are basically pressure-sensitive switches that complete an electrical circuit during an impact of sufficient G force. The electrical signal from these sensors is sent to the electronic diagnostic monitor which then completes the circuit and inflates the airbag(s).

Electronic diagnostic monitor

The electronic diagnostic monitor supplies the current to the airbag system in the event of the collision, even if battery power is cut off. It checks this system every time the vehicle is started, causing the "AIR BAG" light to go on then off, if the system is operating properly. If there is a fault in the system, the light will go on and stay on, flash, or the dash will make a beeping sound. If this happens, the vehicle should be taken to your dealer immediately for service.

Before bringing the vehicle to the dealer, observe the light. If it is blinking, record the number of blinks, which represent a two-digit code. For example, three blinks, followed by a pause, then two blinks, would indicate a code 32. This information will be helpful to the technician, especially for intermittent codes.

Disabling the system

Whenever working in the vicinity of the steering wheel, steering column or near other components of the airbag system, the system should be disarmed. To do this, perform the following steps:

a) *Turn the ignition switch to Off.*

b) *Detach the cable from the negative battery terminal, then detach the positive cable. Wait two minutes for the electronic module backup power supply to be depleted.*

Enabling the system

a) *Turn the ignition switch to the Off position.*

b) *Connect the positive battery cable first, then connect the negative cable.*

31 Wiring diagrams - general information

Refer to illustration 31.3

Since it isn't possible to include all wiring diagrams for every model year covered by this manual, the following diagrams are those that are typical and most commonly needed.

Prior to troubleshooting any circuit, check the fuses and circuit breakers to make sure they're in good condition. Make sure the battery is fully charged and check the cable connections (see Chapter 1). Make sure all connectors are clean, with no broken or loose terminals.

Refer to the accompanying table for the wire color codes applicable to your vehicle **(see illustration)**.

BL	Blue	P	Purple
PK		PK	Pink
BK	Black	R	Red
BR	Brown	T	Tan
DB	Dark blue	W	White
DG	Dark green	Y	Yellow
GN	Green	(H)	Hash
LB	Light blue	(D)	Dot
LG	Light green		
N	Natural		
O	Orange		

31.3 Wiring diagram color codes

12

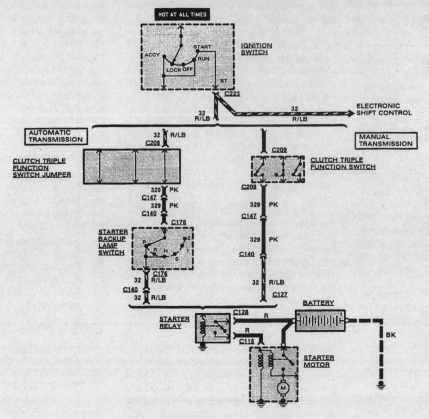

Typical starting system

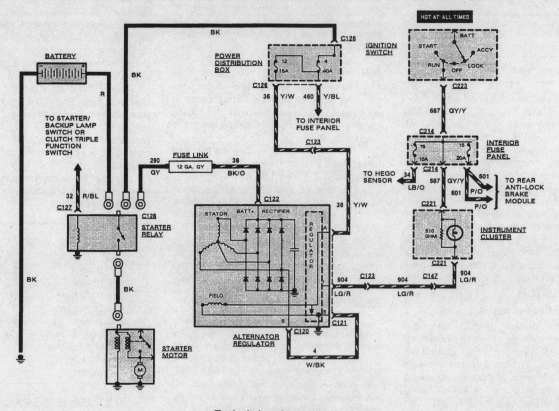

Typical charging system

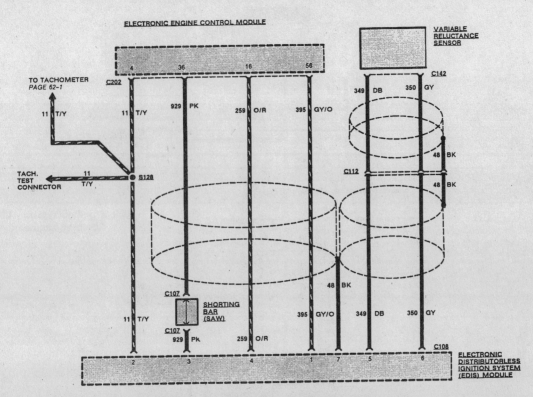

Typical ignition system - 4.0L pushrod V6 engine (1 of 2)

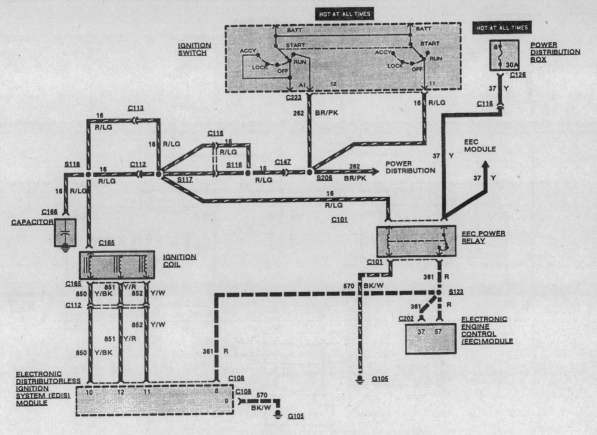

Typical ignition system - 4.0L pushrod V6 engine (2 of 2)

12

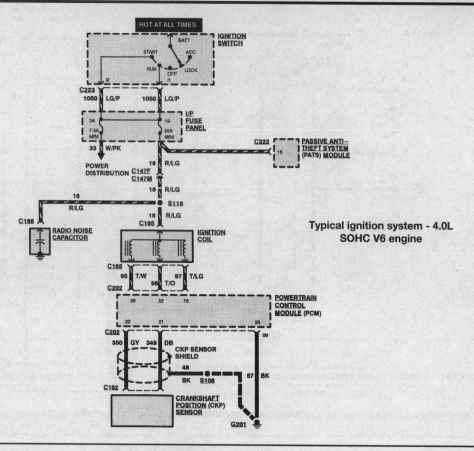

Typical ignition system - 4.0L
SOHC V6 engine

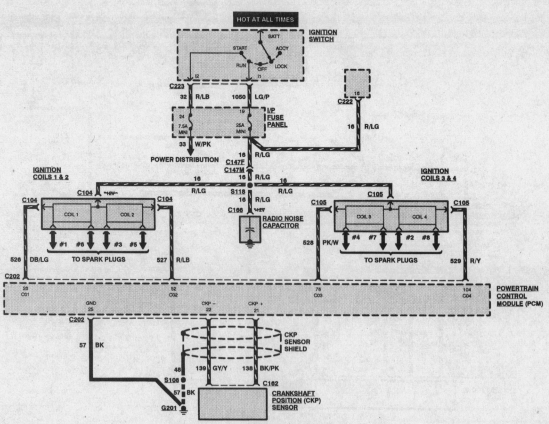

Typical ignition system - 5.0L V8 engine

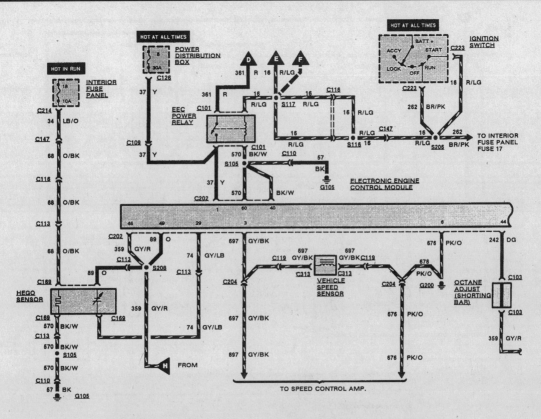

Typical engine control system - 4.0L pushrod V6 engine (1 of 4)

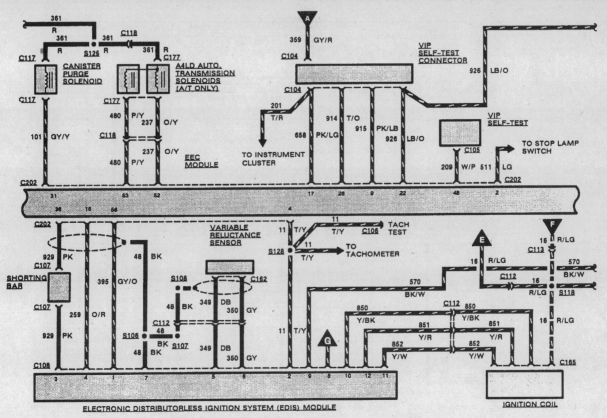

Typical engine control system - 4.0L pushrod V6 engine (2 of 4)

12

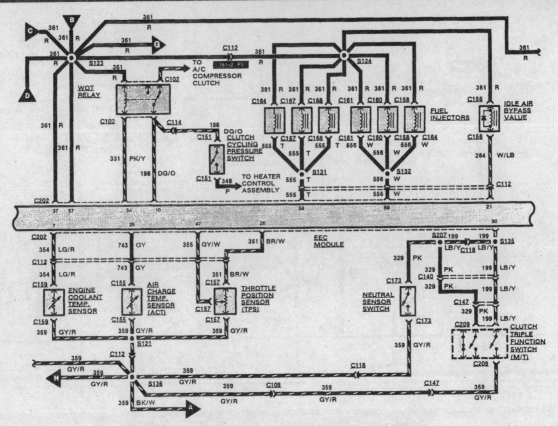

Typical engine control system - 4.0L pushrod V6 engine (3 of 4)

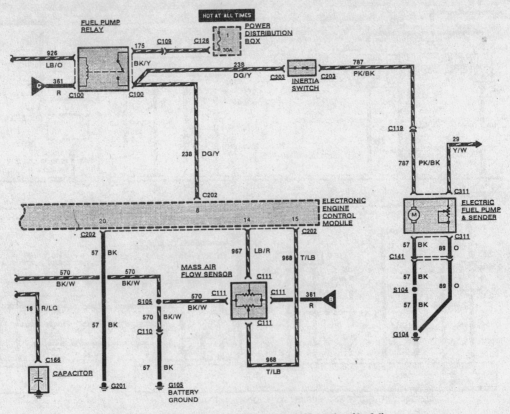

Typical engine control system - 4.0L pushrod V6 engine (4 of 4)

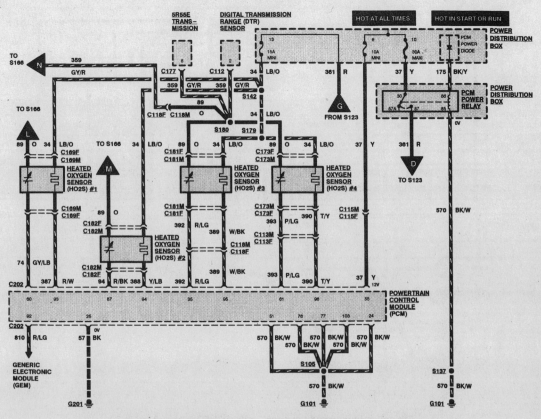

Typical engine control system - 4.0L SOHC V6 engine (1 of 7)

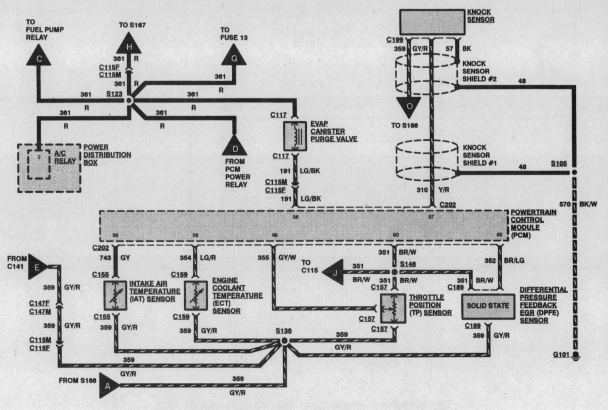

Typical engine control system - 4.0L SOHC V6 engine (2 of 7)

12

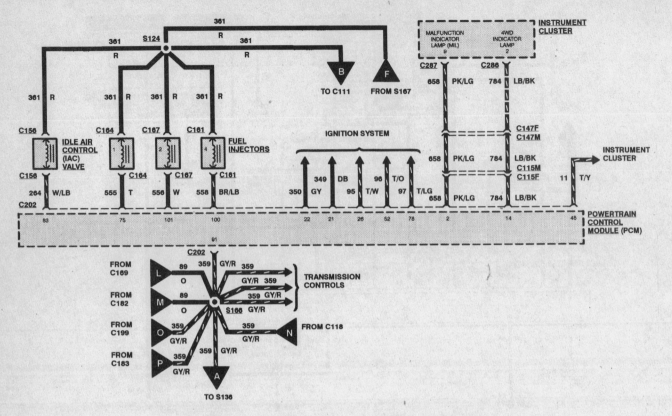

Typical engine control system - 4.0L SOHC V6 engine (3 of 7)

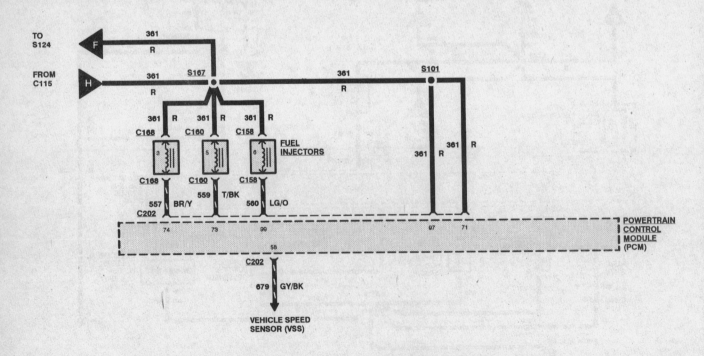

Typical engine control system - 4.0L SOHC V6 engine (4 of 7)

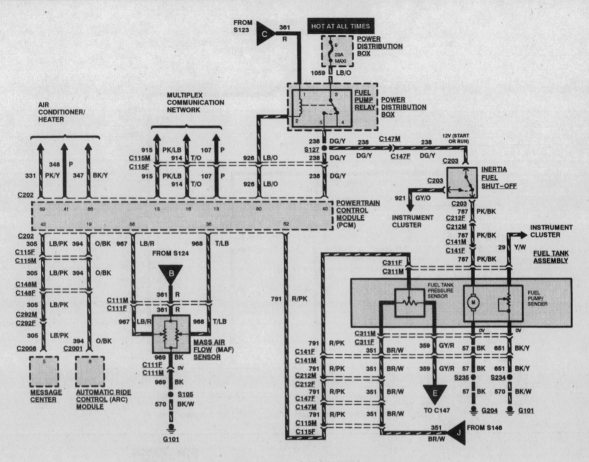

Typical engine control system - 4.0L SOHC V6 engine (5 of 7)

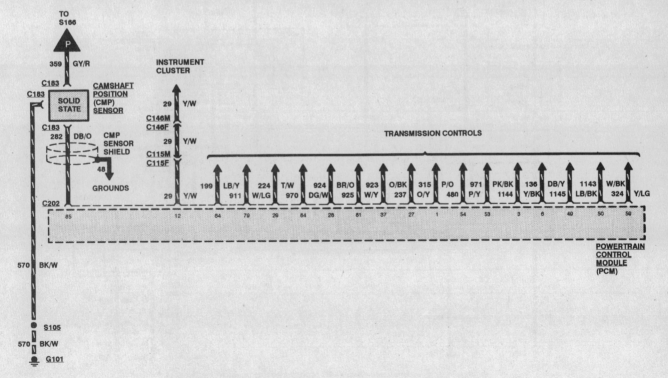

Typical engine control system - 4.0L SOHC V6 engine (6 of 7)

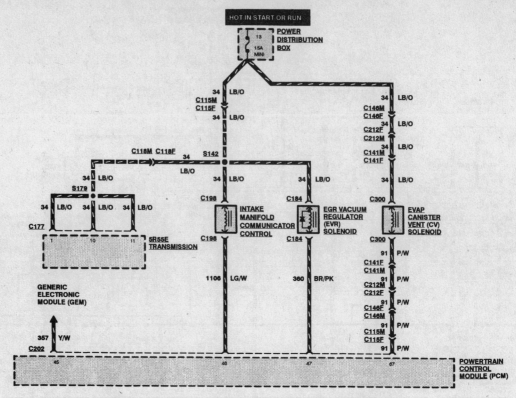

Typical engine control system - 4.0L SOHC V6 engine (7 of 7)

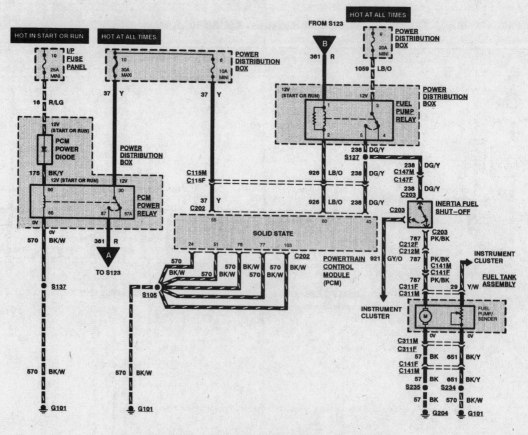

Typical engine control system - 5.0L V8 engine (1 of 7)

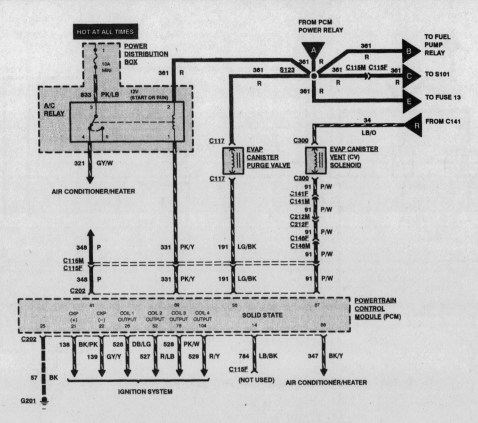

Typical engine control system - 5.0L V8 engine (2 of 7)

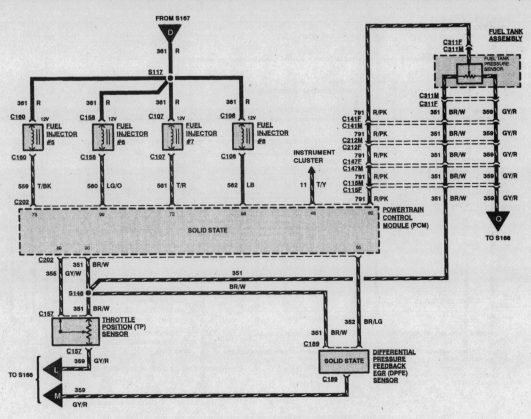

Typical engine control system - 5.0L V8 engine (3 of 7)

12

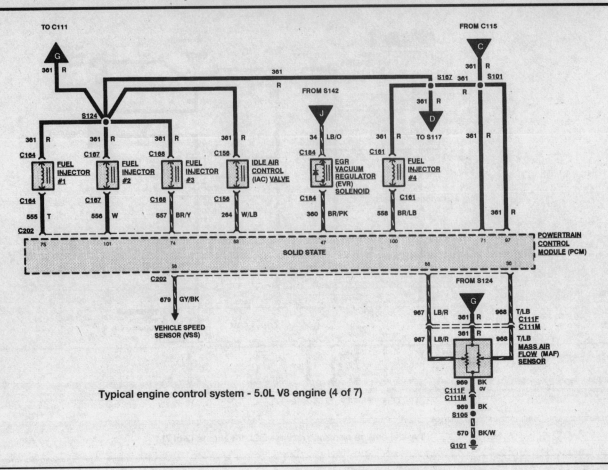

Typical engine control system - 5.0L V8 engine (4 of 7)

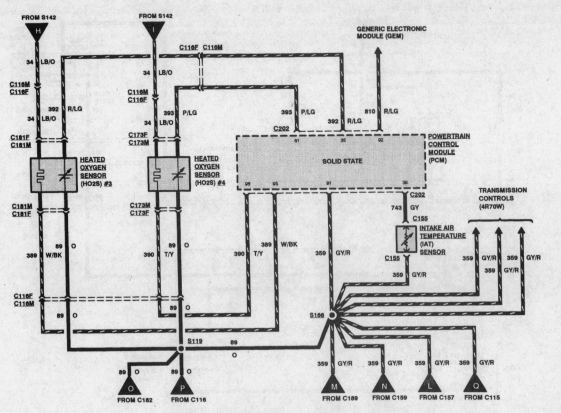

Typical engine control system - 5.0L V8 engine (5 of 7)

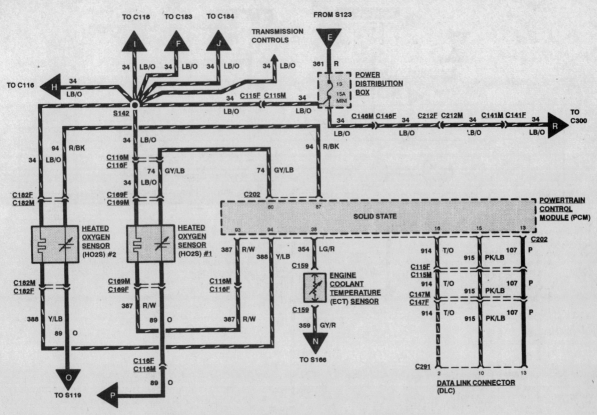

Typical engine control system - 5.0L V8 engine (6 of 7)

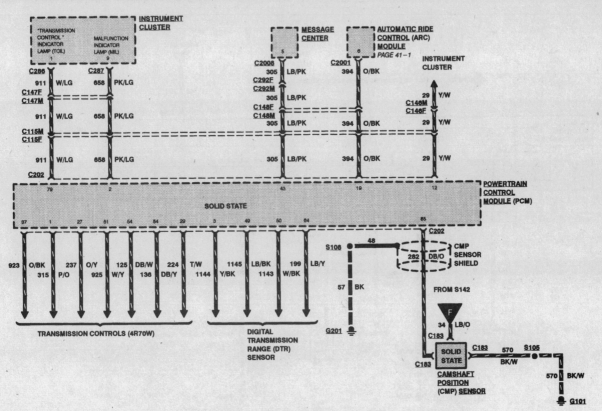

Typical engine control system - 5.0L V8 engine (7 of 7)

12

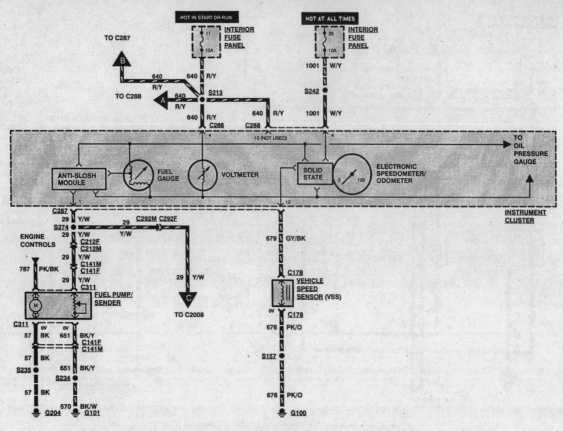

Typical instrument panel gauges and warning light system - with analog gauges (1 of 7)

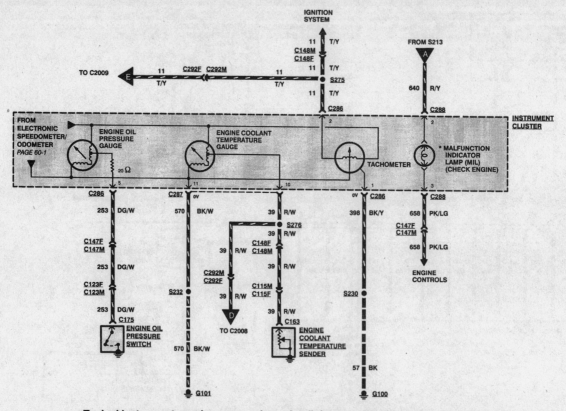

Typical instrument panel gauges and warning light system - with analog gauges (2 of 7)

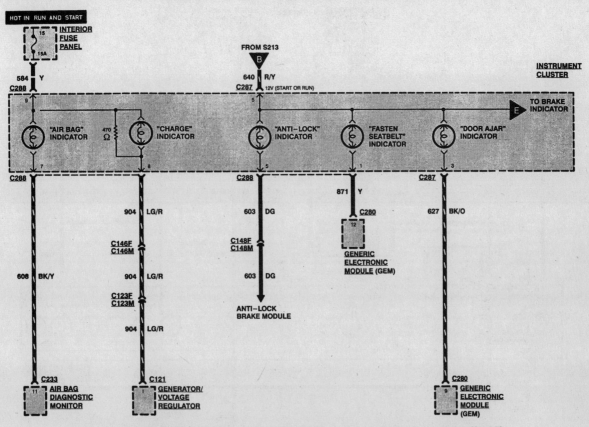

Typical instrument panel gauges and warning light system - with analog gauges (3 of 7)

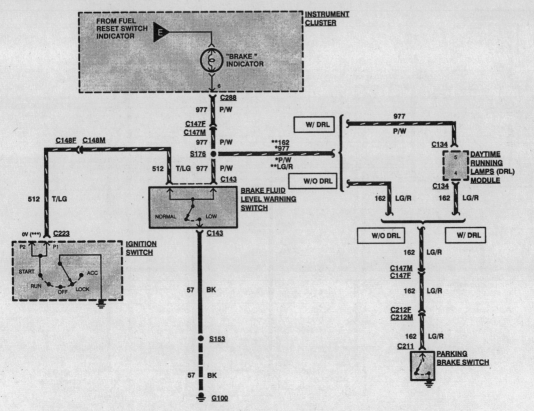

Typical instrument panel gauges and warning light system - with analog gauges (4 of 7)

12

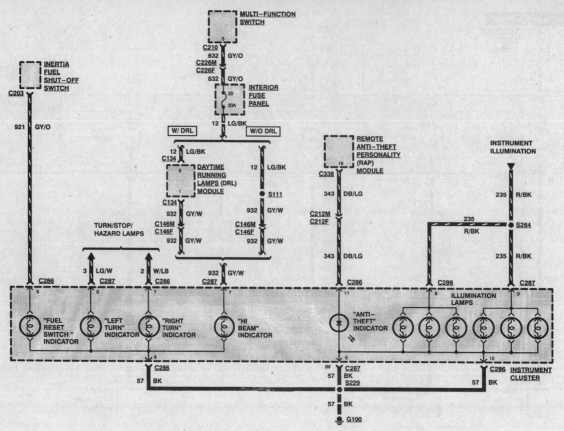

Typical instrument panel gauges and warning light system - with analog gauges (5 of 7)

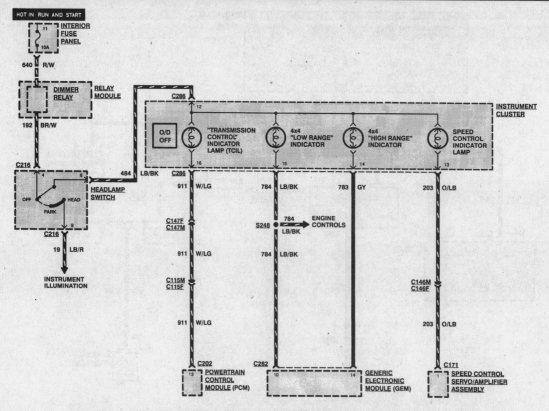

Typical instrument panel gauges and warning light system - with analog gauges (6 of 7)

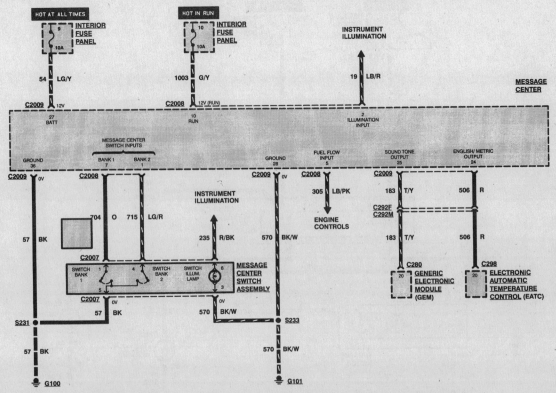

Typical instrument panel gauges and warning light system - with analog gauges (7 of 7)

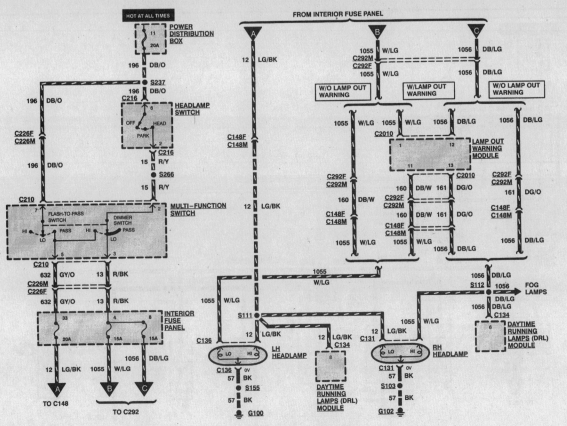

Typical headlight system

12

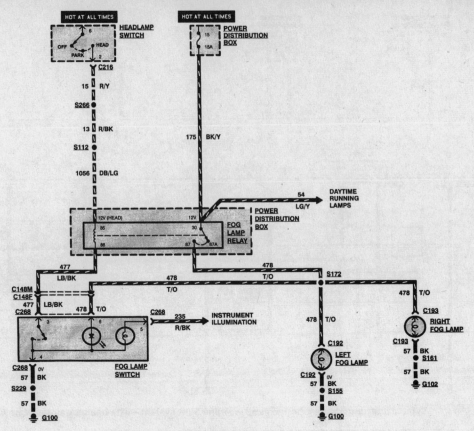

Typical fog light system

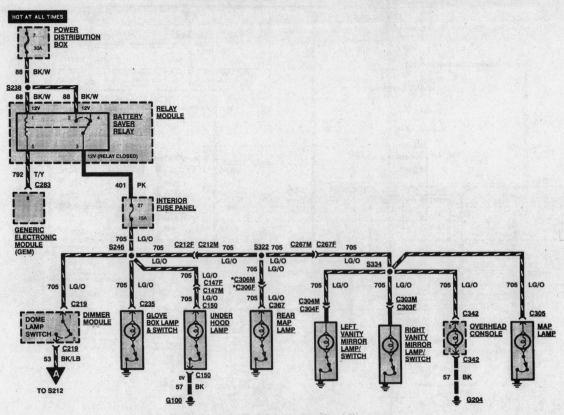

Typical interior lighting system (1 of 2)

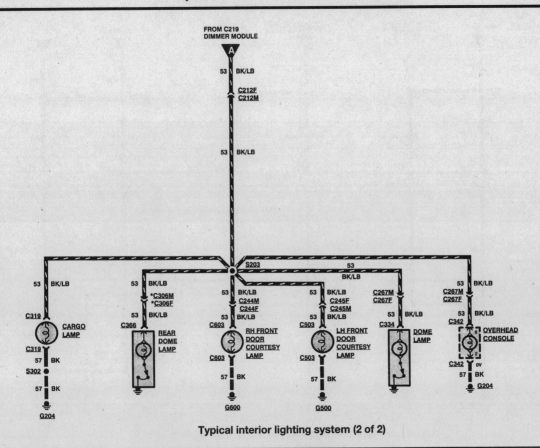

Typical interior lighting system (2 of 2)

Typical turn signal, hazard light and brake light systems (1 of 3)

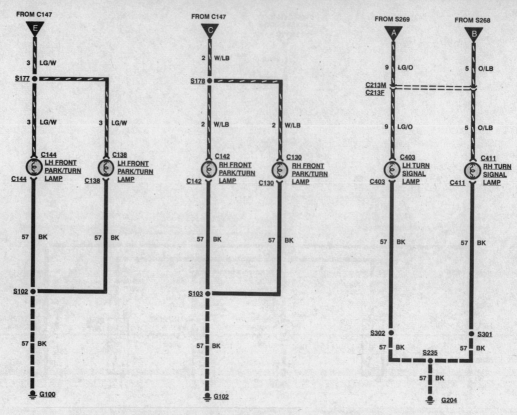

Typical turn signal, hazard light and brake light systems (2 of 3)

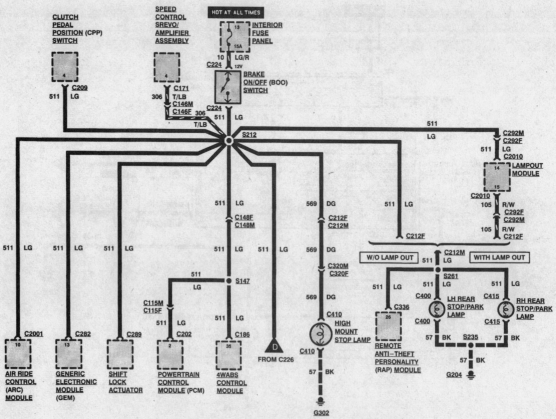

Typical turn signal, hazard light and brake light systems (3 of 3)

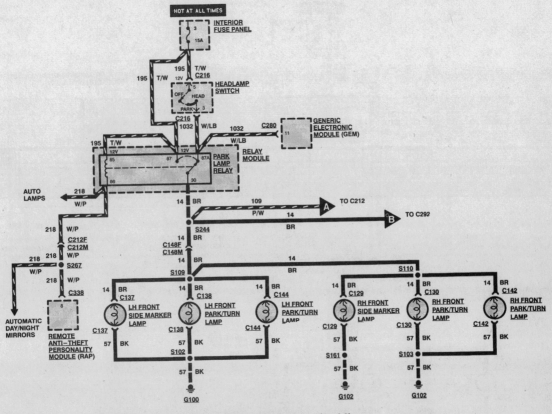

Typical exterior lighting system (1 of 2)

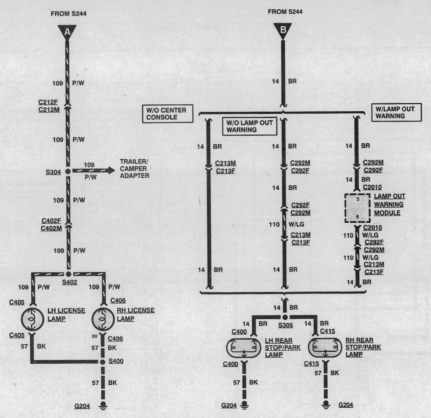

Typical exterior lighting system (2 of 2)

12

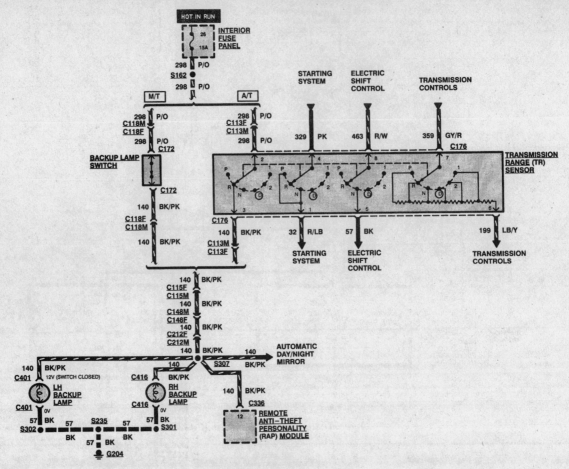

Typical back-up light system

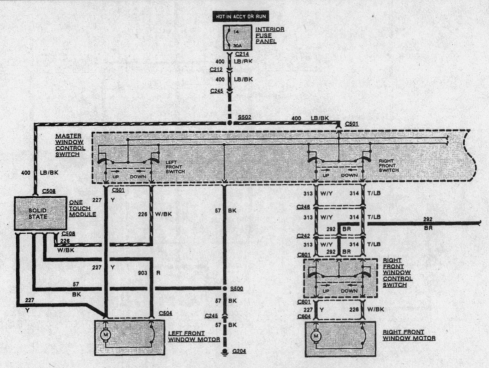

Typical power window system - 1994 and earlier (1 of 2)

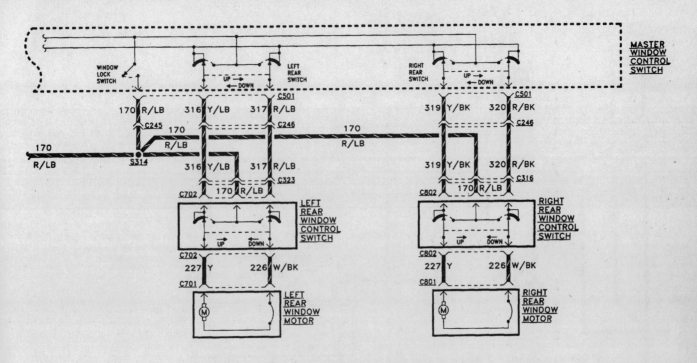

Typical power window system - 1994 and earlier (2 of 2)

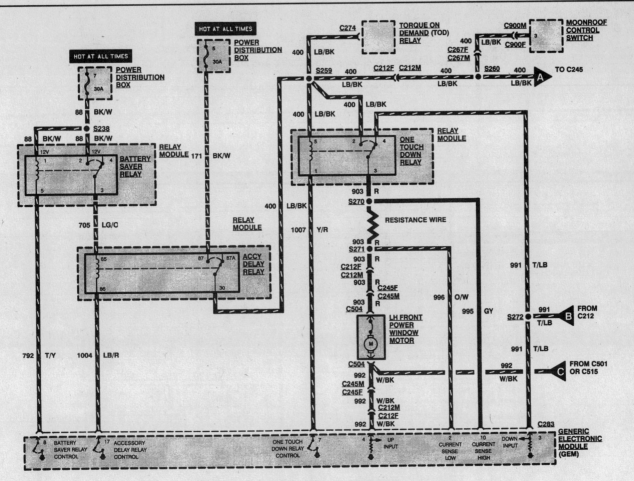

Typical power window system - 1995 and later (1 of 3)

12

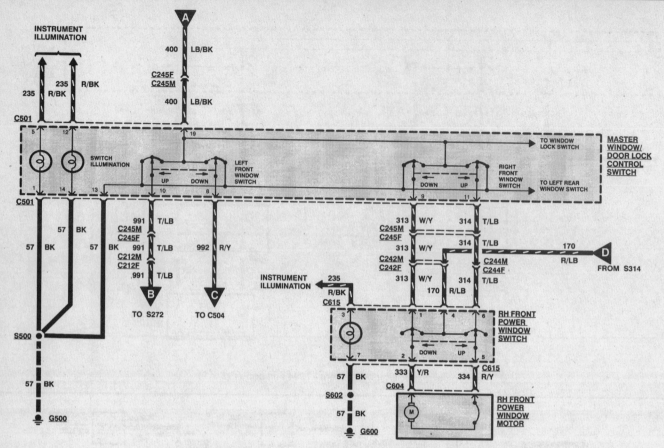

Typical power window system - 1995 and later (2 of 3)

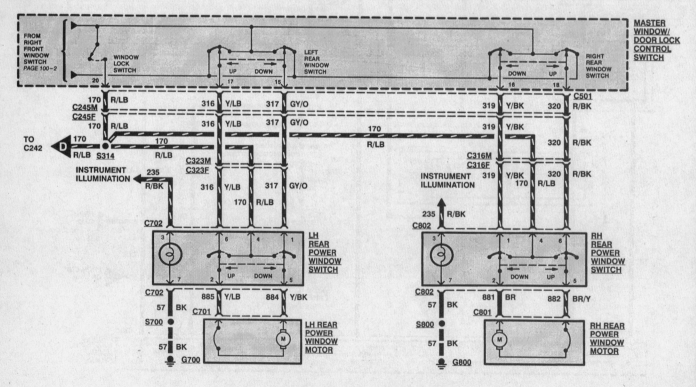

Typical power window system - 1995 and later (3 of 3)

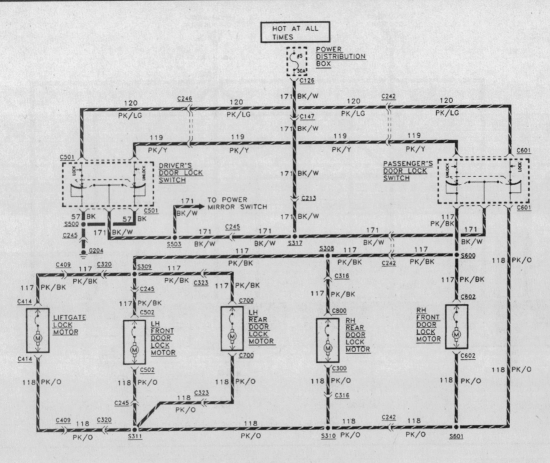

Typical power door lock system - 1994 and earlier

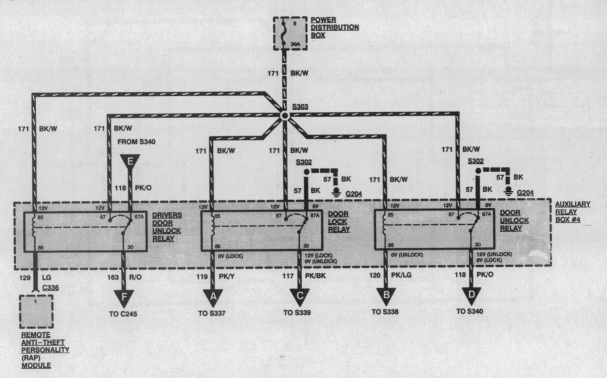

Typical power door lock system - 1995 and later (1 of 3)

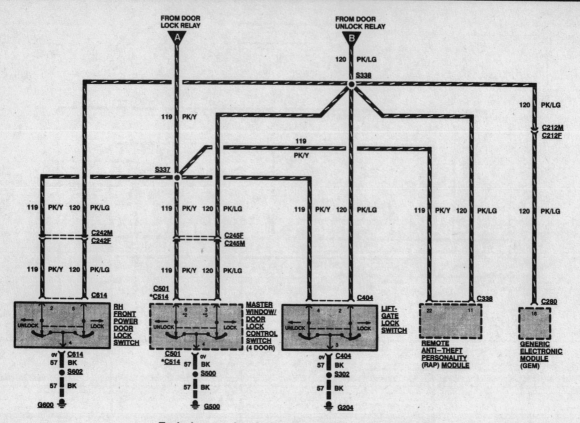

Typical power door lock system - 1995 and later (2 of 3)

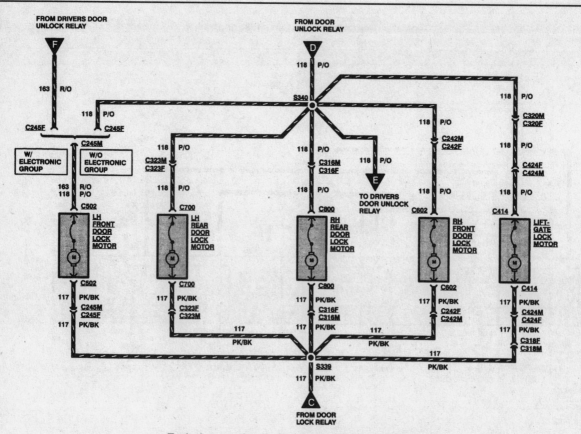

Typical power door lock system - 1995 and later (3 of 3)

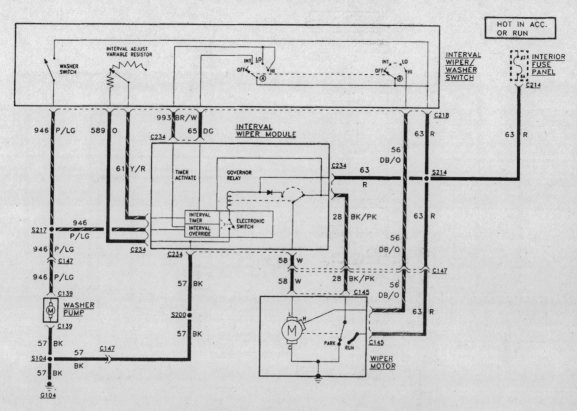

Typical windshield wiper and washer system - 1994 and earlier

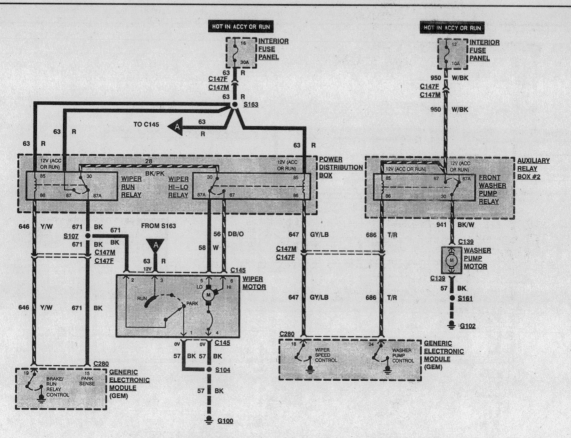

Typical windshield wiper and washer system - 1995 and later

12

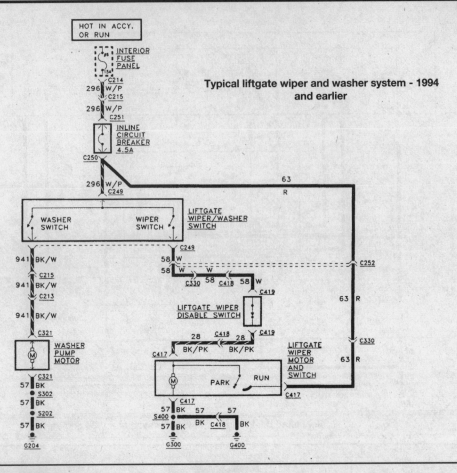

Typical liftgate wiper and washer system - 1994
and earlier

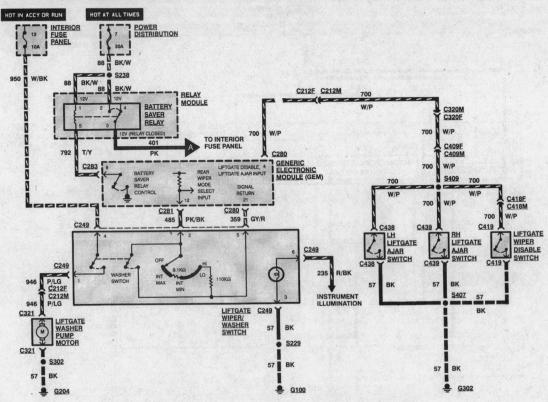

Typical liftgate wiper and washer system - 1995 and later (1 of 2)

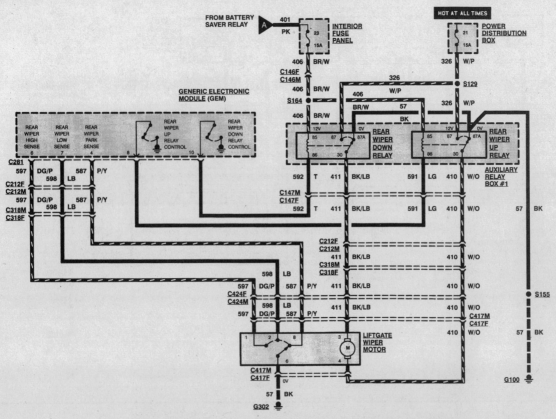

Typical liftgate wiper and washer system - 1995 and later (2 of 2)

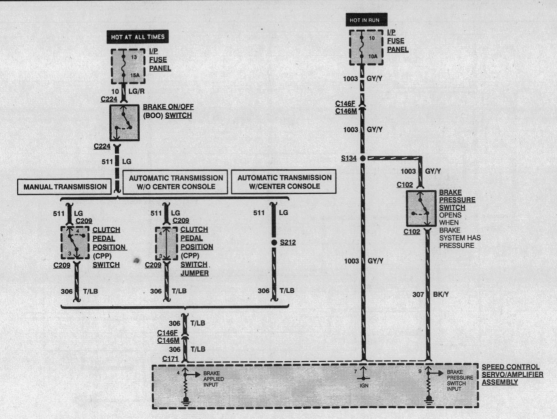

Typical cruise control system (1 of 2)

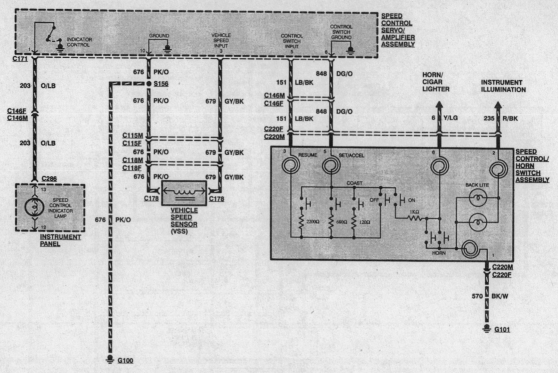

Typical cruise control system (2 of 2)

Cruise control circuit

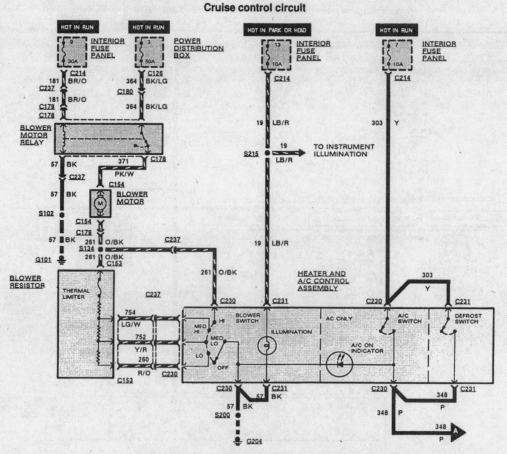

Typical heating and air conditioning system (1 of 2)

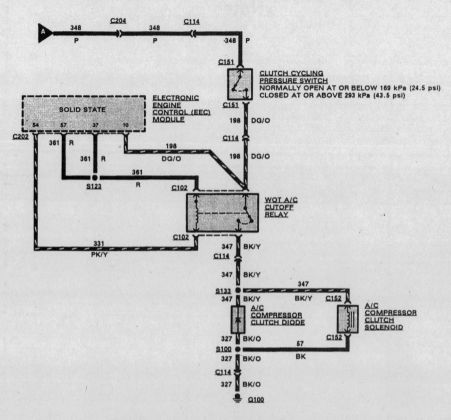

Typical heating and air conditioning system (2 of 2)

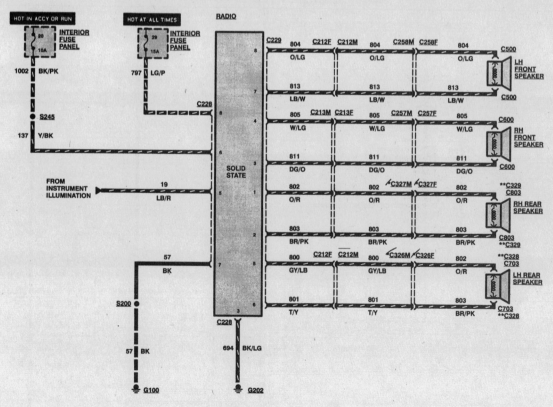

Typical base model audio system

12

Notes

Index

Haynes Automotive Manuals

NOTE: New manuals are added to this list on a periodic basis. If you do not see a listing for your vehicle, consult your local Haynes dealer for the latest product information.

ACURA
*12020 Integra '86 thru '89 & Legend '86 thru '90

AMC
Jeep CJ - see JEEP (50020)
14020 Concord/Hornet/Gremlin/Spirit '70 thru '83
14025 (Renault) Alliance & Encore '83 thru '87

AUDI
15020 4000 all models '80 thru '87
15025 5000 all models '77 thru '83
15026 5000 all models '84 thru '88

AUSTIN
Healey Sprite - see MG Midget (66015)

BMW
*18020 3/5 Series '82 thru '92
*18021 3 Series except 325iX models '92 thru '97
18025 320i all 4 cyl models '75 thru '83
18035 528i & 530i all models '75 thru '80
18050 1500 thru 2002 except Turbo '59 thru '77

BUICK
Century (FWD) - see GM (38005)
*19020 Buick, Oldsmobile & Pontiac Full-size (Front wheel drive) '85 thru '98
Buick Electra, LeSabre and Park Avenue; Oldsmobile Delta 88 Royale, Ninety Eight and Regency; Pontiac Bonneville
19025 Buick Oldsmobile & Pontiac Full-size (Rear wheel drive)
Buick Estate '70 thru '90, Electra'70 thru '84, LeSabre '70 thru '85, Limited '74 thru '79 Oldsmobile Custom Cruiser '70 thru '90, Delta 88 '70 thru '85, Ninety-eight '70 thru '84 Pontiac Bonneville '70 thru '81, Catalina '70 thru '81, Grandville '70 thru '75, Parisienne '83 thru '86
19030 Mid-size Regal & Century '74 thru '87
Regal - see GENERAL MOTORS (38010)
Skyhawk - see GM (38030)
Skylark - see GM (38020, 38025)
Somerset - see GENERAL MOTORS (38025)

CADILLAC
*21030 Cadillac Rear Wheel Drive '70 thru '93
Cimarron, Eldorado & Seville - see GM (38015, 38030)

CHEVROLET
10305 Chevrolet Engine Overhaul Manual
*24010 Astro & GMC Safari Mini-vans '85 thru '93
24015 Camaro V8 all models '70 thru '81
24016 Camaro all models '82 thru '92
Cavalier - see GM (38015)
Celebrity - see GM (38005)
24017 Camaro & Firebird '93 thru '97
24020 Chevelle, Malibu, El Camino '69 thru '87
24024 Chevette & Pontiac T1000 '76 thru '87
Citation - see GENERAL MOTORS (38020)
*24032 Corsica/Beretta all models '87 thru '96
24040 Corvette all V8 models '68 thru '82
*24041 Corvette all models '84 thru '96
24045 Full-size Sedans Caprice, Impala, Biscayne, Bel Air & Wagons '69 thru '90
24046 Impala SS & Caprice and Buick Roadmaster '91 thru '96
Lumina '90 thru '94 - see GM (38010)
24048 Lumina & Monte Carlo '95 thru '98
Lumina APV - see GM (38038)
24050 Luv Pick-up all 2WD & 4WD '72 thru '82
24055 Monte Carlo all models '70 thru '88
Monte Carlo '95 thru '98 - see LUMINA
24059 Nova all V8 models '69 thru '79
*24060 Nova/Geo Prizm '85 thru '92
*24064 Pick-ups '67 thru '87 - Chevrolet & GMC, all V8 & in-line 6 cyl, 2WD & 4WD '67 thru '87; Suburbans, Blazers & Jimmys '67 thru '91
*24065 Pick-ups '88 thru '98 - Chevrolet & GMC, all full-size models '88 thru '98; Blazer & Jimmy '92 thru '94; Suburban '92 thru '98; Tahoe & Yukon '95 thru '98
*24070 S-10 & GMC S-15 Pick-ups '82 thru '93
24071 S-10, Gmc S-15 & Blazer '94 thru '96
*24075 Sprint & Geo Metro '85 thru '94
*24080 Vans - Chevrolet & GMC '68 thru '96

CHRYSLER
10310 Chrysler Engine Overhaul Manual
*25015 Chrysler Cirrus, Dodge Stratus, Plymouth Breeze, '95 thru '98
*25020 Full-size Front-Wheel Drive '88 thru '93
K-Cars - see DODGE Aries (30008)
Laser - see DODGE Daytona (30030)
25025 Chrysler LHS, Concorde & New Yorker, Dodge Intrepid, Eagle Vision, '93 thru '97
*25030 Chrysler/Plym. Mid-size '82 thru '95
Rear-wheel Drive - see DODGE (30050)

DATSUN
28005 200SX all models '80 thru '83
28007 B-210 all models '73 thru '78
28009 210 all models '79 thru '82
28012 240Z, 260Z & 280Z Coupe '70 thru '78
28014 280ZX Coupe & 2+2 '79 thru '83
300ZX - see NISSAN (72010)
28016 310 all models '78 thru '82
28018 510 & PL521 Pick-up '68 thru '73
28020 510 all models '78 thru '81
28022 620 Series Pick-up all models '73 thru '79
720 Series Pick-up - see NISSAN (72030)
28025 810/Maxima all gas models, '77 thru '84

DODGE
400 & 600 - see CHRYSLER (25030)
*30008 Aries & Plymouth Reliant '81 thru '89
30010 Caravan & Ply. Voyager '84 thru '95
*30011 Caravan & Ply. Voyager '96 thru '98
30012 Challenger/Plymouth Saporro '78 thru '83
Challenger '67-'76 - see DART (30025)
30016 Colt/Plymouth Champ '78 thru '87
*30020 Dakota Pick-ups all models '87 thru '96
30025 Dart, Challenger/Plymouth Barracuda & Valiant 6 cyl models '67 thru '76
*30030 Daytona & Chrysler Laser '84 thru '89
Intrepid - see Chrysler (25025)
*30034 Dodge & Plymouth Neon '95 thru '97
*30035 Omni & Plymouth Horizon '78 thru '90
30040 Pick-ups all full-size models '74 thru '93
*30041 Pick-ups all full-size models '94 thru '96
*30045 Ram 50/D50 Pick-ups & Raider and Plymouth Arrow Pick-ups '79 thru '93
30050 Dodge/Ply./Chrysler RWD '71 thru '89
*30055 Shadow/Plymouth Sundance '87 thru '94
*30060 Spirit & Plymouth Acclaim '89 thru '95
*30065 Vans - Dodge & Plymouth '71 thru '96

EAGLE
Talon - see MITSUBISHI Eclipse (68030)
Vision - see CHRYSLER (25025)

FIAT
34010 124 Sport Coupe & Spider '68 thru '78
34025 X1/9 all models '74 thru '80

FORD
10355 Ford Automatic Transmission Overhaul
10320 Ford Engine Overhaul Manual
*36004 Aerostar Mini-vans '86 thru '96
Aspire - see FORD Festiva (36030)
*36006 Contour/Mercury Mystique '95 thru '98
36008 Courier Pick-up all models '72 thru '82
36012 Crown Victoria & Mercury Grand Marquis '88 thru '96
*36016 Escort/Mercury Lynx '81 thru '90
*36020 Escort/Mercury Tracer '91 thru '96
Expedition - see FORD Pick-up (36059)
*36024 Explorer & Mazda Navajo '91 thru '95
36028 Fairmont & Mercury Zephyr '78 thru '83
36030 Festiva & Aspire '88 thru '97
36032 Fiesta all models '77 thru '80
36036 Ford & Mercury Full-size,
Ford LTD & Mercury Marquis ('75 thru '82); Ford Custom 500,Country Squire, Crown Victoria & Mercury Colony Park ('75 thru '87); Ford LTD Crown Victoria & Mercury Gran Marquis ('83 thru '87)
36040 Granada & Mercury Monarch '75 thru '80
36044 Ford & Mercury Mid-size,
Ford Thunderbird & Mercury Cougar ('75 thru '82); Ford LTD & Mercury Marquis ('83 thru '86); Ford Torino,Gran Torino, Elite, Ranchero pick-up, LTD II, Mercury Montego, Comet, XR-7 & Lincoln Versailles ('75 thru '86)
36048 Mustang V8 all models '64-1/2 thru '73
36049 Mustang II 4 cyl, V6 & V8 '74 thru '78
36050 Mustang & Mercury Capri incl. Turbo Mustang, '79 thru '93; Capri, '79 thru '86
*36051 Mustang all models '94 thru '97
36054 Pick-ups and Bronco '73 thru '79
*36058 Pick-ups and Bronco '80 thru '96
*36059 Pick-ups, Expedition & Lincoln Navigator '97 thru '98
36062 Pinto & Mercury Bobcat '75 thru '80
36066 Probe all models '89 thru '92
*36070 Ranger/Bronco II all models '83 thru '92
*36071 Ford Ranger '93 thru '97 &
Mazda Pick-ups '94 thru '97
*36074 Taurus & Mercury Sable '86 thru '95
*36075 Taurus & Mercury Sable '96 thru '98
*36078 Tempo & Mercury Topaz '84 thru '94
36082 Thunderbird/Mercury Cougar '83 thru '88
*36086 Thunderbird/Mercury Cougar '89 and '97
36090 Vans all V8 Econoline models '69 thru '91
*36094 Vans full size '92 thru '95
*36097 Windstar Mini-van '95 thru '98

GENERAL MOTORS
*10360 GM Automatic Transmission Overhaul
*38005 Buick Century, Chevrolet Celebrity, Olds Cutlass Ciera & Pontiac 6000 all models '82 thru '96
*38010 Buick Regal, Chevrolet Lumina, Oldsmobile Cutlass Supreme & Pontiac Grand Prix front wheel drive '88 thru '95
*38015 Buick Skyhawk, Cadillac Cimarron, Chevrolet Cavalier, Oldsmobile Firenza Pontiac J-2000 & Sunbird '82 thru '94
*38016 Chevrolet Cavalier & Pontiac Sunfire '95 thru '98
38020 Buick Skylark, Chevrolet Citation, Olds Omega, Pontiac Phoenix '80 thru '85
38025 Buick Skylark & Somerset, Olds Achieva, Calais & Pontiac Grand Am '85 thru '95
38030 Cadillac Eldorado & Oldsmobile Toronado '71 thru '85, Seville '80 thru '85, Buick Riviera '79 thru '85
*38035 Chevrolet Lumina APV, Oldsmobile Silhouette & Pontiac Trans Sport '90 thru '95
General Motors Full-size Rear-wheel Drive - see BUICK (19025)

GEO
Metro - see CHEVROLET Sprint (24075)
Prizm - see CHEVROLET (24060) or TOYOTA (92036)
*40030 Storm all models '90 thru '93
Tracker - see SUZUKI Samurai (90010)

GMC
Safari - see CHEVROLET ASTRO (24010)
Vans & Pick-ups - see CHEVROLET

HONDA
42010 Accord CVCC all models '76 thru '83
42011 Accord all models '84 thru '89
42012 Accord all models '90 thru '93
*42013 Accord all models '94 thru '95
42020 Civic 1200 all models '73 thru '79
42021 Civic 1300 & 1500 CVCC '80 thru '83
42022 Civic 1500 CVCC all models '75 thru '79
42023 Civic all models '84 thru '91
42024 Civic & del Sol '92 thru '95
Passport - see ISUZU Rodeo (47017)
*42040 Prelude CVCC all models '79 thru '89

HYUNDAI
*43015 Excel all models '86 thru '94

ISUZU
Hombre - see CHEVROLET S-10 (24071)
*47017 Rodeo '91 thru '97, Amigo '89 thru '94, Honda Passport '95 thru '97
*47020 Trooper '84 thru '91, Pick-up '81 thru '93

JAGUAR
*49010 XJ6 all 6 cyl models '68 thru '86
*49011 XJ6 all models '88 thru '94
*49015 XJ12 & XJS all 12 cyl models '72 thru '85

JEEP
*50010 Cherokee, Comanche & Wagoneer Limited all models '84 thru '96
50020 CJ all models '49 thru '86
*50025 Grand Cherokee all models '93 thru '98
*50029 Grand Wagoneer & Pick-up '72 thru '91
*50030 Wrangler all models '87 thru '95

LINCOLN
Navigator - see FORD Pick-up (36059)
59010 Rear Wheel Drive all models '70 thru '96

MAZDA
61010 GLC (rear wheel drive) '77 thru '83
61011 GLC (front wheel drive) '81 thru '85
*61015 323 & Protegé '90 thru '97
*61016 MX-5 Miata '90 thru '97
*61020 MPV all models '89 thru '94
Navajo - see FORD Explorer (36024)
61030 Pick-ups '72 thru '93
Pick-ups '94 on - see Ford (36071)
*61035 RX-7 all models '79 thru '85
*61036 RX-7 all models '86 thru '91
61040 626 (rear wheel drive) '79 thru '82
*61041 626 & MX-6 (front wheel drive) '83 thru '91

MERCEDES-BENZ
63012 123 Series Diesel '76 thru '85
*63015 190 Series 4-cyl gas models, '84 thru '88
63020 230, 250 & 280 6 cyl sohc '68 thru '72
63025 280 123 Series gas models '77 thru '81
63030 350 & 450 all models '71 thru '80

MERCURY
See FORD Listing

MG
66010 MGB Roadster & GT Coupe '62 thru '80
66015 MG Midget & Austin Healey Sprite Roadster '58 thru '80

MITSUBISHI
*68020 Cordia, Tredia, Galant, Precis & Mirage '83 thru '93
*68030 Eclipse, Eagle Talon & Plymouth Laser '90 thru '94
*68040 Pick-up '83 thru '96, Montero '83 thru '93

NISSAN
72010 300ZX all models incl. Turbo '84 thru '89
*72015 Altima all models '93 thru '97
*72020 Maxima all models '85 thru '91
*72030 Pick-ups '80 thru '96, Pathfinder '87 thru '95
72040 Pulsar all models '83 thru '86
72050 Sentra all models '82 thru '94
*72051 Sentra & 200SX all models '95 thru '98
72060 Stanza all models '82 thru '90

OLDSMOBILE
*73015 Cutlass '74 thru '88
For other OLDSMOBILE titles, see BUICK, CHEVROLET or GENERAL MOTORS listing.

PLYMOUTH
For PLYMOUTH titles, see DODGE.

PONTIAC
79008 Fiero all models '84 thru '88
79018 Firebird V8 models except Turbo '70 thru '81
79019 Firebird all models '82 thru '92
For other PONTIAC titles, see BUICK, CHEVROLET or GENERAL MOTORS listing.

PORSCHE
*80020 911 Coupe & Targa models '65 thru '89
80025 914 all 4 cyl models '69 thru '76
80030 924 all models incl. Turbo '76 thru '82
*80035 944 all models incl. Turbo '83 thru '89

RENAULT
Alliance, Encore - see AMC (14020)

SAAB
*84010 900 including Turbo '79 thru '88

SATURN
*87010 Saturn all models '91 thru '96

SUBARU
89002 1100, 1300, 1400 & 1600 '71 thru '79
*89003 1600 & 1800 2WD & 4WD '80 thru '94

SUZUKI
*90010 Samurai/Sidekick/Geo Tracker '86 thru '96

TOYOTA
92005 Camry all models '83 thru '91
*92006 Camry all models '92 thru '96
92015 Celica Rear Wheel Drive '71 thru '85
*92020 Celica Front Wheel Drive '86 thru '93
92025 Celica Supra all models '79 thru '92
92030 Corolla all models '75 thru '79
92032 Corolla rear wheel drive models '80 thru '87
*92035 Corolla front wheel drive models '84 thru '92
*92036 Corolla & Geo Prizm '93 thru '97
92040 Corolla Tercel all models '80 thru '82
92045 Corona all models '74 thru '82
92050 Cressida all models '78 thru '82
92055 Land Cruiser Series FJ40, 43, 45 & 55 '68 thru '82
*92056 Land Cruiser Series FJ60, 62, 80 & FZJ80 '80 thru '82
*92065 MR2 all models '85 thru '87
92070 Pick-up all models '69 thru '78
*92075 Pick-up all models '79 thru '95
*92076 Tacoma '95 thru '98,
4Runner '96 thru '98, T100 '93 thru '98
*92080 Previa all models '91 thru '95
92085 Tercel all models '87 thru '94

TRIUMPH
94007 Spitfire all models '62 thru '81
94010 TR7 all models '75 thru '81

VW
96008 Beetle & Karmann Ghia '54 thru '79
96012 Dasher all gasoline models '74 thru '81
*96016 Rabbit, Jetta, Scirocco, & Pick-up gas models '74 thru '91 & Convertible '80 thru '92
*96017 Golf & Jetta '93 thru '97
96020 Rabbit, Jetta, Pick-up diesel '77 thru '84
96030 Transporter 1600 all models '68 thru '79
96035 Transporter 1700, 1800, 2000 '72 thru '79
96040 Type 3 1500 & 1600 '63 thru '73
96045 Vanagon air-cooled models '80 thru '83

VOLVO
97010 120, 130 Series & 1800 Sports '61 thru '73
97015 140 Series all models '66 thru '74
*97020 240 Series all models '76 thru '93
97025 260 Series all models '75 thru '82
*97040 740 & 760 Series all models '82 thru '88

TECHBOOK MANUALS
10205 Automotive Computer Codes
10210 Automotive Emissions Control Manual
10215 Fuel Injection Manual, 1978 thru 1985
10220 Fuel Injection Manual, 1986 thru 1996
10225 Holley Carburetor Manual
10230 Rochester Carburetor Manual
10240 Weber/Zenith/Stromberg/SU Carburetor
10305 Chevrolet Engine Overhaul Manual
10310 Chrysler Engine Overhaul Manual
10320 Ford Engine Overhaul Manual
10330 GM and Ford Diesel Engine Repair
10340 Small Engine Repair Manual
10345 Suspension, Steering & Driveline
10355 Ford Automatic Transmission Overhaul
10360 GM Automatic Transmission Overhaul
10405 Automotive Body Repair & Painting
10410 Automotive Brake Manual
10415 Automotive Detailing Manual
10420 Automotive Eelectrical Manual
10425 Automotive Heating & Air Conditioning
10430 Automotive Reference Dictionary
10435 Automotive Tools Manual
10440 Used Car Buying Guide
10445 Welding Manual
10450 ATV Basics

SPANISH MANUALS
98903 Reparación de Carrocería & Pintura
98905 Códigos Automotrices de la Computadora
98910 Frenos Automotriz
98915 Inyección de Combustible 1986 al 1994
99040 Chevrolet & GMC Camionetas '67 al '87
99041 Chevrolet & GMC Camionetas '88 al '95
99042 Chevrolet Camionetas Cerradas '68 al '95
99055 Dodge Caravan/Ply. Voyager '84 al '95
99075 Ford Camionetas y Bronco '80 al '94
99077 Ford Camionetas Cerradas '69 al '91
99083 Ford Modelos de Tamaño Grande '75 al '87
99088 Ford Modelos de Tamaño Mediano '75 al '86
99091 Ford Taurus & Mercury Sable '75 al '95
99095 GM Modelos de Tamaño Grande '70 al '90
99100 GM Modelos de Tamaño Mediano '70 al '88
99110 Nissan Camionetas '80 al '96, Pathfinder '80 al '95
99118 Nissan Sentra '82 al '94
99125 Toyota Camionetas y 4-Runner '79 al '95

Listings shown with an asterisk (*) indicate model coverage as of this printing. These titles will be periodically updated to include later model years - consult your Haynes dealer for more information.

Nearly 100 Haynes motorcycle manuals also available

5-98

Haynes North America, Inc., 861 Lawrence Drive, Newbury Park, CA 91320 • (805) 498-6703